Universitext

Universitext is a series of textbooks that presents material from a wide variety of mathematical disciplines at master's level and beyond. The books, often well class-tested by their author, may have an informal, personal, or even experimental approach to their subject matter. Some of the most successful and established books in the series have evolved through several editions, always following the evolution of teaching curricula, into very polished texts.

Thus as research topics trickle down into graduate-level teaching, first textbooks written for new, cutting-edge courses may find their way into *Universitext*.

Thomas Alazard

Analysis and Partial Differential Equations

 Springer

Thomas Alazard
École Normale Supérieure Paris-Saclay
Université Paris-Saclay
Gif-sur-Yvette, France

The original submitted manuscript has been translated into English. The translation was done using artificial intelligence. A subsequent revision was performed by the author(s) to further refine the work and to ensure that the translation is appropriate concerning content and scientific correctness. It may, however, read stylistically different from a conventional translation.

ISSN 0172-5939   ISSN 2191-6675   (electronic)
Universitext
ISBN 978-3-031-70908-1   ISBN 978-3-031-70909-8   (eBook)
https://doi.org/10.1007/978-3-031-70909-8

Mathematics Subject Classification (2020): 54-01, 35-01, 42-01, 46-01

Translation from the French language edition: "Analyse et équations aux dérivées partielles" by Thomas Alazard, © EDP Sciences 2023. Published by EDP Sciences. All Rights Reserved.

This Springer imprint is published by the registered company Springer Nature Switzerland AG
The registered company address is: Gewerbestrasse 11, 6330 Cham, Switzerland

If disposing of this product, please recycle the paper.

To Adèle and Charles

Preface

This book is based on the notes from several different courses given successively at the École Normale Supérieure in France (ENS Paris and ENS Paris-Saclay) and at Tsinghua University. I have added nothing in this book that was not taught, and all the proofs that appear in this book were written on the board. I have tried to offer a demanding but nevertheless accessible exposition of graduate courses.

The book is divided into five parts, whose contents are described below. It aims to give a complete immersion in the advanced concepts of Mathematical Analysis and revolves around four domains, covering Functional Analysis, Harmonic Analysis, Microlocal Analysis, and the Analysis of Partial Differential Equations. Each chapter is constructed to provide readers with an understanding of fundamental concepts while guiding them through the detailed proofs of some significant theorems.

Part I - Functional Analysis

The first chapter is about the foundational aspects of **topological vector spaces**. The discussion includes the study of finite-dimensional spaces, semi-norms, and Banach spaces, with a focus on the space of continuous functions. To conclude, we give applications of the Stone–Weierstrass theorem to study the universal approximation property for artificial neural networks.

The second chapter explores many **fixed-point theorems**, beginning with a review of differential calculus to prepare for local inversion and Cauchy–Lipschitz theorems, Caristi's and Ekeland's theorems. We also discuss Brouwer's fixed-point theorem, Brouwer's domain invariance theorem, and Nash's implicit function theorem.

In the third chapter, we study **Hilbert spaces** and their properties as well as the Hahn–Banach theorem, a powerful tool in functional analysis to prove the existence of linear forms. We address weak convergence and convergence in the weak-* topology. Finally, the Banach–Alaoglu theorem concludes this section. These results allow extracting subsequences converging in a weak sense even in infinite-dimensional vector spaces. We discuss also the role of **duality and convexity** in functional analysis.

Part II - Harmonic Analysis

In the first chapter of this part, we prove the convergence of **Fourier series** of square-integrable functions. The study continues with an examination of pointwise and uniform convergence of Fourier series.

The next chapter generalizes the study of Fourier series and introduces the **Fourier transform**, a powerful tool in mathematical analysis. By exploring the Schwartz space, we lay the theoretical foundations for understanding tempered distributions and the Littlewood–Paley decomposition.

We then give an in-depth study of **convolution**. We begin with the rigorous definition of the convolution product and then address the Hardy–Littlewood maximal function, the pointwise convergence of an approximation of the identity, the Hardy–Littlewood–Sobolev inequality, and other Sobolev inequalities. Finally, we explore Calderón's reconstruction formula, wavelets, and Wiener's theorem.

We then study **Sobolev spaces**, which are fundamental function spaces. After a general presentation, we examine the Poincaré inequality and its application to the Poisson problem. The study continues with Sobolev spaces defined on arbitrary open sets. We then analyze the interaction between Fourier analysis and Sobolev spaces, proving Sobolev embeddings and establishing a characterization of these spaces in terms of Littlewood–Paley decomposition.

This part concludes with a chapter on **harmonic functions**. We begin by introducing the mean value property, which reveals many of the characteristics of these functions. Next, we examine the fundamental solution of the Laplacian, and the regularity of harmonic functions is also discussed.

Part III - Microlocal Analysis

This part is devoted to the study of pseudo-differential operators, symbolic calculus, and hyperbolic equations. This complex set of results and techniques paves the way to a deeper understanding of the singularities of a function.

The first chapter introduces the analysis of **pseudo-differential operators**, an important class of operators generalizing differential operators. We begin by discussing this notion and examining the continuity of pseudo-differential operators on the Schwartz space and on L^2.

The next chapter contains the main results about the **symbolic calculus**, a fruitful approach to manipulating differential and pseudo-differential operators. We study oscillatory integrals, which allow to define the symbols of the adjoint and the composition of pseudo-differential operators. We then address numerous applications of this method.

As a preparation for the study of the propagation of singularities, we then study **hyperbolic equations**, which play an important role in the study of partial differential equations. In this chapter we study transport equations as well as pseudo-differential hyperbolic equations, proving the existence and the uniqueness of the solutions to the Cauchy problem.

The final chapter of this part examines **microlocal singularities**. We define the notion of wave front set and prove a version of the Hörmander theorem of propagation of singularities. The chapter concludes with a discussion of associated nonlinear problems.

Part IV - Partial Differential Equations

This part, which is more challenging, concerns the analysis of partial differential equations. This is a very broad subject and I have chosen to provide complete proofs of a selection of major theorems: resolution of the Calderón problem, De Giorgi's theorem, as well as a Strichartz–Bourgain inequality.

The first chapter is an introduction to the study of the **Calderón Problem**. We begin with a detailed introduction and then examine the density of products of harmonic functions. The study continues with an analysis of equations with variable coefficients before presenting a famous result due to Sylvester and Uhlmann.

The second chapter introduces **De Giorgi's Theorem**, a major advance in the study of partial differential equations. We give a self-contained presentation, explaining the role of subsolutions and convex nonlinear transformations. The Moser's iterations argument and Harnack inequality are then examined, which allow to prove the Hölder regularity of solutions to elliptic equations.

The next chapter is dedicated to **Schauder's Theorem**. We begin with a discussion of the use of local averages to study Hölder regularity, presenting Campanato's theorem. Schauder's theorem is then proved in a general setting, for variational solutions, emphasizing its importance in the context of elliptic equations and in particular to prove the regularity of minimal surfaces.

The final chapter introduces the analysis of **dispersive estimates**, an ubiquitous theme in the study of partial differential equations. We begin with a description of many recent results about the Schrödinger equation. Next, we prove a Strichartz–Bourgain estimate for the Korteweg-De Vries equation.

Part V - Recap and Solutions to the exercises

An appendix contains reminders about general topology and Lebesgue spaces. Exercises complement this presentation and propose to prove numerous famous results. Solutions to some exercises are given in this part. These are indicated by "Exercise (solved)" in the main text. Additional applications and solved problems can be found in the book [1] written with Claude Zuily.

Acknowledgments

I would like to thank Claude Sabbah for all his numerous advices for the French version of this book, which greatly contributed to improving the presentation. I also thank Rémi Lodh who provides me with a decisive assistance for the preparation of the English version.

I warmly thank Cécile Huneau, Isabelle Tristani and Irène Waldspurger, who gave exercises classes on the Microlocal Analysis and PDE parts of this course, for allowing me to use many of their statements and corrections. I would also like to thank Arthur Leclaire, Ayman Rimah, and Rémi Tesson, who gave exercises classes on the parts related to Harmonic Analysis and Functional Analysis, as well as Charles Arnal, Malo Jézequel, Ayman Rimah, and Haocheng Yang for their comments on a preliminary version.

I would like to thank Karim Boulabiar, Nicolas Burq, Didier Bresch, Patrick Gérard, Lingbing He, Guy Métivier, Laure Quivy, Alain Trouvé, Pin Yu and Claude Zuily for the many discussions we have had on the teaching of Analysis.

This book is the result of lectures given over a period of twelve years at various locations. In preparing them, I have used numerous books and lecture notes. I have endeavored to cite them in the text and apologize for any omissions due to oversight. I acknowledge in advance the readers who send me suggestions or corrections.

Bures-sur-Yvette,   *Thomas Alazard*
July 2024

Contents

Part II Harmonic Analysis

Part III Microlocal Analysis

Part IV Analysis of Partial Differential Equations

Part I
Functional Analysis

Chapter 1
Topological Vector Spaces

In 1850, Weierstrass gave the first precise ("epsilon-delta") definition of the concept of the limit of a function at a point. He used the notion of arbitrarily small numbers and the notation lim introduced by L'Huilier and used by Cauchy in his Analysis course. He introduced the notion of a neighborhood of a point as well as the notation $|x|$ for the absolute value of x. Bolzano also contributed to proving Analysis results that were considered intuitively true, such as the intermediate value theorem (see [31, 172]).

Later, it seemed natural to consider the existence of limits of sequences of functions, and to study the notion of continuity of certain mappings that act, not on numbers, but on functions. The most important concept is that of metric space, where the notions of distance and norm generalize the absolute value. This generalization was essential for studying the convergence of trigonometric series that Fourier introduced in 1812 to study the heat equation [52]. It also allowed Riemann to solve the Dirichlet problem by a minimization method (given an open set Ω of $\mathbb{R}^n$, the Dirichlet problem consists in finding a harmonic function $u: \Omega \to \mathbb{R}$ whose value on the boundary $\partial\Omega$ is a given function). These works led to the emergence of a language, known as general topology, which allows us to speak rigorously about limits and continuity in a very general abstract framework. The main definitions are recalled in the appendix to Chapter 17.

The notions of topology are fundamental in functional analysis, which is the branch of analysis that studies function spaces; one of the main objectives being to solve linear equations whose unknown is a function (think of the study of partial differential equations). To study such equations, the most natural framework is that of infinite-dimensional vector spaces with a topology compatible with their algebraic structure: these are the topological vector spaces that we will study in this chapter.

We will begin this chapter with reminders about normed spaces. We will then be interested in spaces whose topology is not defined by a norm, but by a family of semi-norms. We will see the properties specific to finite-dimensional spaces and then study Banach's results on the continuity of linear maps. We will also study several important results concerning the space of continuous functions and the universal approximation property for artificial neural networks.

1.1 Normed Vector Spaces

Let us start with some reminders on the notions of norm and distance.

Definition 1.1 A *distance* on a set X is a mapping

$$d: X \times X \longrightarrow [0, +\infty)$$

which satisfies the following three properties: for all x, y, z in X,

1. (separation) $d(x, y) = 0$ if and only if $x = y$;
2. (symmetry) $d(x, y) = d(y, x)$;
3. (triangle inequality) $d(x, z) \leqslant d(x, y) + d(y, z)$.

The pair (X, d) is a *metric space*. Given a point x in X and a positive real number r, we call the *open ball* centered at x with radius r the set

$$B(x, r) := \{y \in X : d(x, y) < r\}.$$

By definition, the *closed ball* centered at x with radius r is the set

$$B_f(x, r) := \{y \in X : d(x, y) \leqslant r\}.$$

Definition 1.2 Let E be a vector space over the field $\mathbb{K} = \mathbb{R}$ or $\mathbb{C}$. A *semi-norm* on E is a mapping $\rho: E \to [0, +\infty)$ which satisfies the following two properties: for all λ in $\mathbb{K}$ and all x, y in E,

1. (*homogeneity*) $\rho(\lambda x) = |\lambda| \rho(x)$;
2. (*triangle inequality*) $\rho(x + y) \leqslant \rho(x) + \rho(y)$.

A semi-norm is a *norm* if in addition $\rho(x) = 0$ implies $x = 0$ (note that the property of homogeneity implies that $\rho(0) = 0$).

Norms are usually denoted $\|\cdot\|$, a pair $(E, \|\cdot\|)$ is called a *normed vector space*. The mapping d defined on $E \times E$ by $d(x, y) = \|x - y\|$ is then a distance on E.

The fact that we can associate a distance with a norm allows us to equip a normed vector space with a topology. Given a vector space E equipped with a norm $\|\cdot\|$, we will always assume that E is equipped with the topology induced by the distance $d(x, y) = \|x - y\|$. A neighborhood basis (see Definition 1.22) for this topology is given by the collection of open balls $B(x, r)$ with $x \in E$ and $r > 0$, defined by

$$B(x, r) = \{y \in E : \|x - y\| < r\}.$$

Proposition 1.3 *Consider a normed vector space* $(E, \|\cdot\|)$. *The operations of vector addition* $(x, y) \mapsto x + y$ *and scalar multiplication* $(\lambda, x) \mapsto \lambda x$ *are continuous.*

Proof Let us show that the addition $A: E \times E \to E$, with $A(x, y) = x + y$, is continuous. Let U be an open set in E. We want to show that $A^{-1}(U)$ is open. For this, consider $(x, y) \in A^{-1}(U)$ so that $x + y$ belongs to U. Since U is open, there exists $\delta > 0$ such that $B(x + y, \delta) \subset U$. From the triangle inequality, we deduce that $u + v$ belongs to $A^{-1}(U)$ for all $u \in B(x, \delta/2)$ and all $v \in B(y, \delta/2)$. This proves that $B(x, \delta/2) \times B(y, \delta/2)$ is included in $A^{-1}(U)$. As the product of two balls is open for the product topology, we have shown that the preimage of an open set is an open set, which proves that A is continuous.

We prove in the same way that multiplication by a scalar is continuous. $\qquad \square$

Let us recall that a homeomorphism is a continuous bijection whose inverse is continuous.

Corollary 1.4 *Translations and dilations (by a scalar $\lambda \neq 0$) are homeomorphisms.*

Proof Let $u \in E$ and let T_u be the translation by u, defined by $T_u(x) = x + u$. Then T_u is continuous (because vector addition is continuous and the mapping $v \mapsto (u, v)$ is continuous from E to $E \times E$). Moreover, T_u is invertible and its inverse, which is T_{-u}, is continuous. This shows that the translation by u is a homeomorphism. We proceed in the same way to study the dilation $x \mapsto \lambda x$. $\qquad \square$

Two norms $\|\cdot\|_1$ and $\|\cdot\|_2$ are *equivalent* if there exist two positive constants c_1 and c_2 such that, for all $x \in E$,

$$c_1 \|x\|_1 \leqslant \|x\|_2 \leqslant c_2 \|x\|_1.$$

In this case, it is easily verified that the topologies induced by the distances $d_1(x, y) = \|x - y\|_1$ and $d_2(x, y) = \|x - y\|_2$ are the same.

Continuous linear maps play a fundamental role, so we provide several simple criteria to characterize them.

Proposition 1.5 *Let $(E, \|\cdot\|_E)$ and $(F, \|\cdot\|_F)$ be two normed vector spaces and $T: E \to F$ a linear map. The following properties are equivalent:*

1. *T is continuous;*
2. *T is continuous at 0;*
3. *T is bounded on a neighborhood of the origin, which means that there exists a neighborhood V of the origin such that $\sup_{x \in V} \|T(x)\|_F < +\infty$.*

Proof Let us show that (3) implies (2). To show that T is continuous at 0, as $T(0) = 0$, it suffices to show that, for any ball $B_F(0, r)$ in F, there exists a ball $B_E(0, \rho)$ in E such that $B_E(0, \rho)$ is included in $T^{-1}(B_F(0, r))$. As T is bounded on a neighborhood of the origin by hypothesis, there exist $R > 0$ and $M > 0$ such that $\|T(u)\|_F \leqslant M$ for all u satisfying $\|u\|_E \leqslant R$. Let us then set $\rho = (rR)/(2M)$ and suppose that x belongs to $B_E(0, \rho)$. Then $u = (2M/r)x$ satisfies $\|u\|_E \leqslant R$, therefore $\|T(u)\|_F \leqslant M$. But $T(u) = (2M/r)T(x)$, so that $\|T(x)\|_F \leqslant r/2$, which shows that $x \in T^{-1}(B_F(0, r))$ and proves the desired result.

Note that (1) trivially implies (2). Conversely, let us show that (2) implies (1). For this, we recall that the translation by any vector is a homeomorphism according to Corollary 1.4. Therefore, if T is continuous at 0, it is continuous at any point, which implies that T is continuous on E.

Finally, property (2) directly implies (3). It suffices to observe that there exists a ball centered at 0 and of radius $r > 0$ included in the preimage of $B_F(0, 1)$. $\qquad \square$

We denote by $\mathcal{L}(E, F)$ the set of continuous linear maps from E to F. This is a vector space. We define a norm on this vector space by setting

$$\|T\| := \sup_{x \neq 0} \frac{\|T(x)\|_F}{\|x\|_E}.$$

An important special case is where $E = F$. In this case, we can compose two continuous linear maps T_1 and T_2 to obtain another continuous linear map (commonly written $T_1 T_2$ instead of $T_1 \circ T_2$). We verify that

$$\|T_1 T_2\| \leqslant \|T_1\| \, \|T_2\|.$$

Another important special case is where F is the field $\mathbb{K}$ equipped with the topology induced by the absolute value if $\mathbb{K} = \mathbb{R}$ or by the modulus if $\mathbb{K} = \mathbb{C}$. We denote this space by $E' = \mathcal{L}(E, \mathbb{K})$, and call it the topological dual of E (the dual space is often alternatively denoted by E^*; however in this book we prefer the notation E'). We will see in Chapter 3 that a very fruitful point of view is to study the dual space. This will allow us to introduce a dual notion of convergence in the sense of norms, which corresponds to a notion of convergence in the sense of coordinates.

Definition 1.6 A normed vector space that is complete for the distance induced by the norm is called a *Banach space*.

Proposition 1.7 *Let E be any normed vector space and F a Banach space. Then $\mathcal{L}(E, F)$ is a Banach space. In particular, the topological dual of any normed vector space is a Banach space.*

Proof Let $(T_n)_{n \in \mathbb{N}}$ be a Cauchy sequence in $(\mathcal{L}(E, F), \|\cdot\|)$. For all $x \in E$, we have

$$\left\| T_n(x) - T_p(x) \right\|_F \leqslant \left\| T_n - T_p \right\| \cdot \|x\|_E, \tag{1.1}$$

so $(T_n(x))_{n \in \mathbb{N}}$ is a Cauchy sequence in F. As F is complete by hypothesis, this sequence converges to an element denoted $T(x)$. By linearity of the limit, we verify that $E \ni x \mapsto T(x) \in F$ is a linear map. Moreover, the sequence $(\|T_n\|)_{n \in \mathbb{N}}$ is also a Cauchy sequence in $\mathbb{R}$, so it is bounded. Taking the limit in

$$\|T_n(x)\| \leqslant \|T_n\| \, \|x\|,$$

we find that $\|T(x)\| \leqslant M \|x\|$ with $M = \sup_{n \in \mathbb{N}} \|T_n\|$, which implies that T is bounded on a neighborhood of the origin and therefore continuous.

To conclude, let us show that $(T_n)_{n \in \mathbb{N}}$ converges to T in $(\mathcal{L}(E, F), \|\cdot\|)$. Given $\varepsilon > 0$, there exists an integer N such that the right-hand side of (1.1) is smaller than $\varepsilon \|x\|$ when $n \geqslant N$ and $p \geqslant N$. By letting p tend to $+\infty$, we deduce that $\|T_n(x) - T(x)\|_F \leqslant \varepsilon \|x\|_E$, for all $n \geqslant N$ and all x in E. In particular, $\|T_n - T\| \leqslant \varepsilon$, which proves the desired result. $\qquad \square$

1.2 Convex, Bounded, Balanced Sets

The notion of a topological vector space generalizes that of a normed vector space.

Definition 1.8 A *topological vector space* is a $\mathbb{K}$-vector space E, with $\mathbb{K} = \mathbb{R}$ or $\mathbb{C}$, equipped with a separated topology such that the operations of vector addition $(x, y) \mapsto x + y$ and scalar multiplication $(\lambda, x) \mapsto \lambda x$ are continuous.

It follows that translations and dilations (by a scalar $\lambda \neq 0$) are homeomorphisms (the proof is exactly the same as that of Corollary 1.4).

When studying topological vector spaces, the main concepts that come into play are the following three properties.

Definition 1.9 Let E be a topological vector space over the field $\mathbb{K} = \mathbb{R}$ or $\mathbb{C}$, and let A be a subset of E.

1. We say that A is bounded if, for every neighborhood V of 0, there exists $s > 0$ such that $A \subset tV$ for all scalars t satisfying $|t| \geqslant s$.
2. We say that A is convex if, for every pair (x, y) of elements of A, the segment $[x, y] = \{\lambda x + (1 - \lambda)y : \lambda \in [0, 1]\}$ is included in A.
3. We say that A is balanced if $\lambda A \subset A$ for every scalar $\lambda \in \mathbb{K}$ satisfying $|\lambda| \leqslant 1$.

Proposition 1.10 *Let A be a subset of a topological vector space E. Then the following properties are satisfied:*

1. $\overline{A} = \bigcap_V (A + V)$ *where the intersection is taken over all neighborhoods V of the origin;*
2. $\overline{A} + \overline{B} \subset \overline{A + B}$;
3. *every neighborhood V of the origin contains a balanced neighborhood;*
4. *for every open neighborhood V of the origin, there exists an open neighborhood U of the origin such that $\overline{U} + \overline{U} \subset V$;*
5. *every open neighborhood V of the origin contains an open balanced U neighborhood of the origin such that $\overline{U} + \overline{U} \subset V$;*
6. *if A is bounded, then $\overline{A}$ is bounded.*

Proof 1. Recall that $x \in \overline{A}$ if and only if every neighborhood of x intersects A. Let $x \in \overline{A}$ and V be a neighborhood of the origin. Then $x - V$ is a neighborhood of x because the map $y \mapsto x - y$ is a homeomorphism. Therefore $x - V$ intersects A and we deduce that x belongs to $A + V$. Conversely, suppose that x belongs to $\bigcap_V (A + V)$ and consider a neighborhood W of x. Since $x - W$ is a neighborhood of 0, x belongs to $A + (\{x\} - W)$, which implies that there exist $a \in A$ and $w \in W$ such that $x = a + (x - w)$. We deduce that $a = w$, so $A \cap W \neq \emptyset$, which proves that $x \in \overline{A}$.

2. Let $x \in \overline{A} + \overline{B}$. Let us show that $x \in \overline{A + B}$. It suffices according to point (1) to prove that, for every neighborhood V of the origin, we have $x \in A + B + V$. Note for this that, by continuity of addition, there exist two open neighborhoods of the origin, V_1 and V_2, such that $V_1 + V_2 \subset V$. We conclude by observing that $x \in A + V_1 + B + V_2 \subset A + B + V$.

3. Let V be an open neighborhood of the origin. By continuity of scalar multiplication at 0, there exists $\delta > 0$ and a neighborhood W of 0 in E such that $tx \in V$ for
   $|t| \leqslant \delta$ and $x \in W$. Then $U = \bigcup_{|t|\leqslant\delta} tW$ is an open neighborhood of the origin
   that is balanced and contained in V.
4. Consider a neighborhood V of the origin. As we have already said, by continuity
   of addition, there exist two open neighborhoods of the origin, V_1 and V_2, such that
   $V_1 + V_2 \subset V$. Let $W = V_1 \cap V_2$ so that $W + W \subset V$. The same argument implies that
   there exists an open neighborhood of the origin such that $U + U \subset W$. Then, point
   (2) implies that $\overline{U} + \overline{U} \subset \overline{U + U}$ and point (1) implies that $\overline{U + U} \subset (U + U) + W$.
   But $U + U + W \subset W + W \subset V$, which concludes the proof.
5. This is a direct consequence of (3) and (4).
6. Suppose that A is bounded and consider an open neighborhood V of 0. We have
   seen that there exists a balanced open set W such that $W + W \subset V$. Since A is
   bounded, there exists $s > 0$ such that $A \subset tW$ for all scalars t satisfying $|t| \geqslant s$.
   Moreover, point (1) implies that $\overline{A} \subset A + W$. Since W is balanced, it follows that
   $\overline{A} \subset tW + W \subset t(W + W) \subset tV$ if $|t| \geqslant s + 1$, which shows that $\overline{A}$ is bounded. $\square$

Definition 1.11 Let $(E, \mathcal{T})$ be a topological vector space. We say that E is normable
if there exists a norm $\|\cdot\|$ on E such that the topology induced by this norm coincides
with $\mathcal{T}$.

Proposition 1.12 *A topological vector space is normable if and only if there exists
a convex and bounded neighborhood of the origin.*

Proof Let E be a topological vector space over the field $\mathbb{K} = \mathbb{R}$ or $\mathbb{C}$. If E is
normable, then the open ball centered at 0 and of radius 1 is a convex and bounded
neighborhood of the origin. Conversely, suppose there exists a convex and bounded
neighborhood C of the origin.

Lemma 1.13 *There exists a convex and balanced neighborhood U of the origin such
that $U \subset C$.*

Proof Consider $V = \bigcap_{|\lambda|=1} \lambda C$ and let U be the interior of V. We verify that U is
convex (because V is convex) and balanced. $\square$

We then define the Minkowski functional of U by

$$\mu(x) = \inf\{t > 0 : x \in tU\}.$$

The idea is as follows: if $(E, \|\cdot\|)$ is a normed space, then

$$\|x\| = \inf\{t > 0 : x \in tB(0, 1)\}.$$

The following lemma states that μ is precisely a norm and that U corresponds to the
unit ball for this norm.

Lemma 1.14 *The Minkowski functional μ is a norm on E.*

Proof Let $I(x) = \{t > 0 : x \in tU\}$. This set is non-empty because U is a neighborhood of the origin and thus, by continuity of scalar multiplication, εx belongs to U for $|\varepsilon|$ small enough, which is equivalent to $x \in tU$ for t large enough. Now, let us consider an element t from $I(x)$. There exists then $a \in U$ such that $x = ta$. For all $s > t$, by writing

$$x = s\left(\frac{t}{s}a + \left(1 - \frac{t}{s}\right)0\right),$$

and noting that the vector inside the parentheses belongs to U by convexity, we see that $s \in I(x)$. In particular, $I(x)$ is a half-line: $I(x) = (\mu(x), +\infty)$ or $I(x) = [\mu(x), +\infty)$.
  Let us show that μ satisfies the triangle inequality:

$$\mu(x + y) \leqslant \mu(x) + \mu(y).$$

Consider two vectors x and y and two real numbers s and t such that $s > \mu(x)$ and $t > \mu(y)$. It follows that $x \in sU$ and $y \in tU$. As before, by writing

$$x + y = (s + t)\left(\frac{s}{s+t}\left(s^{-1}x\right) + \frac{t}{s+t}\left(t^{-1}y\right)\right),$$

we deduce that $x + y$ belongs to $(s + t)U$ by convexity of U. This implies that $\mu(x + y) \leqslant s + t$. By taking the lower bound in s then in t, we deduce the desired inequality $\mu(x + y) \leqslant \mu(x) + \mu(y)$.
  Next, we directly verify that $\mu(\lambda x) = \lambda\mu(x)$ for all $\lambda > 0$. Then, using the fact that U is balanced ($\lambda U \subset U$ for all scalar $\lambda \in \mathbb{K}$ of modulus smaller than 1), we find that

$$\mu(\lambda x) = |\lambda|\,\mu(x).$$

It remains to verify that $\mu(x) = 0$ if and only if $x = 0$. We clearly have $\mu(0) = 0$. Conversely, we consider x in E such that $\mu(x) = 0$ and show that $x = 0$. Suppose by contradiction that $x \neq 0$. As the singleton $\{0\}$ is closed, we deduce that there exists a neighborhood V of 0 that does not contain x. As U is bounded, there exists $r > 0$ such that $U \subset rV$. In particular, we deduce that rx does not belong to U (otherwise rx would belong to rV and x would belong to V). This implies that $\mu(x) \geqslant r^{-1}$. □

  Let us conclude by showing that the topology induced by the norm μ is that of E. For this, note that the collection of sets $\{rU : r > 0\}$ forms a local basis of the origin. Indeed, for every neighborhood V of the origin, there exists $t > 0$ such that $U \subset tV$. We deduce that $t^{-1}U \subset V$. Moreover, rU is exactly the ball of center 0 and radius r for the norm μ. So the two local basis of the origin coincide. □

1.3 Finite-Dimensional Spaces

Let us recall two fundamental and elementary results in the study of finite-dimensional spaces: (i) the closed unit ball is compact and (ii) all norms are equivalent. We will study in this section two extensions of these results.

1.3.1 Compactness and Finite Dimension

Theorem 1.15 (Riesz) *Let E be a normed vector space. If the closed unit ball is compact, then E is finite-dimensional.*

Proof We will argue by contraposition and show that if E is infinite-dimensional then the unit ball is not compact. For this, we will use the following lemma.

Lemma 1.16 *Let F be a closed linear subspace of E, different from E. Then there exists a vector $u \in E$ such that $\|u\|_E = 1$ and $\mathrm{dist}(u, F) \geqslant 1/2$.*

Proof Let an element x belong to $E \setminus F$. Since F is a closed set, we have $\mathrm{dist}(x, F) > 0$. So $2 \, \mathrm{dist}(x, F) > \mathrm{dist}(x, F)$ and by definition of the distance from a point to a set, we deduce that there exists y in F such that $\|x - y\| \leqslant 2 \, \mathrm{dist}(x, F)$. Let us set

$$u = \frac{1}{\|x - y\|} (x - y).$$

Then u is clearly of norm 1 and moreover, for all z in F, we have

$$\|u - z\| = \frac{1}{\|x - y\|} \left\| x - y - \|x - y\| z \right\| \geqslant \frac{1}{\|x - y\|} \, \mathrm{dist}(x, F) \geqslant \frac{1}{2},$$

where we used the fact that $y + \|x - y\| z$ belongs to F and the fact that $\|x - y\| \leqslant 2 \, \mathrm{dist}(x, F)$ by definition of y.   □

We will use this lemma to construct a sequence $(u_n)_{n \in \mathbb{N}}$ of vectors of norm $\|u_n\| = 1$ satisfying the following property:

$$\forall (n, m) \in \mathbb{N} \times \mathbb{N}, \quad n \neq m \implies \|u_n - u_m\| \geqslant \frac{1}{2}. \tag{1.2}$$

For this, we choose a first vector u_0 of norm 1 then we construct the sequence by induction, the existence of u_n following from Lemma 1.16 applied to the subspace F_n generated by the vectors defined in the previous steps.

Then the sequence $(u_n)_{n \in \mathbb{N}}$ has no subsequential limit, which proves that the space E is not compact.   □

1.3.2 Uniqueness of Topology in Finite-Dimensional Spaces

A Hausdorff space, also called a separated space, is a topological space $(X, \mathcal{T})$ where, for any two distinct points, there exist neighbourhoods of each that are disjoint from each other (see Section 17.2.2). We then say that the topology $\mathcal{T}$ is separated.

Theorem 1.17 *Let E be a finite-dimensional vector space over the field $\mathbb{K} = \mathbb{R}$ or $\mathbb{K} = \mathbb{C}$. There exists a unique separated topology that equips E with the structure of a topological vector space.*

Proof Let d denote the dimension of E and choose an arbitrary base $\beta = (\beta_1, \ldots, \beta_d)$ of E. We define the norm $\|\cdot\|$ by

$$
x = \sum_{i=1}^{d} x_i \beta_i \implies \|x\| = \left(\sum_{i=1}^{d} |x_i|^2 \right)^{1/2}.
$$

Consider a separated topology τ compatible with the algebraic operations. We will denote by E_n the pair $(E, \|\cdot\|)$, and by E_τ the pair (E, τ). We will show that the identity mapping I is a homeomorphism from E_n to E_τ.

(a) We begin by proving that I is continuous from E_n to E_τ.

Let U be any neighborhood of 0 in E_τ. Then the mapping

$$
E_\tau \times \cdots \times E_\tau \ni (x_1, \ldots, x_d) \longmapsto x_1 + \cdots + x_d \in E_\tau
$$

is continuous because τ is compatible with the addition of E, by assumption.

The continuity of this mapping at 0 implies the existence of d open neighborhoods of 0, $V_i \in \tau$, such that $V_1 + \cdots + V_d \subset U$. Let us set

$$
W = \bigcap_{i=1}^{d} V_i \qquad \text{which satisfies} \qquad W + \cdots + W \subset U. \tag{1.3}
$$

Since the set W is an open neighborhood of 0 in E_τ, the continuity of the multiplication by a scalar implies the existence of d positive reals $\varepsilon_1, \ldots, \varepsilon_d$ such that:

$$
\forall i \in \{1, \ldots, d\}, \forall \xi \in \mathbb{K}, \quad [|\xi| < \varepsilon_i] \implies [\xi \beta_i \in W]. \tag{1.4}
$$

Let $\varepsilon = \min\{\varepsilon_i : 1 \leqslant i \leqslant d\}$, then the ball of E_n centered at 0 and of radius ε is included in U according to (1.3) and (1.4). Therefore I is continuous at 0, and by linearity I is continuous from E_n to E_τ.

(b) We now prove that I is continuous from E_τ to E_n.

Let V be a neighborhood of 0 in E_n. Then V contains an open ball $B_\rho = \{x \in E : \|x\| < \rho\}$. Denote by S the boundary of this ball. It is a closed and bounded set in a finite-dimensional normed vector space, so it is a compact subset of E_n (see Lemma 1.18 below). By continuity of I from E_n to E_τ proven above, S is a compact subset of E_τ and, τ being separated, S is a closed set of τ. As $0 \notin S$, there exists $W \in \tau$ such that $S \cap W = \varnothing$. By continuity of scalar multiplication at 0, there exist

$\delta > 0$ and a neighborhood W' of 0 such that $tx \in W$ for $|t| \leqslant \delta$ and $x \in W'$. Therefore

$$U = \bigcup_{|t| \leqslant \delta} tW' \subset W.$$

We claim that $U \subset V$. Indeed, suppose there exists $x \in U \setminus V$, in particular $\|x\| \geqslant \rho$. Let $\lambda = \rho/\|x\|$ and $y = \lambda x$, then $|\lambda| \leqslant 1$ and $x \in U$ implies that $y \in U$. On the other hand, $\|y\| = \rho$, therefore $y \in S$, which contradicts $S \cap W = \emptyset$. We have indeed shown that I is continuous at 0, and by linearity continuous from E_τ to E_n.

The map I is therefore a homeomorphism, which shows $E_n = E_\tau$.

To conclude the proof, it remains to prove the following lemma, which was used above.

Lemma 1.18 *Let $\mathcal{K}$ be a closed and bounded subset of E_n. Then $\mathcal{K}$ is compact.*

Proof We choose a basis $\beta = (\beta_1, \ldots, \beta_d)$ of E arbitrarily, and we define the linear isomorphism:

$$I : E \longrightarrow \mathbb{K}^d, \qquad \sum_{i=1}^{d} x_i \beta_i \longmapsto (x_1, \ldots, x_d).$$

By definition

$$\|x\| = \left(\sum_{i=1}^{d} |x_i|^2 \right)^{1/2},$$

so I is an isometric isomorphism, therefore a homeomorphism. In particular, $I(\mathcal{K})$ is closed and bounded in $\mathbb{K}^d$ therefore compact, and by continuity of I^{-1}, $\mathcal{K}$ is a compact subset of E_n. $\qquad\square$

This concludes the proof of the theorem. $\qquad\square$

Corollary 1.19 *All norms on a finite-dimensional space X are equivalent.*

Proof Consider two norms $\|\cdot\|_1$ and $\|\cdot\|_2$ on a finite-dimensional space X. According to the previous theorem, the identity mapping I is a homeomorphism from $(X, \|\cdot\|_1)$ to $(X, \|\cdot\|_2)$. By continuity of I, and its inverse, at 0, we deduce the existence of two constants δ_1 and δ_2 such that:

$$\|x\|_1 \leqslant \delta_1 \implies \|x\|_2 \leqslant 1, \tag{1.5}$$

$$\|x\|_2 \leqslant \delta_2 \implies \|x\|_1 \leqslant 1. \tag{1.6}$$

The homogeneity property of norms applied to relations (1.5) and (1.6) implies

$$\delta_2 \|x\|_1 \leqslant \|x\|_2 \leqslant \frac{1}{\delta_1} \|x\|_1,$$

which is the desired result of equivalence of the two norms. $\qquad\square$

1.4 Semi-Norms

In this section, we will be interested in topological vector spaces whose topology is not induced by a norm, but by a family of semi-norms. We will not develop the general theory of these spaces but rather consider the simplest case, which will be sufficient to describe all the spaces we will need in this book.[1]

Definition 1.20

1. A graded family of semi-norms is a countable family $\{\rho_n\}_{n\in\mathbb{N}}$ of semi-norms satisfying

$$\rho_0(f) \leqslant \rho_1(f) \leqslant \rho_2(f) \leqslant \cdots$$

   for all f in E.
2. A family $\{\rho_n\}_{n\in\mathbb{N}}$ of semi-norms is separating if and only if

$$x = 0 \iff \forall n \in \mathbb{N}, \ \rho_n(x) = 0.$$

In the following, we only consider separating graded families of semi-norms.

Example 1.21 Several examples are in order concerning this definition.

1. If $(E, \|\cdot\|_E)$ is a normed vector space and $\rho_n = \|\cdot\|_E$ for all n, then $\{\rho_n\}_{n\in\mathbb{N}}$ is a graded family of semi-norms.
2. Consider a compact $K \subset \mathbb{R}^d$ and denote by $C_K^\infty(\mathbb{R}^d)$ the space of functions of class $C^\infty(\mathbb{R}^d)$ with support in K. Then we define a graded family of semi-norms by

$$\rho_n(f) = \max_{|\alpha| \leqslant n} \sup_{x\in K} \left| \partial_x^\alpha f(x) \right|.$$

   These semi-norms are even norms but, in general, we still say 'semi-norm' in practice. The reason for this is that the natural topology on $C_K^\infty(\mathbb{R}^d)$ (given by Proposition 1.23 below) is not that of a normed vector space.
3. Let $k \in \mathbb{N}$ and let Ω be an open set of $\mathbb{R}^d$. We want to define a graded family of semi-norms on $E = C^k(\Omega)$. For this, we consider an exhaustive sequence of compact sets covering Ω (by definition, this is a sequence $(K_n)_{n\in\mathbb{N}}$ of increasing compact sets, such that $K_n \subset \text{int}(K_{n+1})$ and satisfying $\bigcup_{n\in\mathbb{N}} K_n = \Omega$). We obtain such a sequence by setting

$$K_n = \{x \in \Omega : \text{dist}(x, \partial\Omega) \geqslant 1/n\} \cap \overline{B(0, n)}.$$

   Then we can define a graded family of semi-norms by

$$\rho_n(f) = \max_{|\alpha| \leqslant k} \sup_{x\in K_n} \left| \partial_x^\alpha f(x) \right|.$$

---

[1] For a detailed exposition of the classical results on topological vector spaces, we refer to the books of Bony [12], Brezis [15], Hörmander [79], Rudin [164] and Zuily [200, 201].

4. Let us denote by $L_{loc}^p(\Omega)$ the space of functions whose restriction to any bounded open set belongs to $L^p(\Omega)$. Then we define a graded family of semi-norms by

$$\rho_n(f) = \|f\|_{L^p(K_n)}.$$

Let us recall the notion of a neighborhood basis.

Definition 1.22 Consider a topological space X equipped with a topology $\mathcal{T}$ and a collection $\mathcal{B}$ of subsets of X. We say that:

- $\mathcal{B}$ is a basis of the topology $\mathcal{T}$ if every open set of the topology is a union of elements of $\mathcal{B}$.
- If $\mathcal{B}$ is a collection of subsets of X that is a basis for the topology $\mathcal{T}$, then we say that $\mathcal{T}$ is the topology induced by $\mathcal{B}$ (we verify the uniqueness of the topology induced by $\mathcal{B}$).
- $\mathcal{B}$ is a neighborhood basis of a point x_0 if for every neighborhood V of x_0, there exists a $U \in \mathcal{B}$ such that $x_0 \in U \subset V$.

Proposition 1.23 *Consider a vector space E equipped with a separating and graded family of semi-norms $\{\rho_n\}_{n \in \mathbb{N}}$. Given $x \in E$ and $\varepsilon > 0$, we introduce*

$$B_n(x, \varepsilon) = \{y \in E : \rho_n(y - x) < \varepsilon\},$$

and equip E with the topology induced by the collection

$$\mathcal{B} = \{B_n(x, \varepsilon) : n \in \mathbb{N}, \ x \in E, \ \varepsilon > 0\}.$$

Then we have the following properties:

1. *For all $x_0 \in E$, $\mathcal{B}_{x_0} = \{B_n(x_0, \varepsilon) : n \in \mathbb{N}, \ \varepsilon > 0\}$ is a neighborhood basis of x_0.*
2. *E is a topological vector space.*
3. *The convergence of a sequence $(x_j)_{j \in \mathbb{N}}$ towards x is equivalent to:*

$$\forall n \in \mathbb{N}, \quad \lim_{j \to +\infty} \rho_n(x_j - x) = 0.$$

4. *This topology is metrizable and it is induced by the distance*

$$d(x, y) = \sum_{n \in \mathbb{N}} 2^{-n} \frac{\rho_n(x - y)}{1 + \rho_n(x - y)}.$$

Proof 1. Consider an element x_0 of E and a neighborhood V of x_0. Then V contains an open set U that contains x_0. By definition of the topology induced by a basis, V contains a ball: there exist $n \in \mathbb{N}$, $x_1 \in E$ and $\varepsilon > 0$ such that $x_0 \in B_n(x_1, \varepsilon) \subset U$. In particular, we have $\rho_n(x_0 - x_1) < \varepsilon$. Let us then set $\delta = (\varepsilon - \rho_n(x_0 - x_1))/2$. Then, for every $x \in B_n(x_0, \delta)$, using the triangle inequality, we see that

$$\rho_n(x - x_1) \leqslant \rho_n(x - x_0) + \rho_n(x_0 - x_1) < \delta + \rho_n(x_0 - x_1)$$

$$\leqslant \frac{1}{2}(\varepsilon - \rho_n(x_0 - x_1)) + \rho_n(x_0 - x_1) \leqslant \frac{1}{2}\varepsilon + \frac{1}{2}\rho_n(x_0 - x_1) < \varepsilon.$$

This proves that $B_n(x_0, \delta) \subset B_n(x_1, \varepsilon)$ and therefore that V contains an element of $\mathcal{B}_{x_0}$.

2. Let $(x_0, y_0) \in E \times E$ and let V be a neighborhood of $x_0 + y_0$. According to point (1), there exist $n \in \mathbb{N}$ and $\varepsilon > 0$ such that $B_n(x_0 + y_0, \varepsilon) \subset V$. So, for all x in $B_n(x_0, \varepsilon/2)$ and for all y in $B_n(y_0, \varepsilon/2)$, we have

$$x + y \in B_n(x_0 + y_0, \varepsilon) \subset V$$

according to the triangle inequality. This shows that the vector addition $(x, y) \mapsto x + y$ is continuous at every point of $E \times E$, hence continuous.

3. Consider a sequence $(x_j)_{j \in \mathbb{N}}$ and a point $x \in E$. The convergence of the sequence $(x_j)_{j \in \mathbb{N}}$ towards x means by definition that, for every neighborhood V of x, there exists N such that x_j belongs to V for all j greater than N. By applying this result with $V = B_n(x, 1/k)$, it follows that $\rho_n(x_j - x)$ converges to 0 as j tends to $+\infty$, for all n.

Conversely, suppose that, for all n, $\rho_n(x_j - x)$ converges to 0 as j tends to $+\infty$. Consider a neighborhood V of x. Then, according to the first point, there exist $n \in \mathbb{N}$ and $\varepsilon > 0$ such that $B_n(x, \varepsilon) \subset V$. Since $\rho_n(x_j - x)$ converges to 0, it follows that x_j belongs to $B_n(x, \varepsilon) \subset V$ for sufficiently large j. Therefore, $(x_j)_{j \in \mathbb{N}}$ converges to x.

4. Let us show that the function d is a distance. If $d(x, x) = 0$ then $x = 0$ because the family $\{\rho_n\}_{n \in \mathbb{N}}$ is separating. Next, note that $d(x, y) = d(y, x)$ because $\rho_n(x - y) = \rho_n(y - x)$.

To verify the triangle inequality for d, it suffices to show that for each integer n, the function $(x, y) \mapsto \rho_n(x - y)/(1 + \rho_n(x, y))$ satisfies the triangle inequality. For this, note that the function $t \mapsto t/(1 + t)$ is increasing on $\mathbb{R}^+$, therefore

$$\frac{\rho_n(x - y)}{1 + \rho_n(x - y)} \leqslant \frac{\rho_n(x - z) + \rho_n(z - y)}{1 + \rho_n(x - z) + \rho_n(z - y)}.$$

Next, we verify that for all t_1, t_2 in $\mathbb{R}^+$, we have

$$\frac{t_1 + t_2}{1 + t_1 + t_2} \leqslant \frac{t_1}{1 + t_1} + \frac{t_2}{1 + t_2},$$

which demonstrates that d is a distance.

It remains to show that the topology $\mathcal{T}$ of E coincides with the topology induced by d. We must therefore show that the identity is a homeomorphism from $(E, \mathcal{T})$ to (E, d). For this, it suffices to show that every element $B_n(x_0, \varepsilon)$ of the basis $\mathcal{B}$ contains a ball for the distance d, and vice versa. Let us denote by $B(x, \varepsilon)$ the ball of center x and radius ε for the distance d. Then we have $B(x_0, 2^{-n-1}\varepsilon) \subset B_n(x_0, \varepsilon)$ because $d(x, x_0) < 2^{-n-1}\varepsilon$ directly implies that

$$2^{-n}\frac{\varepsilon}{2} > 2^{-n}\frac{\rho_n(x - x_0)}{1 + \rho_n(x - x_0)},$$

from which we deduce that $\rho_n(x - x_0) < \varepsilon/(2 - 2\varepsilon) \leqslant \varepsilon$ for $\varepsilon \leqslant 1/2$.

Conversely, consider a ball $B(x_0, \varepsilon)$ for the distance d. Choose N such that $2^{-N} < \varepsilon/2$ and suppose that $x \in B_N(x_0, \varepsilon/2)$. Then, as $\rho_0 \leqslant \rho_1 \leqslant \cdots \leqslant \rho_N$, we have, using the growth of the function $t \mapsto t/(1+t)$ on $\mathbb{R}^+$,

$$d(x, x_0) \leqslant \sum_{0 \leqslant n \leqslant N} 2^{-n} \frac{\rho_N(x - x_0)}{1 + \rho_N(x - x_0)} + \sum_{n > N} 2^{-n},$$

from which

$$d(x, x_0) \leqslant 2\rho_N(x - x_0) + \frac{\varepsilon}{2} < \varepsilon,$$

which demonstrates that $B_N(x_0, \varepsilon/2) \subset B(x_0, \varepsilon)$. This completes the proof that the topology induced by d is the same as that of E. □

Definition 1.24 Consider a topological vector space whose topology is induced by a graded family of semi-norms $\{\rho_n\}_{n \in \mathbb{N}}$. We say that E is a *Fréchet space* if it is complete for the distance

$$d(x, y) = \sum_{n \in \mathbb{N}} 2^{-n} \frac{\rho_n(x - y)}{1 + \rho_n(x - y)}.$$

Proposition 1.25 *The local spaces $L^p_{\text{loc}}(\Omega)$, $C^k(\Omega)$ and $C^\infty_K(\Omega)$, equipped with the families of semi-norms introduced in Example 1.21, are Fréchet spaces.*

We skip the proof of this proposition, which is a corollary of the classical fact that the corresponding spaces of functions restricted to K_n are complete metric spaces.

1.5 Banach Spaces

In this section, we will prove fundamental results that rely on fine uses of Baire's theorem. Let us recall the following definition.

Definition 1.26 We say that a topological space is a *Baire space* if every countable intersection of open dense sets is a dense subset.

Remark 1.27 It is equivalent to say that a topological space is a Baire space if every countable union of closed sets with empty interior is a set with empty interior.

Theorem 1.28 (Baire) *Every complete metric space is a Baire space.*

This theorem is proved in the recap on General Topology (Chapter 17). To use this result, the idea is often to find a representation of the entire space as a countable union of closed sets. Then one generally gets the desired result by using the fact that one of these closed sets has a non-empty interior. A simple demonstration of this principle is in the proof that the dimension of a Banach space is either finite or uncountable.

Proposition 1.29 *Let E be a Banach space. Suppose there exists a set $I \subset \mathbb{N}$ and a family $\{e_i\}_{i \in I}$ such that $E = \mathrm{Vect}\{e_i : i \in I\}$. Then I is finite.*

Proof Let us denote by E_n the vector space generated by

$$\{e_i : i \in I \text{ and } i \leqslant n\}.$$

Then E_n is a finite-dimensional vector space and therefore it is closed. Since E is the union of the sets E_n, Baire's theorem implies that one of the sets E_n, denoted E_{n_0}, has a non-empty interior. Consequently, E_{n_0} contains a ball $B(a, r)$. Since $a \in E_{n_0}$, by translation, we can reduce to the case where $a = 0$. But then, if $B(0, r) \subset E_{n_0}$, by homogeneity, we see that E_{n_0} contains any ball $B(0, \lambda r)$, where λ is any positive real number. This implies that $E = E_{n_0}$ and hence I is finite. □

Let E and F be two normed vector spaces. Recall that we denote by $\mathcal{L}(E, F)$ the set of continuous linear maps from E to F. This is a vector space, which we can equip with a norm[2]

$$\|T\|_{\mathcal{L}(E,F)} := \sup_{x \neq 0} \frac{\|Tx\|_F}{\|x\|_E}.$$

Theorem 1.30 (Banach–Steinhaus) *Let E be a Banach space and F a normed vector space. Consider an arbitrary family $\{T_\alpha\}_{\alpha \in A}$ of continuous linear maps from E to F that is pointwise bounded in the sense that*

$$\forall x \in E, \quad \sup_{\alpha \in A} \|T_\alpha x\|_F < +\infty.$$

Then this family is bounded in $\mathcal{L}(E, F)$:

$$\sup_{\alpha \in A} \|T_\alpha\|_{\mathcal{L}(E,F)} < +\infty.$$

Proof For every integer $k \in \mathbb{N}$, let us introduce

$$E_k = \{x \in E : \forall \alpha \in A, \ \|T_\alpha x\|_F \leqslant k\}.$$

By hypothesis, each element x of E belongs to one of these sets and therefore E can be written as the union of the sets E_k. Moreover, each set E_k is closed because it is an intersection of closed sets: we have

$$E_k = \bigcap_{\alpha \in A} \{x \in E : \|T_\alpha x\|_F \leqslant k\},$$

and each of the sets above is closed because the maps $x \mapsto \|T_\alpha x\|_F$ are continuous due to the continuity of the maps T_α and the continuity of the norm. Thus, as the union of the E_k has a non-empty interior, Baire's theorem 1.28 implies that at least

---

[2] From now on, we will simply write $T x$ for the image of x under a linear mapping, instead of $T(x)$ (by analogy with the notation Ax for a vector x and a matrix A).

one of the sets E_k, let us denote it E_K, has a non-empty interior. Suppose that $B(a,r) \subset E_K$ and consider $x \in E$ with $x \neq 0$. Then the element y defined by

$$y = a + \frac{r}{2\|x\|_E} x$$

belongs to the ball $B(a,r)$. We then obtain the desired result by observing that

$$
\begin{aligned}
\|T_\alpha x\|_F &= \left\| T_\alpha \left(\frac{2\|x\|_E}{r} (y-a) \right) \right\|_F \\
&\leqslant \frac{2\|x\|_E}{r} \left(\|T_\alpha y\|_F + \|T_\alpha a\|_F \right) \\
&\leqslant \frac{4K}{r} \|x\|_E,
\end{aligned}
$$

where we first used the triangle inequality and then the fact that a and y belong to $B(a,r) \subset E_K$ to obtain the second inequality.   □

Remark 1.31 Exercise 1.3 proposes an elementary proof of the Banach–Steinhaus theorem, which does not rely on Baire's theorem.

Corollary 1.32 *Let E and F be two Banach spaces. Consider a sequence $(T_n)_{n \in \mathbb{N}}$ of continuous linear maps that converges simply: for every $x \in E$, the sequence $(T_n x)_{n \in \mathbb{N}}$ converges towards a limit, denoted Tx. Then T is a continuous linear map.*

Proof The fact that T is linear is obvious. The hypothesis implies that

$$\forall x \in E, \quad \sup_{n \in \mathbb{N}} \|T_n x\|_F < +\infty.$$

The Banach–Steinhaus theorem implies that there exists a constant C such that, for every $x \in E$, $\|T_n x\|_F \leqslant C\|x\|_E$. We can take the limit in this inequality to obtain that $\|Tx\|_F \leqslant C\|x\|_E$, which implies that T is continuous.   □

We will conclude the study of the Banach–Steinhaus theorem by showing how to extend this result to the framework of topological vector spaces.

Definition 1.33 Let X and Y be two topological vector spaces, and let $\mathcal{F}$ be a family of continuous linear maps from X to Y.

1. We say that the family $\mathcal{F}$ is *pointwise bounded* if, for every x in X, the set $\{f(x) : f \in \mathcal{F}\}$ of images of x under the elements of $\mathcal{F}$ is bounded in Y.
2. We say that $\mathcal{F}$ is equicontinuous if, for every neighborhood $V \subset Y$ of 0, there exists a neighborhood $U \subset X$ of 0 such that $f(U) \subset V$ for every f in $\mathcal{F}$.

Theorem 1.34 (Banach–Steinhaus) *Let X be a Fréchet space, Y a topological vector space and let $\mathcal{F}$ be a family of continuous linear maps from X to Y. If $\mathcal{F}$ is pointwise bounded, then $\mathcal{F}$ is equicontinuous.*

Proof Let V be an open neighborhood of 0 in Y. We want to show that there exists an open neighborhood $U \subset X$ of 0 such that

$$\forall f \in \mathcal{F}, \quad f(U) \subset V. \tag{1.7}$$

We have seen in Proposition 1.10 that there exists a balanced open neighborhood W such that $\overline{W} + \overline{W} \subset V$. Let us define

$$E := \bigcap_{f \in \mathcal{F}} f^{-1}(\overline{W}).$$

By continuity of the elements of $\mathcal{F}$, E is an intersection of closed sets and therefore E is closed. The goal is to show that E has a non-empty interior. Indeed, if we know that there exists an open set U included in E, then the desired property (*cf.* (1.7)) will be verified. For this, we will apply Baire's theorem (see Theorem 1.28) by first showing that

$$X = \bigcup_{n \geqslant 1} nE.$$

To obtain this equality, consider the sets $\mathcal{F}_x = \{f(x) : f \in \mathcal{F}\}$. For every x in X, the set $\mathcal{F}_x$ is bounded by the hypothesis made on $\mathcal{F}$. Therefore, there exists an integer n_x such that $\mathcal{F}_x$ is included in $n_x W$. Also, $f(x)$ belongs to $n_x W$, for every f in $\mathcal{F}$. Thus, $x \in f^{-1}(n_x W)$, for every f in $\mathcal{F}$, and as f is linear

$$x \in n_x f^{-1}(W), \quad \forall f \in \mathcal{F}.$$

This implies $x \in n_x E$ and we indeed have $X = \bigcup_{n \geqslant 1} nE$.

We can then use Baire's theorem to obtain that E has a non-empty interior. Consider $x_0 \in E^\circ$ and define $U := x_0 - E^\circ$, which is a neighborhood of 0. For $f \in \mathcal{F}$, we conclude that

$$f(U) = f(x_0) - f(E^\circ) \subset \overline{W} - \overline{W} = \overline{W} + \overline{W} \subset V.$$

We used the fact that $\overline{W}$ is balanced because W is balanced. $\qquad\square$

We will now study another famous result of Banach.

Theorem 1.35 (Banach isomorphism) *Let E and F be two Banach spaces and $T : E \to F$ a continuous, bijective linear map from E to F. Then the inverse T^{-1} is continuous.*

Proof We need to show that there exists a constant C such that, for every y in F,

$$\left\|T^{-1}y\right\|_E \leqslant C \left\|y\right\|_F .$$

By linearity, this is equivalent to showing that

$$T^{-1}\big(B_F(0,1)\big) \subset B_E(0,C).$$

However, since T is bijective, we have $T\left(T^{-1}(B_F(0,1))\right) = B_F(0,1)$. Also, we only need to prove that there exists $C > 0$ such that $B_F(0,1) \subset T(B_E(0,C))$. It is of course equivalent to show that there exists $\delta > 0$ such that

$$B_F(0,\delta) \subset T(B_E(0,1)). \tag{1.8}$$

Let us start by proving the following lemma.

Lemma 1.36 *If there exists a constant $c > 0$ such that*

$$B_F(0,c) \subset \overline{T(B_E(0,1))},$$

then $B_F(0,c/2) \subset T(B_E(0,1))$.

Proof Let $y \in B_F(0,c)$. Then y belongs to $\overline{T(B_E(0,1))}$. Consequently, there exists $y_0 \in T(B_E(0,1))$ such that $\|y - y_0\|_F \leqslant c/2$. By homogeneity, the lemma's assumption implies that $B_F(0,c/2) \subset \overline{T(B_E(0,1/2))}$. We can therefore find $y_1 \in T(B_E(0,1/2))$ such that $\|y - y_0 - y_1\|_F \leqslant c/4$. By proceeding recursively, we define a sequence $(y_n)_{n\in\mathbb{N}}$ such that

$$y_n \in T\left(B_E(0,2^{-n})\right) \quad \text{and} \quad \|y - y_0 - \cdots - y_n\|_F \leqslant 2^{-n-1}c.$$

Let $x_n \in B_E(0,2^{-n})$ such that $y_n = Tx_n$. As $\|x_n\|_E < 2^{-n}$, the series $\sum x_n$ converges normally and therefore converges because E is a Banach space. The limit of this series, denoted z, is of norm strictly smaller than 2 and satisfies $y = Tz$. We have shown that $B_F(0,c) \subset T(B_E(0,2))$, which implies the desired result by linearity. $\square$

From the above, to conclude the proof, it remains only to show that there exists $c > 0$ such that $B_F(0,c) \subset \overline{T(B_E(0,2))}$; we then obtain the desired result (1.8) with $\delta = c/4$. For this, we will apply Baire's theorem 1.28 with the sequence of closed sets F_n defined by $F_n = \overline{T(B_E(0,n))}$. As T is surjective, we have

$$F = \bigcup_{n\in\mathbb{N}} F_n.$$

Baire's theorem implies that one of the closed sets has a non-empty interior. As $F_n = n\overline{T(B_E(0,1))}$, we deduce that there exists an element $y_0 \in F$ and a positive number c such that $B_F(y_0,c) \subset \overline{T(B_E(0,1))}$. By symmetry, $-y_0$ also belongs to $\overline{T(B_E(0,1))}$. It follows that $B_F(0,c) \subset \overline{T(B_E(0,2))}$. This concludes the proof of the theorem. $\square$

Corollary 1.37 (Equivalence of norms) *Let E be a vector space equipped with two norms $\|\cdot\|_1$ and $\|\cdot\|_2$. We assume that E is a Banach space for these two norms. Suppose further that there exists a constant C such that*

$$\forall x \in E, \quad \|x\|_1 \leqslant C\|x\|_2.$$

Then there exists a constant C' such that $\|x\|_2 \leqslant C'\|x\|_1$ for all $x \in E$.

Proof It suffices to apply Theorem 1.35 to the identity mapping, from $(E, \|\cdot\|_2)$ into $(E, \|\cdot\|_1)$. □

Definition 1.38 Consider two normed vector spaces E and F and a linear map $T: E \to F$. By definition, the graph of T is the subset $G(T)$ of $E \times F$ defined by

$$G(T) = \{(x, y) \in E \times F : y = Tx\}.$$

Theorem 1.39 (Closed Graph Theorem) *Let E and F be two Banach spaces and $T: E \to F$ a linear map. Then T is continuous if and only if its graph is closed in $E \times F$.*

Proof Suppose that T is continuous. Then $G(T)$ is a closed set. Indeed, if (x, y) belongs to the closure of $G(T)$, then there exists a sequence $(x_n)_{n \in \mathbb{N}}$ converging to x and such that $(Tx_n)_{n \in \mathbb{N}}$ converges to y. By continuity we have $y = Tx$ and therefore $(x, y) \in G(T)$. This demonstrates that $G(T) = \overline{G(T)}$.

It remains to show that if the graph of T is closed, then T is continuous. For this, we introduce the map $N: E \to [0, +\infty)$ defined by

$$N(x) = \|x\|_E + \|Tx\|_F.$$

Then we easily verify that this is a norm on E. Let us show that E, equipped with this new norm, is a Banach space. Consider a Cauchy sequence $(x_n)_{n \in \mathbb{N}}$ for the norm N. Then $(x_n)_{n \in \mathbb{N}}$ is a Cauchy sequence in $(E, \|\cdot\|_E)$ and $(Tx_n)_{n \in \mathbb{N}}$ is a Cauchy sequence in $(F, \|\cdot\|_F)$. Therefore, there exist $x \in E$ and $y \in F$ such that $\|x_n - x\|_E$ and $\|Tx_n - y\|_F$ converge to 0 as n tends to $+\infty$. As the graph of T is closed by hypothesis, we deduce that $y = Tx$. This demonstrates that $N(x_n - x)$ tends to 0 as n tends to $+\infty$, and therefore E is a Banach space for the norm N.

As we trivially have $\|x\|_E \leqslant N(x)$, it follows from Corollary 1.37 that there exists a constant C such that $N(x) \leqslant C\|x\|_E$ for all x in E, which implies that $\|Tx\|_F \leqslant C\|x\|_E$. This proves that T is continuous. □

1.6 The Space of Continuous Functions

In this section, we will consider a central example in the study of Banach spaces: the space of continuous and bounded functions, equipped with the uniform norm. It is classical that this is a Banach space. We will study the compact subsets of this space (the Arzelà–Ascoli theorem) and exploit its algebraic structure to study the existence of dense sub-algebras (a central example, treated by Weierstrass, concerns the case of polynomials; we can also study the trigonometric polynomials of Fourier). Finally, we will see subtle links, related to the works of Baire and Banach, between continuity and uniform convergence.

1.6.1 The Arzelà–Ascoli Theorem

Definition 1.40 Let $(X, \mathcal{T})$ be a topological space, (Y, ρ) be a metric space and $\mathcal{F}$ be a family of functions $f \colon X \to Y$. Given an element x of X, we denote $\mathcal{V}(x)$ the set of neighborhoods of x. By definition, $\mathcal{F}$ is a family of equicontinuous functions as soon as

$$\forall x \in X, \ \forall \varepsilon > 0, \ \exists O \in \mathcal{V}(x) \ / \ \forall z \in O, \ \forall f \in \mathcal{F}, \qquad \rho(f(x), f(z)) < \varepsilon.$$

Theorem 1.41 (Arzelà–Ascoli) *Let* $(X, \mathcal{T})$ *be a separable topological space*, (Y, ρ) *a metric space and* $\mathcal{F} = \{f_n \colon X \to Y\}_{n \in \mathbb{N}}$ *a sequence of equicontinuous functions. If the closure of the set* $\{f_n(x) : n \in \mathbb{N}\}$ *is compact for all* $x \in X$, *then:*

1. *there exists a subsequence of* $\mathcal{F}$ *that converges pointwise to a continuous function* $f \colon X \to Y$;
2. *the convergence is uniform on all compact sets.*

Proof Let us denote by $\mathcal{D} = \{x_n : n \in \mathbb{N}\}$ a countable dense subset in X. By hypothesis, the closure of the set $\{f_n(x_0) : n \in \mathbb{N}\}$ is a compact set in Y. As Y is a metric space, we can extract a subsequence of $(f_n(x_0))_{n \in \mathbb{N}}$, denoted $(f_{0,n}(x_0))_{n \in \mathbb{N}}$, which converges. We then successively construct a sequence $(f_{j+1,n}(x_{j+1}))_{n \in \mathbb{N}}$ from $(f_{j,n}(x_{j+1}))_{n \in \mathbb{N}}$ which converges. Then, for all $j \geqslant 1$, we consider, by diagonal extraction, the sequence $(f_{n,n}(x_j))_{n \geqslant j}$. This is a subsequence of $(f_{j,n}(x_j))_{n \in \mathbb{N}}$ so it converges. We set $g_n = f_{n,n}$. By construction, the sequence $(g_n)_{n \in \mathbb{N}}$ converges pointwise on $\mathcal{D}$.

We will now show that the sequence $(g_n)_{n \in \mathbb{N}}$ converges pointwise on X. For this, it is enough to show that, for any element $x \in X$, the sequence $(g_n(x))_{n \in \mathbb{N}}$ is Cauchy in Y. Indeed, $\{g_n(x) : n \in \mathbb{N}\}$ is included in $\overline{\{f_n(x) : n \in \mathbb{N}\}} \subset Y$, which is a compact set and therefore complete. Let us fix $\varepsilon > 0$. Since $\mathcal{F}$ is equicontinuous, there exists a neighborhood O of x such that

$$\forall n \in \mathbb{N}, \ \forall z \in O, \quad \rho(g_n(x), g_n(z)) < \varepsilon/3.$$

Since $\mathcal{D}$ is dense in X, there exists $x_i \in \mathcal{D}$ such that $x_i \in O$. Finally, by convergence of $(g_n(x_i))_{n \in \mathbb{N}}$, there is N such that, for $m, n \geqslant N$, we have $\rho(g_n(x_i), g_m(x_i)) < \varepsilon/3$. We deduce that:

$$\rho(g_n(x), g_m(x)) \leqslant \rho(g_n(x), g_n(x_i)) + \rho(g_n(x_i), g_m(x_i)) + \rho(g_m(x_i), g_m(x))$$
$$\leqslant \varepsilon/3 + \varepsilon/3 + \varepsilon/3.$$

This shows that $(g_n(x))_{n \in \mathbb{N}}$ is a Cauchy sequence and therefore converges. Let us denote by $f(x)$ the limit of this sequence.

Let us show that f is continuous.[3] Let x be in X and $\varepsilon > 0$. There exists a neighborhood $O \subset X$ of x such that, for every integer n,

$$\forall z \in O, \quad \rho(f_n(x), f_n(z)) < \varepsilon/3.$$

Then, for every integer n, and every z in O, we trivially have the inequality $\rho(g_n(x), g_n(z)) < \varepsilon/3$. Moreover, by pointwise convergence, there exists N such that if $n \geqslant N$ then $\rho(f(x), g_n(x)) \leqslant \varepsilon/3$ and $\rho(f(y), g_n(y)) \leqslant \varepsilon/3$. We use the triangle inequality again to write that, for every y in O,

$$\rho(f(x), f(y)) \leqslant \rho(f(x), g_n(x)) + \rho(g_n(x), g_n(y)) + \rho(g_n(y), f(y))$$
$$\leqslant \varepsilon/3 + \varepsilon/3 + \varepsilon/3.$$

This shows that f is continuous at x.

Now let K be a compact set of X. Let us show that the convergence is uniform on K. The property of equicontinuity implies that, for every element x of X, there exists a neighborhood O_x of x such that,

$$\forall z \in O_x, \ \forall f \in \mathcal{F}, \quad \rho(f(x), f(z)) < \varepsilon.$$

Then $(O_x)_{x \in K}$ forms an open cover of K, from which we can extract a finite subcover indexed by $x_1, \ldots, x_M$. We can therefore find an N such that for every $n \geqslant N$ and for every $i \in \{1, \ldots, M\}$, $\rho(g_n(x_i), g(x_i)) < \varepsilon/3$. Now, for every $x \in K$:

$$\rho(g_n(x), g(x)) \leqslant \rho(g_n(x), g_n(x_i)) + \rho(g_n(x_i), g(x_i)) + \rho(g(x_i), g(x))$$
$$< \varepsilon/3 + \varepsilon/3 + \varepsilon/3 = \varepsilon.$$

This implies the desired result and concludes the proof of the theorem.   □

1.6.2 The Stone–Weierstrass Theorem

The space of continuous functions has the property of being an algebra. It is natural to seek to determine if certain remarkable subalgebras are dense in this space. The main example concerns the study of the approximation of continuous functions by polynomial functions. There are two fundamental results. The first, due to Weierstrass, states that any continuous function on a compact interval can be approximated by algebraic polynomials $P(x) = \sum_{n=0}^{N} a_n x^n$. The second shows that a periodic continuous function can be approximated by trigonometric polynomials, of the form $P(x) = c_0 + \sum_{n=0}^{N}(a_n \cos(nx) + b_n \sin(nx))$. We will state and prove these results at the end of this section. We will start by proving a general result due to Stone [178].

---

[3] In general, continuity is not preserved by passing to the pointwise limit. For example, the sequence $f_n \colon x \mapsto x^n$, on $[0, 1]$, converges pointwise to a discontinuous function at 1.

Theorem 1.42 (Stone–Weierstrass) *Let X be a compact topological space. Let us equip the space $C(X;\mathbb{R})$ of real-valued continuous functions with the uniform norm, $\|f\| = \sup_{x\in X}|f(x)|$. Consider a unitary subalgebra A of $C(X;\mathbb{R})$ (that is, a subset of $C(X;\mathbb{R})$ that contains the constant functions and is stable under addition and multiplication: if f, g are in A then $f + g$ and fg are in A). We also assume that A separates the points of X in the sense that for all x, y in X with $x \neq y$, there exists $f \in A$ such that $f(x) \neq f(y)$. Then A is dense in $C(X;\mathbb{R})$.*

Proof The proof relies on three different ideas. The first concerns the resolution of the problem of approximating the absolute value function by polynomials.

Lemma 1.43 *For all $a > 0$, there exists a sequence of polynomials $(p_n)_{n\in\mathbb{N}}$ which converges uniformly to the absolute value function on $[-a, a]$.*

Proof By the elementary change of variables $u(x) \mapsto u(x/a)$, we reduce to the case $a = 1$. We will start by constructing an auxiliary sequence of polynomials that converges uniformly to the square root function on the interval $[0, 1]$. Consider for this the sequence of functions $P_n \colon [0, 1] \to \mathbb{R}$ defined by:

$$P_0 = 0 \quad \text{and} \quad P_{n+1}(x) = P_n(x) + \frac{1}{2}(x - P_n(x)^2). \tag{1.9}$$

Then P_n is a polynomial function such that $0 \leqslant P_n(x)$. Let us show by induction that $P_n(x) \leqslant \sqrt{x}$ for all x in $[0, 1]$. For this, we write

$$P_{n+1}(x) - P_n(x) = \frac{1}{2}(\sqrt{x} - P_n(x))(\sqrt{x} + P_n(x)),$$

then we use the induction hypothesis to study the right-hand side. We find that the first factor $\sqrt{x} - P_n(x)$ is positive or zero and that the second $\sqrt{x} + P_n(x)$ is bounded by 2, hence the desired result:

$$P_{n+1}(x) \leqslant P_n(x) + (\sqrt{x} - P_n(x)) = \sqrt{x}.$$

Now, returning to the definition of the sequence (1.9), we see that the property $0 \leqslant P_n(x) \leqslant \sqrt{x}$ implies that the sequence $(P_n(x))_{n\in\mathbb{N}}$ is increasing and bounded and therefore convergent. The limit $P(x)$ satisfies the equation

$$P(x) = P(x) + \frac{1}{2}(x - P(x)^2),$$

hence $P(x) = \sqrt{x}$. Then, we use a classical lemma of Dini (see Lemma 1.46) to show that the sequence $(P_n)_{n\in\mathbb{N}}$ converges uniformly to its limit $x \mapsto \sqrt{x}$ on $[0, 1]$.

Next, we set $p_n(x) = P_n(x^2)$ to obtain a sequence of polynomials which converges uniformly to the absolute value function on the symmetric interval $[-1, 1]$, which concludes the proof.   □

The second idea of the proof is that $\overline{A}$ satisfies a particular stability property given by the following result.

Lemma 1.44 *If f and g are two elements of $\overline{A}$, then $\max\{f, g\}$ and $\min\{f, g\}$ belong to $\overline{A}$.*

Proof Let f, g be in $\overline{A}$. Then $f + g$ and $f - g$ belong to $\overline{A}$. To prove this lemma, we will further show that $\overline{A}$ is stable under the absolute value, which means that if f belongs to A, then so does $|f|$. We will deduce the desired result from the elementary identities

$$\max\{x, y\} = \frac{x + y + |x - y|}{2}, \quad \min\{x, y\} = \frac{x + y - |x - y|}{2}.$$

Consider a function f belonging to $\overline{A}$ and show that $|f| \in \overline{A}$. There exists a sequence $(f_p)_{p \in \mathbb{N}}$ that converges to f uniformly on X. Since X is compact, $f(X)$ is bounded in $\mathbb{R}$, from which we deduce that there exists $a > 0$ such that $f_p(X) \subset [-a, a]$ for all $p \in \mathbb{N}$.

We showed in Lemma 1.43 that there exists a sequence of polynomials $(p_n)_{n \in \mathbb{N}}$ that converges to the absolute value $|\cdot|$ uniformly on $[-a, a]$. Then, for all $\varepsilon > 0$, we can find two indices n and p such that $\|f_p - f\| \leqslant \varepsilon/2$ and $\sup_{t \in [-a,a]} |p_n(t) - |t|| \leqslant \varepsilon/2$. It directly follows that

$$\big\|p_n(f_p) - |f|\big\| \leqslant \big\|p_n(f_p) - |f_p|\big\| + \big\||f_p| - |f|\big\|$$
$$\leqslant \sup_{t \in [-a,a]} |p_n(t) - |t|| + \big\|f_p - f\big\| \leqslant \varepsilon.$$

Since A is an algebra, $p_n(f_p)$ belongs to A, which implies that $|f|$ belongs to $\overline{A}$. $\square$

It remains then to explain how this property of being stable under passage to the absolute value intervenes. This is the object of the last lemma.

Lemma 1.45 *Let X be a compact topological space that contains at least two elements. Let H be a subset of $C(X; \mathbb{R})$ satisfying the two following conditions:*

1. *for all u, v in H, the functions $\max\{u, v\}$ and $\min\{u, v\}$ are in H;*
2. *for any pair of distinct points of X, if α_1 and α_2 are two real numbers, then there exists a $u \in H$ such that $u(x_1) = \alpha_1$ and $u(x_2) = \alpha_2$.*

Then H is dense in $C(X; \mathbb{R})$.

Proof Let $f \in C(X; \mathbb{R})$ and $\varepsilon > 0$. Fix a point $x \in X$. For all $y \neq x$, there exists a $v_y \in H$ such that $v_y(x) = f(x)$ and $v_y(y) = f(y)$. Let

$$O_y = \{z \in X : v_y(z) > f(z) - \varepsilon\}.$$

For all $y \in X$, by continuity of the function v_y, the set O_y is open and contains y and x, so $X = \bigcup_{y \neq x} O_y$. By compactness, we can extract a finite subcover: $X = \bigcup_{j=1}^{r} O_{y_j}$, with $y_j \neq x$ for all j. Set $u_x = \max\{v_{y_1}, \ldots, v_{y_r}\}$. This function satisfies $u_x \in H$ and moreover

$$u_x(x) = f(x) \quad \text{and} \quad \forall x' \in X, \ u_x(x') > f(x') - \varepsilon.$$

We now vary x and set, for each $x \in X$,

$$\Omega_x = \{x' \in X : u_x(x') < f(x') + \varepsilon\}.$$

Thus, Ω_x is an open set of X by continuity of u_x. Moreover, Ω_x contains x. We can use the compactness of X again to obtain a finite number of points such that $X = \bigcup_{i=1}^{p} \Omega_{x_i}$. Finally, let $u = \min\{u_{x_1}, \ldots, u_{x_p}\}$. Then $u \in H$ and, for all $x \in X$, we have

$$f(x) - \varepsilon < u(x) < f(x) + \varepsilon.$$

This demonstrates that $\|f - u\| < \varepsilon$, which concludes the proof. $\square$

The Stone–Weierstrass theorem is obtained from the previous lemmas as follows. First, if X is reduced to a single element then the result is trivial because $C(X; \mathbb{R})$ consists of constant functions, which are in A by hypothesis. If, now, X contains at least two elements, Lemma 1.44 shows that $H = \overline{A}$ satisfies the first hypothesis of Lemma 1.45. It remains only to verify that $H = \overline{A}$ satisfies the second hypothesis. For this, let us consider x_1, x_2 in X and two real numbers α_1 and α_2. Since A separates the points, there exists f in A such that $f(x_1) \neq f(x_2)$. We then set

$$u(x) = \alpha_1 + (\alpha_2 - \alpha_1)\frac{f(x) - f(x_1)}{f(x_2) - f(x_1)}.$$

Since A is an algebra that contains constant functions, we verify that the function u belongs to A (thus to $\overline{A}$) and satisfies the desired property. This concludes the proof of the Stone–Weierstrass theorem. $\square$

To be complete, we recall Dini's lemma, which we used in the proof of the Stone–Weierstrass theorem.

Lemma 1.46 (Dini) *Let $I = [a, b]$ be a compact interval of $\mathbb{R}$. Consider a sequence of continuous functions $f_n : I \rightarrow \mathbb{R}$ with $n \in \mathbb{N}$, which is increasing in the sense that $f_n \leqslant f_{n+1}$. If the sequence converges pointwise to a continuous function $f \in C(I; \mathbb{R})$, then it converges uniformly to f.*

Proof Let $\varepsilon > 0$. For every integer n we set $\Omega_n = \{x \in I : f_n(x) > f(x) - \varepsilon\}$. The sets Ω_n are open because the functions f_n and f are continuous. Moreover, they form an exhaustive increasing sequence of I, because the sequence $(f_n)_{n \in \mathbb{N}}$ is increasing and converges pointwise to f. By the Borel–Lebesgue property, there exists an integer m such that $I = \Omega_m$, and therefore for all $x \in I$, $f_m(x) > f(x) - \varepsilon$. By increasing convergence, we have the other inequality: $f_m(x) \leqslant f(x)$. Thus, $\sup_{x \in I} |f(x) - f_m(x)| \leqslant \varepsilon$, which implies that the convergence is uniform. $\square$

Corollary 1.47 (Complex Stone–Weierstrass) *Let X be a compact metric space. Let us equip the space $C(X; \mathbb{C})$ of continuous functions with complex values with the uniform norm, $\|f\| = \sup_{x \in X} |f(x)|$. Consider a sub-algebra A of $C(X; \mathbb{C})$ that is unitary, stable under conjugation (if $f \in A$ then $\overline{f}$ belongs to A) and separates the points of X. Then A is dense in $C(X; \mathbb{C})$.*

Proof Note that if $f \in A$ then $\mathrm{Re}\, f$ and $\mathrm{Im}\, f$ belong to A. Let us set $H = A \cap C(X; \mathbb{R})$. Then H is a sub-algebra of $C(X; \mathbb{R})$ that is unitary and separates the points (if $x \neq y$ and if $f \in A$ is such that $f(x) \neq f(y)$, then either $\mathrm{Re}\, f$ is suitable, or $\mathrm{Im}\, f$ is suitable). Therefore, H is dense in $C(X; \mathbb{R})$. This implies the desired result by decomposition into real and imaginary parts. □

We can now deduce some fundamental density results.

Corollary 1.48 (Density of algebraic polynomials) *Let f be a continuous function on a compact interval $[a, b] \subset \mathbb{R}$, with complex values. Then, for every $\varepsilon > 0$, there exists a polynomial P such that, for every $x \in [a, b]$, we have $|f(x) - P(x)| < \varepsilon$.*

Proof It is easily verified that the set of polynomials is a unitary sub-algebra, stable under conjugation, and separates the points (the identity function $x \mapsto x$ is a polynomial that trivially separates the points). The desired result is therefore a corollary of the complex-valued version of the Stone–Weierstrass theorem. □

Definition 1.49 A trigonometric polynomial is a function $P \colon \mathbb{R}^d \to \mathbb{C}$ of the form

$$P(x) = \sum_{|n| \leqslant N} c_n e^{in \cdot x}$$

with $N \in \mathbb{N}$, $n = (n_1, \ldots, n_d) \in \mathbb{N}^d$, $c_n \in \mathbb{C}$ and $n \cdot x = n_1 x_1 + \cdots + n_d x_d$.

Corollary 1.50 (Density of trigonometric polynomials) *Consider a function $f \colon \mathbb{R}^d \to \mathbb{C}$ continuous and 2π-periodic with respect to each variable. For all $\varepsilon > 0$, there exists a trigonometric polynomial P such that $\sup_{x \in \mathbb{R}^d} |f(x) - P(x)| < \varepsilon$.*

Proof Let $\mathbb{S}^1$ be the circle of complex numbers of modulus 1, $\mathbb{T}^d = \mathbb{S}^1 \times \cdots \times \mathbb{S}^1$, the product of d copies of $\mathbb{S}^1$ (which is a compact topological space), and consider the algebra $C(\mathbb{T}^d; \mathbb{C})$ of continuous functions $f \colon \mathbb{T}^d \to \mathbb{C}$. Let A be the sub-algebra formed by functions of the form $P(z) = \sum_{|n| \leqslant N} c_n z^n$ where $z^n = z_1^{n_1} \cdots z_d^{n_d}$. Then A is a sub-algebra of $C(\mathbb{T}^d; \mathbb{C})$, unitary, stable under conjugation, and separates the points (indeed, if $x \neq y$, then there exists an index $1 \leqslant k \leqslant d$ such that $P_k(x) \neq P_k(y)$ for the map $P_k(z) = z_k$). The Stone–Weierstrass theorem (in the complex-valued version) implies that A is dense. Now consider $f \colon \mathbb{R}^d \to \mathbb{C}$ continuous and 2π-periodic with respect to each variable. Then we can define a function $F \colon \mathbb{T}^d \to \mathbb{C}$ by $F(e^{ix_1}, \ldots, e^{ix_d}) = f(x_1, \ldots, x_d)$ and apply the previous result. □

1.6.3 Continuity and Pointwise Convergence

Proposition 1.51 *Let X be a compact topological space. Suppose that the space $C(X; \mathbb{R})$ is complete for a norm $\|\cdot\|$ that satisfies the following property: if $\lim_{n \to +\infty} \|f_n\| = 0$, then $(f_n(x))_{n \in \mathbb{N}}$ converges to 0 for all $x \in X$. Then the norm $\|\cdot\|$ is equivalent to the norm $\|f\|_\infty = \sup_{x \in X} |f(x)|$.*

Proof We denote by E the pair $(C(X;\mathbb{R}), \|\cdot\|)$ and given $x \in X$, we denote by Λ_x the linear map defined by

$$\Lambda_x(f) := f(x).$$

Let us prove that for all x in X the map Λ_x is continuous. For this, we will use the sequential characterization of continuity given by the following result.

Lemma 1.52 *Let (Y_1, d_1) and (Y_2, d_2) be two metric spaces and $y \in Y_1$. A map $T : Y_1 \to Y_2$ is continuous at the point y if and only if it is sequentially continuous (which means that for any sequence $(y_n)_{n\in\mathbb{N}}$ that converges to y in Y_1, the sequence $(T(y_n))_{n\in\mathbb{N}}$ converges to $T(y)$ in Y_2).*

Proof If T is continuous at the point y then, for all $\varepsilon > 0$, there exists $\delta > 0$ such that

$$\forall x \in Y_1, \quad d_1(x,y) < \delta \implies d_2(T(x), T(y)) < \varepsilon.$$

We directly deduce that if $(y_n)_{n\in\mathbb{N}}$ converges to y in Y_1, then the sequence $(T(y_n))_{n\in\mathbb{N}}$ converges to $T(y)$. Conversely, suppose that T is not continuous at point y. Then, there exists $\varepsilon > 0$ and for every integer $n \in \mathbb{N}$ there exists $x_n \in Y_1$ such that $d(x_n, y) < 1/n$ and $d_2(T(x_n), T(y)) \geqslant \varepsilon$. This directly implies that T is not sequentially continuous. □

Let us show that for every $x \in X$ the map Λ_x is continuous. It is sufficient by linearity to show that Λ_x is continuous at 0. According to the previous lemma, it is therefore sufficient to show that Λ_x is sequentially continuous at 0. Let us then consider a sequence $(f_n)_{n\in\mathbb{N}}$ such that $\|f_n\|$ converges to 0 when n tends to $+\infty$. By hypothesis, the sequence $(f_n)_{n\in\mathbb{N}}$ also converges pointwise to 0. In particular, $(f_n(x))_{n\in\mathbb{N}}$ also converges to 0. By definition of Λ_x, this implies that the sequence $(\Lambda_x(f_n))_{n\in\mathbb{N}}$ converges to 0.

Let us note then that the family $\mathcal{F} := \{\Lambda_x : x \in X\}$ is pointwise bounded. Indeed, for every x in X and every f belonging to $C(X;\mathbb{R})$,

$$|\Lambda_x(f)| = |f(x)| \leqslant \sup_{y\in X} |f(y)| < +\infty,$$

where we used the fact that $f(X)$ is a compact subset of $\mathbb{R}$ to obtain the last inequality.

Thus, the Banach–Steinhaus theorem ensures that the family $\mathcal{F}$ is uniformly bounded in the space of continuous linear functions from E to $\mathbb{R}$. This implies the existence of a constant C such that:

$$\forall x \in X, \ \forall f \in E, \quad |f(x)| = |\Lambda_x(f)| \leqslant \|\Lambda_x\|_{\mathcal{L}(E,\mathbb{R})} \|f\| \leqslant C \|f\|.$$

Therefore,

$$\|f\|_\infty := \sup_{x\in X} |f(x)| \leqslant C \|f\|.$$

We conclude the proof by using the completeness of E and Corollary 1.37, which states that two norms on a Banach space are equivalent as soon as one dominates the other. □

1.7 Applications to Neural Networks

In this section, we introduce the concept of Artificial Neural Networks and explore some applications of the Stone–Weierstrass theorem within this context.

Some context

Ada Lovelace, a mathematician and English writer, published the first algorithm in 1843, a century before the construction of the first computer in 1945 (a part of this gigantic machine is preserved at the University of Pennsylvania). It is noteworthy that she foresaw applications beyond pure calculation. Her vision of the computer was that of a universal calculator, far superior to a purely numerical calculator, capable of manipulating general symbols.

One of the great successes of this approach was the resolution of intellectually challenging problems for humans but relatively simple for computers because they are easy to formally describe (think of chess and the supercomputer Deep Blue, developed in 1990, capable of evaluating 200 million positions per second). This marked the early developments of artificial intelligence. The current challenges in this field are of a different nature, involving algorithmic solutions to problems that the human brain solves intuitively (unlike chess), but in a way that is challenging to formalize due to a vast amount of subjective knowledge (think, for example, of word recognition in a manuscript or face recognition in images).

In 1957, psychologist Frank Rosenblatt, an engineer at the Cornell Aeronautical Laboratory, published an article ([162]) introducing a class of artificial neural networks inspired by the functioning of the brain and receptors of biological systems. His goal was to build a device with functions similar to those of humans, such as perception, recognition, concept formation, and the ability to generalize from experience.

These Artificial Neural Networks allow finding solutions to questions that the human brain handles with apparent simplicity, using mechanisms that are still poorly understood. Consider an example from everyday life, such as recognizing a specific object in a series of images. Suppose, for simplicity, that these images are in black and white, each composed of n pixels (recall that the pixel is the basic unit used to measure the definition of a digital image; a digital photo contains several million pixels). Each pixel is associated with a grayscale level ranging from 0 to 1. Now, consider a function $f: [0, 1]^n \to [0, 1]$ that associates an image with the probability that it contains the target object. We cannot analytically describe such a function f, which is the difficulty of this problem. However, algorithms based on artificial neural networks today can approximate this function f with greater accuracy than a human operator could achieve.

The objective of this section is to provide a definition of artificial neural networks, particularly in the simplest case, and to explore the so-called approximation property, which, in broad terms, posits that the function f mentioned earlier can be approximated by the output of an artificial neural network.

Formal Neuron by McCulloch and Pitts

Neurons are nerve cells that respond to stimuli and convert them into a bioelectric signal, which they transmit. Stimuli are received at the level of dendrites (several thousand on average), then analyzed at the cell body, which emits a signal transmitted by the axon (unique) to synaptic zones (also thousands on average). There are approximately 10^{11} neurons in the human brain, and the connections between these neurons form a network of formidable complexity.

Since Rosenblatt's seminal paper, numerous studies have focused on networks composed of formal neurons, inspired by the functioning of the brain and biological neurons. A formal neuron is an automaton that transforms its inputs into an output according to precise rules. McCulloch and Pitts proposed the first binary model in 1943 ([132]), having multiple inputs $(x_j)_{1 \leqslant j \leqslant n}$ and a single output, all equal to either 0 or 1. This neuron calculates a weighted sum of these inputs, i.e., the dot product

$$\Sigma = w \cdot x = w_1 x_1 + \cdots + w_n x_n$$

where $w = (w_1, \ldots, w_n) \in \mathbb{R}^n$ is a vector modeling synaptic weights (that is to say, modeling the excitatory or inhibitory action of a given synapse). Then the neuron compares this sum Σ to a threshold value $b \in \mathbb{R}$ (called the bias). It returns the value 1 or 0 depending on whether the sum Σ is greater or smaller than the threshold.

The formal neuron by McCulloch and Pitts is a function $f : \{0, 1\}^n \rightarrow \{0, 1\}$ of the form

$$f(x) = H(w \cdot x - b),$$

where H is the Heaviside function (i.e., the indicator function of $[0, +\infty)$).

General Formal Neuron

Definition 1.53 (*i*) A formal neuron is any function $f : \mathbb{R}^n \rightarrow \mathbb{R}$ (with $n \in \mathbb{N}^*$ arbitrary) that can be written in the following form:

$$f(x) = \rho(w \cdot x - b),$$

where $w \in \mathbb{R}^n$, $b \in \mathbb{R}$, and $\rho : \mathbb{R} \rightarrow \mathbb{R}$ is an arbitrary given function.

(*ii*) The function ρ is called the activation function of the neuron, and b is called the bias.

In practice, the most commonly used activation functions are:

1. the sigmoid $\rho_1(t) = \dfrac{1}{1 + e^{-t}}$,

2. the hyperbolic tangent $\rho_2(t) = \dfrac{e^t - e^{-t}}{e^t + e^{-t}}$,

3. the function $\rho_3(t) = \max\{0, t\}$ (called ReLU).

Neural Network with One Hidden Layer

A neural network is a set of formal neurons with inputs being the outputs of other formal neurons. We will focus on the simplest case to describe, that of a network called a "network with one hidden layer."

The scenario is that of a network with n inputs and a single output. These n inputs are connected to N formal neurons, all having the same activation function $\rho \colon \mathbb{R} \to \mathbb{R}$. Each of these neurons produces an output $\beta_j := \rho(w_j \cdot x - b_j)$, $1 \leqslant j \leqslant N$, and the output of the network is given by the weighted sum $\alpha_1 \beta_1 + \cdots + \alpha_N \beta_N$.

Definition 1.54 Fix an integer $n \in \mathbb{N}^*$ and an activation function $\rho \colon \mathbb{R} \to \mathbb{R}$.

(*i*) A neural network with one hidden layer is a function $f \colon \mathbb{R}^n \to \mathbb{R}$ of the following form:

$$f(x) = \sum_{j=1}^{N} \alpha_j \rho(w_j \cdot x - b_j) \quad \text{with } N \geqslant 1 \quad \text{and} \quad \begin{cases} (\alpha_1, \ldots, \alpha_N) \in \mathbb{R}^N, \\ (w_1, \ldots, w_N) \in (\mathbb{R}^n)^N, \\ (b_1, \ldots, b_N) \in \mathbb{R}^N. \end{cases}$$

(*ii*) We denote $\Sigma_n(\rho)$ the set of functions obtained in this way, i.e., the vector space

$$\Sigma_n(\rho) = \text{vect}\{\rho(w \cdot x - b) : w \in \mathbb{R}^n, \ b \in \mathbb{R}\}.$$

Approximation results

We will prove a surprising result stating that any continuous function $f \colon [0,1]^n \to \mathbb{R}$ can be approximated by a neural network with one hidden layer, regardless of the function ρ (under the sole assumption that ρ is not a polynomial).

Let $n \geqslant 1$. Denote by $C([0,1]^n; \mathbb{R})$ the space of continuous real-valued functions, equipped with the uniform norm $\|f\|_{C([0,1]^n;\mathbb{R})} = \sup_{x \in [0,1]^n} |f(x)|$.

Examples

We start by studying the case of two specific activation functions, $\rho = \sin$ or $\rho = \exp$.

Proposition 1.55 *For any integer $n \geqslant 1$, the set $\Sigma_n(\sin)$ is dense in $C([0,1]^n; \mathbb{R})$.*

Proof Let $f \in C([0,1]^n; \mathbb{R})$. We extend f to a function $F \colon [-1,1]^n \to \mathbb{R}$ defined by $F(x_1, \ldots, x_n) = f(|x_1|, \ldots, |x_n|)$. We obtain a function F that is also continuous and takes identical values on opposite faces pairs of the hypercube $[-1,1]^n$. We can then extend it to $\mathbb{R}^n$ by periodicity, setting $\widetilde{F}(x_1, \ldots, x_n) = F(y_1, \ldots, y_n)$ where y is the unique element of $(-1,1]^n$ such that $x_k - y_k \in 2\mathbb{Z}$ for all $1 \leqslant k \leqslant n$. Then introduce the function $\widetilde{f} \colon \mathbb{R}^n \to \mathbb{R}$ defined by

$$\widetilde{f}(x_1, \ldots, x_n) = \widetilde{F}(x_1/\pi, \ldots, x_n/\pi).$$

The function $\widetilde{f}$ is 2π-periodic with respect to each variable. Consequently, by the density of trigonometric polynomials in the space of continuous and 2π-periodic functions (see Corollary 1.50), we deduce that, for all $\varepsilon > 0$, there exists a trigonometric polynomial P such that $\sup_{x \in \mathbb{R}^n} \left| \widetilde{f}(x) - P(x) \right| < \varepsilon$. By construction of $\widetilde{f}(x)$, this implies that

$$\sup_{x \in [0,1]^n} |f(x) - P(\pi x)| < \varepsilon.$$

Now observe that $\sin(\theta + \pi/2) = \cos(\theta)$. Consequently, $\Sigma_n(\sin)$ contains all functions of the form $x \mapsto \sum_{0 \leqslant |k| \leqslant p} a_k \cos(k \cdot x) + b_k \sin(k \cdot x)$ where $k \in \mathbb{Z}^n$. Since $\Sigma_n(\sin)$ is stable by dilation of the variable (if $P = P(x)$ belongs to $\Sigma_n(\sin)$ then so does the function $x \mapsto P(\pi x)$) we obtain the wanted result. $\square$

Proposition 1.56 *Let $n \geqslant 1$. Given $w \in \mathbb{R}^n$, denote by e_w the function $x \mapsto \exp(w \cdot x)$. Then the set $\mathcal{A} = \mathrm{vect}\{e_w : w \in \mathbb{R}^n\}$ is dense in $C([0,1]^n; \mathbb{R})$. It directly follows that the set $\Sigma_n(\exp)$ is dense in $C([0,1]^n; \mathbb{R})$.*

Remark 1.57 Note the difference with the case of the sine function treated earlier: the space $\mathrm{vect}\{x \mapsto \sin(w \cdot x) : w \in \mathbb{R}^n\}$ is not dense in $C([0,1]^n; \mathbb{R})$ because it only contains odd functions.

Proof As the product of two exponentials is an exponential, it is easily verified that $\mathcal{A}$ is a unitary subalgebra of $C([0,1]^n; \mathbb{R})$. According to the Stone–Weierstrass theorem (see Theorem 1.42), to prove that $\mathcal{A}$ is dense in $C([0,1]^n; \mathbb{R})$, it suffices to demonstrate that $\mathcal{A}$ separates points in $[0,1]^n$. Consider two distinct points $x_1 \in [0,1]^n$ and $x_2 \in [0,1]^n$. Then the function $f(x) = \exp((x_1 - x_2) \cdot x)$ belongs to $\mathcal{A}$ and we have $f(x_1)/f(x_2) = \exp((x_1 - x_2)^2) \neq 1$, so $f(x_1) \neq f(x_2)$. This concludes the proof. $\square$

Universal Approximation

We now study the universal approximation theorem of Hornik, Stinchcombe and White ([82, 81]) and Cybenko [41]. This result states that networks with a single hidden layer can approximate any continuous functions on a compact domain.

Theorem 1.58 *Let $n \geqslant 1$ and consider a function $\rho \colon \mathbb{R} \to \mathbb{R}$ of class C^∞. Then the set $\Sigma_n(\rho)$ is dense in $C([0,1]^n; \mathbb{R})$ if and only if ρ is not a polynomial.*

Remark 1.59 i) Kolmogorov [104] was the first to study the possibility of representing functions of several variables by means of the superpositions of functions of a smaller number of variables.

ii) This result holds for arbitrary continuous functions ρ not necessarily smooth (see [120]). For this extension and a general introduction to a broader set of results, we refer to the survey article by Pinkus [154] and the Lectures notes by Wolf [197].

Proof First, suppose that ρ is a polynomial of degree N. Then all elements of $\Sigma_n(\rho)$ are polynomials of degree N. Since the space of polynomials of degree N is a finite-dimensional vector space, hence closed, we conclude that $\Sigma_n(\rho)$ cannot be dense in $C([0,1]^n; \mathbb{R})$.

We now have to prove that $\Sigma_n(\rho)$ is dense when ρ is not a polynomial. We begin by observing that one can reduce the analysis to the case $n = 1$. To do this, recall that Proposition 1.56 implies that, for any $\varepsilon > 0$, there exists an integer $N \geqslant 1$, scalars $\alpha_j \in \mathbb{R}$ and vectors $w_j \in \mathbb{R}^n$ for $1 \leqslant j \leqslant N$, so that

$$\sup_{x \in [0,1]^n} \left| f(x) - \sum_{j=1}^{N} \alpha_j \exp(w_j \cdot x) \right| \leqslant \varepsilon.$$

Set

$$m = \sqrt{n} \times \sup_{1 \leqslant j \leqslant N} |w_j|, \quad a = \sup_{1 \leqslant j \leqslant N} |\alpha_j|$$

Notice that, by the Cauchy–Schwarz inequality, if $x \in [0,1]^n$ then $|w_j \cdot x| \leqslant |w_j| |x| \leqslant m$. Now, assume that one can find an integer $M \geqslant 1$ and three families of real numbers β_k, ω_k and b_k for $1 \leqslant k \leqslant M$ so that

$$\sup_{y \in [-m,m]} \left| \exp(y) - \sum_{k=1}^{M} \beta_k \rho(\omega_k y - b_k) \right| \leqslant \frac{\varepsilon}{Na}. \tag{1.10}$$

By the triangle inequality, this will imply that

$$\sup_{x \in [0,1]^n} \left| f(x) - \sum_{j=1}^{N} \alpha_j \sum_{k=1}^{M} \beta_k \rho(\omega_k (w_j \cdot x) - b_k) \right| \leqslant 2\varepsilon.$$

We thus see that, to conclude the proof of Theorem 1.58, it will suffice to prove the approximation property (1.10). We will prove in fact a stronger result, namely the following result.

Lemma 1.60 *Assume that $\rho \colon \mathbb{R} \to \mathbb{R}$ is of class C^∞ and is not a polynomial. Then, for any $m > 0$, the set of functions*

$$E = \text{vect}\{[-m,m] \ni x \mapsto \rho(wx - b) : w \in \mathbb{R}, \ b \in \mathbb{R}\}$$

is dense in $C([-m,m]; \mathbb{R})$.   □

Proof Let $t > 0$. Note that the function Δ_t defined by

$$\Delta_t(x) = \frac{1}{t} \left(\rho((w+t)x - b) - \rho(wx - b) \right)$$

belongs to E (as it is a linear combination of two functions of the form $x \mapsto \rho(w'x - b)$). Then, by writing

$$\Delta_t(x) = \frac{1}{t}\int_0^t x\rho'((w+\tau)x - b)\,d\tau = \int_0^1 x\rho'((w+st)x - b)\,ds$$

$$= x\rho'(wx - b) + \int_0^1 (x\rho'((w+st)x - b) - x\rho'(wx - b))\,ds,$$

it is easily verified that

$$\sup_{x\in[0,1]} |\Delta_t(x) - x\rho'(wx - b)| \text{ tends to 0 as } t \text{ tends to 0.}$$

This shows that the function defined by $\phi^1_{w,b}(x) = x\rho'(wx - b)$ belongs to $\overline{E}$ for every $(w, b) \in \mathbb{R}^2$.

We can iterate the previous reasoning: by writing that $\frac{1}{t}(\phi^1_{w+t,b} - \phi^1_{w,b})$ belongs to $\overline{E}$ (as it is a linear combination of two elements from the vector space $\overline{E}$), by taking the limit as t goes to 0, we deduce that the function defined by $\phi^2_{w,b}(x) = x^2\rho''(wx - b)$ belongs to $\overline{E}$ for every w, b in $\mathbb{R}$. By repeating this argument, we deduce successively that the function $x \mapsto x^k\rho^{(k)}(wx - b)$ belongs to $\overline{E}$ for every w, b in $\mathbb{R}$, where $\rho^{(k)}$ denotes the k-th derivative of ρ. Since ρ is not a polynomial, for every $k \in \mathbb{N}$, we can choose $w = 0$ and $b = b_k$ such that $\rho^{(k)}(-b_k) \neq 0$. We conclude that

$$\text{vect}\{x \mapsto x^k \ ; \ k \in \mathbb{N}\} \subset \overline{E}.$$

It follows that $\overline{E}$ contains all polynomials. Then, by density of polynomials in $C([-m, m]; \mathbb{R})$ (according to the Weierstrass theorem, see Corollary 1.48), we conclude that $\overline{E} = C([-m, m]; \mathbb{R})$, which is the desired result. □

This concludes the proof of Theorem 1.58. □

1.8 Exercises

Exercise (solved) 1.1 Consider an infinite-dimensional normed vector space E. Using the sequence introduced in the proof of Lemma 1.16, show that there exists a continuous function $F: E \to E$ that is continuous and unbounded on the closed unit ball.

Exercise (solved) 1.2 Let A be an unbounded subset of $\mathbb{R}$. Prove that if $f: A \to \mathbb{R}$ is a uniform limit of polynomial functions on $\mathbb{R}$, then f is a polynomial (which explains why it is necessary to work on a bounded set to prove the Stone–Weierstrass theorem).

Exercise (solved) 1.3 The aim of this exercise is to provide an elementary proof of the Banach–Steinhaus theorem, due to Alan Sokal [175]. Consider a Banach space $(E, \|\cdot\|_E)$ and a normed space $(F, \|\cdot\|_F)$. We denote by $\mathcal{L}(E, F)$ the space of continuous linear maps from E to F, equipped with the norm $\|T\|_{\mathcal{L}(E,F)} = \sup_{\|x\|_E=1} \|Tx\|_F$.

1. Let $T \in \mathcal{L}(E, F)$. Show that, for all $x \in E$ and all $\xi \in E$, we have

$$\|T\xi\|_F \leqslant \max\{\|T(x + \xi)\|_F, \|T(x - \xi)\|_F\}.$$

2. Deduce that, for all $x \in E$ and all $r > 0$,

$$\sup_{\{x' \in E : \|x' - x\|_E = r\}} \|Tx'\|_F \geqslant \|T\|_{\mathcal{L}(E,F)} r.$$

3. Consider a family $\{T_\alpha\}_{\alpha \in A}$ of maps $T_\alpha \in \mathcal{L}(E, F)$ that is pointwise bounded (that is, for all $x \in E$, we have $\sup_{\alpha \in A} \|T_\alpha x\|_F < +\infty$). Suppose by contradiction that $\{T_\alpha\}_{\alpha \in A}$ is not bounded in $\mathcal{L}(E, F)$ and consider $(\alpha_n)_{n \in \mathbb{N}}$ such that $\|T_{\alpha_n}\|_{\mathcal{L}(E,F)} \geqslant 4^n$. Show that there exists a sequence $(x_n)_{n \in \mathbb{N}}$ of E such that

$$x_0 = 0, \quad \|x_n - x_{n-1}\|_E = 3^{-n}, \quad \|T_{\alpha_n} x_n\|_F \geqslant \frac{2}{3} 3^{-n} \|T_{\alpha_n}\|_{\mathcal{L}(E,F)}.$$

Deduce the Banach–Steinhaus theorem.

Exercise 1.4 Let E be a Banach space and denote by E' its dual. Consider a linear map $T : E \to E'$ such that, for all $(x, y) \in E \times E$,

$$\langle Tx, y \rangle = \langle Ty, x \rangle.$$

Show that T is continuous.

Exercise 1.5 Let E be a Banach space. Consider a continuous linear map $T : E \to E$ satisfying the following property: for all $x \in E$, there exists $m_x \in \mathbb{N}^*$ such that $T^{m_x}(x) = 0$. Show that $\sup_{n \in \mathbb{N}} \|a_n T^n\| < +\infty$ for any sequence $(a_n)_{n \in \mathbb{N}}$ of real numbers, then deduce that there exists an integer n such that $T^n = 0$.

Exercise 1.6 The aim of this exercise is to show that any inclusion between two complete linear subspaces continuously included in $L^1_{\text{loc}}(\Omega)$ is in fact a continuous inclusion (in the sense that the canonical embedding is continuous).

More precisely, consider an open set $\Omega \subset \mathbb{R}^n$ and consider two linear subspaces E and F of the space $L^1_{\text{loc}}(\Omega)$. We suppose that E and F are endowed with norms $\|\cdot\|_E$, $\|\cdot\|_F$, which make them complete and such that any sequence $(f_n)_{n \in \mathbb{N}}$ that converges to 0 for $\|\cdot\|_E$ (resp. for $\|\cdot\|_F$) converges to 0 in $L^1_{\text{loc}}(\Omega)$ in the following sense: for all $\varphi \in C_0^\infty(\Omega)$,

$$\int f_n(x)\varphi(x)\, dx \xrightarrow[n \to +\infty]{} 0.$$

Suppose that $E \subset F$. Prove that there exists $C > 0$ such that $\|u\|_F \leqslant C\|u\|_E$ for all $u \in E$. (Hint: use the closed graph theorem).

Exercise 1.7 (Schauder's continuation principle) We consider a Banach space $(B, \|\cdot\|_B)$ and a normed vector space $(V, \|\cdot\|_V)$. Consider two continuous linear operators $L_0 : B \to V$ and $L_1 : B \to V$.

For t in $[0, 1]$, we define

$$L_t = (1 - t)L_0 + tL_1.$$

We suppose that there exists a constant $A > 0$ such that

$$\forall t \in [0, 1], \ \forall u \in B, \quad \|u\|_B \leqslant A \|L_t u\|_V.$$

1. Suppose that L_s is surjective for a certain $s \in [0, 1]$. Show that L_s is bijective and that its inverse is a continuous linear map satisfying

$$\|L_s^{-1}\|_{V \to B} \leqslant A.$$

2. Let $f \in V$ and let $s \in [0, 1]$ such that L_s is surjective. Observe that for all $t \in [0, 1]$,

$$f = L_t u \iff f = L_s u + (t - s)(L_1 - L_0)u.$$

Introduce a map $T_{s,t} : B \to B$ depending on f and s, t and satisfying the two properties:

(i) $f = L_t u \iff u = T_{s,t}(u),$

and

(ii) $T_{s,t}$ is a contraction if $|t - s| < \delta = \dfrac{1}{A(\|L_0\|_{B \to V} + \|L_1\|_{B \to V})}$.

3. Deduce that L_t is surjective for all $t \in [0, 1]$ such that $|t - s| \in [0, \delta)$. Then show that if L_0 is surjective then L_1 is surjective.

Chapter 2
Fixed Point Theorems

In this chapter, we address the question of solving an equation $\Phi(x) = 0$ from the point of view of Analysis. The unknown x may belong to an infinite-dimensional vector space. We will see several situations in which we can guarantee the existence of a solution and also obtain the solution as the limit of a sequence defined iteratively. We will prove several fundamental results: the Banach fixed point theorem, the local inversion theorem, the Cauchy–Lipschitz theorem, Brouwer's theorem and the domain invariance theorem. We will also address the study of Ekeland's variational principle and Nash's implicit function theorem, which play a key role in many research fields.

2.1 Differential Calculus Reminders

Let E and F be two real normed vector spaces and let U be an open set of E. Consider a function $f \colon U \to F$ and a point $a \in U$. We say that f is differentiable at the point a in the sense of Fréchet if there exists a continuous linear map $L \colon E \to F$ and a function $\varepsilon \colon E \to F$ such that

$$f(x) = f(a) + L(x - a) + \|x - a\|_E\, \varepsilon(x - a) \quad \text{with} \quad \lim_{\|h\|_E \to 0} \varepsilon(h) = 0.$$

The existence of L depends on the choice of the norm $\|\cdot\|_E$. Such a linear map L is necessarily unique and is called the differential of f at a, denoted $\mathrm{d}f(a)$ or $\mathrm{d}_a f$ or $f'(a)$. By abuse, we will simply say differentiable instead of differentiable in the sense of Fréchet.

Suppose that f is differentiable at every point of U. We then denote by $\mathrm{d}f$ the function $a \mapsto \mathrm{d}f(a)$, called the differential of f. If the function $\mathrm{d}f$ is continuous from U to $\mathcal{L}(E, F)$, then we say that f is of class C^1 on U and we write $f \in C^1(U)$. If the function $\mathrm{d}f$ is differentiable at every point a of U, then we say that f is twice differentiable on U and we denote by $\mathrm{d}^2 f$ the obtained function. If this function is continuous from U to $\mathcal{L}(E, \mathcal{L}(E, F))$, then we say that f belongs to the space $C^2(U)$

© The Author(s), under exclusive license to Springer Nature Switzerland AG 2024
T. Alazard, *Analysis and Partial Differential Equations*, Universitext,
https://doi.org/10.1007/978-3-031-70909-8_2

of functions of class C^2 on U. By induction, we define more generally the notion of a function of class C^k for every integer k. We will say that f belongs to the space $C^\infty(U)$ of functions of class C^∞ on U if f is of class C^k for every k. Finally, given a closed set $K \subset U$, we will say that f is of class C^k on K if there exists an open set V such that $K \subset V \subset U$ and such that f belongs to $C^k(V)$.

Let us recall the following version of the mean value theorem.

Theorem 2.1 *Let $f : U \subset E \to F$ be a differentiable function on a convex open set U. Suppose there exists a constant C such that*

$$\forall a \in U, \quad \|df(a)\|_{\mathcal{L}(E,F)} \leqslant C.$$

Then, for every pair (x, y) in $U \times U$, we have $\|f(x) - f(y)\|_F \leqslant C \|x - y\|_E$.

2.2 Banach's Fixed Point Theorem

Let us start with the fundamental example of solving $\Phi(u) = 0$ in the case where $\Phi - I$ is a contraction, where $I : x \mapsto x$ denotes the identity and where a contraction is defined as follows.

Definition 2.2 Let (E, d) be a metric space and k a positive real number. We say that a function $f : E \to E$ is k-Lipschitzian if, for every pair (x, y) in $E \times E$,

$$d(f(x), f(y)) \leqslant kd(x, y).$$

We say that f is a contraction if it is k-Lipschitzian for a certain $k \in [0, 1)$.

Theorem 2.3 *Let E be a complete metric space and $f : E \to E$ a contraction. There exists a unique fixed point $x^* \in E$, that is to say a solution to $f(x^*) = x^*$. Moreover, every sequence $(x_n)_{n \in \mathbb{N}}$ of elements of E satisfying $x_{n+1} = f(x_n)$ converges to x^*.*

Proof Let $x_0 \in E$ and let $(x_n)_{n \in \mathbb{N}}$ be the sequence defined by $x_{n+1} = f(x_n)$. Then $d(x_{m+1}, x_m) \leqslant kd(x_m, x_{m-1})$, therefore

$$d(x_{m+1}, x_m) \leqslant k^m d(x_1, x_0).$$

As $x_{n+p} - x_n = x_{n+1} - x_n + \cdots + x_{n+p} - x_{n+p-1}$, we deduce

$$d(x_{n+p}, x_n) \leqslant (k^n + \cdots + k^{n+p-1})d(x_1, x_0) \leqslant k^n \frac{1}{1-k} d(x_1, x_0).$$

Since $k < 1$, this implies that the sequence $(x_n)_{n \in \mathbb{N}}$ is Cauchy. As E is a complete space, this sequence converges towards an element x^*. To show that x^* is a fixed point of f we will use the previous inequality $d(x_{m+1}, x_m) \leqslant k^m d(x_1, x_0)$, which implies that $d(f(x_m), x_m) \leqslant k^m d(x_1, x_0)$. As $k < 1$ and f is continuous, we can take the limit in this inequality to deduce that $d(f(x^*), x^*) = 0$, which shows that x^* is a fixed point of f. $\square$

2.3 The Local Inversion Theorem

In this section, we will prove the local inversion theorem for Banach spaces[1].

Definition 2.4 Consider two normed spaces B_1, B_2 and open sets $U \subset B_1, V \subset B_2$. We say that a function $f : U \to V$ is a C^k-diffeomorphism, with $k \in \mathbb{N} \cup \{\infty\}$, if:

- f is of class C^k;
- f is a bijection from U to V;
- the inverse function f^{-1} is of class C^k.

Theorem 2.5 *Let $f : U \to B_2$ be a C^1 function from an open set U of a Banach space B_1 to a Banach space B_2. If $df(x_0)$ is an isomorphism from B_1 to B_2 then f is a C^1 diffeomorphism from a neighborhood of x_0 to a neighborhood of $f(x_0)$.*

Remark 2.6 Furthermore, it can be shown that if f is injective and if for all x in U the differential $df(x)$ is a continuous isomorphism, then $f(U)$ is open and the inverse bijection, from $f(U)$ to U, is of class C^1.

Proof We start by reducing to the case $B_1 = B_2$, $x_0 = 0$, $f(x_0) = x_0$ and $df(x_0) = I$, where we denote by I the identity mapping. For this, we replace U by the set $\widetilde{U}$ of elements x such that $x_0 + x$ belongs to U, and we replace f by the function

$$\widetilde{f}(x) = (df(x_0))^{-1}(f(x_0 + x) - f(x_0)).$$

Let us introduce $\varphi(x) = x - f(x)$. The differential $d\varphi(0)$ is equal to 0 at 0 so there exists an $r > 0$ such that $\overline{B}_r$ is included in U and such that the norm of the differential of φ is always less than $1/2$ on this ball. We introduce $W = B_{r/2}$ and $V = B_r \cap f^{-1}(W)$. Let us show that f is bijective from V to W.

Surjectivity

Let $y \in W$. We look for $x \in V$ such that $y = f(x)$. For this, we write the equation $y = f(x)$ in the form

$$x = h(x) \quad \text{with} \quad h(x) = y + \varphi(x) = x + y - f(x),$$

and we look for a *fixed point* of h.

Theorem 2.1 implies that φ is $1/2$-Lipschitz on $\overline{B}_r$. Thus, $\|\varphi(x)\| \leqslant r/2$ for all $x \in \overline{B}_r$. For all $y \in W = B_{r/2}$, we have $\|y\| < r/2$ so h sends $\overline{B}_r$ into B_r by the triangle inequality. Moreover, h, like φ, is $1/2$-Lipschitz. So h has a fixed point x in $\overline{B}_r$ (according to the Fixed Point Theorem 2.3). We verify that x belongs to B_r because $x = h(x)$. Similarly, we have $x \in f^{-1}(W)$ because $f(x) = y$. This proves that for all $y \in W$, we can find $x \in V$ such that $f(x) = y$.

---

[1] We refer to [3, 47, 105] for an historical account of the subject as well as for various proofs of this and related results.

Injectivity

For all $(x_1, x_2) \in V \times V$,

$$\|x_1 - x_2\| = \|\varphi(x_1) + f(x_1) - \varphi(x_2) - f(x_2)\|$$
$$\leqslant \frac{1}{2}\|x_1 - x_2\| + \|f(x_1) - f(x_2)\|$$

so

$$\|x_1 - x_2\| \leqslant 2\|f(x_1) - f(x_2)\|, \qquad (2.1)$$

which implies that $f: V \to W$ is injective.

Regularity

First, we observe that $df(x) = I - d\varphi(x)$ and $d\varphi(x)$ is of norm less than $1/2 < 1$ for all x in V. Therefore $df(x)$ has a bounded inverse according to a classical result proved below (see Lemma 2.7), and its inverse is given by $\sum_{n \in \mathbb{N}} (d\varphi(x))^n$, of norm smaller than 2. Let us then show that f^{-1} is differentiable and that its differential is the inverse of $df(f^{-1}(x))$. For this, let us fix $y \in W$, let $x = f^{-1}(y)$ and set $L = (df(f^{-1}(y)))^{-1}$. We want to show that

$$\|f^{-1}(y + z) - f^{-1}(y) - Lz\| = o(\|z\|). \qquad (2.2)$$

For this, we introduce h such that $x + h = f^{-1}(y + z)$. Then

$$\|f^{-1}(y + z) - f^{-1}(y) - Lz\| = \|x + h - x - L(f(x + h) - f(x))\|$$
$$= \|L(f(x + h) - f(x) - L^{-1}h)\|$$
$$\leqslant 2\|f(x + h) - f(x) - L^{-1}h\|,$$

because $\|L\|_{\mathcal{L}(E)}$ is bounded by 2. Now, we use $L^{-1} = df(x)$ to conclude

$$\|f^{-1}(y + z) - f^{-1}(y) - Lz\| \leqslant 4\|h\|.$$

We can then use the inequality (2.1) to conclude that $\|h\| \leqslant 2\|z\|$, which finishes the proof of (2.2).

Since f^{-1} is differentiable, it is continuous and $x \mapsto (df(f^{-1}(x)))^{-1}$ is continuous by composition of continuous functions. $\square$

It remains to prove the following result used in the proof above.

Lemma 2.7 (Neumann Series) *Let B be a Banach space and suppose $T \in \mathcal{L}(B)$ satisfies $\|T\| < 1$. Then $I - T$ is invertible and its inverse is given by*

$$(I - T)^{-1} = \sum_{n=0}^{\infty} T^n.$$

Proof The proof relies on the fact that $\mathcal{L}(B)$ is a Banach space (since B is one) and on the fact that the operator norm on $\mathcal{L}(B)$ satisfies the following inequality: $\|T_1 T_2\|_{\mathcal{L}(B)} \leqslant \|T_1\|_{\mathcal{L}(B)} \|T_2\|_{\mathcal{L}(B)}$.

Let us consider the sum $S_n = I + T + \cdots + T^n$. Then $(I - T)S_n = I - T^{n+1}$ converges towards I because $\|T^{n+1}\|_{\mathcal{L}(B)} \leqslant \|T\|_{\mathcal{L}(B)}^{n+1}$ and $\|T\|_{\mathcal{L}(B)} < 1$ by hypothesis. Moreover, the series S_n converges normally and therefore converges because $\mathcal{L}(B)$ is a Banach space. This classical result is proved directly in the following way: since

$$\|S_{n+m} - S_n\|_{\mathcal{L}(B)} \leqslant \sum_{j=n+1}^{n+m} \|T^j\|_{\mathcal{L}(B)} \leqslant \sum_{j=n+1}^{n+m} \|T\|_{\mathcal{L}(B)}^j \leqslant \sum_{j=n+1}^{+\infty} \|T\|_{\mathcal{L}(B)}^j,$$

and since the right-hand side of the preceding inequality converges towards 0, the sequence $(S_n)_{n \in \mathbb{N}}$ is Cauchy and consequently it is a convergent sequence in the Banach space $\mathcal{L}(B)$.   □

2.4 The Cauchy–Lipschitz Theorem

The Cauchy–Lipschitz Theorem, also known as the Picard–Lindelöf theorem, gives a condition which ensures that a system of differential equations has a unique solution. More precisely, consider a function

$$f : (t, x) \in \mathbb{R} \times \mathbb{R}^n \mapsto f(t, x) \in \mathbb{R}^n.$$

The variable t is called the time variable (which corresponds to applications in Physics) and f is often called a time-dependent vector field (in simple terms, a vector field assigns a vector to each point in a given space; these vectors typically represent physical quantities such as force or velocity whose direction and magnitude may vary from point to point in the space). Given a vector $y_0 \in \mathbb{R}^n$ (called the initial data), we seek an interval $[-T, T] \subset \mathbb{R}$ and a function $y \in C^1([-T, T]; \mathbb{R}^n)$ satisfying $y(0) = y_0$ and solution to

$$y'(t) = f(t, y(t)), \quad \forall t \in (-T, T), \quad \text{where} \quad y'(t) = \frac{dy}{dt}(t).$$

Theorem 2.8 Let $n \geqslant 1$ and $f \in C^1(\mathbb{R} \times \mathbb{R}^n; \mathbb{R}^n)$. Then, for all $y_0 \in \mathbb{R}^n$, there exists $T > 0$ such that the system of differential equations $y' = f(t, y)$ has a unique solution $y \in C^1([-T, T]; \mathbb{R}^n)$ satisfying $y(0) = y_0$.

Proof We follow a very elegant proof found in [3]. Introduce a parameter $T > 0$. Note that y is a solution of

$$y'(t) = f(t, y(t)), \quad y|_{t=0} = y_0, \tag{2.3}$$

if and only if the function $z(\tau) = y(T\tau) - y_0$ satisfies

$$z'(\tau) = Tf(T\tau, z(\tau) + y_0), \quad z|_{\tau=0} = 0. \tag{2.4}$$

Therefore, if we can find $T > 0$ such that there exists a solution z defined over a time interval $[-1, 1]$ we will have a solution to the initial problem, defined over a time interval $[-T, T]$.

Let us introduce the spaces

$$B_0 = C^0([-1, 1]; \mathbb{R}^n) \quad \text{and} \quad B_1 = \{z \in C^1([-1, 1]; \mathbb{R}^n) : z(0) = 0\}.$$

These are Banach spaces for the norms

$$\|u\|_{B_0} := \sup_{[-1,1]} |u(\tau)|, \quad \|u\|_{B_1} := \sup_{[-1,1]} |u(\tau)| + \sup_{[-1,1]} |u'(\tau)|,$$

where $|\cdot|$ denotes any norm on $\mathbb{R}^n$. Let us also introduce the functional $\Phi \colon \mathbb{R} \times B_1 \to \mathbb{R} \times B_0$ defined by

$$\Phi(T, z) = (T, v) \quad \text{where} \quad v(\tau) = z'(\tau) - Tf(T\tau, z(\tau) + y_0).$$

Then Φ is a C^1-function and the differential at $(0, 0)$ is given by the map

$$d\Phi(0, 0) \cdot (T, h) = (T, u) \quad \text{with} \quad u(\tau) = h'(\tau) - Tf(0, y_0).$$

This map is a linear isomorphism of $\mathbb{R} \times B_1$ onto $\mathbb{R} \times B_0$, whose inverse is the map L defined by $L(T, v) = (T, w)$, where $w(\tau) = T\tau f(0, y_0) + \int_0^\tau v(s)\,ds$.

We can then apply the local inversion theorem, which implies that Φ is a C^1-diffeomorphism of a neighborhood $U \subset \mathbb{R} \times B_1$ of $(0, 0)$ onto $\Phi(U)$. Since $(0, 0)$ belongs to U, $\Phi((0, 0))$ belongs to $\Phi(U)$. By definition of Φ, we have $\Phi((0, 0)) = (0, 0)$. Moreover, $\Phi(U)$ is open, because it is the preimage of the open set U under the continuous map Φ^{-1}. In particular, there exists $T_1 > 0$ such that the pair $(T, 0)$ belongs to $\Phi(U)$ for all $T \in (0, T_1]$. Therefore, there exists a pair (T', z) in $\mathbb{R} \times B_1$ such that $\Phi((T', z)) = (T, 0)$. We deduce that $T' = T$ and that z is a solution of equation (2.4). Then, as we explained at the beginning of the proof, the function $y(t) = y_0 + z(t/T)$ is a solution of (2.3) over the time interval $[-T, T]$.

It remains to show uniqueness. Suppose that y_1 and y_2 are two solutions defined on $[-T, T]$. Let, for $j \in \{1, 2\}$, $z_{j,T}(\tau) = y_j(T\tau) - y_0$. Note that

$$\|z_{j,T}\|_{B_1} \leqslant 2T \sup_{[-T,T]} |y_j'(t)|.$$

Also, there exists $T_2 > 0$ such that if $T \in (0, T_2]$ then $(T, z_{1,T})$ and $(T, z_{2,T})$ belong to the open set U constructed in the previous step. Since $z_{j,T}$ is a solution of $\Phi(T, z_{j,T}) = (T, 0)$, and since Φ is a bijection from U onto its image, it follows that $z_{1,T} = z_{2,T}$ and therefore $y_1 = y_2$.

By taking $T = \min\{T_1, T_2\}$, we obtain the existence and uniqueness of a solution defined over the time interval $[-T, T]$. This concludes the proof. $\qquad \square$

2.5 The Caristi and Ekeland Theorems

In this section, we will prove a famous result of Ekeland [48] which guarantees the existence of nearly optimal solutions to certain optimization problems, without making any compactness assumptions. As we will see, this result allows for a very quick demonstration of another fixed point theorem, that of Caristi [29].

Theorem 2.9 (Ekeland's Variational Principle) *Let (E, d) be a complete metric space and let $F \colon E \to \mathbb{R}$ be a continuous function, bounded from below ($\inf_E F > -\infty$). Consider $\varepsilon > 0$ and $x \in E$ such that*

$$F(x) \leqslant \inf_E F + \varepsilon.$$

For every $\delta > 0$, there exists $y \in E$ such that

(i) $F(y) \leqslant F(x)$;

(ii) $d(x, y) \leqslant \delta$;

(iii) $\forall z \in E \smallsetminus \{y\}, \quad F(z) > F(y) - \dfrac{\varepsilon}{\delta} d(y, z).$

Proof We can assume that $\varepsilon = 1 = \delta$ (by replacing the distance d with the distance $\delta(x, y) = d(x, y)/\delta$, and F by the function $\tilde{F}(x) = F(x)/\varepsilon$).

We define a sequence $(x_n)_{n \in \mathbb{N}}$ by induction starting from $x_0 = x$. Assume x_n is known. Then there are two cases:

1. Either, for all $z \in E \smallsetminus \{x_n\}$, we have $F(z) > F(x_n) - d(x_n, z)$, and then we set $x_{n+1} = x_n$.
2. Or there exists $z \neq x_n$ such that $F(z) \leqslant F(x_n) - d(x_n, z)$. In which case we consider the set S_n of elements satisfying this condition and we choose $x_{n+1} \in S_n$ such that

$$F(x_{n+1}) - \inf_{S_n} F \leqslant \frac{1}{2}\left(F(x_n) - \inf_{S_n} F\right). \tag{2.5}$$

We will see that the sequence $(x_n)_{n \in \mathbb{N}}$ is Cauchy. For this, note that in both cases of the previous alternative, we have

$$d(x_n, x_{n+1}) \leqslant F(x_n) - F(x_{n+1}),$$

so that by telescopic summation, for any pair of integers $0 \leqslant n \leqslant p$, we have

$$d(x_n, x_p) \leqslant F(x_n) - F(x_p). \tag{2.6}$$

However, the sequence $(F(x_n))_{n \in \mathbb{N}}$ is decreasing (by construction) and bounded from below (because F is), so it converges. The sequence $(F(x_n))_{n \in \mathbb{N}}$ is therefore Cauchy, which proves that $(x_n)_{n \in \mathbb{N}}$ is Cauchy. We denote by y the limit of the sequence $(x_n)_{n \in \mathbb{N}}$.

It is immediate that $F(y) \leqslant F(x)$ because the sequence $(F(x_n))_{n \in \mathbb{N}}$ is decreasing.

Moreover, for any integer p, by applying (2.6) with $n = 0$, we have

$$d(x, x_p) \leqslant F(x) - F(x_p) \leqslant F(x) - \inf_E F \leqslant 1 \qquad \text{(by hypothesis on } x\text{)}.$$

By taking the limit as p tends to $+\infty$, we verify that $d(x, y) \leqslant 1$.

It remains to show property (iii). Suppose it is not true and there exists $z \in E$ such that $F(z) < F(y) - d(y, z)$. We will combine this inequality with the one obtained by taking the limit as p tends to $+\infty$ in (2.6), that is:

$$\forall n \in \mathbb{N}, \quad F(y) = \lim_{p \to +\infty} F(x_p) \leqslant F(x_n) - d(x_n, y).$$

So, using the triangle inequality, we find that, for all $n \in \mathbb{N}$,

$$F(z) < F(y) - d(y, z) \leqslant F(x_n) - d(x_n, y) - d(y, z) \leqslant F(x_n) - d(x_n, z),$$

which implies that $z \in S_n$ for all $n \in \mathbb{N}$. We can then use inequality (2.5) to write

$$2F(x_{n+1}) - F(x_n) \leqslant \inf_{S_n} F \leqslant F(z).$$

Taking the limit, it follows that $F(y) \leqslant F(z)$, which contradicts the definition of z.$\square$

One can deduce from Ekeland's Variational Principle a famous result by Caristi.

Theorem 2.10 (Caristi's Fixed Point) *Let (E, d) be a complete metric space and let $\phi \colon E \to [0, +\infty)$ be a continuous function. For any continuous function $f \colon E \to E$ satisfying*

$$\forall x \in E, \quad d(x, f(x)) \leqslant \phi(x) - \phi(f(x)), \tag{2.7}$$

there exists a fixed point y (such that $y = f(y)$).

Proof We apply Theorem 2.9 to the function ϕ, with $\delta = 1$ and $\varepsilon = 1/2$ and x any point such that $\phi(x) \leqslant \inf_E \phi + 1/2$. We deduce the existence of $y \in E$ such that,

$$\forall z \in E, \quad \phi(z) \geqslant \phi(y) - \frac{1}{2}d(y, z).$$

By applying this inequality with $z = f(y)$, we get

$$\phi(y) - \phi(f(y)) \leqslant \frac{1}{2}d(y, f(y)),$$

so that the assumption (2.7) leads to $d(y, f(y)) \leqslant \frac{1}{2}d(y, f(y))$. This implies that $y = f(y)$.$\qquad\square$

Now, following Lieberman's classical book [124], we prove a corollary of Caristi's fixed point theorem which is very useful for solving the Cauchy problem for certain nonlinear partial differential equations.

We first need a definition. Let E and B be Banach spaces and $f: E \to B$ a mapping such that the limit

$$df(u) \cdot \psi = \lim_{\varepsilon \to 0} \frac{1}{\varepsilon}\big(f(u + \varepsilon\psi) - f(u)\big)$$

exists for all $u \in E$ and all $\psi \in E$. Then we say that df is the Gateaux derivative of f and that $df(u) \cdot \psi$ is the Gateaux derivative of f at u in the direction ψ.

Proposition 2.11 *Let E and B be Banach spaces and $f: E \to B$ a mapping such that $f(E)$ is closed in B. Assume that f has a Gateaux derivative and that, for all $u \in U$, there exists $\psi \in E$ such that*

$$df(u) \cdot \psi + f(u) = 0.$$

Then there exists $u \in E$ such that $f(u) = 0$.

Proof Introduce $F = f(E)$. This is a complete metric space since $f(E)$ is closed by assumption. We want to prove that $0 \in F$. Assume by contradiction that this is not the case. Consider a point $v \in F$ and let $u \in E$ be such that $v = f(u)$. By assumption there exists $\psi \in E$ such that $df(u) \cdot \psi + f(u) = 0$. Moreover, according to the definition of the Gateaux derivative and the assumption that $f(u) = v \neq 0$, there exists $\varepsilon > 0$ such that

$$\|f(u + \varepsilon\psi) - f(u) - \varepsilon df(u) \cdot \psi\| \leqslant \frac{\varepsilon}{2}\|f(u)\|.$$

Therefore

$$\|f(u + \varepsilon\psi) - (1 - \varepsilon)f(u)\| \leqslant \frac{\varepsilon}{2}\|f(u)\|. \qquad (2.8)$$

Once ψ and ε have been so determined, we set $g(v) = f(u + \varepsilon\psi)$. We claim that the map $g: F \to F$ has a fixed point. Let us admit this claim and conclude the proof. To do so, notice that if $g(v) = v$ then the inequality (2.8) implies that

$$\|v - (1 - \varepsilon)v\| \leqslant \frac{\varepsilon}{2}\|v\|,$$

which in turn immediately implies that $v = 0$ which is a contradiction.

We now have to prove that g admits a fixed point. It follows from (2.8) and the triangle inequality that

$$\|f(u + \varepsilon\psi) - f(u)\| \leqslant \frac{3\varepsilon}{2}\|f(u)\| \quad \text{and} \quad \|f(u + \varepsilon\psi)\| \leqslant \left(1 - \frac{\varepsilon}{2}\right)\|f(u)\|.$$

Consequently,

$$\|f(u + \varepsilon\psi) - f(u)\| \leqslant 3\big(\|f(u)\| - \|f(u + \varepsilon\psi)\|\big).$$

This reads $\|g(v) - v\| \leqslant 3\big(\|v\| - \|g(v)\|\big)$ which means that we are in position to apply Caristi's fixed point theorem with the function $\phi(x) = 3\|x\|$.   □

2.6 Brouwer's Fixed Point Theorem

Given a vector $x = (x_1, \ldots, x_n)$ in $\mathbb{R}^n$, we denote by $|x|$ the Euclidean norm defined by $|x|^2 = x_1^2 + \cdots + x_n^2$. The aim of this section is to prove the following fundamental result:

Theorem 2.12 (Brouwer) *Let $n \geqslant 1$ and $\overline{B} = \{x \in \mathbb{R}^n : |x| \leqslant 1\}$ be the closed unit ball in $\mathbb{R}^n$. Any continuous function $\psi : \overline{B} \to \overline{B}$ has a fixed point.*

Corollary 2.13 *Let $C \subset \mathbb{R}^n$ be a set homeomorphic to $\overline{B}$. Then any continuous function $\phi : C \to C$ has a fixed point.*

Proof Consider a homeomorphism $\kappa : C \to \overline{B}$ (that is, a continuous and bijective function, whose inverse $\kappa^{-1} : \overline{B} \to C$ is also continuous). Then the function $\psi = \kappa \circ \phi \circ \kappa^{-1} : \overline{B} \to \overline{B}$ is continuous, therefore has a fixed point $y \in \overline{B}$ according to Theorem 2.12. Then $x = \kappa^{-1}(y) \in C$ is a fixed point of ϕ.   □

Remark 2.14 In particular, we will deduce in the solved Exercise 2.2 that Theorem 2.12 remains true if we replace the Euclidean norm with any norm.

Proof (of Theorem 2.12) We begin by showing that the theorem will be a simple consequence of the following lemma.

Lemma 2.15 *Consider a continuous function $\theta : \overline{B} \to \overline{B}$ such that $\theta(x) = x$ for x belonging to the sphere $\mathbb{S}^{n-1} = \partial \overline{B}$. Then $\overline{B} = \theta(\overline{B})$.*

Proof (of the theorem from the lemma) We argue by contradiction. Suppose that $\psi(x) \neq x$ for all x in $\overline{B}$. Consider x in $\overline{B}$ and denote by D_x the half-line originating from $\psi(x)$ that passes through x:

$$D_x = \{\psi(x) + \lambda(x - \psi(x)) : \lambda > 0\}.$$

This half-line intersects the sphere $\mathbb{S}^{n-1}$ at a unique point, denoted $\theta(x)$. Then we easily verify that the function $\theta : \overline{B} \to \mathbb{S}^{n-1}$ is continuous and moreover $\theta(x) = x$ for all $x \in \mathbb{S}^{n-1}$. Then Lemma 2.15 implies that $\overline{B} = \theta(\overline{B})$, which is absurd because $\theta(\overline{B}) \subset \mathbb{S}^{n-1}$ by construction.   □

It remains to prove Lemma 2.15. There are many proofs of this result. The one presented below is due to Peter Lax [111] and is based on the formula for change of variables. We begin by recalling the following statement.

Theorem 2.16 (Change of variables in an integral) *Let $\kappa : U \to V$ be a diffeomorphism of class C^1 (see Definition 2.4). Then, for any Borel function $f : V \to \mathbb{R}_+$, we have*

$$\int_V f(y)\, dy = \int_U f(\kappa(x))\, |J(x)|\, dx,$$

where $J(x) = \det(\kappa'(x)) = \det(\partial \kappa_j / \partial x_i)$ is the Jacobian of κ.

Peter Lax's very clever idea is to show the formula for change of variables for a multiple integral, without assuming that the change of variables is a diffeomorphism.

Lemma 2.17 (Peter Lax) *Let* $f: \mathbb{R}^n \to \mathbb{R}$ *be a continuous function with compact support and let* $\varphi: \mathbb{R}^n \to \mathbb{R}^n$ *be a function of class* C^2 *that coincides with the identity outside the unit ball. Then:*

$$\int_{\mathbb{R}^n} f(\varphi(x))J(x)\,dx = \int_{\mathbb{R}^n} f(y)\,dy, \tag{2.9}$$

where $J = \det\left(\partial\varphi_j/\partial x_i\right)$ *is the Jacobian of* φ.

Remark 2.18

1. Note that the integral $\int_{\mathbb{R}^n} f(\varphi(x))J(x)\,dx$ involves the Jacobian of φ and not the absolute value of the Jacobian.
2. The key point is that we do not assume that φ is a diffeomorphism. If φ was a diffeomorphism, then this result would be a consequence of Theorem 2.16. Indeed, if φ is a diffeomorphism, the function $J = \det\left(\partial\varphi_j/\partial x_i\right)$ does not vanish on $\mathbb{R}^n$ therefore it keeps a constant sign. Since J is equal to 1 outside the ball B (because φ is the identity outside of B), we deduce that J is a positive function and therefore that $J = |J|$.
3. Peter Lax also showed (*cf.* [112]) that one can deduce Theorem 2.16 from this lemma.

Proof Let $f: \mathbb{R}^n \to \mathbb{R}$ be a continuous function with compact support. Given $y = (y_1, \dots, y_n)$, we introduce

$$g(y) = \int_{-\infty}^{y_1} f(s, y_2, \dots, y_n)\,ds.$$

Fix $c \geqslant 1$ large enough so that supp f is included in the cube $Q = [-c, c]^n$. Then we have $f(\varphi(x)) = 0$ if $x \notin Q$ and

$$\int_{\mathbb{R}^n} f(\varphi(x))J(x)\,dx = \int_Q (\partial_{y_1} g)(\varphi(x))J(x)\,dx.$$

The proof relies on elementary properties of the determinant function. We start by noting that

$$\frac{\partial g}{\partial y_1}(\varphi(x))J(x) = \frac{\partial g}{\partial y_1}(\varphi(x))\det(\nabla\varphi_1, \dots, \nabla\varphi_n)$$
$$= \det\big(\nabla(g(\varphi(x)), \nabla\varphi_2, \dots, \nabla\varphi_n)\big),$$

where we used the fact that, for all k different from 1, we have

$$\det(\nabla\varphi_k, \nabla\varphi_2, \dots, \nabla\varphi_n) = 0.$$

Then, by expanding the determinant $\det(\nabla(g(\varphi(x)), \nabla\varphi_2, \dots, \nabla\varphi_n)$ with respect to the first column, we get

$$\det(\nabla(g(\varphi)), \nabla\varphi_2, \ldots, \nabla\varphi_n) = M_1 \partial_1(g(\varphi)) + \cdots + M_n \partial_n(g(\varphi)),$$

where we denoted by M_j the minor of the term $\partial_{x_j}(g(\varphi))$. Therefore,

$$\int_{\mathbb{R}^n} f(\varphi) J \, dx = - \int_Q g(\varphi) \, (\partial_1 M_1 + \cdots + \partial_n M_n) \, dx + B, \qquad (2.10)$$

where the term B is given by:

$$B = \int_Q \left(\partial_1 \left(M_1 g(\varphi) \right) + \cdots + \partial_n \left(M_n g(\varphi) \right) \right) dx.$$

The key point (due to Kronecker) comes from the following cancellation

$$\partial_1 M_1 + \cdots + \partial_n M_n = 0. \qquad (2.11)$$

Let us prove this identity. For this, we will use the formal identity

$$\partial_1 M_1 + \cdots + \partial_n M_n = \det(\nabla, \nabla\varphi_2, \ldots, \nabla\varphi_n),$$

which is also obtained by expanding the determinant in the right-hand side with respect to the first column. Then, if $n = 2$, we directly have

$$\partial_1 M_1 + \partial_2 M_2 = \det \begin{pmatrix} \partial_1 & \partial_1 \varphi_2 \\ \partial_2 & \partial_2 \varphi_2 \end{pmatrix} = \partial_1 \partial_2 \varphi_2 - \partial_2 \partial_1 \varphi_2 = 0,$$

where the last cancellation comes from Schwarz's theorem (symmetry of the Hessian matrix) which applies because we assumed that φ is of class C^2. Let us now consider the case $n \geqslant 3$. Let us denote by A the matrix $A = (\nabla, \nabla\varphi_2, \ldots, \nabla\varphi_n)$. We use the definition of the determinant from permutations[2] to write,

$$\det(\nabla, \nabla\varphi_2, \ldots, \nabla\varphi_n) = \sum_{\sigma \in \mathfrak{S}_n} \varepsilon(\sigma) \prod a_{\sigma(j)j} = \sum_{\sigma \in \mathfrak{S}_n} \varepsilon(\sigma) \partial_{\sigma(1)} \left(\prod_{j=2}^n \partial_{\sigma(j)} \varphi_j \right)$$

$$= \sum_{\sigma \in \mathfrak{S}_n} \varepsilon(\sigma) (\partial_{\sigma(1)} \partial_{\sigma(2)} \varphi_2)(\partial_{\sigma(3)} \varphi_3) \cdots (\partial_{\sigma(n)} \varphi_n) + \cdots$$

$$+ \sum_{\sigma \in \mathfrak{S}_n} \varepsilon(\sigma) (\partial_{\sigma(2)} \varphi_2) \cdots (\partial_{\sigma(n-1)} \varphi_{n-1})(\partial_{\sigma(1)} \partial_{\sigma(n)} \varphi_n).$$

---

[2] Let $\mathfrak{S}_n$ denote the symmetric group of $E = \{1, \ldots, n\}$, that is the set of bijective functions from E to E, called permutations, equipped with the law of composition of functions. A transposition is a particular permutation that exchanges two elements while leaving the others unchanged; we then denote by (k, ℓ) the transposition that exchanges element k with element ℓ. Any permutation σ can be written as a product of transpositions. Such an expression is not unique. However, we can define an invariant: the parity of the number of terms of such a product depends only on the permutation. We then speak of even or odd permutations, and we define the signature $\varepsilon(\sigma)$ of a permutation σ as follows: $\varepsilon(\sigma) = 1$ if σ is even and $\varepsilon(\sigma) = -1$ otherwise.

Let us show that the $n - 1$ terms of the right-hand side above are all equal to 0. To fix our ideas, we study the first of these n terms:

$$\Pi := \sum_{\sigma \in \mathfrak{S}_n} \varepsilon(\sigma)(\partial_{\sigma(1)}\partial_{\sigma(2)}\varphi_2)(\partial_{\sigma(3)}\varphi_3)\dots(\partial_{\sigma(n)}\varphi_n).$$

Let $\tau \in \mathfrak{S}_n$ be any permutation. As the map $\sigma \mapsto \sigma \circ \tau$ is a bijection of the symmetric group to itself, we verify that

$$\Pi = \sum_{\sigma \in \mathfrak{S}_n} \varepsilon(\sigma \circ \tau)(\partial_{\sigma\circ\tau(1)}\partial_{\sigma\circ\tau(2)}\varphi_2)(\partial_{\sigma\circ\tau(3)}\varphi_3)\cdots(\partial_{\sigma\circ\tau(n)}\varphi_n).$$

We apply this with the transposition $\tau = (1,2)$ given by $\tau(1) = 2$ and $\tau(2) = 1$. Then we have $\varepsilon(\sigma \circ \tau) = -\varepsilon(\sigma)$ and according to Schwarz's theorem $\partial_{\sigma\circ\tau(1)}\partial_{\sigma\circ\tau(2)}\varphi_2 = \partial_{\sigma(1)}\partial_{\sigma(2)}\varphi_2$. Moreover, $\sigma \circ \tau(k) = \sigma(k)$ for $k \geq 3$. It follows that $\Pi = -\Pi$, which proves that $\Pi = 0$. We conclude that the identity (2.10) leads to

$$\int_{\mathbb{R}^n} f(\varphi)J\,dx = \int_Q \left(\partial_1(M_1 g(\varphi)) + \cdots + \partial_n(M_n g(\varphi))\right)dx.$$

Note that $g(x) = 0$ if $x = (x_1,\dots,x_n)$ is such that $|x_k| \geq c$ for a certain index k such that $2 \leq k \leq n$. By combining this remark with the fact that $\varphi(x) = x$ if $|x| \geq 1$, we deduce that, for any index k such that $2 \leq k \leq n$,

$$\int_Q \partial_k(M_k g(\varphi))\,dx = 0.$$

We deduce that

$$\int_{\mathbb{R}^n} f(\varphi)J\,dx = \int_Q \partial_1(M_1 g(\varphi))\,dx. \tag{2.12}$$

Note that

$$\int_Q \partial_1(M_1 g(\varphi))\,dx = \int_{[-c,c]^{n-1}} \theta(x_2,\dots,x_n)\,dx_2\cdots dx_n \qquad \text{where}$$
$$\theta(x_2,\dots,x_n) = (M_1 g(\varphi))(c,x_2,\dots,x_n) - (M_1 g(\varphi))(-c,x_2,\dots,x_n).$$

By definition, M_1 is the determinant of the square matrix $(\partial\varphi_j/\partial x_i)_{2 \leq i,j \leq n}$. As $\varphi(\pm c, x_2,\dots,x_n) = (\pm c, x_2,\dots,x_n)$ because $c \geq 1$ and because $\varphi(y) = y$ if $y \notin B$ (by hypothesis), we deduce that $M_1(\pm c, x_2,\dots,x_n) = 1$. Moreover, by definition of g and using the hypothesis supp $f \subset [-c,c]^n$, we have

$$g(\varphi)(-c,x_2,\dots,x_n) = 0 \quad \text{and} \quad g(\varphi)(c,x_2,\dots,x_n) = \int_{-\infty}^{\infty} f(t,x_2,\dots,x_n)\,dt.$$

We have therefore shown that $\int_Q \partial_1(M_1 g(\varphi))\,dx = \int_{\mathbb{R}^n} f(y)\,dy$. This concludes the proof in view of (2.12). $\qquad\square$

We are now in a position to prove Lemma 2.15. For this, we will proceed in two stages. We start by studying a weak form of Lemma 2.15 where we assume that the function is of class C^2 (and not just continuous).

Lemma 2.19 *Consider $\varphi \colon \mathbb{R}^n \to \mathbb{R}^n$ of class C^2 and such that $\varphi(x) = x$ if $|x| \geqslant 1$. Then we have $\overline{B} \subset \varphi(\overline{B})$.*

Proof By contradiction, suppose there exists an element y_0 of $\overline{B}$ that does not belong to $\varphi(\overline{B})$. As φ coincides with the identity outside the open ball, we verify that y_0 belongs to B. As $\overline{B}$ is compact, the image set $\varphi(\overline{B})$ is also compact and therefore closed. We deduce that the complement $\mathbb{R}^n \setminus \varphi(\overline{B})$ is open so there exists $r > 0$ such that the ball $B(y_0, r)$ is included in $B \cap (\mathbb{R}^n \setminus \varphi(\overline{B}))$. The idea is then to use the identity

$$\int_{\mathbb{R}^n} f(\varphi(x)) J(x)\, dx = \int_{\mathbb{R}^n} f(y)\, dy,$$

with $f \in C_0^\infty(\mathbb{R}^n)$ chosen such that supp f is included in $B(y_0, r)$. For such a function f we find that $\int f(\varphi(x)) J(x)\, dx = 0$, which is absurd as soon as $\int f \neq 0$.   □

End of the proof of Lemma 2.15

We start by extending $\theta \colon \overline{B} \to \overline{B}$ to a continuous function $\varphi \colon \mathbb{R}^n \to \mathbb{R}^n$ by setting $\varphi(x) = \theta(x)$ if $x \in \overline{B}$ and $\varphi(x) = x$ if $|x| \geqslant 1$. Next, we will use a convolution method[3] to approximate φ by functions φ_ε of class C^2, satisfying $\varphi_\varepsilon(x) = x$ if $|x| \geqslant 1 + \varepsilon$. For this, we introduce the functions

$$\varphi_\varepsilon(x) = \frac{1}{\varepsilon^n} \int_{\mathbb{R}^n} \rho\left(\frac{x - y}{\varepsilon}\right) \varphi(y)\, dy,$$

where ρ is a C^∞ function with support in the unit ball, satisfying $\rho(y) = \rho(-y)$, such that $\rho \geqslant 0$ and $\int_{\mathbb{R}^n} \rho(y)\, dy = 1$. For all $\varepsilon > 0$, φ_ε is of class C^∞ by differentiation under the integral sign.

Let us show that $\varphi_\varepsilon(x) = x$ if $|x| \geqslant 1 + \varepsilon$. For this, by an elementary change of variables, we start by writing that

$$\varphi_\varepsilon(x) = \frac{1}{\varepsilon^n} \int_{\mathbb{R}^n} \rho\left(\frac{y}{\varepsilon}\right) \varphi(x - y)\, dy.$$

Consider x such that $|x| \geqslant 1 + \varepsilon$. If y/ε belongs to the support of ρ, then $|y| \leqslant \varepsilon$ and therefore $|x - y| \geqslant 1$ from the triangle inequality. It follows that

$$\varphi_\varepsilon(x) = \frac{1}{\varepsilon^n} \int_{\mathbb{R}^n} \rho\left(\frac{y}{\varepsilon}\right)(x - y)\, dy = \frac{1}{\varepsilon^n} \int_{\mathbb{R}^n} \rho\left(\frac{y}{\varepsilon}\right) x\, dy = x,$$

where we used the fact that $\int \rho(y)\, dy = 1$ and $\int y\rho(y)\, dy = 0$ (since ρ is even). Moreover, it follows from classical results (see the proof of Theorem 6.8) that these

---

[3] The convolution will be studied in detail in Chapter 6.

functions converge uniformly towards φ in the sense that

$$\lim_{\varepsilon \to 0} \sup_{x \in \mathbb{R}^n} |\varphi_\varepsilon(x) - \varphi(x)| = 0.$$

Finally, we introduce the function

$$\widetilde{\varphi}_\varepsilon(x) = \frac{1}{1 + \varepsilon} \varphi_\varepsilon((1 + \varepsilon)x),$$

so that $\widetilde{\varphi}_\varepsilon : \mathbb{R}^n \to \mathbb{R}^n$ is of class C^2 and satisfies $\widetilde{\varphi}_\varepsilon(x) = x$ if $|x| \geqslant 1$.

Now, let us consider an element y from the ball $\overline{B}$. According to Lemma 2.19, there exists an $\widetilde{x}_\varepsilon$ in $\overline{B}$ such that $\widetilde{\varphi}_\varepsilon(\widetilde{x}_\varepsilon) = y$. By compactness of $\overline{B}$, there exists a sequence $(\widetilde{x}_{\varepsilon_k})_{k \in \mathbb{N}}$ that converges towards an element $x \in \overline{B}$. By taking the limit, we verify that $\varphi(x) = y$.

This concludes the proof of Lemma 2.15 and therefore that of Brouwer's fixed point theorem.   □

Brouwer's fixed point theorem has many applications (see [139, 55, 151]). We will see in the next section a (difficult) application to the proof of Brouwer's domain invariance theorem. We show here how to use this fixed point theorem to prove a famous result on matrices with non-negative coefficients.

Corollary 2.20 (Perron–Frobenius Theorem) *Let $n \geqslant 1$ and let $A = (a_{ij})_{1 \leqslant i, j \leqslant n}$ be a matrix such that $a_{ij} \in [0, +\infty)$ for all i, j with $1 \leqslant i, j \leqslant n$. Then A admits an eigenvalue $\lambda \in [0, +\infty)$.*

Proof If A is not invertible, then 0 is an eigenvalue.

If A is invertible, we introduce the function $f : \mathbb{R}^n \setminus \{0\} \to \mathbb{R}^n$ defined by $f(x) = Ax/|Ax|_1$, where we used the notation $|(y_1, \ldots, y_n)|_1 = |y_1| + \cdots + |y_n|$. We also consider the set

$$C = \{(x_1, \ldots, x_n) \in [0, 1]^n : x_1 + \cdots + x_n = 1\}.$$

We verify that C is a convex compact of $\mathbb{R}^n$, which is homeomorphic[4] to the closed unit ball $\overline{B_{n-1}}$ of $\mathbb{R}^{n-1}$, that is

$$\overline{B_{n-1}} = \{(X_1, \ldots, X_{n-1}) \in \mathbb{R}^{n-1} : X_1^2 + \cdots + X_{n-1}^2 \leqslant 1\}.$$

Now, we observe that $f(C) \subset C$ by construction of f, using the fact that, if the coordinates of x are positive, then Ax is also a vector with positive coordinates (since the coefficients of A are positive). Therefore, Brouwer's theorem (see Corollary 2.13) implies that $f|_C$ admits a fixed point $x \in C$, which implies that $|Ax|_1$ is an eigenvalue of A.   □

---

[4] We do not detail here the proof that C is homeomorphic to $\overline{B_{n-1}}$; we refer to the solved Exercise 2.2 where a similar construction is proposed.

2.7 Invariance of Domain

The purpose of this section is to prove the domain invariance theorem due to Brouwer.[5]

Theorem 2.21 (Invariance of Domain)

1. *Consider an integer $n \geq 1$ and an open set $U \subset \mathbb{R}^n$. If $f : U \to \mathbb{R}^n$ is continuous and injective, then the image set $f(U)$ is open.*
2. *Consider two integers $n, m \geq 1$. If $\mathbb{R}^n$ is homeomorphic to $\mathbb{R}^m$ then $n = m$.*
3. *Consider an integer $n \geq 1$. Let $U \subset \mathbb{R}^n$ be open and let $f : U \to \mathbb{R}^n$ be a continuous and injective function. Then $f : U \to f(U)$ is a homeomorphism.*

Remark 2.22

- There exists $f : (0, 1) \to (0, 1)^2$ which is bijective and continuous but whose inverse is not continuous. For example, the function defined by:

$$f(x) = (y, z) \quad \text{where} \quad y = 0.d_1 d_3 d_5 \ldots \quad \text{and} \quad z = 0.d_2 d_4 d_6 \ldots .$$

where

$$x = 0.d_1 d_2 \ldots = \sum_{j=1}^{\infty} \frac{d_j}{10^j} \quad \text{with} \quad d_j \in \{0, \ldots, 9\}.$$

This function is bijective and continuous but its inverse is not continuous. Indeed, the numbers $a = 0.1$ and $b_p = 0.099 \ldots 9$, containing p copies of the digit 9, are arbitrarily close, while the distance between the numbers $f^{-1}(a, a) = 0.11$ and $f^{-1}(b_p, b_p) = 0.0099 \ldots 9$ is at least 0.1).
- The result is false in infinite-dimensional spaces. For example, $\tau : \ell^\infty(\mathbb{N}) \to \ell^\infty(\mathbb{N})$ defined by

$$\tau(x_0, \ldots, x_n, \ldots) = (0, x_0, \ldots, x_n, \ldots).$$

This function is continuous and injective, but the set $\tau(\ell^\infty(\mathbb{N}))$ is not open.
- There exists a continuous and injective function $f : \mathbb{R} \to \mathbb{R}^2$ whose image is not open (for example $f(t) = (t, 0)$). This shows that we must assume that the dimensions of the starting and arriving spaces are the same.

Proof To prove the domain invariance theorem, we will use the following lemma.

Lemma 2.23 *Let $\overline{B}$ be the closed ball $\overline{B(0, 1)}$. If $f : \overline{B} \to \mathbb{R}^n$ is continuous and injective, then $f(0)$ belongs to the interior of $f(\overline{B})$.*

Let us momentarily assume this lemma and prove the theorem.

---

[5] This result was announced by Poincaré. It was first proved by Brouwer [17]. We will give here a very simple demonstration due to Kulpa [108]. Terence Tao revisits this demonstration in his course notes (see [183, §6.2]) and we will follow his version of Kulpa's demonstration (see also the text by Mawhin [130] and the many references it contains).

Proof of statement (1). Consider $y_0 \in f(U)$. There exists an element x_0 of U such that $f(x_0) = y_0$. Let $\varepsilon > 0$ be sufficiently small, so that $x_0 + \varepsilon \overline{B} \subset U$. Consider then the function $F: \overline{B} \to \mathbb{R}^n$ defined by $F(x) = f(x_0 + \varepsilon x)$. This is a continuous and injective function by the hypotheses on f. According to Lemma 2.23, $y_0 = F(0)$ belongs to the interior of $F(\overline{B})$. But $F(\overline{B}) \subset f(U)$ so $f(U)$ is a neighborhood of y_0. This proves that $f(U)$ is open.

Proof of statement (2). Suppose, by contradiction, that there exists a homeomorphism $f: \mathbb{R}^n \to \mathbb{R}^m$ with $m \neq n$. If necessary, by exchanging n and m, we can assume that $m < n$. Let $E_m = \mathbb{R}^m \times \{0\}^{n-m}$ and consider the function $p: \mathbb{R}^m \to \mathbb{R}^n$ defined by $p: (x_1, \ldots, x_m) \mapsto (x_1, \ldots, x_m, 0, \ldots, 0)$. Then the function $F: \mathbb{R}^n \to \mathbb{R}^n$, defined by $F(x) = p(f(x))$, is continuous and injective as the composition of two continuous and injective functions. So $F(\mathbb{R}^n)$ is open according to statement (1). This is absurd because $F(\mathbb{R}^n) \subset E_m$ and no ball of $\mathbb{R}^n$ is included in E_m.

Proof of statement (3). The function f is bijective from U to $f(U)$. Let us denote by $g = f^{-1}: f(U) \to U$ its inverse. Consider an open subset V of U. Then V is an open subset of $\mathbb{R}^n$. As $g^{-1}(V) = f(V)$, the result of statement (1) implies that $g^{-1}(V)$ is open. This shows that $g = f^{-1}$ is continuous, therefore f is a homeomorphism from U to $f(U)$.

We now come to the main part of the proof.

Proof of Lemma 2.23

We will use the following result.

Theorem 2.24 (Tietze) *Consider a metric space X and a real number $M > 0$. For any continuous function $f: A \to [-M, M]$ defined on a closed subset A of X, there exists a continuous function $g: X \to [-M, M]$ such that $g|_A = f$.*

Proof The proof relies on the following lemma.

Lemma 2.25 *Let $f: A \to [-M, M]$ be a continuous function on a closed subset A of X. There exists a continuous function $\widetilde{f}: X \to [-M/3, M/3]$ such that*

$$\forall x \in A, \quad \left| f(x) - \widetilde{f}(x) \right| \leq \frac{2M}{3}. \tag{2.13}$$

Proof Introduce the two closed sets

$$F_1 = f^{-1}([-M, -M/3]) \quad \text{and} \quad F_2 = f^{-1}([M/3, M]).$$

Note that these two sets are disjoint. Observe that the function $X \ni x \mapsto \text{dist}(x, F_1) + \text{dist}(x, F_2) \in \mathbb{R}_+$ does not vanish. Indeed, if $\text{dist}(x, F_1) = 0$, then $x \in F_1$ because F_1 is closed, so $x \notin F_2$ by hypothesis and it follows that $\text{dist}(x, F_2) > 0$ because F_2 is closed. We can then pose

$$\widetilde{f}(x) = -\frac{M}{3} \frac{\text{dist}(x, F_2)}{\text{dist}(x, F_1) + \text{dist}(x, F_2)} + \frac{M}{3} \frac{\text{dist}(x, F_1)}{\text{dist}(x, F_1) + \text{dist}(x, F_2)}.$$

This function is continuous, with values in $[-M/3, M/3]$. It equals $-M/3$ on F_1 and $M/3$ on F_2, and we deduce (2.13) from the triangle inequality by separately studying the cases $x \in F_1, x \in F_2$ and $x \in A \setminus (F_1 \cup F_2)$.   □

Using this lemma, we construct by induction a sequence $(f_k)_{k \in \mathbb{N}^*}$ of continuous functions $f_k : X \to \mathbb{R}$ such that

$$\forall x \in A, \quad \left| f(x) - \sum_{k=1}^{n} f_k(x) \right| \leq (2/3)^n M, \tag{2.14}$$

$$\forall x \in X, \quad |f_k(x)| \leq (2/3)^k \frac{M}{2}. \tag{2.15}$$

According to (2.15), the series $\sum f_k$ converges normally, therefore uniformly, and its sum $g = \sum_{k \in \mathbb{N}^*} f_k$ satisfies $|g| \leq M$ and coincides with f on A according to (2.14). This concludes the proof of Theorem 2.24.   □

By hypothesis, the function $f : \overline{B} \to f(\overline{B})$ is a continuous bijection. We want to show that $f(0)$ belongs to the interior of $f(\overline{B})$.

Consider the function $g : f(\overline{B}) \to \overline{B}$ defined by $g(x) = f^{-1}(x)$. Let us show that it is continuous. Let F be a closed set of $\overline{B}$. Then F is a compact set of $\mathbb{R}^n$, so $g^{-1}(F) = f(F)$ is compact because it is the image of a compact set by a continuous function. In particular, $g^{-1}(F)$ is closed, which shows that g is continuous. We then use the Tietze theorem 2.24 to obtain the existence of a continuous function $G : \mathbb{R}^n \to \mathbb{R}^n$ that extends g (we apply the Tietze theorem to each coordinate of g). By construction, we have

$$G(f(0)) = g(f(0)) = f^{-1}(f(0)) = 0,$$

so G vanishes at some point. The idea of the proof is to construct a small perturbation of G that never vanishes, then deduce a contradiction from it.

Suppose that $f(0)$ does not belong to the interior of $f(\overline{B})$. By continuity of G, there exists $\varepsilon > 0$ such that

$$|G(y)| < \frac{1}{3}, \qquad \forall y \in B(f(0), 2\varepsilon).$$

Moreover, there exists c in $\mathbb{R}^n$ such that

$$|c - f(0)| < \varepsilon \quad \text{and} \quad c \notin f(\overline{B}).$$

Then $|G(y)| < 1/3$ if $y \in \overline{B(c, \varepsilon)}$. We set:

$$\Sigma = \Sigma_1 \cup \Sigma_2 \quad \text{with} \quad \Sigma_1 = \{y \in f(\overline{B}) : |y - c| \geq \varepsilon\} \quad \text{and} \quad \Sigma_2 = \partial B(c, \varepsilon).$$

The function G does not vanish on Σ_1. Indeed, if $y \in f(\overline{B})$, then $G(y) = g(y) = f^{-1}(y)$ and so $G(y) = 0$ implies $y = f(0)$. However, if $y \in \Sigma_1$, we have $y \in f(\overline{B})$ and $y \neq f(0)$.

Since Σ_1 is compact, G is bounded below on Σ_1 by a positive constant. There exists a $\delta > 0$ such that

$$\inf_{y \in \Sigma_1} |G(y)| > \delta. \tag{2.16}$$

We can of course assume that $\delta \leqslant 1/3$. It follows that G does not vanish on Σ_1. We will show that we can perturb G to obtain a function that also does not vanish on Σ_2.

Lemma 2.26 *There exists a continuous function $P \colon \mathbb{R}^n \to \mathbb{R}^n$ such that*

$$\sup_{y \in f(\overline{B})} |P(y) - G(y)| \leqslant \delta, \tag{2.17}$$

and which does not vanish on $\Sigma_1 \cup \Sigma_2$.

Proof Since $f(\overline{B})$ is compact, we can use the Stone–Weierstrass theorem (see Corollary 1.48) to deduce the existence of a function $Q \colon \mathbb{R}^n \to \mathbb{R}^n$ whose coordinates are polynomials and which satisfies

$$\sup_{y \in f(\overline{B})} |Q(y) - G(y)| \leqslant \frac{\delta}{2}. \tag{2.18}$$

Let us now show that the image set $Q(\Sigma_2)$ has measure zero. For this, note that for any real number $\mu > 0$, we can cover the sphere Σ_2 with a finite collection of balls $\Sigma_2 \subset \bigcup_{j=1}^N B(x_j, r_j)$ such that

$$\sum_{1 \leqslant j \leqslant N} |B(x_j, r_j)| \leqslant \mu,$$

where $|B(x_j, r_j)|$ is the n-dimensional Lebesgue measure of the ball $B(x_j, r_j)$. Since Q is of class C^1, it is Lipschitz on bounded parts according to the mean value theorem (see Theorem 2.1). Also, there exists a constant $K > 0$ such that $|Q(B(x_j, r_j))| \leqslant K |B(x_j, r_j)|$. Then, by writing

$$Q(\Sigma_2) \subset \bigcup_{j=1}^N Q(B(x_j, r_j)),$$

we obtain the inequality $|Q(\Sigma_2)| \leqslant K\mu$. Since this inequality is true for all $\mu > 0$, we indeed find that $Q(\Sigma_2)$ has measure zero. In particular, the ball $B(0, \delta/2)$ is not included in $Q(\Sigma_2)$, which shows that there exists a vector $e \in B(0, \delta/2)$ such that $e \notin Q(\Sigma_2)$. We then consider the function $P \colon \mathbb{R}^n \to \mathbb{R}^n$ defined by $P(x) = Q(x) - e$. This function satisfies (2.17) by the triangle inequality and moreover does not vanish on Σ_2 by construction.

Finally, it directly follows from (2.16) and the triangle inequality that P does not vanish on Σ_1. $\qquad\square$

Let us introduce the function $\widetilde{G}: f(\overline{B}) \to \mathbb{R}^n$ defined by

$$\widetilde{G}(y) = P\left(c + \max\left\{\frac{\varepsilon}{|y - c|}, 1\right\}(y - c)\right).$$

Since c does not belong to $f(\overline{B})$ by assumption, the function $\widetilde{G}$ is well defined and it is continuous on $f(\overline{B})$. Moreover, for all $y \in f(\overline{B})$, we have

$$z := c + \max\left\{\frac{\varepsilon}{|y - c|}, 1\right\}(y - c) \in \Sigma.$$

Indeed, if $|y - c| \geqslant \varepsilon$, then $z = y$ and therefore $z \in \Sigma_1$. If $|y - c| < \varepsilon$, then $|z - c| = \varepsilon$ and therefore $z \in \Sigma_2$. In both cases we verify that z belongs to $\Sigma = \Sigma_1 \cup \Sigma_2$. We deduce that $\widetilde{G}(y) \neq 0$ for all $y \in f(\overline{B})$ because the function P does not vanish on Σ.

If $y \in f(\overline{B})$ with $|y - c| \geqslant \varepsilon$, then

$$\left|G(y) - \widetilde{G}(y)\right| = |G(y) - P(y)| \leqslant \delta.$$

If $y \in f(\overline{B})$ and $|y - c| < \varepsilon$, then

$$|G(y)| < \frac{1}{3} \quad \text{and} \quad |G(z)| < \frac{1}{3}.$$

Indeed, $|G(x)| \leqslant 1/3$ for all x in the closed ball $\overline{B}(c, \varepsilon)$. Therefore

$$|\widetilde{G}(y)| = |P(z)| \leqslant |P(z) - G(z)| + |G(z)| \leqslant \delta + \frac{1}{3}.$$

It follows that

$$|\widetilde{G}(y) - G(y)| \leqslant |G(y)| + |\widetilde{G}(y)| \leqslant \frac{2}{3} + \delta \leqslant 1.$$

It follows that, for all $y \in f(\overline{B})$, we have

$$\left|G(y) - \widetilde{G}(y)\right| \leqslant 1.$$

Let us then consider the function $h: \overline{B} \to \mathbb{R}^n$ defined by

$$h(x) = G(f(x)) - \widetilde{G}(f(x)).$$

This function is continuous and $|h(x)| \leqslant 1$ for all $x \in \overline{B}$. Therefore h is in fact a continuous function of the unit ball $\overline{B}$ into itself. According to Brouwer's theorem 2.12, h has a fixed point. Since $G(f(x)) = x$ if $x \in \overline{B}$, it follows that there exists $x \in \overline{B}$ such that $\widetilde{G}(f(x)) = 0$. This is absurd because we have seen that $\widetilde{G}$ does not vanish on $f(\overline{B})$.

This concludes the proof of the lemma, and therefore the proof of the domain invariance theorem. $\square$

2.8 Nash's Theorem

We will study a simple version of Nash's theorem, which is essentially an extension of the local inversion theorem to the framework of Fréchet spaces.

We start by giving a counterexample, taken from Hamilton's article [67], that shows that the local inversion theorem is not true in a Fréchet space. Consider the operator $P\colon C^\infty([-1,1]) \to C^\infty([-1,1])$ defined by

$$P(f) = f - xf\frac{\mathrm{d}f}{\mathrm{d}x}.$$

The operator P is regular, at least in the sense where

$$DP(f)g = \lim_{\varepsilon \to 0} \frac{1}{\varepsilon}(P(f + \varepsilon g) - P(f)) = g - xg\frac{\mathrm{d}f}{\mathrm{d}x} - xf\frac{\mathrm{d}g}{\mathrm{d}x}.$$

For $f = 0$, we have $DP(0)g = g$ so $DP(0) = I$. As $P(0) = 0$, if the local inversion theorem were true in this setting, the image of P should contain a neighborhood of 0. Let us show that this is not the case. Consider the sequence

$$g_n = 1/n + x^n/n!$$

which converges to 0 in $C^\infty([-1,1])$. We will show that g_n does not belong to the image of P for all $n > 1$. More generally, let us show that $G_n = 1/n + b_n x^n$ does not belong to the image of P for all $n > 1$.

Suppose that $P(f) = 1/n + b_n x^n$ and consider the formal series expansion of f:

$$f = a_0 + a_1 x + a_2 x^2 + \cdots.$$

Then

$$P(f) = a_0 + (1 - a_0)a_1 x + (a_2 - a_1^2 - 2a_0a_2)x^2 + (a_3 - 3a_1a_2 - 3a_0a_3)x^3 + \cdots$$

can be written in the form $P(f) = a_0 + \alpha_1 x + \cdots + \alpha_k x^k + \cdots$ with

$$\alpha_1 = (1 - a_0)a_1, \quad \alpha_k = (1 - ka_0)a_n + Q_k(a_1, \ldots, a_{k-1}) \text{ for } k \geqslant 2,$$

where Q_k satisfies $Q_k(0) = 0$. Necessarily, $a_0 = 1/n$ and therefore $a_0 \neq 1$ if $n > 1$. Then $(1 - a_0)a_1 = 0$, so $a_1 = 0$. We then show by induction that $a_k = 0$ for $k < n$. Then $\alpha_n = (1 - na_0)a_n$ and therefore $\alpha_n = 0$ because $a_0 = 1/n$. It is thus impossible to impose $\alpha_n = b_n$ and therefore to solve $P(f) = 1/n + b_n x^n$.

Remark 2.27 The problem comes from the fact that $DP(f)$ can be non-invertible for arbitrarily small f. Indeed,

$$DP(1/n)x^k = (1 - k/n)x^k$$

so $DP(1/n)x^n = 0$. Even if $DP(0)$ is the identity, $DP(f)$ can be non-invertible for f arbitrarily close to 0. In the assumptions of the Nash–Moser theorem, we will assume that the differential is invertible in *a neighborhood* of the considered point. Note the contrast with the local inversion theorem, which ensures that we can solve a nonlinear equation from the resolution of a single linear equation.

For simplicity, we will prove the Nash–Moser theorem in the simplest context.[6] The context is given by a scale of Banach spaces $(X^\sigma, \|\cdot\|_\sigma)_{\sigma \geqslant 0}$. We will think of σ as a parameter measuring the regularity of a function, that is, the number of derivatives that we control in a certain norm (for example $X^k = C^k(\mathbb{R})$ for $k \in \mathbb{N}$). We will see in this book several scales of Banach spaces, the most important in applications being those of Sobolev and Hölder spaces.

Definition 2.28 We say that a family of Banach spaces $(X^\sigma, \|\cdot\|_\sigma)_{\sigma \geqslant 0}$ is a scale of Banach spaces if the following two properties are satisfied:

(P1) For all σ', σ such that $0 \leqslant \sigma' \leqslant \sigma < \infty$,

$$X^\infty \subset X^\sigma \subset X^{\sigma'} \subset X^0, \quad X^\infty := \bigcap_{\sigma \geqslant 0} X_\sigma,$$

and $\|\cdot\|_{\sigma'} \leqslant \|\cdot\|_\sigma$.

(P2) There exists an approximate identity: there is a family $(S(N))_{N \geqslant 1}$ of regularizing linear operators $S(N) \colon X^0 \to X^\infty$ such that

$$\lim_{N \to +\infty} \|S(N)u - u\|_0 = 0, \quad \forall u \in X_0,$$

and, for all $s, t \geqslant 0$,

$$\|u - S(N)u\|_s \leqslant C(t)N^{-t}\|u\|_{s+t},$$
$$\|S(N)u\|_{s+t} \leqslant C(t)N^t\|u\|_s.$$

It is interesting to note that under this single assumption we can prove an interpolation inequality.

Lemma 2.29 *Let λ_1, λ_2 be such that $0 \leqslant \lambda_1 \leqslant \lambda_2$ and let $\alpha \in [0, 1]$. There exists a constant A such that, for all $u \in X_{\lambda_2}$,*

$$\|u\|_\lambda \leqslant A\|u\|_{\lambda_1}^{1-\alpha}\|u\|_{\lambda_2}^\alpha, \quad \lambda = (1 - \alpha)\lambda_1 + \alpha\lambda_2.$$

---

[6] The Nash–Moser theorem comes into play in many problems from different fields: geometry (isometric embedding), dynamical systems (stability in celestial mechanics), the study of partial differential equations (infinite-dimensional Hamiltonian systems of equations). The common goal is the resolution of a nonlinear equation from the resolution of linear equations, using a quadratic iterative scheme. We will deal with the simplest situation and, for a detailed exposition of the methods to be used to solve the general cases, we refer to the original articles by Kolmogorov [103], Nash [147], Arnold [7], Moser [142, 143], Zehnder [199], and Herman [73], as well as Alinhac–Gérard [3], Craig [40], Ghys [59], Hamilton [67], Hörmander [77], Kuksin [107], Nirenberg [150], Poschel [156] and Wayne [195].

Proof Let $N > 0$. We can decompose u in the form $u = S(N)u + (I - S(N))u$ and deduce that

$$\|u\|_\lambda \leqslant \|S(N)u\|_\lambda + \|(I - S(N))u\|_\lambda \leqslant CN^{\lambda-\lambda_1}\|u\|_{\lambda_1} + CN^{-(\lambda_2-\lambda)}\|u\|_{\lambda_2}.$$

We obtain the desired result by optimizing this inequality. □

Consider a scale of Banach spaces (X^σ) and a function $u \mapsto \Phi(u)$ with a domain of definition X^m for a certain $m \geqslant 0$ and an image contained in X^0. We want to solve the equation $\Phi(u) = 0$, assuming that we know a good approximate solution, that is, an element $u_0 \in X^\infty$ such that $\Phi(u_0)$ is sufficiently small in X^k with k large. The Nash–Moser method is an elaboration of the Newton scheme that allows us to solve the equation $\Phi(u) = 0$ in certain situations where the inverse of $\Phi'(0)$ is not bounded. This method constructs a solution u of $\Phi(u) = 0$ as the limit of a sequence $(u_n)_{n\in\mathbb{N}}$ defined by

$$u_{n+1} = u_n - S(N_n)\Phi'(u_n)^{-1}\Phi(u_n),$$

for some $N_n \in \mathbb{N}$. The idea is that the very rapid convergence of the scheme allows us to compensate for the fact that the inverse of $\Phi'(u_n)$ is not bounded.

The main assumption is that Φ is of class C^2 (or $C^{1,\alpha}$ with $\alpha > 0$) and that the linearized equation

$$\lim_{t\to 0} \frac{1}{t}\left(\Phi(u + tv) - \Phi(u)\right) = \Phi'(u)v = g$$

has approximate solutions, not only for $u = 0$ but for all u in a neighborhood of 0.

Assumption 2.30 Throughout the rest of this section, we fix $m \geqslant 0$ and consider a function $u \mapsto \Phi(u)$ with a domain of definition X^m and an image contained in X^0. We assume that there exist constants $K_j \geqslant 1$ $(1 \leqslant j \leqslant 4)$ and $\tau \geqslant 0$ such that for all $s \geqslant 0$ the following properties are satisfied:

- (C^0 condition) $\Phi : X^{s+m} \to X^s$ and

$$\|\Phi(u)\|_s \leqslant K_1(1 + \|u\|_{s+m}), \quad \forall u \in X^\infty; \tag{2.19}$$

- ($C^{1,1}$ condition) $\Phi : X^{s+m} \to X^s$ is differentiable and

$$\|\Phi'(u)h\|_s \leqslant K_2\|h\|_{s+m}, \quad \forall u, h \in X^\infty, \tag{2.20}$$

and the quadratic part

$$Q(u, u') = \Phi(u') - \Phi(u) - \Phi'(u)[u' - u]$$

is estimated by

$$\|Q(u, u')\|_s \leqslant K_3\|u' - u\|_{s+m}^2, \quad \forall u, u' \in X^\infty; \tag{2.21}$$

- (inversion of the differential with loss of derivative) for all $u \in X^\infty$, there exists a linear operator $L(u): X^\tau \to X^0$ satisfying

$$\Phi'(u)\big(L(u)h\big) = h, \quad \forall h \in X^\infty$$

and such that

$$\|L(u)h\|_s \leqslant K_4 \|h\|_{s+\tau}, \quad \forall h \in X^\infty. \tag{2.22}$$

Theorem 2.31 (Nash) *Let $s_0 > m + \tau$. If $\|\Phi(0)\|_{s_0+\tau}$ is small enough then there exists a solution $u \in X^{s_0}$ of the equation $\Phi(u) = 0$.*

Remark 2.32 We can allow the constants K_1, K_2, K_3, K_4 in the assumption 2.30 to depend on certain norms of the unknown, provided that the estimates are *tame* in the sense given by Hamilton [67], that is: let F be a graded Fréchet space and let $P: U \subset F \to F$ be a function. We say that P satisfies a tame estimate of degree r and base b if

$$\|P(f)\|_n \leqslant C(1 + \|f\|_{n+r})$$

for all $f \in U$ and all $n \geqslant b$ (with a constant C that may depend on n). We say that P is a tame function if P is defined on an open set and is continuous and satisfies a tame estimate in a neighborhood of each point.

Proof We will construct by induction a sequence $(u_n)_{n \in \mathbb{N}}$ such that

$$\|\Phi(u_n)\|_{s_0-m} < M_n^{-1}, \tag{2.23}$$

$$\|u_{n+1} - u_n\|_{s_0} < M_n^{-1}, \tag{2.24}$$

where M_n is a rapidly increasing sequence such that $M_{n+1} = M_n^\gamma$ for some $\gamma > 1$. The second estimate implies that $(u_n)_{n \in \mathbb{N}}$ is a Cauchy sequence in X^{s_0} and therefore it converges. The first estimate and the continuity of Φ imply that the limit is a solution of $\Phi(u) = 0$.

Starting from $u_0 = 0$, we will define the sequence $(u_n)_{n \in \mathbb{N}}$ by solving, approximately, the equation

$$\Phi'(u_n)(u_{n+1} - u_n) + \Phi(u_n) = 0.$$

This means that at each step, we use a regularizing operator to compensate for the fact that $\Phi'(u)$ is not invertible due to a possible loss of derivative.

Let $2 < N_0 < N_1 < \cdots$ be a rapidly increasing sequence given by

$$N_n := \exp(\lambda \chi^n), \quad N_{n+1} = N_n^\chi, \quad \chi := \frac{3}{2},$$

with λ large enough depending on m, τ, K_j, s_0, which we will choose later. We define

$$v_n := -L(u_n)\Phi(u_n)$$

$$u_{n+1} := u_n + S(N_{n+1})v_n$$

so that $u_n \in X^\infty$ and thus $\Phi(u_n) \in X^\infty$ for all $n \geqslant 0$.

We want to estimate

$$\varepsilon_n := \|\Phi(u_n)\|_{s_0-m}.$$

As $\Phi(u_n) = \Phi'(u_n)L(u_n)\Phi(u_n)$, we have

$$\Phi(u_{n+1}) = \Phi(u_n) + \Phi'(u_n)(u_{n+1} - u_n) + Q(u_n, u_{n+1})$$
$$= \Phi'(u_n)(I - S(N_{n+1}))L(u_n)\Phi(u_n) + Q(u_n, u_{n+1}).$$

This identity and the estimates

$$\|\Phi'(u_n)h\|_{s_0-m} \leqslant K_2 \|h\|_{s_0},$$
$$\|Q(u_n, u_{n+1})\|_{s_0-m} \leqslant K_3 \|u_{n+1} - u_n\|_{s_0}^2,$$

give us

$$\|\Phi(u_{n+1})\|_{s_0-m} \leqslant K_2 \|(I - S(N_{n+1}))L(u_n)\Phi(u_n)\|_{s_0}$$
$$+ K_3 \|S(N_{n+1})L(u_n)\Phi(u_n)\|_{s_0}^2.$$

Thus, for all $\beta \geqslant 0$,

$$\|\Phi(u_{n+1})\|_{s_0-m} \leqslant CN_{n+1}^{-\beta} \|L(u_n)\Phi(u_n)\|_{s_0+\beta}$$
$$+ CN_{n+1}^{2m+2\tau} \|L(u_n)\Phi(u_n)\|_{s_0-m-\tau}^2,$$

with $C = C(\beta, s_0, m, \tau, K_2, K_3)$. In the following, we will denote by C several different constants that only depend on β, s_0, m, τ, K_j. We will sometimes simply write $A \lesssim B$ to indicate that there exists a constant C that only depends on β, s_0, m, τ, K_j such that $A \leqslant CB$. Note that the parameter β will be chosen depending on s_0, m, τ, K_j.

Using the estimate for $L(u_n)$, we then find that

$$\varepsilon_{n+1} \lesssim N_{n+1}^{-\beta} \|\Phi(u_n)\|_{s_0+\beta+\tau} + N_{n+1}^{2m+2\tau} \varepsilon_n^2.$$

The idea is that the super-fast convergence of the Newton scheme compensates for the $N_{n+1}^{2m+2\tau}$ factor that comes from the unboundedness of the Fréchet derivative and its inverse. We will show that if β and λ are large enough then the first term satisfies

$$\|\Phi(u_n)\|_{s_0+\beta+\tau} \leqslant N_n^{\beta}, \tag{2.25}$$

so that $N_{n+1}^{-\beta} \|\Phi(u_n)\|_{s_0+\beta+\tau}$ tends to 0 quickly because N_n/N_{n+1} tends to 0 quickly. We prove (2.25) by induction on n. For $n = 0$, the condition (2.25) is that

$$\|\Phi(0)\|_{s_0+\beta+\tau} \leqslant e^{\lambda\beta}, \tag{2.26}$$

which is satisfied for λ large enough because $\|\Phi(0)\|_{s_0+\beta+\tau} \leqslant K_1$.

Now suppose (2.25) holds at rank $n - 1$ with $n > 0$. As

$$\|\Phi(u_n)\|_{s_0+\beta+\tau} \leqslant K_1(1 + \|u_n\|_{s_0+\beta+\tau+m}),$$

the triangle inequality implies that

$$\|\Phi(u_n)\|_{s_0+\beta+\tau} \leqslant K_1\Big(1 + \sum_{k=1}^{n} \|u_k - u_{k-1}\|_{s_0+\beta+\tau+m}\Big).$$

It follows that

$$\|\Phi(u_n)\|_{s_0+\beta+\tau} \leqslant K_1\Big(1 + \sum_{k=1}^{n} \|S(N_k)L(u_{k-1})\Phi(u_{k-1})\|_{s_0+\beta+\tau+m}\Big)$$

$$\leqslant K_1\Big(1 + \sum_{k=1}^{n} C N_k^{\tau+m} \|L(u_{k-1})\Phi(u_{k-1})\|_{s_0+\beta}\Big)$$

$$\lesssim 1 + \sum_{k=1}^{n} N_k^{\tau+m} \|\Phi(u_{k-1})\|_{s_0+\beta+\tau}.$$

The induction hypothesis implies

$$\|\Phi(u_n)\|_{s_0+\beta+\tau} \lesssim 1 + \sum_{k=1}^{n} N_k^{\tau+m} N_{k-1}^{\beta}.$$

Since

$$\frac{N_{k+1}}{N_n} \leqslant \frac{N_{n-1}}{N_n} \leqslant \Big(\frac{1}{N_{n-1}}\Big)^{\chi-1} \leqslant 2^{-n},$$

we have

$$\|\Phi(u_n)\|_{s_0+\beta+\tau} \lesssim (1 + n2^{-n})N_n^{\tau+m} N_{n-1}^{\beta} \leqslant C N_n^{\tau+m} N_{n-1}^{\beta}.$$

Now, we verify that, if λ and β are large enough, namely such that:

$$\beta > \frac{\chi(\tau+m)}{\chi-1}, \qquad \lambda \geqslant \frac{\log(C)}{(\chi-1)\beta - \chi(m+\tau)}, \tag{2.27}$$

then

$$C N_n^{\tau+m} N_{n-1}^{\beta} = C \exp\big(\lambda(\chi\tau + \chi m + \beta)\chi^{n-1}\big) \leqslant N_n^{\beta} = \exp(\lambda\beta\chi^n).$$

This proves (2.25).

It follows that

$$\varepsilon_{n+1} \leqslant C N_{n+1}^{-\beta} N_n^{\beta} + C N_{n+1}^{2m+2\tau} \varepsilon_n^2,$$

with C depending only on $\beta, m\tau, s_0, K_j$. If the following inequalities are true

$$\frac{\chi-1}{\chi}\beta > \nu > \frac{2\chi}{2-\chi}(m+\tau), \qquad \lambda \geqslant \frac{\log(C)}{(2-\chi)\nu - 2\chi(m+\tau)},$$

and

$$\lambda \geqslant \frac{\log(C)}{(\chi - 1)\beta - \chi\nu},$$

then we can show that

$$\varepsilon_n \leqslant N_n^{-\nu}$$

provided $\varepsilon_0 \leqslant N_0^{-\nu}$. By definition of ε_0, the last condition is satisfied as soon as $\|\Phi(0)\|_{s_0-m} \leqslant e^{-\lambda\nu}$. This concludes the proof. $\qquad\qquad\Box$

2.9 Exercises

Exercise (solved) 2.1 Let E be a Banach space and let $F \colon [0, +\infty) \times E \to E$ be a continuous function satisfying the following property: there exists $C > 0$ such that,

$$\forall t \in [0, +\infty), \ \forall (x, y) \in E \times E, \quad \|F(t, x) - F(t, y)\|_E \leqslant C \, \|x - y\|_E \, .$$

The goal of this problem is to provide two proofs of the fact that, for every u_0 in E, there exists a unique function $u \in C^1([0, +\infty); E)$ which is a solution of

$$\frac{du}{dt} = F(t, u), \quad u|_{t=0} = u_0.$$

1. We are looking for a solution u of the equation $\Phi(u) = u$ with

$$\Phi(u) = u_0 + \int_0^t F(s, u(s)) \, ds.$$

Given $T > 0$, we let $X_T = C^0([0, T]; E)$. Show that Φ is a contraction of X_T into X_T for T small enough.
2. Deduce that, if T is small enough, then for every u_0 in E, there exists a unique function $u \in C^1([0, T]; E)$ which is a solution of

$$\frac{du}{dt} = F(t, u), \quad u|_{t=0} = u_0.$$

Then deduce the existence of a solution defined for all time by piecing together solutions defined over time intervals of length T.
3. We want to provide another argument that directly yields a global existence result in time. Given a parameter $\lambda > 0$, we introduce the space of functions with at most exponential growth of factor λ:

$$X = \left\{ u \in C^0([0, +\infty); E) : \sup_{t \in [0, +\infty)} e^{-\lambda t} \|u(t)\|_E < +\infty \right\}.$$

Verify that this is a Banach space for the norm

$$\|u\|_X = \sup_{t \in [0,+\infty)} e^{-\lambda t} \|u(t)\|_E.$$

Let u belong to X. Show that $\Phi(u)$ also belongs to X. Show furthermore that for every u and v in X, we have

$$\|\Phi(u) - \Phi(v)\|_X \leqslant \frac{C}{\lambda} \|u - v\|_X.$$

Conclude. (The idea of using an exponential factor to solve the Cauchy problem was introduced by Peyser [153].)

Exercise (solved) 2.2 (Brouwer's Fixed Point Theorem) We define $\overline{B} = \{x \in \mathbb{R}^n : \|x\| \leqslant 1\}$ where $\| \cdot \|$ is the Euclidean norm, defined by $\|x\|^2 = x_1^2 + \cdots + x_n^2$. We have proved (see Section 2.6) the following result: every continuous function of $\overline{B}$ into $\overline{B}$ has a fixed point. A natural question is to determine if the previous result is true if we replace the Euclidean norm with another norm.

1. Now consider a set $K \subset \mathbb{R}^n$ homeomorphic to the unit ball of $\mathbb{R}^n$. Show that every continuous function of K into K has a fixed point.
2. Let C be a non-empty convex compact set of $\mathbb{R}^n$ with $0 \in \overset{\circ}{C}$. Show that the function $\mu: \mathbb{R}^n \to [0,+\infty)$ defined by

$$\mu(x) = \inf\{t > 0 : x/t \in C\}$$

satisfies the following properties:

   (a) $\exists r, R > 0, \forall x \in \mathbb{R}^n, \|x\|/R \leqslant \mu(x) \leqslant \|x\|/r$;
   (b) $\forall \lambda \geqslant 0, \mu(\lambda x) = \lambda \mu(x)$;
   (c) $\forall x, y \in \mathbb{R}^n, \mu(x + y) \leqslant \mu(x) + \mu(y)$;
   (d) μ is continuous;
   (e) $x \in C$ if and only if $\mu(x) \leqslant 1$.

3. Show that C is homeomorphic to the unit ball of $\mathbb{R}^n$ and deduce that every continuous function of C into C has a fixed point.
4. Consider a norm N on $\mathbb{R}^n$. Show that

$$\overline{B} = \{x \in \mathbb{R}^n : N(x) \leqslant 1\}$$

is a convex compact set and that 0 belongs to the interior of $\overline{B}$. Infer from the above that Brouwer's theorem remains true if we replace the Euclidean norm with any norm.

Chapter 3
Hilbertian Analysis, Duality and Convexity

This chapter introduces several essential concepts in functional analysis and establishes the link between duality and geometry. We begin with a simple and powerful theory: the study of Hilbert spaces. These are the spaces for which we have an analogue of Pythagoras' theorem, so we can use the methods of Euclidean geometry in infinite-dimensional spaces. We will study the properties of orthogonality and convexity and see how they intervene in the study of the topological dual.

One of the objectives of this chapter is to introduce the point of view of duality, which consists in studying a topological vector space by studying the continuous linear forms operating on it. We can think of a linear form as a coordinate, for example the function $(x_1, \ldots, x_n) \mapsto x_1$ from $\mathbb{R}^n$ to $\mathbb{R}$. In infinite-dimensional spaces, it is non-trivial to prove even the existence of a non-zero continuous linear form. In the case of a Hilbert space H, we will prove the Riesz–Fréchet theorem, which allows us to identify the space of continuous linear forms on H with the space H itself: every continuous linear form can be represented by a scalar product with a given vector. This result is fundamental because it is an existence theorem. Indeed, it is often used to guarantee the existence of a solution to an equation whose unknown is a function (we will later give an application of this principle to the resolution of the Dirichlet problem). We will also see in the Hilbertian framework a notion of weak convergence: instead of studying convergence in the sense of the norm, we will be interested in sequences that converge coordinate by coordinate. This is a very fruitful point of view that allows us to study the compactness of bounded parts, in a weak sense, even in infinite-dimensional spaces.

The following sections concern the extension of this point of view to Banach spaces. In this general setting there is no analogue of the Riesz–Fréchet theorem that allows us to study the dual. The key theorem here will be the Hahn–Banach theorem. This is an abstract result from which we will deduce the existence of numerous linear forms. This will lead us to study weak convergence in Banach spaces. The notion of convexity will also play an important role in this part.

T. Alazard, *Analysis and Partial Differential Equations*, Universitext,
https://doi.org/10.1007/978-3-031-70909-8_3

3.1 Introduction to Hilbert Spaces

In this section, we will introduce the concept of Hilbert space and prove the main
results: the Cauchy–Schwarz inequality, projection onto a convex set, the Riesz–
Fréchet representation theorem and weak compactness of the unit ball.[1]

Throughout this section, $\mathbb{K}$ will denote either $\mathbb{R}$ or $\mathbb{C}$ and we will denote by $\bar{z}$ the
complex conjugate of z; we will also use the notation $\bar{z}$ when z is a real number. We
want to consider real or complex Hilbert spaces and we will write the definitions
and proofs so that they apply equally well to both cases.

3.1.1 Scalar Product

Consider a $\mathbb{K}$-vector space H. By definition, a scalar product is a map from $H \times H$
to $\mathbb{K}$, denoted $(\cdot, \cdot)$, such that, for all $x, y, z \in H$ and all $\lambda, \mu \in \mathbb{K}$,

1. $(x, x) \geqslant 0$ with equality if and only if $x = 0$;
2. $(x, y) = \overline{(y, x)}$;
3. $(\lambda x + \mu y, z) = \lambda(x, z) + \mu(y, z)$.

From (2) and (3) it follows that $(z, \lambda x + \mu y) = \bar{\lambda}(z, x) + \bar{\mu}(z, y)$.

If $\mathbb{K} = \mathbb{R}$, then of course $\overline{(y, x)} = (y, x)$ and likewise $\bar{\lambda} = \lambda$ (as we mentioned,
we write the definitions and proofs so that they apply equally well to the real and
complex case).

Theorem 3.1 (Cauchy–Schwarz Inequality) *Suppose that $(\cdot, \cdot)$ is a scalar product
on any $\mathbb{K}$-vector space H. For all $x, y \in H$,*

$$|(x, y)| \leqslant \sqrt{(x, x)}\sqrt{(y, y)}.$$

Proof Given any element x of H, we will denote $\sqrt{(x, x)}$ by $\|x\|$. Let $\lambda \in \mathbb{K}$ be of
modulus 1 and let $(x, y) \in H \times H$. We verify that

$$0 \leqslant \left\| x\,\|y\| - \lambda\|x\|y \right\|^2 = 2\|x\|^2\,\|y\|^2 - \|x\|\,\|y\|\left((x, \lambda y) + (\lambda y, x)\right)$$
$$\leqslant 2\|x\|\,\|y\|\left(\|x\|\,\|y\| - \mathrm{Re}(x, \lambda y)\right).$$

The Cauchy–Schwarz inequality is trivial if $|(x, y)| = 0$. If $|(x, y)| \neq 0$, we can
set $\lambda = (x, y)/|(x, y)|$ and then $\mathrm{Re}(x, \lambda y) = |(x, y)|$ and we deduce that $\|x\|\,\|y\| -
|(x, y)| \geqslant 0$. $\square$

---

[1] The theory of Hilbert spaces plays a fundamental role in mathematics and physics. We refer to
the books by Brezis [15], Landsman [110], Lax [113], Reed and Simon [159, 158], as well as
the lecture notes by Golse, Laszlo, Pacard and Viterbo [61], Landsman [109], Robert [161] and
Saint-Raymond [165].

Corollary 3.2 *Let $(\cdot,\cdot)$ be a scalar product on H. Then the map $\|\cdot\| : H \to [0,+\infty)$ defined by $\|x\| = \sqrt{(x,x)}$ is a norm on H. This norm is called the norm induced by the scalar product. Furthermore, this norm satisfies the so-called parallelogram law: for all x, y in H, we have*

$$\|x + y\|^2 + \|x - y\|^2 = 2\|x\|^2 + 2\|y\|^2. \tag{3.1}$$

Proof By definition of the scalar product, we directly have that $\|x\| = 0$ if and only if $x = 0$. Similarly, $\|\lambda x\| = |\lambda|\,\|x\|$ for all $\lambda \in \mathbb{K}$ and all $x \in H$. It remains to verify the triangle inequality. For this, we observe that

$$\|x + y\|^2 = (x + y, x + y) = \|x\|^2 + \|y\|^2 + 2\,\mathrm{Re}(x,y),$$

then we use the Cauchy–Schwarz inequality to obtain

$$\|x + y\|^2 \leqslant \|x\|^2 + \|y\|^2 + 2\|x\|\,\|y\| = (\|x\| + \|y\|)^2,$$

from which we get $\|x + y\| \leqslant \|x\| + \|y\|$. This proves that $\|x\|$ is a norm.

The identity (3.1) is obtained directly by writing

$$\|x + y\|^2 + \|x - y\|^2 = (x + y, x + y) + (x - y, x - y),$$

and expanding the right-hand side. $\qquad\square$

Note that the scalar product is continuous from $(H, \|\cdot\|)$ into $\mathbb{K}$. This is directly deduced from the Cauchy–Schwarz inequality.

Definition 3.3 By definition, a Hilbert space is a $\mathbb{K}$-vector space, equipped with a scalar product $(\cdot,\cdot)$, which is complete for the associated norm (defined by $\|x\| = \sqrt{(x,x)}$).

Example 3.4 The three most important examples of Hilbert spaces are the following:

- the space $\mathbb{R}^n$ equipped with the Euclidean norm $|x|^2 = \sum_{1 \leqslant i \leqslant n} x_i^2$;
- the space $\ell^2(\mathbb{N};\mathbb{C})$ of square summable complex-valued sequences $(u_n)_{n \in \mathbb{N}}$, i.e. such that $\|u\|_{\ell^2}^2 = \sum_{n \in \mathbb{N}} |u_n|^2 < +\infty$;
- the space $L^2(\Omega;\mathbb{C})$ of square integrable functions (quotiented by the equivalence relation of almost everywhere equality).

Proposition 3.5 *Consider a Hilbert space H. Any subset $A \subset H$ that is convex and closed admits a unique element of minimal norm.*

Remark 3.6 It is in the proof of this result that we use the fact that every Cauchy sequence converges.

Proof Consider a sequence $(x_n)_{n \in \mathbb{N}}$ of elements of A such that $\|x_n\|$ converges to $\delta := \inf_{x \in A} \|x\|$. As A is convex, for all integers n, m, the midpoint $(x_n + x_m)/2$ belongs to A and therefore $\|(x_n + x_m)/2\| \geqslant \delta$.

The parallelogram law (3.1) implies that

$$\|(x_n - x_m)/2\|^2 + \delta^2 \leqslant \|(x_n - x_m)/2\|^2 + \|(x_n + x_m)/2\|^2$$

$$\leqslant \frac{1}{2} \|x_n\|^2 + \frac{1}{2} \|x_m\|^2.$$

The right-hand side of the inequality converges to δ^2 when n and m tend to $+\infty$. We deduce that $\lim_{n,m\to+\infty} \|x_n - x_m\| = 0$. Thus, the sequence $(x_n)_{n\in\mathbb{N}}$ is Cauchy, and hence converges by completeness of H. The limit belongs to A because A is closed. This proves the existence of an element of minimal norm, and uniqueness is deduced again from the parallelogram law.   □

3.1.2 Orthogonality

Let H be a Hilbert space. We say that two vectors x, y are orthogonal if $(x, y) = 0$. We then write $x \perp y$. We denote by $\{x\}^\perp$ the set of vectors that are orthogonal to x. More generally, if F is any subspace of H, we set

$$F^\perp = \{x \in H : \forall y \in F, \ (x, y) = 0\}.$$

This is a closed linear subspace. Moreover, from the continuity of $y \mapsto (x, y)$ we easily deduce that $F^\perp = (\overline{F})^\perp$.

Theorem 3.7

1. *Let F be a closed linear subspace of H. There exists a linear map $P_F : H \to F$, called the orthogonal projection, such that $\|x - P_F(x)\| = \mathrm{dist}(x, F) = \inf_{y\in F} \|x - y\|$.*
2. *We have $x - P_F(x) \in F^\perp$.*

Proof 1. Let $x \in H$. Let $A_x = F - \{x\} = \{y - x : y \in F\}$. Then A_x is a convex set. Proposition 3.5 implies that there exists a unique vector $z \in A_x$ of minimal norm. We define $P_F(x)$ by $P_F(x) = z + x$, so that $P_F(x) \in F$ and

$$\|x - P_F(x)\| = \|z\| = \inf_{y\in F} \|y - x\| = \mathrm{dist}(x, F).$$

2. Let $y \in F$ be non-zero. We want to show that $x - P_F(x)$ is orthogonal to y. For this, consider $\lambda \in \mathbb{K}$ (to be chosen later) and observe that, by definition of $P_F(x)$ *via* a minimization argument,

$$\|P_F(x) - x - \lambda y\|^2 \geqslant \|P_F(x) - x\|^2 .$$

By expanding the left-hand side, we deduce that

$$|\lambda|^2 \|y\|^2 - 2\,\mathrm{Re}(P_F(x) - x, \lambda y) \geqslant 0.$$

We then choose $\lambda \neq 0$ such that

$$|\lambda|^2 \|y\|^2 - 2\operatorname{Re}(P_F(x) - x, \lambda y) = -|\lambda|\,|(P_F(x) - x, y)|$$

(we will verify that this is possible) to deduce that

$$|(P_F(x) - x, y)| = 0.$$

This implies that y is orthogonal to $P_F(x) - x$. □

We now state an immediate corollary of Theorem 3.7, whose interest is that it contains that part of the theorem that we will often use later.

Corollary 3.8

1. *If F is a closed linear subspace, then $H = F \oplus F^\perp$.*
2. *If F is any linear subspace of H, then $(F^\perp)^\perp = \overline{F}$.*
3. *Let $F \subset H$ be any linear subspace. Then F is dense if and only if $F^\perp = \{0\}$.*

We now come to a fundamental result that allows us to identify the topological dual of a Hilbert space H with H itself.

Definition 3.9 We denote by H' the topological dual of H. It is the set of continuous linear forms $\varphi \colon H \to \mathbb{K}$.

Theorem 3.10 (Riesz–Fréchet) *Let H be a Hilbert space. For every $\varphi \in H'$, there exists a unique $f \in H$ such that*

$$\forall v \in H, \quad \varphi(v) = (v, f).$$

Proof Given $f \in H$, we denote by $\Theta_f \colon H \to \mathbb{K}$ the function defined by $\Theta_f(v) = (v, f)$. The Cauchy–Schwarz inequality implies that $|\Theta_f(v)| = |(v, f)| \leqslant \|v\|\|f\|$ which shows that Θ_f is a continuous linear form on H. Let $\varphi \in H'$. We want to show that there exists $f \in H$ such that $\varphi = \Theta_f$. If $\varphi = 0$, then the result is trivially verified with $f = 0$.

Suppose that $\varphi \neq 0$ and introduce $F = \ker \varphi$, which is a closed subspace by continuity of φ. The previous corollary implies that $H = F \oplus F^\perp$. Let us show that $F^\perp$ is of dimension 1. For this, consider two non-zero vectors x, y in $F^\perp$. As $F \cap F^\perp = \{0\}$ and $x \neq 0$ by hypothesis, we have $x \notin F = \ker \varphi$, which implies that $\varphi(x) \neq 0$. Now, we have

$$\varphi\left(y - \frac{\varphi(y)}{\varphi(x)} x\right) = 0,$$

which implies that the vector $y - \frac{\varphi(y)}{\varphi(x)} x$ belongs to $\ker \varphi = F$. Moreover, this vector belongs to $F^\perp$ because x and y are in the subspace $F^\perp$. Reusing that $F \cap F^\perp = \{0\}$, we deduce that this vector is equal to 0, which proves that $F^\perp$ is of dimension 1.

To show that there exists $f \in F^\perp$ such that $\varphi = \Theta_f$, it remains just to choose f in $F^\perp$ such that $\varphi(f) = \|f\|^2$. Indeed, with this choice, φ and Θ_f vanish and therefore coincide on F. Moreover, these two functions also coincide on $F^\perp$ because they are equal at the point $f \in F^\perp$, and $F^\perp$ is of dimension 1. Therefore $\varphi = \Theta_f$ on $H = F \oplus F^\perp$. $\square$

Definition 3.11 Consider a sequence $(x_n)_{n\in\mathbb{N}}$ of elements of a Hilbert space H. We say that this sequence converges weakly to $x \in H$ if, for every continuous linear form $\varphi \in H'$, we have

$$\lim_{n\to+\infty} \varphi(x_n - x) = 0.$$

According to the previous theorem, it is equivalent to say that $(x_n)_{n\in\mathbb{N}}$ converges weakly to x if, for all $y \in H$, we have

$$\lim_{n\to+\infty} (y, x_n) = (y, x).$$

We then write $x_n \rightharpoonup x$.

Remark 3.12 Strong convergence implies weak convergence:

$$\lim_{n\to+\infty} \|x_n - x\| = 0 \implies x_n \rightharpoonup x.$$

Moreover, the Banach–Steinhaus theorem implies that every weakly convergent sequence is bounded (see Proposition 3.41 for a generalization of this remark).

The two results that follow explain the fundamental interest of weak convergence: there are sequences that admit weakly convergent subsequences but which do not admit a strongly convergent subsequence.

Proposition 3.13 *Consider an infinite-dimensional vector space H equipped with a scalar product (we say that H is a pre-Hilbert space). Then the closed unit ball is not compact.*

Proof This is a particular case of Theorem 1.15. Let us give a direct proof using the principle of orthonormalization (recalled below in Proposition 3.15). This principle allows us to construct a sequence $(e_n)_{n\in\mathbb{N}}$ of elements of H such that $(e_n, e_m) = 1$ if $n = m$ and 0 otherwise. Then, if $n \neq m$, we have $\|e_n - e_m\|^2 = \|e_n\|^2 + \|e_m\|^2 = 2$. This sequence therefore does not admit a subsequence that is Cauchy. $\square$

Theorem 3.14 *Let H be a Hilbert space. Then, every bounded sequence $(x_n)_{n\in\mathbb{N}}$ of elements of H has a weakly convergent subsequence.*

Proof From the Cauchy–Schwarz inequality, the sequence with general term (x_0, x_n) is bounded in $\mathbb{K}$ and therefore it has a convergent subsequence. By reusing this argument, we construct by induction a sequence $(\varphi_k)_{k\in\mathbb{N}}$ of strictly increasing functions $\varphi_k : \mathbb{N} \to \mathbb{N}$ such that, for all $k \in \mathbb{N}$, the sequence $\big((x_k, x_{\varphi_0 \circ \cdots \circ \varphi_k(n)})\big)_{n\in\mathbb{N}}$ converges in $\mathbb{K}$.

We define $y_n = x_{\varphi_0 \circ \cdots \circ \varphi_n(n)}$ and we will show that this subsequence converges weakly. By linearity, for all v in the vector space E generated by $\{x_n : n \in \mathbb{N}\}$, the sequence of general term (v, y_n) converges to an element $U(v)$ of $\mathbb{K}$. We verify that U is a linear form on $\overline{E}$ with values in $\mathbb{K}$, which is bounded because $|(v, y_n)| \leqslant M\|v\|$, where $M = \sup \|x_n\|$. The space $\overline{E}$ equipped with the scalar product of H is a Hilbert space, which allows us to use the Riesz representation theorem in this space to conclude that there exists $x \in \overline{E}$ such that $U(v) = \lim(v, y_n) = (v, x)$ for all $v \in \overline{E}$.

Moreover, if $v \in \overline{E}^{\perp}$, then $v \in E^{\perp}$, from which $(v, y_n) = 0$. Therefore, for all $v \in \overline{E} \oplus \overline{E}^{\perp}$, we have

$$\lim_{n \to +\infty} (v, y_n) = (v, x).$$

Since $\overline{E}$ is closed, we have $H = \overline{E} \oplus \overline{E}^{\perp}$ and the above shows that the subsequence $(y_n)_{n \in \mathbb{N}}$ converges weakly towards x. □

3.2 Hilbert Basis

A sequence $(e_n)_{n \in \mathbb{N}}$ of elements of a Hilbert space H is called an orthonormal system if and only if

$$(e_n, e_m) = \delta_n^m \qquad \forall n, m \in \mathbb{N},$$

where δ_n^m is 1 if $n = m$ and 0 otherwise.

Proposition 3.15 (Gram–Schmidt Orthonormalization) *Let H be a vector space equipped with a scalar product. Consider a family $(u_n)_{n \in \mathbb{N}}$ of linearly independent vectors. Then there exists an orthonormal system $(e_n)_{n \in \mathbb{N}}$ such that, for all $N \in \mathbb{N}$,*

$$\mathrm{Vect}\{e_0, \ldots, e_N\} = \mathrm{Vect}\{u_0, \ldots, u_N\}.$$

Proof We set $e_0 = u_0/\|u_0\|$ and define the following elements by induction, so that

$$e_n = v_n/\|v_n\| \quad \text{where} \quad v_n = u_n - (u_n, e_0)e_0 - \cdots - (u_n, e_{n-1})e_{n-1}.$$

We verify that e_n is orthogonal to $\mathrm{Vect}\{e_0, \ldots, e_{n-1}\}$. □

Lemma 3.16 (Bessel's Inequality) *Consider an orthonormal system $(e_n)_{n \in \mathbb{N}}$ and an element f of H. Then*

$$\sum_{n=0}^{\infty} |(f, e_n)|^2 \leqslant \|f\|^2.$$

Proof Let $S_N f = \sum_{n=0}^{N} (f, e_n)e_n$. We have

$$\|S_N f\|^2 = \sum_{0 \leqslant n_1, n_2 \leqslant N} (f, e_{n_1}) \overline{(f, e_{n_2})} (e_{n_1}, e_{n_2}) = \sum_{n=0}^{N} |(f, e_n)|^2.$$

From this, we deduce that

$$(f, S_N f) = \sum_{n=0}^{N} (f, (f, e_n)e_n) = \sum_{n=0}^{N} |(f, e_n)|^2 = \|S_N f\|^2.$$

The Cauchy–Schwarz inequality implies that $\|S_N f\|^2 \leqslant \|f\| \, \|S_N f\|$, from which we conclude that $\|S_N f\| \leqslant \|f\|$. ▫

Theorem 3.17 *Consider a separable Hilbert space H. The following properties are equivalent:*

1. *the vector space generated by the $(e_n)_{n \in \mathbb{N}}$ is dense in H;*
2. *for all $f \in H$, $\|f\|^2 = \sum_{n=0}^{+\infty} |(f, e_n)|^2$;*
3. *for all $f \in H$, the series $\sum (f, e_n)e_n$ converges to f;*
4. *if $f \in H$ satisfies $(f, e_n) = 0$ for all $n \in \mathbb{N}$, then $f = 0$.*

Proof The implications (3) $\implies$ (1) and (3) $\implies$ (4) are trivial. Let us prove that (2) $\implies$ (3). For this, we use the already seen identity $(f, S_N f) = \|S_N f\|^2$ to deduce that

$$\|f - S_N f\|^2 = \|f\|^2 + \|S_N f\|^2 - 2\operatorname{Re}(f, S_N f) = \|f\|^2 - \|S_N f\|^2,$$

which implies

$$\left\| f - \sum_{n=0}^{N} (f, e_n)e_n \right\|^2 = \|f\|^2 - \sum_{n=0}^{N} |(f, e_n)|^2. \tag{3.2}$$

So $f - \sum_{n=0}^{N}(f, e_n)e_n$ converges to 0 if $\sum_{n=0}^{N} |(f, e_n)|^2$ converges to $\|f\|^2$.

Consider the implication (1) $\implies$ (2). Recall that $\|S_N f\| \leqslant \|f\|$ for all $f \in H$. Let E be the vector space generated by $(e_n)_{n \in \mathbb{N}}$. Let $\varepsilon > 0$ and let $f' \in E$ such that $\|f - f'\| < \varepsilon$. For N large enough we have $S_N f' = f'$. Moreover,

$$\|S_N f - S_N f'\| = \|S_N (f - f')\| \leqslant \|f - f'\| \leqslant \varepsilon,$$

so

$$\|S_N f - f\| \leqslant \|S_N f - S_N f'\| + \|S_N f' - f'\| + \|f' - f\| \leqslant \varepsilon + 0 + \varepsilon,$$

so $(f - S_N f)$ converges to 0. Now, we can take the limit in (3.2) and we get $\|f\|^2 = \sum_{n=0}^{+\infty} |(f, e_n)|^2$, which concludes the proof of (1) $\implies$ (2).

Let us show that (4) $\implies$ (3) (this is where we use the fact that H is complete). Let $a_n = (f, e_n)$ and $f_p = \sum_{n=1}^{p} a_n e_n$. Bessel's inequality leads to $(a_n) \in \ell^2$. Now, for $m > p$ we have $\|f_m - f_p\|^2 = \sum_{n=p+1}^{m} |a_n|^2$ and so (f_p) is a Cauchy sequence and converges to an element denoted f'. But then (considering the partial sums and taking the limit), we find that $(f', e_n) = a_n$ for all n, which implies that $(f - f', e_n) = 0$ for all n. We deduce that $f = \sum_{n=1}^{\infty} a_n e_n$, which concludes the proof. ▫

3.3 The Hahn–Banach Theorem

We now propose to extend some of the results on Hilbert spaces to Banach spaces. We will no longer have the Riesz–Fréchet theorem which allows us to study the dual of a Hilbert space. In fact, in a Banach space, it is already non-trivial to show the existence of non-zero linear forms. The key theorem here will be the Hahn–Banach theorem, which is a result concerning the existence of linear forms. We will begin with an analytic version that allows the extension of linear forms. We will then present a geometric form that concerns the separation of convex sets by means of a hyperplane (which is the kernel of a linear form).

3.3.1 Analytic Version

Throughout this section, $\mathbb{K}$ will denote either $\mathbb{R}$ or $\mathbb{C}$. Recall that if $(E, \|\cdot\|)$ is a $\mathbb{K}$-normed vector space, we denote its topological dual by E'. This is the set of continuous linear forms $\ell: E \to \mathbb{K}$. Then E' is a $\mathbb{K}$-vector space, equipped with the norm

$$\|\ell\|_{E'} = \sup_{\substack{x \in E \\ \|x\|=1}} |\ell(x)| = \sup_{\substack{x \in E \\ x \neq 0}} \frac{|\ell(x)|}{\|x\|}.$$

Theorem 3.18 (Analytic Form) *Let* $(E, \|\cdot\|)$ *be a* $\mathbb{K}$-*normed vector space with* $\mathbb{K} = \mathbb{R}$ *or* $\mathbb{K} = \mathbb{C}$, $F \subset E$ *a linear subspace and* $f: F \to \mathbb{K}$ *a continuous linear form. Then there exists a continuous linear form* $g: E \to \mathbb{K}$ *that extends* f *(so that* $g|_F = f$*) and such that*

$$\|g\|_{E'} = \sup_{\substack{x \in E \\ \|x\|=1}} |g(x)| = \sup_{\substack{x \in F \\ \|x\|=1}} |f(x)| = \|f\|_{F'}.$$

Proof For simplicity, we prove this result in the case where E is a separable $\mathbb{K}$-normed vector space (that is, it admits a countable dense subset). The general case is not more difficult, except that the proof uses the axiom of choice (we refer to the classical textbook by Brézis [15] which contains many results centering around the Hahn–Banach theorem).

Real Case

Suppose that $\mathbb{K} = \mathbb{R}$. Consider a family $\{e_n\}_{n \in \mathbb{N}^*}$ dense in E. We introduce an increasing family of linear subspaces defined by:

$$F_0 = F, \quad F_n = \text{Vect}\{F \cup \{e_1, \ldots, e_n\}\}.$$

The sequence $(F_n)_{n \in \mathbb{N}}$ is increasing but not strictly increasing in general. Let us show that we can construct by induction a sequence of continuous linear forms $f_n \colon F_n \to \mathbb{R}$ that satisfy

$$f_0 = f, \quad f_n|_{F_{n-1}} = f_{n-1} \quad \text{and} \quad \|f_n\|_{F_n'} = \|f\|_{F'}.$$

We then define g by $g(x) = f_n(x)$ if $x \in F_n$ and we extend g by density over the entire space E thanks to a classical result on the extension of uniformly continuous functions (see Lemma 17.7).

Let us now show how to construct the sequence $(f_n)_{n \in \mathbb{N}}$. Suppose we have constructed the sequence up to rank $n-1$. We explain how to construct f_n, distinguishing two cases. If $F_n = F_{n-1}$, we set $f_n = f_{n-1}$. If $F_n \neq F_{n-1}$ then $e_n \notin F_{n-1}$ and we have a unique decomposition of an element u of F_n in the form $u = x + t e_n$ with $x \in F_{n-1}$ and $t \in \mathbb{R}$. We look for f_n in the form

$$f_n(x + t e_n) = f_{n-1}(x) + t a_n \quad \text{with } a_n \in \mathbb{R}.$$

Then f_n is indeed a linear form, which coincides with f_{n-1} on F_{n-1}. To conclude, it is enough to show that we can choose a_n such that $\|f_n\|_{F_n'} = \|f\|_{F'}$.

We have $\|f_n\|_{F_n'} \geqslant \|f\|_{F'}$ because f_n coincides with f on $F_0 = F$. It remains to show that $\|f_n\|_{F_n'} \leqslant \|f\|_{F'}$. More precisely, we need to show that there exists a_n such that, for all $x \in F_{n-1}$ and all $t > 0$,

$$f_{n-1}(x) + t a_n \leqslant \|f\|_{F'} \|x + t e_n\|,$$
$$f_{n-1}(x) - t a_n \leqslant \|f\|_{F'} \|x - t e_n\|.$$

If we divide by $t > 0$ and replace x with $u = x/t$, it suffices to show that there exists $a_n \in \mathbb{R}$ such that

$$f_{n-1}(u) + a_n \leqslant \|f\|_{F'} \|u + e_n\|,$$
$$f_{n-1}(u) - a_n \leqslant \|f\|_{F'} \|u - e_n\|.$$

The existence of $a_n \in \mathbb{R}$ satisfying the two previous inequalities for all $u \in F_{n-1}$ is equivalent to the inequality $m_n \leqslant M_n$, where

$$M_n := \inf_{u \in F_{n-1}} \left\{ \|f\|_{F'} \|u + e_n\| - f_{n-1}(u) \right\},$$
$$m_n := \sup_{v \in F_{n-1}} \left\{ f_{n-1}(v) - \|f\|_{F'} \|v - e_n\| \right\}.$$

To see that $m_n \leqslant M_n$, we start by writing

$$|f_{n-1}(u) + f_{n-1}(v)| = |f_{n-1}(u + v)| \leqslant \|f_{n-1}\|_{F_{n-1}'} \|u + v\|$$
$$\leqslant \|f_{n-1}\|_{F_{n-1}'} (\|u + e_n\| + \|v - e_n\|).$$

As $\|f_{n-1}\|_{F'_{n-1}} = \|f\|_{F'}$ by the induction hypothesis, we deduce that

$$f_{n-1}(v) - \|f\|_{F'}\|v - e_n\| \leqslant \|f\|_{F'}\|u + e_n\| - f_{n-1}(u).$$

We then show that $m_n \leqslant M_n$ by taking the supremum of the left-hand side and the infimum of the right-hand side, which concludes the induction and thus the proof.

Complex case

Now let $(E, \|\cdot\|)$ be a $\mathbb{C}$-normed vector space. Consider a linear subspace $F \subset E$ and a continuous linear form $f\colon F \to \mathbb{C}$. As $f(ix) = if(x)$, we have

$$\forall x \in E, \quad \operatorname{Im} f(x) = -(\operatorname{Re} f)(ix). \tag{3.3}$$

This identity shows that $\operatorname{Re} f$ completely determines f, which will allow us to deduce the complex case from the real case. For this, note that the space E is trivially equipped with the structure of an $\mathbb{R}$-normed vector space for which the linear map $\operatorname{Re} f\colon F \to \mathbb{R}$ is continuous. From the above, there exists a continuous linear form $g\colon E \to \mathbb{R}$ that extends $\operatorname{Re} f$ and such that

$$\forall x \in E, \quad |g(x)| \leqslant \|\operatorname{Re} f\|_{F'}\|x\| \leqslant \|f\|_{F'}\|x\|. \tag{3.4}$$

We then define the function $h\colon E \to \mathbb{C}$ by

$$h(x) = g(x) - ig(ix).$$

We verify that it is a $\mathbb{C}$-linear map. As g is real-valued, we have $\operatorname{Re} h = g$ and the identity (3.3) implies that h extends f. To conclude, let us show that

$$\forall x \in E, \quad |h(x)| \leqslant \|f\|_{F'}\|x\|. \tag{3.5}$$

As $\operatorname{Re} h(x) = g(x)$, inequality (3.4) implies that $|\operatorname{Re} h(x)| \leqslant \|f\|_{F'}\|x\|$ for all $x \in E$. In particular, for all $\lambda \in \mathbb{C}$ with $|\lambda| = 1$, we have

$$|\operatorname{Re} h(\lambda x)| \leqslant \|f\|_{F'}\|x\|.$$

We then obtain (3.5) by applying this inequality with $\lambda = \overline{h(x)}/|h(x)|$ (note that if $|h(x)| = 0$ then (3.5) is true). $\qquad\square$

Remark 3.19 In the case where E is a $\mathbb{R}$-vector space, we have actually shown a stronger result. Specifically, consider an $\mathbb{R}$-vector space E and a function $p\colon E \to \mathbb{R}_+$ such that, for all $(x, y) \in E \times E$ and all $\lambda \in [0, +\infty)$,

$$p(x + y) \leqslant p(x) + p(y), \quad p(\lambda x) = \lambda p(x).$$

Then, for any subspace F and any linear form $f\colon F \to \mathbb{R}$ satisfying $f(x) \leqslant p(x)$ for all $x \in F$, there exists a linear form $\widetilde{f}\colon E \to \mathbb{R}$ that extends f and satisfies $\widetilde{f} \leqslant p$.

We will now present several applications of the Hahn–Banach theorem.

Proposition 3.20 *Let $(E, \|\cdot\|_E)$ be a $\mathbb{K}$-normed vector space. For any non-zero element u in E, there exists a continuous linear form $\ell \in E'$ such that $\|\ell\|_{E'} = 1$ and $\ell(u) = \|u\|_E$. In particular,*

$$\|u\|_E = \sup_{\substack{\ell \in E' \\ \|\ell\|_{E'}=1}} |\ell(u)|.$$

Moreover, the map $J : E \to (E')'$ defined by $J(u)(\ell) = \ell(u)$ is an isometry from E into $(E')'$.

Notation 3.21 We simply denote by E'' the dual of E', called the bi-dual of E.

Proof Let $F = \mathbb{K}u$ and define $f : F \to \mathbb{K}$ by $f(\lambda u) = \lambda \|u\|_E$. Then f is a continuous linear form (since $|f(\lambda u)| \leqslant \|\lambda u\|_E$). According to the Hahn–Banach theorem, we can extend f to a continuous linear form $\ell : E \to \mathbb{K}$ that satisfies $\|\ell\|_{E'} = \|f\|_{F'} = 1$. Moreover, $\ell(u) = f(u) = \|u\|_E$. This directly implies that

$$\|J(u)\|_{E''} = \sup_{\substack{\ell \in E' \\ \|\ell\|_{E'}=1}} |J(u)(\ell)| = \sup_{\substack{\ell \in E' \\ \|\ell\|_{E'}=1}} |\ell(u)| = \|u\|_E,$$

which concludes the proof.   □

Corollary 3.22 *Let E be a $\mathbb{K}$-normed vector space with $\mathbb{K} = \mathbb{R}$ or $\mathbb{K} = \mathbb{C}$. Consider a subset $B \subset E$. If $\phi(B)$ is bounded for every continuous linear form $\phi \in E'$, then B is bounded.*

Proof Consider the family of maps $T_b : E' \to \mathbb{K}$ defined by $T_b(\phi) = \phi(b)$ with $b \in B$. As E' is a Banach space for any normed vector space E, we can apply the Banach–Steinhaus theorem (see Theorem 1.30). The assumption that $\phi(B)$ is bounded for all $\phi \in E'$ implies that the family $\{T_b\}_{b \in B}$ is pointwise bounded and therefore bounded according to Banach–Steinhaus. This means that there exists a $\Lambda > 0$ such that

$$\forall b \in B, \quad \|T_b\|_{E''} \leqslant \Lambda.$$

It follows that,

$$\forall \phi \in E', \quad |\phi(b)| = |T_b(\phi)| \leqslant \Lambda \|\phi\|_{E'}.$$

However, according to Proposition 3.20, for all $b \in B$, there exists $\ell \in E'$ such that $\ell(b) = \|b\|_E$ and $\|\ell\|_{E'} = 1$. It follows that, for all $b \in B$, we have $\|b\|_E \leqslant \Lambda$, which proves that B is bounded.   □

Proposition 3.23 *Let E be a normed vector space over $\mathbb{K}$ and F a closed linear subspace. Consider an element u of E not belonging to F. Then there exists a continuous linear form $\ell \in E'$ that vanishes on F and such that $\ell(u) = 1$.*

Proof Consider the quotient space E/F equipped with the quotient norm. By definition, the elements of E/F are the equivalence classes for the following equivalence relation: $x \sim y$ if and only if $x - y \in F$.

We denote by $\dot{v}$ the equivalence class of v, which is therefore $v + F = \{v + f : f \in F\}$. This is a vector space for the following operations:

$$\dot{v} + \dot{w} = \overbrace{v + w}^{\cdot}, \qquad \lambda\dot{v} = \overbrace{\lambda v}^{\cdot} .$$

We equip E/F with the norm $\|\dot{v}\|_{E/F} = \inf\{\|x\| : x \in \dot{v}\}$. Then we verify (exercise) that this is indeed a norm and that

$$\|\dot{v}\|_{E/F} = \inf_{f \in F} \|v + f\|_E = \inf_{f \in F} \|v - f\|_E = \mathrm{dist}(v, F) \leqslant \|v\|_E,$$

where we used that 0 belongs to F to obtain the last inequality.

Now introduce the linear form $f : \mathbb{K}\dot{u} \to \mathbb{K}$ defined by $f(\lambda\dot{u}) = \lambda$. As $u \notin F$ we have $\|\dot{u}\|_{E/F} \neq 0$, which allows us to write

$$\left|f(\lambda\dot{u})\right| = |\lambda| = \frac{1}{\|\dot{u}\|_{E/F}} \|\lambda\dot{u}\|_{E/F} .$$

This shows that f is continuous. Then, using the Hahn–Banach theorem, we deduce that we can extend f to a continuous linear form $g : E/F \to \mathbb{K}$. Then we define $\ell : E \to \mathbb{K}$ by $\ell(v) = g(\dot{v})$. We verify that ℓ is continuous by writing

$$|\ell(v)| \leqslant \|g\|_{(E/F)'} \|\dot{v}\|_{E/F} \leqslant \|g\|_{(E/F)'} \|v\|_E,$$

where we used the inequality $\|\dot{v}\|_{E/F} \leqslant \|v\|_E$, which was proved above. Moreover, $\ell(u) = g(\dot{u}) = f(\dot{u}) = f(1 \cdot \dot{u}) = 1$ and $\ell(v) = g(\dot{0}) = 0$ for all $v \in F$. $\qquad\square$

Corollary 3.24 *Let E be a $\mathbb{K}$-normed vector space and F a linear subspace. Then F is dense in E if and only if the only continuous linear form that vanishes on F is the null form.*

Proof This is an immediate consequence of the previous result. $\qquad\square$

We will now study a result whose proof allows us to illustrate the Hahn–Banach theorem and the Stone–Weierstrass theorem.

Proposition 3.25 *For $a > 1$, we define $f_a(x) = 1/(x - a)$. Consider a sequence $(a_n)_{n \in \mathbb{N}}$ of real numbers $a_n > 1$ such that $\lim_{n \to +\infty} a_n = +\infty$. We will show that*

$$V := \mathrm{Vect}\{f_{a_n} : n \in \mathbb{N}\}$$

is dense in the space of continuous functions $C([0, 1]; \mathbb{R})$ equipped with the norm of uniform convergence.

Proof Consider a linear form μ, continuous on $C([0, 1]; \mathbb{R})$ and such that $\langle \mu, f_{a_n} \rangle = 0$ for all $n \in \mathbb{N}$. If we can show that $\mu = 0$, then the desired result will follow from the density criterion deduced from the Hahn–Banach theorem. Let $\Theta_k : x \mapsto x^k$, then the series of functions $\sum_{k \geqslant 0} a_n^{-k} \Theta_k$ converges normally on $[0, 1]$ towards $-a_n f_{a_n}$.

We deduce by continuity of μ that

$$\sum_{k \in \mathbb{N}} \frac{\langle \mu, \Theta_k \rangle}{a_n^k} = 0. \tag{3.6}$$

Let $\varphi(z) = \sum_{k \in \mathbb{N}} \langle \mu, \Theta_k \rangle z^k$. The sequence $(\langle \mu, \Theta_k \rangle)_{k \in \mathbb{N}}$ is bounded by $\|\mu\|$, so φ is holomorphic on the open unit disk. Moreover, according to (3.6), $\varphi(1/a_n) = 0$ for all $n \in \mathbb{N}$. But the sequence $(1/a_n)_{n \in \mathbb{N}}$ converges towards 0, so the principle of isolated zeros implies that $\varphi = 0$. Therefore, $\langle \mu, \Theta_k \rangle = 0$ for all $k \in \mathbb{N}$. Thus, μ vanishes on the dense subspace of polynomial functions, so it vanishes on $C([0, 1]; \mathbb{R})$, which concludes the proof. $\square$

3.3.2 Geometric Version

In the case of a real vector space, we can also show:

Theorem 3.26 (Hahn–Banach geometric form) *Let E be a real normed vector space and A, B two convex, non-empty and disjoint subsets.*

1. *Assume that A is open. Then there exists a closed affine hyperplane that separates A and B in the broad sense. This means that there exists a non-zero continuous linear form $f : E \to \mathbb{R}$ such that*

$$\sup_A f \leqslant \inf_B f.$$

2. *Assume that A is closed and B is compact. Then there exists a closed affine hyperplane that separates A and B in the strict sense. This means that there exists a non-zero continuous linear form $f : E \to \mathbb{R}$ such that*

$$\sup_A f < \inf_B f.$$

Proof 1. Assume that $B = \{u\}$ and that A is a convex open set containing 0. We introduce the Minkowski functional of A,

$$\mu(x) = \inf\{t > 0 : x \in tA\}.$$

Then, as we have seen in the proof of Lemma 1.14, μ is sub-additive and homogeneous:

$$\mu(x + y) \leqslant \mu(x) + \mu(y), \quad \mu(\lambda x) = \lambda \mu(x) \qquad (x, y \in E, \; \lambda \geqslant 0),$$

and furthermore $A = \mu^{-1}([0, 1))$.

We then consider the linear form $f : \mathbb{R}u \to \mathbb{R}$ defined by $f(tu) = t$. Let us show that $f(tu) \leqslant \mu(tu)$. It is obvious if $t \leqslant 0$ because $f(tu) = t \leqslant 0$ while $\mu(tu) \geqslant 0$.

For $t \geqslant 0$, we use the fact that u does not belong to A to get that $\mu(u) \geqslant 1$, therefore

$$f(tu) = t \leqslant t\mu(u) = \mu(tu).$$

By applying the version of the Hahn–Banach theorem given in Remark 3.19, we can extend f to a linear form on E satisfying $f(x) \leqslant \mu(x)$ for all x in E. Then, for all x in A we have

$$f(x) \leqslant \mu(x) < 1,$$

therefore $\sup_A f \leqslant f(u)$.

In the general case, we choose $a \in A$, $b \in B$ and we set

$$C = A - B - a + b.$$

Then C is open, convex and contains the origin. The hypothesis $A \cap B = \varnothing$ implies that $u = b - a \notin C$. We then apply the previous case.

2. Let us consider $\varepsilon > 0$ and introduce the sets

$$A_\varepsilon = \{x \in E : \mathrm{dist}(x, A) < \varepsilon\}, \quad B_\varepsilon = \{x \in E : \mathrm{dist}(x, B) \leqslant \varepsilon\}.$$

Let us show that $A_\varepsilon \cap B_\varepsilon = \varnothing$ for ε small enough. Indeed, if this is false then there exists a sequence $(x_n)_{n \in \mathbb{N}}$ in E such that $\mathrm{dist}(x_n, A) \leqslant 1/n$ and $\mathrm{dist}(x_n, B) \leqslant 1/n$. By definition of the distance from a point to a set, we deduce that there exists a sequence $(y_n)_{n \in \mathbb{N}}$ of points from B such that $\|x_n - y_n\| \leqslant 2/n$. As B is compact, we can extract a subsequence $(y_{\theta(n)})_{n \in \mathbb{N}}$ from $(y_n)_{n \in \mathbb{N}}$ that converges towards an element y from B. However, according to the triangle inequality, the distance from $x_{\theta(n)}$ to y is bounded by

$$\|x_{\theta(n)} - y_{\theta(n)}\| + \|y_{\theta(n)} - y\|,$$

therefore $(x_{\theta(n)})_{n \in \mathbb{N}}$ converges towards y. As A is closed we deduce that $y \in A$. This is absurd because $A \cap B = \varnothing$. Therefore there exists $\varepsilon > 0$ such that $A_\varepsilon \cap B_\varepsilon = \varnothing$. Moreover, we verify (exercise) that A_ε and B_ε are convex sets and that A_ε is open. We can therefore apply point (1) to deduce that there exists a non-zero $f : E \to \mathbb{R}$ such that $\sup_{A_\varepsilon} f \leqslant \sup_{B_\varepsilon} f$. Then

$$f(u + \varepsilon w) \leqslant f(v + \varepsilon w'), \qquad \forall u \in A, \forall v \in B, \forall w, w' \in B_E(0, 1).$$

As

$$\sup_{w \in B_E(0,1)} f(w) = \|f\|, \qquad \inf_{w \in B_E(0,1)} f(w) = -\|f\|,$$

it follows that, for all u in A and v in B,

$$f(u) + \frac{\varepsilon}{2}\|f\| \leqslant f(v) - \frac{\varepsilon}{2}\|f\|,$$

which implies the desired result since $\|f\| > 0$. $\qquad\qquad\square$

3.4 Lebesgue Spaces

Let us recall that we have already seen that every Hilbert space can be identified with its dual.

Theorem 3.27 *Consider a Hilbert space H equipped with the scalar product $(\cdot, \cdot)$. For every continuous linear form $\varphi \in H'$, there exists a unique $g \in H$ such that*

$$\forall u \in H, \quad \varphi(u) = (u, g).$$

The fundamental example is that of the space $L^2(\Omega; \mathbb{R})$ of square-summable functions with real values. In this case, the previous result implies that for any continuous linear form on $L^2(\Omega)$, there exists $f \in L^2(\Omega)$ such that

$$\forall u \in L^2(\Omega), \quad \varphi(u) = \int_\Omega f(x)u(x)\, dx.$$

We now recall a classical result on Lebesgue spaces[2] $L^p(\Omega)$ which generalizes the previous discussion (*cf.* [42, 123, 163, 182] for its proof).

Theorem 3.28 *Let Ω be an open set of $\mathbb{R}^n$. Let $p \in [1, +\infty)$ and $p' = p/(p-1)$ be the conjugate exponent of p (if $p = 1$ then $p' = +\infty$). For any continuous linear form $\Lambda: L^p(\Omega) \to \mathbb{R}$, there exists a unique element $f \in L^{p'}(\Omega)$ such that, for all $u \in L^p(\Omega)$,*

$$\Lambda(u) = \int_\Omega f(x)u(x)\, dx.$$

Furthermore, $\|\Lambda\|_{(L^p)'} = \|f\|_{L^{p'}}$.

Remark 3.29

1. We deduce that the map $f \mapsto \Theta_f$ defined by $\Theta_f(u) = \int_\Omega fu\, dx$ is a bijective isometry of $L^{p'}(\Omega)$ into $(L^p(\Omega))'$. This result allows us to identify $(L^p(\Omega))'$ and $L^{p'}(\Omega)$, and we commonly say that $L^{p'}(\Omega)$ is the dual of $L^p(\Omega)$.
2. It follows that $L^p(\Omega)$ is a reflexive Banach space for $1 < p < \infty$.
3. This theorem remains true if we replace Ω equipped with the Lebesgue measure by a measure space (X, μ) which is σ-finite (that is, X can be written as the union of a countable sequence of sets of finite measure).

Note that the previous theorem implies that $L^\infty(\Omega)$ is the dual of $L^1(\Omega)$.

The following proposition states that the topological dual of $(L^\infty(\mathbb{R}^n))'$ is strictly larger than $L^1(\mathbb{R}^n)$. To be able to compare these two spaces, we must first embed $L^1(\mathbb{R}^n)$ into $L^\infty(\mathbb{R}^n)$. For this, given a function $f \in L^1(\mathbb{R}^n)$ with real values, we consider the linear form $\Theta_f : L^\infty(\mathbb{R}^n) \to \mathbb{R}$ defined by $\Theta_f(u) = \int_{\mathbb{R}^n} fu\, dx$. Then Θ_f is clearly a continuous linear form.

---

[2] See Chapter 18 for reminders on these spaces.

Proposition 3.30 *There exists a continuous linear form* $\Lambda \colon L^\infty(\mathbb{R}^n) \to \mathbb{R}$ *that is not of the form* Θ_f *with* f *in* $L^1(\mathbb{R}^n)$.

Proof We introduce the subspace

$$V = \{u \in C^0(\mathbb{R}) : u \text{ is bounded on } \mathbb{R} \text{ and } u \text{ has a limit at } +\infty\}.$$

We also introduce the linear map $\ell \colon V \to \mathbb{R}$ defined by $\ell(u) = \lim_{+\infty} u$. We have $|\ell(u)| \leqslant \|u\|_{L^\infty(\mathbb{R})}$. We can therefore use the Hahn–Banach theorem in its analytical form to extend ℓ to a continuous linear form Λ on $L^\infty(\mathbb{R})$. We suppose that we can write $\Lambda = \Theta_f$ with $f \in L^1(\mathbb{R})$ and show that we then obtain an absurdity. For this, let us now give ourselves $\xi \in \mathbb{R}$ and $\varepsilon > 0$. Then the functions $\cos(x\xi)\exp(-\varepsilon x^2)$ and $\sin(-x\xi)\exp(-\varepsilon x^2)$ belong to V. It follows that

$$\int_{\mathbb{R}} f(x)\cos(x\xi)e^{-\varepsilon x^2}\,dx = \Theta_f\left(\cos(x\xi)e^{-\varepsilon x^2}\right)$$

$$= \Lambda\left(\cos(x\xi)e^{-\varepsilon x^2}\right)$$

$$= \ell\left(\cos(x\xi)e^{-\varepsilon x^2}\right) = 0$$

and we have an analogous result by replacing $\cos(x\xi)e^{-\varepsilon x^2}$ with $\sin(-x\xi)e^{-\varepsilon x^2}$. Then

$$\int_{\mathbb{R}} f(x)e^{-ix\cdot\xi - \varepsilon x^2}\,dx = 0.$$

Since $f \in L^1(\mathbb{R})$, we can apply the dominated convergence theorem and let ε tend to 0. We obtain that, for all $\xi \in \mathbb{R}$,

$$\int_{\mathbb{R}} f(x)e^{-ix\cdot\xi}\,dx = 0.$$

To conclude, we use the injectivity theorem of Fourier (which we will see later, see Corollary 5.14) and whose statement we give here: if $f \in L^1(\mathbb{R}^n)$ is such that $\int e^{-i\xi\cdot x} f(x)\,dx = 0$ for all $\xi \in \mathbb{R}^n$, then $f = 0$. It follows that $\ell = 0$, which is absurd. $\qquad\square$

3.5 Weak Convergence, Weak-* Convergence

In this section, $(E, \|\cdot\|_E)$ denotes a normed vector space over $\mathbb{K} = \mathbb{R}$ or $\mathbb{C}$. We denote by E' the topological dual of E; it is the set of continuous linear forms $f \colon E \to \mathbb{K}$. Recall that E' is equipped with the norm $\|f\|_{E'} = \sup_{\|x\|_E \leqslant 1} |f(x)|$ and that it is a Banach space for this norm (let us emphasize that E' is a Banach space for any normed space E; the completeness of E' comes from the fact that $\mathbb{K}$ is complete). We are going to be interested in two topologies on E and E', respectively called the weak topology and the weak-* topology. To define these topologies, we need to introduce the concept of a induced topology.

Induced topology

Let us start with some topology reminders (see Section 17.1.2 of Chapter 17 in the appendix). Consider a set X and a collection $\mathcal{A} \subset \mathcal{P}(X)$ of subsets of X. By definition, the topology induced by $\mathcal{A}$, denoted $\mathcal{T}_{\mathcal{A}}$, is the coarsest topology containing $\mathcal{A}$ (which means it is the smallest topology in terms of inclusion). Then $\mathcal{T}_{\mathcal{A}}$ consists of the empty set, X and all unions of finite intersections of elements of $\mathcal{A}$. The concept of induced topology allows us to define the concept of a topology induced by a family of functions. It is the coarsest topology that makes a given family of functions continuous. More precisely, consider a set X and a family of functions $\mathcal{F} = \{f_\alpha : X \to Y\}_{\alpha \in A}$ from X to a topological space Y equipped with a topology $\mathcal{T}_Y$. By definition, the topology induced by $\mathcal{F}$ on X is equal to the topology $\mathcal{T}_{\mathcal{A}}$ induced by the collection

$$\mathcal{A} := \{f_\alpha^{-1}(U) : \alpha \in A, U \in \mathcal{T}_Y\}.$$

Definition 3.31 Consider a normed vector space E. Denote by E' the topological dual of E and E'' the topological dual of E'.

1. The weak topology on E, denoted $\sigma(E, E')$, is the topology induced by E'.
2. The weak topology on E', denoted $\sigma(E', E'')$, is the topology induced by E''.
3. Given $x \in E$, let us denote by J_x the linear form of E' in $\mathbb{K}$ defined by $J_x(f) = f(x)$. The weak-* topology on E', denoted $\sigma(E', E)$, is the topology induced by the family $\mathcal{F} = \{J_x : x \in E\}$.

Definition 3.32 Consider a normed vector space E. We say that E is reflexive if the map $J : x \mapsto J_x$ is an isomorphism from E to its bi-dual $E'' = (E')'$.

Remark 3.33

1. As the dual of a normed vector space is always complete, a reflexive space is necessarily a Banach space.
2. Note that there exists a Banach space that is isomorphic to its bi-dual without being reflexive (a counter-example due to James). This shows that to define reflexivity, we must involve the map J above.

Given two topologies $\mathcal{T}_1$ and $\mathcal{T}_2$ on the same set X, recall that we say that $\mathcal{T}_2$ is finer than $\mathcal{T}_1$ if $\mathcal{T}_1 \subset \mathcal{T}_2$. It follows directly from the definition of weak topologies that we have the following propositions.

Proposition 3.34 *Let E be a $\mathbb{K}$-normed vector space. The weak topology is less fine than the strong topology and the weak-* topology is less fine than the weak topology on E', that is*

$$\sigma(E, E') \subset \mathcal{T}(\|\cdot\|_E), \quad \sigma(E', E) \subset \sigma(E', E''),$$

where $\mathcal{T}(\|\cdot\|_E)$ denotes the topology induced on E by the normed space structure.

Proposition 3.35 *Let E be a $\mathbb{K}$-normed vector space. Then $(E, \sigma(E, E'))$ and $(E', \sigma(E', E))$ are topological vector spaces.*

Remark 3.36 In general, the topologies $\sigma(E', E'')$ and $\sigma(E', E)$ are different.

Proposition 3.37 *Let E be a $\mathbb{K}$-normed vector space.*

1. *The weak topology $\sigma(E, E')$ is separated.*
2. *The weak-* topology $\sigma(E', E)$ is separated.*

Proof 1. Suppose that $\mathbb{K} = \mathbb{R}$. Consider x, y in E with $x \neq y$. As $y - x \neq 0$, according to the Hahn–Banach theorem (see Proposition 3.20), there exists a linear form $f \in E'$ such that $f(y - x) > 0$. We set $h = (f(x) + f(y))/2$ and introduce

$$U = f^{-1}((-\infty, h)) \quad \text{and} \quad V = f^{-1}((h, +\infty)).$$

By definition of the weak topology, these are two weak open sets. They are disjoint and by construction we have $x \in U$ and $y \in V$.

Now suppose that $\mathbb{K} = \mathbb{C}$. We denote by $E_{\mathbb{R}}$ the space E equipped with the $\mathbb{R}$-normed vector space structure canonically induced by its $\mathbb{C}$-normed vector space structure. As before, there exists a linear form $f \in E_{\mathbb{R}}'$ such that $f(x) < f(y)$. We set $h = (f(x) + f(y))/2$ and introduce $U = f^{-1}((-\infty, h))$ and $V = f^{-1}((h, +\infty))$. Let us show that these are two weak open sets. For this, we introduce the linear map $g : E \to \mathbb{C}$ defined by

$$g(x) = f(x) - if(ix).$$

So we easily verify that $g \in E'$. We deduce that $f = \operatorname{Re} g \in E'$ and therefore that U and V are weak open sets.

2. Consider two distinct linear forms f and f' in E'. Then there exists x in E such that $f(x) \neq f'(x)$. Let $\varepsilon = |f(x) - f'(x)| > 0$ and

$$V := \{g \in E' : |g(x) - f(x)| < \varepsilon/2\}, \ V' := \{g \in E' : |g(x) - f'(x)| < \varepsilon/2\}.$$

These are two open sets for the topology $\sigma(E', E)$, disjoint and such that $f \in V$ and $f' \in V'$. This proves the desired result.   □

Corollary 3.38 *Let E be a $\mathbb{K}$-normed vector space of finite dimension, then the weak, weak-* and strong topologies coincide.*

Remark 3.39 This is in fact a necessary and sufficient condition. Indeed, in infinite-dimensional spaces, the weak topology is not metrizable (see Exercise 3.3).

Proof This is a consequence of Theorem 1.17 and the fact that the weak topology equips X with the structure of a separated topological vector space.   □

3.5.1 Weak Convergence

Definition 3.40 Let $(E, \|\cdot\|_E)$ be a $\mathbb{K}$-normed vector space and $(x_n)_{n \in \mathbb{N}}$ be a sequence of elements of E.

1. We say that $(x_n)_{n \in \mathbb{N}}$ converges strongly to $x \in E$ if $\|x_n - x\|_E$ tends to 0 as n tends to $+\infty$. We most often simply say that the sequence $(x_n)_{n \in \mathbb{N}}$ converges to x and we denote it by $x_n \to x$.
2. We say that $(x_n)_{n \in \mathbb{N}}$ converges weakly to x if, for every f in E', the sequence $(f(x_n))_{n \in \mathbb{N}}$ converges to $f(x)$ in $\mathbb{K}$. We denote it by $x_n \rightharpoonup x$.

Proposition 3.41 *Let $(x_n)_{n \in \mathbb{N}}$ be a sequence of elements of a normed vector space E and let $x \in E$. Then we have the following properties.*

1. *If $(x_n)_{n \in \mathbb{N}}$ converges strongly to x, then $(x_n)_{n \in \mathbb{N}}$ also converges weakly to x.*
2. *The sequence $(x_n)_{n \in \mathbb{N}}$ converges to x for the topology $\sigma(E, E')$ if and only if $(x_n)_{n \in \mathbb{N}}$ converges weakly to x ($x_n \rightharpoonup x$).*
3. *If $(x_n)_{n \in \mathbb{N}}$ converges weakly to x and to x', then $x = x'$.*
4. *If $(x_n)_{n \in \mathbb{N}}$ converges weakly to x, then the sequence $(x_n)_{n \in \mathbb{N}}$ is bounded in E.*

Proof 1. If $x_n \to x$, then $f(x_n) \to f(x)$ for all $f \in E'$ and therefore $x_n \rightharpoonup x$.
2. Suppose that $(x_n)_{n \in \mathbb{N}}$ converges to x for the topology $\sigma(E, E')$. Let $f \in E'$. Since f is continuous for the weak topology (by definition of the weak topology), we have $f(x_n) \to f(x)$. This shows that $x_n \rightharpoonup x$. Conversely, suppose that $x_n \rightharpoonup x$. Consider a neighborhood of x for the weak topology. We can assume that this neighborhood is of the form $U = \bigcap_{1 \leqslant i \leqslant N} f_i^{-1}(V_i)$ where V_i is an open neighborhood of $f_i(x)$. For all i, there exists an integer N_i such that x_n belongs to $f_i^{-1}(V_i)$ for $n \geqslant N_i$. Therefore, for all n greater than the maximum of these N_i, we have $x_n \in U$. This shows that $(x_n)_{n \in \mathbb{N}}$ converges to x for the topology $\sigma(E, E')$.
3. This is a consequence of the fact that the weak topology is separated.
4. Let us introduce the linear map $T_n \colon E' \to \mathbb{K}$ defined by $T_n(f) = f(x_n)$. As $f(x_n)$ converges to $f(x)$ for all f in E', we deduce that the sequence $(T_n(f))_{n \in \mathbb{N}}$ is bounded for all f in E'. This means that $(T_n)_{n \in \mathbb{N}}$ is a pointwise bounded family of linear maps. As E' is a Banach space, we can apply the Banach–Steinhaus theorem 1.30, which implies that this family is bounded: there exists $C > 0$ such that

$$\sup_{n \in \mathbb{N}} |T_n(f)| \leqslant C\|f\|.$$

Therefore, for all $f \in E'$,
$$|f(x_n)| \leqslant C\|f\|. \tag{3.7}$$

Now, we recall that according to the Hahn–Banach theorem (see Proposition 3.20), for all $n \in \mathbb{N}$, there exists an $\ell_n \in E'$ such that $\|\ell_n\|_{E'} = 1$ and $\ell_n(x) = \|x_n\|_E$. By applying (3.7) with $f = \ell_n$, we find that $\|x_n\|_E \leqslant C$ for all $n \in \mathbb{N}$, which is the desired result.   □

Proposition 3.42 *Let E be a real normed vector space and C a subset of E that is convex. Then C is strongly closed if and only if C is weakly closed. In particular, if C is strongly closed and if $(x_n)_{n \in \mathbb{N}}$ is a sequence of elements of C that converges weakly to x, then x belongs to C.*

Proof Any set open for the weak topology is open for the strong topology (by definition of the weak topology). Therefore, any set closed for the weak topology is closed for the strong topology. It thus suffices to show that if C is convex and strongly closed, then C is weakly closed. We will show that the complement of C is weakly open. For this, it suffices to show that $E \setminus C$ is a neighborhood of each of its points. Consider therefore an element x_0 of $E \setminus C$. Then $\{x_0\}$ is a convex compact set (for the strong topology). According to the geometric Hahn–Banach theorem, we can strictly separate $\{x_0\}$ and C: There exists $f \in E'$ such that

$$\sup_{x \in C} f(x) < f(x_0).$$

Let $\varepsilon = f(x_0) - \sup_{x \in C} f(x)$. Then

$$V = \{y \in E : f(y) > f(x_0) - \varepsilon/2\}$$

is an open set for the weak topology that contains x_0 and is disjoint from C. This completes the proof.   □

Proposition 3.43 *Consider two Banach spaces E and F and a linear map $T : E \to F$. Then T is continuous for the strong topologies on E and F if and only if T is continuous for the weak topologies on E and F.*

Proof Assume that T is continuous for the strong topologies. Then the continuity for the weak topologies is a topology exercise left to the reader.

Conversely, assume that T is continuous for the weak topologies. Let us introduce the graph of T:
$$G(T) = \{(x, y) \in E \times F : y = Tx\}.$$

Let us start by showing that this set is weakly closed. To see this, note that $G(T) = \Lambda^{-1}(\{0\})$ where $\Lambda : F \times E \to F$ is defined by

$$\Lambda(y, x) = y - Tx.$$

We have seen that the weak topology is separated. Therefore, singletons are closed for the weak topology. It follows that $G(T)$ is closed for the weak topology as the preimage of a weakly closed set by a continuous function.

To conclude the proof, we then note that $G(T)$ is a convex set (which is immediate). Then, according to Proposition 3.42, it is strongly closed. It follows that T is continuous for the strong topologies by applying the closed graph theorem.   □

3.5.2 Weak-* Convergence

Definition 3.44 Let $(E, \|\cdot\|_E)$ be a $\mathbb{K}$-normed vector space.

1. Let $(f_n)_{n \in \mathbb{N}}$ be a sequence of elements of E'. We say that this sequence converges strongly to $f \in E'$ if $\|f_n - f\|_{E'}$ tends to 0 as n tends to $+\infty$. For simplicity, we often just say that the sequence $(f_n)_{n \in \mathbb{N}}$ converges to f (without specifying strongly) and we write $f_n \to f$.
2. Let $(f_n)_{n \in \mathbb{N}}$ be a sequence of elements of E'. We say that this sequence converges to f in the weak-* sense if, for all x in E, the sequence $(f_n(x))_{n \in \mathbb{N}}$ converges to $f(x)$ in $\mathbb{K}$. We write $f_n \rightharpoonup f$ weak-*.

Proposition 3.45 *Let $(E, \|\cdot\|_E)$ be a $\mathbb{K}$-normed vector space. Consider a sequence $(f_n)_{n \in \mathbb{N}}$ of elements of E' and $f \in E'$. Then the following properties hold:*

1. *if $(f_n)_{n \in \mathbb{N}}$ converges strongly to f in E', then $(f_n)_{n \in \mathbb{N}}$ converges to f in the weak-* sense ;*
2. *$(f_n)_{n \in \mathbb{N}}$ converges to f for the topology $\sigma(E', E)$ if and only if $f_n \rightharpoonup f$ weak-*;*
3. *if $(f_n)_{n \in \mathbb{N}}$ converges to f and to f' in the weak-* sense, then $f = f'$;*
4. *if $(f_n)_{n \in \mathbb{N}}$ converges to f in the weak-* sense, then $(f_n)_{n \in \mathbb{N}}$ is bounded in E'.*

The proof is similar to that of Proposition 3.41. It is left as an exercise.

3.6 The Banach–Alaoglu Theorem

Consider a normed vector space E. A fundamental result, called the Banach–Alaoglu–Bourbaki theorem, states that the closed unit ball of E' is compact for the weak-* topology and that, moreover, if E is separable, then the topology $\sigma(E', E)$ is metrizable so that the unit ball is sequentially weak-* compact. In this section, we will prove this compactness result only in the case where E is separable. In this case, the result is due to Banach, the proof is simpler and this is the most important case for applications.

Theorem 3.46 (weak-* compactness of the closed unit ball) *Suppose that $\mathbb{K} = \mathbb{R}$ or $\mathbb{K} = \mathbb{C}$. Consider a $\mathbb{K}$-normed vector space E that is separable. Any bounded sequence $(f_n)_{n \in \mathbb{N}}$ in E' has a subsequence that converges in the weak-* sense.*

Proof Consider a countable dense sequence $D = \{e_j\}_{j \in \mathbb{N}}$ in E. As the sequence $(f_n(e_0))_{n \in \mathbb{N}}$ is bounded, by hypothesis on the sequence $(f_n)_{n \in \mathbb{N}}$, we can extract a convergent subsequence $(f_{\theta_0(n)}(e_0))_{n \in \mathbb{N}}$. We thus construct, by induction, a sequence of strictly increasing functions $\theta_j : \mathbb{N} \to \mathbb{N}$ such that $(f_{\theta_0 \circ \cdots \circ \theta_j(n)}(e_j))_{n \in \mathbb{N}}$ converges. Using the diagonal extraction principle, we then consider the subsequence $g_n = f_{\theta_0 \circ \cdots \circ \theta_n(n)}$. By construction, the sequence $(g_n(e))_{n \in \mathbb{N}}$ converges for all $e \in D$.

Let $g: D \to \mathbb{K}$ be the limit.

Let e and e' be in D. The hypothesis that the sequence $(f_n)_{n \in \mathbb{N}}$ is bounded in E' implies that there exists $\Lambda > 0$ such that

$$\forall n \in \mathbb{N}, \ \forall x \in E, \quad |f_n(x)| \leqslant \Lambda \|x\|_E.$$

By applying this result with $x = e - e'$ and taking the limit as n tends to $+\infty$, we find that

$$|g(e) - g(e')| \leqslant \Lambda \|e - e'\|_E,$$

which shows that g is Lipschitzian. We can then use Lemma 17.7 to extend g to a Lipschitzian function $\widetilde{g}: E \to \mathbb{K}$.

Recall that weak-* convergence corresponds to pointwise convergence. We have seen that $(g_n)_{n \in \mathbb{N}}$ converges to $\widetilde{g}$ on D. To conclude the proof, we still need to see that $(g_n)_{n \in \mathbb{N}}$ converges pointwise to g on the whole space E and that g is a linear map.

Consider an element x in E and a real number $\varepsilon > 0$. By density of D, there exists $e_j \in D$ such that $\Lambda \|x - e_j\|_E \leqslant \varepsilon/3$. On the other hand, by pointwise convergence on D, if n is large enough, we have $|g_n(e_j) - \widetilde{g}(e_j)| \leqslant \varepsilon/3$ so that

$$|g_n(x) - \widetilde{g}(x)| \leqslant \left|g_n(x) - g_n(e_j)\right| + \left|g_n(e_j) - \widetilde{g}(e_j)\right| + \left|\widetilde{g}(e_j) - \widetilde{g}(x)\right|$$
$$\leqslant \Lambda \left\|x - e_j\right\|_E + \frac{\varepsilon}{3} + \Lambda \left\|e_j - x\right\|_E \leqslant \varepsilon.$$

This shows that $(g_n)_{n \in \mathbb{N}}$ converges in the weak-* sense to $\widetilde{g}$.

It remains just to show that $\widetilde{g}$ is linear. For this, consider $\lambda \in \mathbb{K}$, x, y in E and consider three sequences of elements of D such that $x_n \to x$, $y_n \to y$, $z_n \to \lambda x + y$. Then

$$\widetilde{g}(z_j) - \lambda \widetilde{g}(x_j) - \widetilde{g}(y_j) = \lim_{n \to +\infty} \left\{ g_n(z_j) - \lambda g_n(x_j) - g_n(y_j) \right\}.$$

However, by linearity of g_n, the expression $|g_n(z_j) - \lambda g_n(x_j) - g_n(y_j)|$ is bounded by $\Lambda \|z_j - \lambda x_j - y_j\|_E$ which converges to 0 when j tends to $+\infty$. We deduce that

$$\widetilde{g}(\lambda x + y) - \lambda \widetilde{g}(x) - \widetilde{g}(y) = 0,$$

which is the desired result. $\qquad\qquad\qquad\qquad\qquad\qquad\qquad\qquad\qquad\qquad\quad \square$

A direct corollary of this weak-* compactness result is that the closed unit ball of a reflexive Banach space is weakly compact.

Corollary 3.47 *Suppose that E is a reflexive Banach space whose dual is separable. Then any bounded sequence $(x_n)_{n \in \mathbb{N}}$ in E has a weakly convergent subsequence.*

Proof Let us set $F = E'$. By hypothesis, F is a separable Banach space. Consider a bounded sequence $(x_n)_{n \in \mathbb{N}}$ in E. As the map $J: E \to E''$ is an isometry, the sequence $\phi_n = J_{x_n}$ is bounded in F'. According to the previous theorem, it admits a subsequence $(\phi_{\theta(n)})_{n \in \mathbb{N}}$ which converges in the weak-* sense towards a function ϕ.

This means that for all $f \in F$, we have $\phi_n(f) \to \phi(f)$. But $\phi_{\theta(n)}(f) = J_{x_{\theta(n)}}(f) = f(x_{\theta(n)})$. As E is reflexive, the map J is surjective and therefore there exists $x \in E$ such that $J_x = \phi$. Then $\phi(f) = J_x(f) = f(x)$. We have therefore shown that $(f(x_{\theta(n)}))_{n \in \mathbb{N}}$ converges towards $f(x)$ for all $f \in F = E'$, which means that the sequence $(x_{\theta(n)})_{n \in \mathbb{N}}$ converges weakly towards x.   □

3.7 Exercises

Exercise (solved) 3.1 (Lax–Milgram Theorem)
Let H be a Hilbert space over $\mathbb{R}$ equipped with the scalar product denoted $\langle \cdot , \cdot \rangle$. Let $a \colon H \times H \to \mathbb{R}$ be a continuous bilinear form satisfying the following property:

$$\exists c > 0/ \quad \forall x \in H, \quad a(x, x) \geqslant c\|x\|^2.$$

1. Show that, for all $u \in H$, there exists $Au \in H$ such that:

$$\forall v \in H, \quad a(u, v) = \langle Au, v \rangle.$$

2. Show that A is linear and continuous.
3. Show that $\mathrm{Im}(A)$ is a closed subset of H then deduce that $\mathrm{Im}(A) = H$.
4. Deduce that, for any continuous linear form $\phi \in H'$, there exists a unique $u \in H$ such that:

$$\forall v \in H, \quad a(u, v) = \phi(v).$$

Exercise (solved) 3.2 (Bessel Sequences)
Let H be a real Hilbert space equipped with a scalar product $(\cdot, \cdot)$ and the norm $\|x\| = \sqrt{(x, x)}$.

1. Consider a sequence $(x_n)_{n \in \mathbb{N}}$ in H. We say that this sequence is a "frame" if there exist two positive real numbers A and B such that,

$$\forall x \in H, \quad A\|x\|^2 \leqslant \sum_{n \in \mathbb{N}} |(x, x_n)|^2 \leqslant B\|x\|^2. \qquad (*)$$

(a) Suppose that $(e_n)_{n \in \mathbb{N}}$ is a Hilbert basis of H. State whether the following four families are or are not frames (provide a brief justification each time):

- $E_1 = (e_n)_{n \in \mathbb{N}}$;
- $E_2 = (e_{n+1})_{n \in \mathbb{N}}$;
- $E_3 = (e_0, e_0, e_1, e_1, e_2, e_2, e_3, e_3, \ldots)$ where each element is repeated twice;
- $E_4 = (e_n/(n+1))_{n \in \mathbb{N}}$.

(b) Suppose that $(x_n)_{n \in \mathbb{N}}$ is a frame. Let $V = \mathrm{Vect}\{x_n : n \in \mathbb{N}\}$. Show that V is a dense subspace in H.

2. (a) Let $(x_n)_{n \in \mathbb{N}}$ be a sequence of elements of H such that

$$\forall x \in H, \quad \sum_{n \in \mathbb{N}} |(x, x_n)|^2 < +\infty.$$

Let us introduce the map $U: H \to \ell^2(\mathbb{N})$ defined by

$$U(x) = \big((x, x_n)\big)_{n \in \mathbb{N}}.$$

Show using a general theorem on Banach spaces that the map U is continuous. Deduce that there exists $B > 0$ such that

$$\forall x \in H, \quad \sum_{n \in \mathbb{N}} |(x, x_n)|^2 \leqslant B\|x\|^2. \tag{**}$$

(b) Consider a sequence $(x_n)_{n \in \mathbb{N}}$ in H that satisfies the property (**). Consider a finite set $F \subset \mathbb{N}$ and a sequence $(c_n)_{n \in \mathbb{N}} \in \ell^2(\mathbb{N})$. Show that

$$\left\| \sum_{n \in F} c_n x_n \right\|^2 \leqslant \sup_{y \in H, \, \|y\|=1} \left(\sum_{n \in F} |c_n|^2 \right) \left(\sum_{n \in F} |(y, x_n)|^2 \right).$$

(c) Deduce that

$$\left\| \sum_{n \in F} c_n x_n \right\|^2 \leqslant B \sum_{n \in F} |c_n|^2.$$

Then show that the series $\sum c_n x_n$ converges.

We refer to Christensen's book [35] for a complete treatment of these notions.

Exercise (solved) 3.3 (The weak topology is not metrizable) Let E be an infinite-dimensional $\mathbb{R}$-vector space. We denote by $\sigma(E, E')$ the weak topology on E. By construction, a neighborhood basis of the origin for $\sigma(E, E')$ is given by the family of sets of the form

$$\bigcap_{k=1}^{N} \{y \in E : |\varphi_k(y)| < \varepsilon\} \quad \text{with} \quad N \geqslant 1, \quad \varphi_k \in E' \quad \text{and} \quad \varepsilon > 0.$$

1. Show that, in a infinite-dimensional space, a finite intersection of hyperplanes always contains a line. Deduce that every weak neighborhood of the origin contains a line $\mathbb{R}y$.
2. We assume that the weak topology $\sigma(E, E')$ is metrizable, that is, there exists a distance $d: E \to \mathbb{R}_+$ such that the balls $B_d(x, r) = \{y \in E : d(x, y) < r\}$ form a basis of neighborhoods of $\sigma(E, E')$.
   (a) Show that there exists a sequence $(x_n)_{n \in \mathbb{N}}$ of E satisfying $\|x_n\|_E = n$ and $x_n \to 0$.
   (b) Deduce a contradiction.

Exercise 3.4 Consider the sequence of functions

$$f_n(x) = \frac{1}{1 + (x - n)^2} \qquad (n \in \mathbb{N}, \ x \in \mathbb{R}).$$

1. Verify that this sequence does not converge strongly to 0 in $L^2(\mathbb{R})$.
2. Let $g \in C_0^0(\mathbb{R})$ (the space of continuous functions with compact support, which is dense in $L^2(\mathbb{R})$). Show that

$$\lim_{n \to +\infty} \int_{-\infty}^{\infty} f_n(t) g(t) \, dt = 0.$$

3. Deduce that $(f_n)_{n \in \mathbb{N}}$ converges weakly to 0 in $L^2(\mathbb{R})$.
4. Consider the same questions for the sequence of functions $h_n(x) = \sqrt{n}/(1+n\,|x|)$: show that this sequence does not converge strongly to 0 in $L^2(\mathbb{R})$ but that it converges weakly to 0 in $L^2(\mathbb{R})$.

Exercise 3.5 (Convex optimization)

1. Consider a topological space X. We say that a function $F: X \to \mathbb{R}$ is lower semi-continuous if, for every $\lambda \in \mathbb{R}$, the set

$$F^{-1}((-\infty, \lambda]) = \{x \in X : F(x) \leqslant \lambda\}$$

is closed.
   (a) Let $F: X \to \mathbb{R}$ be lower semi-continuous and (x_n) a sequence of points in X converging to x. Show that

$$\liminf_{n \to +\infty} F(x_n) \geqslant F(x).$$

   (b) Suppose that X is compact and consider a function $F: X \to \mathbb{R}$ that is lower semi-continuous. Show that F reaches its minimum on X.
2. We now assume that E is a Banach space and we consider a function $F: E \to \mathbb{R}$. We say that F is convex if

$$\forall (x, y) \in E \times E, \ \forall \lambda \in [0, 1], \quad F(\lambda x + (1 - \lambda)y) \leqslant \lambda F(x) + (1 - \lambda) F(y).$$

Show that if $F: E \to \mathbb{R}$ is a convex function and lower semi-continuous for the strong topology, then F is also lower semi-continuous for the weak topology. Moreover, if (x_n) converges weakly to x, then

$$\liminf_{n \to +\infty} F(x_n) \geqslant F(x).$$

3. Let E be a reflexive Banach space whose dual is separable. Consider a convex and lower semi-continuous function $F: E \to \mathbb{R}$ and a convex, bounded, closed and non-empty subset $C \subset E$. Show that there exists $x \in C$ such that

$$F(x) = \inf_{c \in C} F(c).$$

Exercise 3.6 (Study of the Kapitza pendulum by weak convergence) The following exercise requires knowledge of Sobolev spaces, which will be studied in Chapter 7.

Let us fix three positive numbers a, b, T. Given $(\alpha, \beta) \in \mathbb{R}^2$ and $\varepsilon \in (0, 1]$, we admit the existence of a function $\theta_\varepsilon \in C^2([0, T])$ satisfying

$$\begin{cases} \theta_\varepsilon''(t) = \left(a + \dfrac{b}{\varepsilon} \cos(t/\varepsilon)\right) \sin(\theta_\varepsilon(t)), & t \geqslant 0, \\ \theta_\varepsilon(0) = \alpha, & \\ \theta_\varepsilon'(0) = \beta. & \end{cases} \tag{$*$}$$

The previous equation governs the dynamics of an inverted pendulum (where the weight is at the top) subjected to a rapid vertical oscillation. If $b = 0$, the null solution is unstable. The aim of this problem is to show that the null solution is stable if $|b|$ is large enough.[3]

1. Show that if θ_ε is a solution of $(*)$ if and only if $\theta_\varepsilon(0) = \alpha$ and

$$\theta_\varepsilon'(t) = \beta + b \sin(t/\varepsilon) \sin(\theta_\varepsilon(t))$$
$$+ \int_0^t \left(a \sin(\theta_\varepsilon(s)) - b \sin(s/\varepsilon) \cos(\theta_\varepsilon(s)) \theta_\varepsilon'(s)\right) ds. \tag{$**$}$$

2. We recall the Gronwall inequality: if $u : [0, T] \to [0, +\infty)$ is a continuous function satisfying

$$u(t) \leqslant A + B \int_0^t u(s) \, ds,$$

then $u(t) \leqslant A e^{Bt}$ for all $t \in [0, T]$.
By applying this to $u_\varepsilon(t) = |\theta_\varepsilon'(t)|$, deduce that there exists a constant C depending only on (a, b, α, β, T) such that,

$$\sup_{\varepsilon \in (0,1]} \max_{t \in [0,T]} |\theta_\varepsilon'(t)| \leqslant C.$$

Infer from this that there exists a constant C', depending only on (a, b, α, β, T), such that

$$\sup_{\varepsilon \in (0,1]} \max_{t \in [0,T]} |\theta_\varepsilon(t)| \leqslant C'.$$

3. Infer that there exists $\theta \in C^0([0, T]) \cap H^1([0, T])$ such that we can extract a subsequence $(\theta_{\varepsilon_n})_{n \in \mathbb{N}}$ where ε_n tends to 0, converging to θ strongly in $C^0([0, T])$ and weakly in $H^1([0, T])$.

---

[3] This result is due to Kapitza [96] and we will present a proof given by Evans and Zhang [50]; we refer to this article for the solution.

4. Let $\psi \in C_0^\infty((0,T))$. Show that

$$\int_0^T \psi(t)\theta'_\varepsilon(t) \sin(t/\varepsilon)\, dt = R_\varepsilon + \int_0^T b\psi(t) \cos^2(t/\varepsilon) \sin(\theta_\varepsilon(t))\, dt,$$

where $R_\varepsilon = O(\varepsilon)$. Hint: use equation $(*)$ and integration by parts.

5. Show that $\cos^2(t/\varepsilon)$ converges weakly to the constant function $1/2$ when ε tends to 0. Infer from this that

$$\theta'_{\varepsilon_n}(t) \sin(t/\varepsilon_n) \longrightarrow \frac{b}{2} \sin(\theta(t)) \quad \text{in } L^2(0,T).$$

6. Using $(**)$, show that the weak derivative of θ is given by

$$\theta'(t) = \beta + \int_0^t \left(a \sin(\theta(s)) - \frac{b^2}{4} \sin(2\theta(s)) \right) ds.$$

7. Show that $\theta \in C^2([0,T])$ and θ satisfies

$$\begin{cases} \theta'' = a \sin(\theta(t)) - \dfrac{b^2}{4} \sin(2\theta(t)), & t \geqslant 0, \\ \theta(0) = \alpha, \\ \theta'(0) = \beta. \end{cases}$$

Infer from this that $(\theta_\varepsilon)_{\varepsilon \in (0,1]}$ converges to θ when ε tends to 0 (and not only the subsequence $(\theta_{\varepsilon_n})_{n \in \mathbb{N}}$).

Part II
Harmonic Analysis

Chapter 4
Fourier Series

Joseph Fourier is one of the most famous mathematicians. For an introduction to his life, his influence, and the late recognition of his importance, we refer to the beautiful texts of Jean-Pierre Kahane [90, 91, 93]. He is mainly known for having introduced the decomposition of a periodic function into an infinite sum of trigonometric functions of frequencies each of which is a multiple of a fundamental frequency. He had conjectured that every function was representable by a convergent trigonometric series, which is incorrect. We will see counterexamples to the pointwise convergence of Fourier series and will show results on the convergence of Fourier series in the context of integrable functions.

4.1 Introduction

We are going to study periodic functions $f : \mathbb{R}^n \to \mathbb{C}$. To lighten the notation, rather than considering arbitrary periods, we will assume that the functions are 2π-periodic with respect to each variable. We can reduce to this situation by a change of variables of the form

$$f(x_1, \ldots, x_n) \longmapsto f(T_1 x_1/(2\pi), \ldots, T_n x_n/(2\pi)).$$

By definition, we say that a function $f : \mathbb{R}^n \to \mathbb{C}$ is 2π-periodic with respect to each variable if

$$f(x + 2\pi e_j) = f(x) \qquad \text{for all } j \text{ such that } 1 \leqslant j \leqslant n,$$

where $(e_1, \ldots, e_n)$ is the canonical basis of $\mathbb{R}^n$. We will simply say in the following that f is periodic. For p in $[1, +\infty]$, we will denote by $\mathcal{L}^p_{\mathrm{per}}(\mathbb{R}^n)$ the space of measurable functions $f : \mathbb{R}^n \to \mathbb{C}$ which are periodic and such that $|f|^p$ is integrable over the cube $[0, 2\pi]^n$ (then $|f|^p$ is integrable over any compact set of $\mathbb{R}^n$ by periodicity). We denote by $L^p_{\mathrm{per}}(\mathbb{R}^n)$ the quotient space for the equivalence relation of almost everywhere equality (see Chapter 18 in the appendix for more details).

© The Author(s), under exclusive license to Springer Nature Switzerland AG 2024   95
T. Alazard, *Analysis and Partial Differential Equations*, Universitext,
https://doi.org/10.1007/978-3-031-70909-8_4

The space $L_{per}^P(\mathbb{R}^n)$ is a Banach space with $\|f\|_{L_{per}^P} = \left(\int_{[-\pi,\pi]^n} |f(x)|^p \, dx\right)^{1/p}$.

In 1812, Joseph Fourier explained how to represent a periodic function $f : \mathbb{R}^n \to \mathbb{C}$ as an infinite sum of sinusoidal functions. This theory allows a periodic function to be approximated by trigonometric polynomials and for this reason it plays a central role in many fields. Recall that a trigonometric polynomial is a function of the form

$$P(x) = \sum_{|k| \leqslant N} a_k e^{ik \cdot x}, \qquad a_k \in \mathbb{C}, \ k \in \mathbb{Z}^n, \tag{4.1}$$

where we have used the notations[1]

$$k \cdot x = k_1 x_1 + \cdots + k_n x_n, \qquad |k| = |k_1| + \cdots + |k_n|.$$

Let e_k denote the function (called the oscillatory exponential) defined by

$$e_k(x) = e^{ik \cdot x} = \exp(ik \cdot x),$$

and introduce the scalar product $(\cdot, \cdot)$ defined on $L_{per}^2(\mathbb{R}^n)$ by

$$(f, g) = \frac{1}{(2\pi)^n} \int_{[0,2\pi]^n} f(x)\overline{g(x)} \, dx.$$

The key point is the orthogonality relation

$$(e_k, e_{k'}) = \delta_k^{k'} \qquad \text{(equal to 1 if } k = k' \text{ and 0 otherwise)},$$

which is verified by a direct calculation. We deduce that the coefficients a_k in (4.1) satisfy

$$a_k = \frac{1}{(2\pi)^n} \int_{[0,2\pi]^n} P(x)e^{-ik \cdot x} \, dx, \qquad \forall k \in \mathbb{Z}^n.$$

This motivates the following definition.

Definition 4.1 Given a function $f \in L_{per}^1(\mathbb{R}^n)$, we define the kth Fourier coefficient of f by

$$\widehat{f}(k) = (f, e_k) = \frac{1}{(2\pi)^n} \int_{[0,2\pi]^n} f(x)e^{-ik \cdot x} \, dx,$$

and we call the Fourier series of f the sequence $(S_N(f))_{N \in \mathbb{N}}$ defined by

$$S_N f(x) = \sum_{|k| \leqslant N} \widehat{f}(k) e^{ik \cdot x}.$$

---

[1] In general, for a vector $x \in \mathbb{R}^n$, the notation $|x|$ denotes the Euclidean norm $|x| = \sqrt{x_1^2 + \cdots + x_n^2}$. However, when we consider multi-indices $k \in \mathbb{Z}^n$ (i.e., vectors whose coordinates are integers), the notation $|k|$ will denote the norm $|k| = |k_1| + \cdots + |k_n|$.

In this chapter, we will study the question of the convergence *in norm* of the Fourier series. This means that we are not interested in pointwise convergence. Let us mention a major result by Carleson [30] which states that the Fourier series of an L^2 function converges pointwise almost everywhere, without extracting a subsequence. For the proof of this result and other results on Fourier series, we refer to the books by Arias de Reya [6], Kahane and Lemarié-Rieusset [94], Katznelson [97], Grafakos [62], Muscalu and Schlag [145, 146] and Zygmund [203].

We will in particular prove the following result.

Theorem 4.2

1. *For all $n \geqslant 1$ and all $f \in L^2_{\mathrm{per}}(\mathbb{R}^n)$,*

$$\lim_{N \to \infty} \|S_N f - f\|_{L^2} = 0.$$

2. *There exists $f \in C^0_{\mathrm{per}}(\mathbb{R})$ such that $\|S_N f - f\|_{L^\infty}$ does not tend towards 0.*
3. *There exists $g \in L^1_{\mathrm{per}}(\mathbb{R})$ such that $\|S_N g - g\|_{L^1}$ does not tend towards 0.*

4.2 Square-Integrable Functions

The simplest and perhaps most important result to know about Fourier series is the following theorem.

Theorem 4.3 *For all $f \in L^2_{\mathrm{per}}(\mathbb{R}^n)$, we have*

$$f = \sum_{k \in \mathbb{Z}^n} \widehat{f}(k) e^{ik \cdot x}$$

with convergence in $L^2_{\mathrm{per}}(\mathbb{R}^n)$, which means that

$$\lim_{N \to +\infty} \|f - S_N f\|_{L^2} = 0 \quad \text{where} \quad S_N f(x) = \sum_{|k| \leqslant N} \widehat{f}(k) e^{ik \cdot x}.$$

Furthermore,

$$\|f\|_{L^2}^2 = \sum_{k \in \mathbb{Z}^n} |\widehat{f}(k)|^2.$$

Conversely, if $c = (c_k) \in \ell^2(\mathbb{Z}^n)$, then the series $\sum_{k \in \mathbb{Z}^n} c_k e^{ik \cdot x}$ converges in $L^2_{\mathrm{per}}(\mathbb{R}^n)$ towards a function f satisfying $\widehat{f}(k) = c_k$.

Proof We have already noted that $(e_k)_{k \in \mathbb{Z}^n}$ is an orthonormal family: $(e_k, e_{k'}) = 0$ if $k \neq k'$ and $\|e_k\|_{L^2} = 1$. Thus, according to the theorem on Hilbert basis (see Theorem 3.17), to prove this result, it is sufficient to show that the vector space $F = \mathrm{vect}\{e_k : k \in \mathbb{Z}^n\}$ is dense in $L^2_{\mathrm{per}}(\mathbb{R}^n)$.

Let $f \in L^2_{\mathrm{per}}(\mathbb{R}^n)$ and $\varepsilon > 0$. As the set $C^0_{\mathrm{per}}(\mathbb{R}^n)$ of continuous and 2π-periodic functions is dense in $L^2_{\mathrm{per}}(\mathbb{R}^n)$, there exists $g \in C^0_{\mathrm{per}}(\mathbb{R}^n)$ such that $\|f - g\|_{L^2} \leqslant \varepsilon$.

Furthermore, we have seen that the Stone–Weierstrass theorem implies that trigono-metric polynomials are dense in $C^0_{per}(\mathbb{R}^n)$. Specifically, Proposition 1.50 implies that there exists $h \in \text{vect}\{e_k : k \in \mathbb{Z}^n\}$ such that $\|g - h\|_{L^\infty} \leqslant \varepsilon$. We deduce that

$$\|f - h\|_{L^2} \leqslant \|f - g\|_{L^2} + \|g - h\|_{L^2} \leqslant \varepsilon + \|g - h\|_{L^\infty} \leqslant 2\varepsilon.$$

This shows that F is dense in $L^2_{per}(\mathbb{R}^n)$, which concludes the proof.   □

The previous proof relies on a property of Hilbert basis and on the density property of trigonometric polynomials. We will give another proof of the density of trigonometric polynomials which relies on an explicit construction (instead of using the Stone–Weierstrass theorem). This more direct method of proof has the advantage of clarifying the roles of convolution kernels in the theory of Fourier series. To simplify the notation, we assume that the dimension n is equal to 1.

Theorem 4.4 (Fejer) *Let* $f \in L^p_{per}(\mathbb{R})$ *with* $1 \leqslant p < \infty$. *Let us define*

$$\sigma_N f(x) = \frac{1}{N+1} \sum_{m=0}^{N} S_m f(x).$$

Then

$$\lim_{N \to +\infty} \int_0^{2\pi} |\sigma_N(f) - f|^p \, dx = 0.$$

If f is a continuous 2π-periodic function, then

$$\lim_{N \to +\infty} \sup_{x \in [0,2\pi]} |\sigma_N(f)(x) - f(x)| = 0.$$

Remark 4.5 Let $f \in L^\infty_{per}(\mathbb{R})$ such that $\lim_{N \to +\infty} \|\sigma_N(f) - f\|_{L^\infty} = 0$. Since $\sigma_N(f)$ is a continuous function, it follows that f is necessarily continuous (because the uniform limit of a sequence of continuous functions is continuous).

Proof Introduce the convolution product $f * g$ of two 2π-periodic functions f and g, defined by:

$$f * g(x) = \int_{-\pi}^{\pi} f(x - y)g(y) \, dy.$$

Then, by using identity $(a - 1)(1 + a + \cdots + a^\ell) = a^{\ell+1} - 1$ with $\ell = 2N$, we verify that

$$S_N f(x) = \frac{1}{2\pi} \sum_{k=-N}^{N} \int_{-\pi}^{\pi} f(y)e^{ik(x-y)} \, dy = \frac{1}{2\pi} \int_{-\pi}^{\pi} D_N(x - y)f(y) \, dy,$$

where D_N is the Nth Dirichlet kernel, defined by

$$D_N(x) = \frac{1}{2\pi} \sum_{k=-N}^{N} e^{ikx} = \frac{1}{2\pi} \frac{\sin((N + \frac{1}{2})x)}{\sin(\frac{1}{2}x)}.$$

Moreover,

$$\sigma_N f(x) = \frac{1}{N+1} \sum_{m=0}^{N} S_m f(x) = F_N * f(x),$$

where F_N is the Nth Fejer kernel, defined by

$$F_N = \frac{1}{N+1} \sum_{m=0}^{N} D_m = \frac{1}{2\pi(N+1)} \frac{\sin^2(\frac{N+1}{2}x)}{\sin^2(\frac{1}{2}x)}. \tag{4.2}$$

Here we used

$$\sum_{m=0}^{N} \sin((m+1/2)x) = \text{Im}\left(\sum_{m=0}^{N} \exp(i(m+1/2)x)\right)$$

and the same algebraic identity $(a-1)(1+a+\cdots+a^\ell) = a^{\ell+1}-1$.

We will use two elementary results on the convolution product. The first lemma is a special case of Young's inequality, which will be studied in Chapter 6.

Lemma 4.6 *Let $p \in [1, +\infty)$. Consider two functions $f \in L^1_{\text{per}}(\mathbb{R})$ and $g \in L^p_{\text{per}}(\mathbb{R})$. Then $f * g \in L^p_{\text{per}}(\mathbb{R})$ and moreover*

$$\|f * g\|_{L^p} \leqslant \|f\|_{L^1} \|g\|_{L^p}.$$

Proof Let I be the interval $[-\pi, \pi]$. By periodicity, using an elementary change of variables, we see that

$$f * g(x) = g * f(x) = \int_I f(x-y)g(y)\,dy.$$

We then use Minkowski's inequality (see Proposition 18.4) to write

$$\left(\int_I |f*g(x)|^p\,dx\right)^{1/p} = \left(\int_I \left|\int_I f(x-y)g(y)\,dy\right|^p dx\right)^{1/p}$$

$$= \left(\int_I \left|\int_I g(x-y)f(y)\,dy\right|^p dx\right)^{1/p}$$

$$\leqslant \int_I \left(\int_I |g(x-y)f(y)|^p\,dx\right)^{1/p} dy \quad \text{(Minkowski)}$$

$$\leqslant \int_I \left(\int_I |g(x-y)|^p\,dx\right)^{1/p} |f(y)|\,dy$$

$$\leqslant \int_I \|g\|_{L^p}\,|f(y)|\,dy = \|g\|_{L^p}\|f\|_{L^1}.$$

This concludes the proof. $\qquad\square$

The second result is related to the notion of approximation of the identity for periodic functions, which will also be studied in Chapter 6.

Definition 4.7 A sequence $(K_N)_{N \in \mathbb{N}}$ of 2π-periodic continuous functions is an approximation of the identity if

1. $\int_{-\pi}^{\pi} K_N(x) \, dx = 1$ for all N;
2. $\sup_N \int_{-\pi}^{\pi} |K_N(x)| \, dx < \infty$;
3. for all $\delta > 0$, we have

$$\lim_{N \to +\infty} \int_{|x| > \delta} |K_N(x)| \, dx = 0.$$

Note that the Dirichlet kernels do not form an approximation of the identity because $\|D_N\|_{L^1}$ tends to $+\infty$ as N tends to $+\infty$ (see Lemma 4.10). On the other hand, it is easy to verify (exercise) that the sequence of Fejer kernels $(F_N)_{N \in \mathbb{N}}$ (see (4.2)) is an approximation of the identity. Therefore, to prove Fejer's theorem, it suffices to prove the following lemma.

Lemma 4.8 *Consider an approximation of the identity $(K_N)_{N \in \mathbb{N}}$. If f belongs to $L^p([-\pi, \pi])$ with $1 \leqslant p < \infty$, or if $p = \infty$ and f is a continuous 2π-periodic function, then*

$$\lim_{N \to +\infty} \|K_N * f - f\|_{L^p([-\pi,\pi])} = 0.$$

To prove this lemma, we start by studying the case $p = \infty$. Since $[-\pi, \pi]$ is compact, if f is continuous then f is uniformly continuous. Therefore, for any $\varepsilon > 0$, there exists $\delta > 0$ such that

$$\sup_{x \in [-\pi,\pi]} \sup_{|y| < \delta} |f(x - y) - f(x)| < \varepsilon.$$

Therefore, using the fact that K_N is a periodic function and the property

$$\int_{-\pi}^{\pi} K_N(y) \, dy = 1,$$

we obtain that

$$
\begin{aligned}
K_N * f(x) - f(x) &= \int_{-\pi}^{\pi} K_N(x - y) f(y) \, dy - f(x) \\
&= \int_{x-\pi}^{x+\pi} K_N(y) f(x - y) \, dy - \int_{-\pi}^{\pi} K_N(y) f(x) \, dy \\
&= \int_{-\pi}^{\pi} K_N(y) (f(x - y) - f(x)) \, dy.
\end{aligned}
$$

Using the other two properties satisfied by the sequence $(K_N)_{N \in \mathbb{N}}$ (see Definition 4.7), we deduce that, if N is large enough then

$$|K_N * f(x) - f(x)| \leqslant \sup_{x \in [-\pi, \pi]} \sup_{|y| < \delta} |f(x - y) - f(x)| \int_{-\pi}^{\pi} |K_N(t)| \, dt$$

$$+ \int_{|y| \geqslant \delta} |K_N(y)| 2 \|f\|_{L^\infty} \, dy$$

$$\leqslant C\varepsilon.$$

Now, consider $f \in L^p_{per}(\mathbb{R})$ and $g \in C^0_{per}(\mathbb{R})$ such that $\|f - g\|_{L^p} < \varepsilon$. Then

$$\|K_N * f - f\|_{L^p} \leqslant \|K_N * (f - g)\|_{L^p} + \|f - g\|_{L^p} + \|K_N * g - g\|_{L^p}$$

$$\leqslant \left(\sup_N \|K_N\|_{L^1} + 1 \right) \|f - g\|_{L^p} + \|K_N * g - g\|_{L^\infty},$$

where we used Young's inequality $\|u * v\|_{L^p} \leqslant \|u\|_{L^1} \|v\|_{L^p}$ and the previous result for $p = \infty$, which also implies that $\|K_N * g - g\|_{L^\infty}$ is arbitrarily small for N large enough. This concludes the proof of the lemma and Fejer's theorem. □

4.3 Pointwise Convergence and Uniform Convergence

4.3.1 Divergent Fourier Series

Proposition 4.9 *There exists a continuous periodic function $f \in C^0_{per}(\mathbb{R})$ whose Fourier series at 0 diverges.*

Proof Recall that the Fourier series of f, denoted $(S_N(f))_{N \in \mathbb{N}}$, satisfies

$$S_N(f)(x) = \int_{-\pi}^{\pi} D_N(x - y) f(y) \, dy, \qquad (4.3)$$

where D_N is the Nth Dirichlet kernel:

$$D_N(x) = \frac{1}{2\pi} \sum_{k=-N}^{N} e^{ikx} = \frac{1}{2\pi} \frac{\sin((N + \frac{1}{2})x)}{\sin(\frac{1}{2}x)}.$$

For all $N \in \mathbb{N}$, we introduce the linear form $\ell_N : C^0_{per}(\mathbb{R}) \to \mathbb{C}$ defined by $\ell_N(f) = S_N(f)(0)$. Then ℓ_N is continuous because

$$|\ell_N(f)| \leqslant \|D_N\|_{L^1} \|f\|_{C^0_{per}(\mathbb{R})} \quad \text{where} \quad \|f\|_{C^0_{per}(\mathbb{R})} = \sup_{x \in [-\pi, \pi]} |f(x)|. \qquad (4.4)$$

By definition of ℓ_N, to show that there exists a function whose Fourier series diverges at 0, it suffices to show that the family $(\ell_N)_{N \in \mathbb{N}}$ is not pointwise bounded (which means showing that there exists $f \in C^0_{per}(\mathbb{R})$ such that $(\ell_N(f))_{N \in \mathbb{N}}$ is not bounded in $\mathbb{C}$). According to the Banach–Steinhaus theorem, it suffices to show that $(\ell_N)_{N \in \mathbb{N}}$ is not uniformly bounded, which means showing that the sequence $(\|\ell_N\|)_{N \in \mathbb{N}}$ diverges. For this, we will show that $\|\ell_N\| = \|D_N\|_{L^1}$ so that the desired result will be a consequence of the following lemma.

Lemma 4.10 *The sequence* $(\|D_N\|_{L^1})_{N \in \mathbb{N}}$ *is not bounded.*

Proof Indeed, we verify that

$$
\|D_N\|_{L^1} = \frac{1}{2\pi} \int_{-\pi}^{\pi} \left| \frac{\sin((N+1/2)x)}{\sin(x/2)} \right| dx = \frac{1}{\pi} \int_0^{\pi} \left| \frac{\sin((N+1/2)x)}{\sin(x/2)} \right| dx
$$

$$
\geqslant \frac{1}{\pi} \int_0^{\pi} \left| \frac{\sin((N+1/2)x)}{x/2} \right| dx = \frac{2}{\pi} \int_0^{(2N+1)\pi/2} \left| \frac{\sin(u)}{u} \right| du,
$$

which implies the announced result because $\int_0^{\infty} |\sin(u)/u|\, du = +\infty$.   □

It remains to show that $\|\ell_N\| = \|D_N\|_{L^1}$. Already, the estimate (4.4) implies that $\|\ell_N\| \leqslant \|D_N\|_{L^1}$. Furthermore, by considering

$$
f_\varepsilon(x) = \frac{D_N(x)}{|D_N(x)| + \varepsilon},
$$

and using the dominated convergence theorem (which applies because D_N is integrable on $[0, 1]$) we obtain $|\ell_N(f)| \geqslant \|D_N\|_{L^1}$, which concludes the proof.   □

4.3.2 Pointwise Convergence

In the previous section, we proved the existence of a continuous function whose Fourier series diverges at 0. For this, we used the fact that the L^1 norm of the Dirichlet kernel D_N diverges when N tends to infinity. We will now exploit the fact that if the function f is a bit more regular, then the oscillations of the Dirichlet kernel do not appear and we can obtain a pointwise convergence result. We will study the normal convergence of Fourier series later.

Definition 4.11 Consider a real number α in $(0, 1]$. A function $f : \mathbb{R} \to \mathbb{C}$ is α-Hölderian (or Hölderian of exponent α) if there exists a $C > 0$ such that, for all x, y in $\mathbb{R}$,

$$
|f(x) - f(y)| \leqslant C|x - y|^\alpha.
$$

We will use the notation

$$
[f]_\alpha = \sup_{\substack{x,y \in \mathbb{R} \\ x \neq y}} \frac{|f(x) - f(y)|}{|x - y|^\alpha}.
$$

Proposition 4.12 *Consider a function f that is 2π-periodic and α-Hölderian for a certain $\alpha \in (0, 1]$. Then the Fourier series of f converges pointwise to f: namely $\lim_{N \to +\infty} |f(x) - S_N(f)(x)| = 0$ for all x in $\mathbb{R}$.*

Remark 4.13 Under the same assumptions, we can show that the Fourier series of f converges uniformly to f:

$$\lim_{N \to +\infty} \sup_{x \in [-\pi, \pi]} |f(x) - S_N(f)(x)| = 0.$$

Proof If necessary, by considering $f(x + a)$, it is sufficient to study the convergence at $x = 0$. Let us show that the partial sums $S_N(f)(0)$ converge to $f(0)$.

Recall that (cf. (4.3))

$$S_N(f)(0) = \frac{1}{2\pi} \int_{-\pi}^{\pi} f(y) \frac{\sin\left(\left(N + \frac{1}{2}\right)y\right)}{\sin\left(\frac{1}{2}y\right)} \, dy.$$

We can then write $f(0) - S_N(f)(0)$ in the form

$$f(0) - S_N(f)(0) = \frac{1}{2\pi} \int_{-\pi}^{\pi} (f(0) - f(y)) \frac{\sin\left(\left(N + \frac{1}{2}\right)y\right)}{\sin\left(\frac{1}{2}y\right)} \, dy$$

$$= \frac{1}{2\pi} \int_{-\pi}^{\pi} \frac{f(0) - f(y)}{\sin\left(\frac{1}{2}y\right)} \sin\left(\left(N + \frac{1}{2}\right)y\right) \, dy.$$

We use the following inequality, deduced from the concavity of sine on $[0, \pi/2]$,

$$\forall x \in \left[0, \frac{\pi}{2}\right], \quad \sin(x) \geqslant \frac{2}{\pi} x,$$

and the fact that f is α-Hölderian, to obtain the existence of $C > 0$ such that

$$\forall y \in [-\pi, \pi], \quad \left|\frac{f(0) - f(y)}{\sin\left(\frac{1}{2}y\right)}\right| \leqslant C |y|^{\alpha - 1}.$$

So the function g defined by

$$g(y) = (f(0) - f(y)) / \sin\left(\tfrac{1}{2}y\right)$$

is locally integrable. According to the Riemann–Lebesgue lemma, we have

$$\lim_{k \to +\infty} \int_{-\pi}^{\pi} g(y) e^{iky} \, dy = 0.$$

Now we get that $S_N(f)(0)$ converges to $f(0)$ by expanding $\sin((N + 1/2)y)$, to deduce that

$$\lim_{n \to +\infty} \int_{-\pi}^{\pi} g(y) \sin\left(\left(N + \frac{1}{2}\right)y\right) dy = 0.$$

This concludes the proof.   □

4.3.3 Uniform Convergence

We now give sufficient conditions that lead to the normal convergence of the Fourier series. We will return to this question in Section 4.5 dedicated to Wiener's theorem.

Proposition 4.14 *Consider a function $f : \mathbb{R} \to \mathbb{C}$ of class C^2 and 2π-periodic. Then its Fourier series converges normally and moreover*

$$\sum_{k \in \mathbb{Z}} |\widehat{f}(k)| \leqslant \frac{\pi^2}{3} \|f''\|_{L^\infty} + \|f\|_{L^\infty}. \tag{4.5}$$

Remark 4.15 Since $\sup_{x \in \mathbb{R}} |\widehat{f}(k) e^{ikx}| = |\widehat{f}(k)|$, the absolute convergence of the Fourier coefficients implies the normal convergence of the Fourier series and therefore its uniform convergence. Furthermore, as Proposition 4.12 implies that there is pointwise convergence of the Fourier series towards f, by uniqueness of the limit, we deduce that the Fourier series of f converges uniformly to f.

Proof By integrating by parts, we directly verify that

$$\widehat{f''}(k) = \frac{1}{2\pi} \int_{-\pi}^{\pi} f''(x) e^{-ikx} \, dx = \frac{ik}{2\pi} \int_{-\pi}^{\pi} f'(x) e^{-ikx} \, dx$$
$$= \frac{(ik)^2}{2\pi} \int_{-\pi}^{\pi} f(x) e^{-ikx} \, dx = -k^2 \widehat{f}(k).$$

We deduce the desired inequality because $\sum_{k \in \mathbb{Z}^*} |k|^{-2} = \pi^2/3$.   □

The following proposition shows that it is sufficient to assume that the functions are α-Hölderian with $\alpha > 1/2$.

Proposition 4.16 *Let $f : \mathbb{R} \to \mathbb{C}$ be a 2π-periodic function and α-Hölderian. If $\alpha > 1/2$, the Fourier series of f converges normally to f:*

$$\sum_{p \in \mathbb{Z}} |\widehat{f}(p)| < +\infty.$$

Proof This is based on a dyadic decomposition.[2] Specifically, we will show that there exists $C > 0$ such that

$$\sum_{2^{n-1} < |p| \leqslant 2^n} |\widehat{f}(p)| \leqslant C 2^{(1-2\alpha)n/2}. \tag{4.6}$$

The assumption $\alpha > 1/2$ implies that the right-hand side is the general term of a convergent series. Thus, the desired result will be a direct consequence of (4.6).

---

[2] This result can be shown more directly, using Sobolev spaces (studied later in this book). The following proof has the merit of providing a first illustration of dyadic packet decomposition methods, which will also be studied in detail later.

Given a parameter $h > 0$, consider the function

$$f_h = f(\cdot + h) - f(\cdot - h).$$

Then the Fourier coefficients of f_h are calculated from those of f:

$$\widehat{f_h}(k) = \frac{1}{2\pi} \int_{-\pi}^{\pi} f(x + h)e^{-ikx}\, dx - \frac{1}{2\pi} \int_{-\pi}^{\pi} f(x - h)e^{-ikx}\, dx$$
$$= (e^{ikh} - e^{-ikh})\widehat{f}(k) = 2i\sin(kh)\widehat{f}(k).$$

Parseval's identity implies that

$$\sum_{k \in \mathbb{Z}} 4\sin^2(kh)\left|\widehat{f}(k)\right|^2 = \int_{-\pi}^{\pi} |f(x + h) - f(x - h)|^2\, dx.$$

Let $n \in \mathbb{N}^*$ and $h = 2^{-n}$. The previous identity implies that

$$\sum_{2^{n-1} < p \leqslant 2^n} 4\sin^2(ph)\left|\widehat{f}(p)\right|^2 \leqslant \sum_{p \in \mathbb{Z}} 4\sin^2(ph)\left|\widehat{f}(p)\right|^2$$
$$\leqslant \int_{-\pi}^{\pi} [f]_\alpha^2 \left|2^{-n+1}\right|^{2\alpha}\, dx \leqslant (2\pi)[f]_\alpha^2 2^{-2\alpha(n-1)}.$$

$$(4.7)$$

For $2^{n-1} < |p| \leqslant 2^n$, we have $1/2 \leqslant |ph| \leqslant 1 \leqslant \pi/2$, therefore

$$\sin^2(ph) \geqslant \left(\frac{2}{\pi} ph\right)^2 \geqslant \frac{1}{\pi^2}.$$

From (4.7), we deduce that

$$\sum_{2^{n-1} < p \leqslant 2^n} \left|\widehat{f}(p)\right|^2 \leqslant C_\alpha 2^{-2\alpha n} \quad \text{where} \quad C_\alpha = \frac{\pi^3}{4} [f]_\alpha^2 2^{2\alpha}.$$

Now, the Cauchy–Schwarz inequality implies that

$$\sum_{p=2^{n-1}+1}^{2^n} \left|\widehat{f}(p)\right| \leqslant \left(\sum_{p=2^{n-1}+1}^{2^n} 1\right)^{1/2} \left(\sum_{p=2^{n-1}+1}^{2^n} \left|\widehat{f}(k)\right|^2\right)^{1/2}$$
$$\leqslant 2^{n/2}\sqrt{C_\alpha} 2^{-n\alpha},$$

which implies the wanted estimate (4.6). $\square$

4.4 Applications of the Plancherel Formula

Recall that for any function f in $L^2_{\mathrm{per}}(\mathbb{R}^n)$, we have the Plancherel formula

$$\|f\|^2_{L^2} = \sum_{k \in \mathbb{Z}^n} |\widehat{f}(k)|^2.$$

We will apply this identity to obtain a Poincaré inequality. Generally, a Poincaré inequality is an inequality that allows us to control the L^2 norm of a function by the L^2 norm of its derivative. The simplest of these inequalities is given by the following lemma.

Lemma 4.17 (Poincaré–Wirtinger Inequality) *Suppose that $u \in C^1(\mathbb{R})$ is T-periodic. If $\int_0^T u(t)\,dt = 0$, then*

$$\left(\int_0^T |u(t)|^2 \, dt \right)^{1/2} \leqslant \frac{T}{2\pi} \left(\int_0^T |u'(t)|^2 \, dt \right)^{1/2}.$$

Proof The proof is easily obtained from the Fourier series decomposition of functions belonging to $L^2(0,T)$. Consider the usual scalar product: $(f,g) = \int_0^T f(t)\overline{g(t)}\,dt$. We then set $e_k(t) = T^{-1/2}\exp(2ik\pi t/T)$. These functions are T-periodic and we have $(e_k, e_\ell) = \delta^\ell_k$. Introduce the Fourier coefficients

$$\widehat{u}_k = (u, e_k) = \frac{1}{\sqrt{T}} \int_0^T u(t)\exp\left(-\frac{2i\pi kt}{T}\right) dt.$$

By integrating by parts, we verify that the Fourier coefficients of u' satisfy

$$\widehat{(u')}_k = \frac{1}{\sqrt{T}} \int_0^T u'(t)\exp\left(-\frac{2i\pi kt}{T}\right) dt = \frac{2i\pi}{T} k\widehat{u}_k.$$

We deduce from the Plancherel formula applied to u and u' that

$$\int_0^T |u(t)|^2 \, dt = \sum_{k \in \mathbb{Z}} |\widehat{u}_k|^2$$

$$= \sum_{k \in \mathbb{Z}^*} |\widehat{u}_k|^2$$

$$\leqslant \sum_{k \in \mathbb{Z}^*} |k|^2 |\widehat{u}_k|^2$$

$$= \left(\frac{T}{2\pi}\right)^2 \int_0^T |u'(t)|^2 \, dt,$$

where we used that $\widehat{u}_0 = T^{-1/2}\int_0^T u(t)\,dt = 0$.   □

As an application, let us prove the following result.

Proposition 4.18 (Yorke's Inequality) *Let* $n \geqslant 2$ *and* $V : \mathbb{R}^n \to \mathbb{R}^n$ *be a class* C^1 *such that* ∇V *is bounded on* $\mathbb{R}^n$. *We set*

$$K = \sup_{x \in \mathbb{R}^n} \left(\sum_{1 \leqslant i, j \leqslant n} (\partial_i V_j(x))^2 \right)^{1/2},$$

where V_j $(1 \leqslant j \leqslant n)$ *are the coordinates of* V. *Suppose there exists a solution to the differential equation*

$$m'(t) = V(m(t))$$

which is T-*periodic and non-constant. Then we have* $T \geqslant 2\pi / K$.

Proof We have

$$\left| m_j''(t) \right| = \left| \nabla V_j(m(t)) \cdot m'(t) \right| \leqslant \left| \nabla V_j(m(t)) \right| \left| m'(t) \right|$$

so $|m''(t)| \leqslant K |m'(t)|$. By integrating, we deduce that

$$\int_0^T |m''(t)|^2 \, dt \leqslant K^2 \int_0^T |m'(t)|^2 \, dt.$$

Furthermore, as $\int_0^T m'(t) \, dt = 0$, we can apply the Poincaré–Wirtinger inequality, which implies that

$$\int_0^T |m'(t)|^2 \, dt \leqslant \frac{T^2}{(2\pi)^2} \int_0^T |m''(t)|^2 \, dt.$$

From this, we deduce $T \geqslant 2\pi / K$. $\square$

4.5 Wiener's Theorem

In this section, we will give an elementary proof[3] of a theorem of Wiener.[4]

Theorem 4.19 (Wiener) *Let* f *be a continuous* 2π-*periodic function with an absolutely convergent Fourier series. Then, if* f *does not vanish, the Fourier series of the function* $1/f$ *is absolutely convergent.*

Recall that we denote by $C^0_{\mathrm{per}}(\mathbb{R})$ the space of continuous and 2π-periodic functions $f : \mathbb{R} \to \mathbb{C}$. It is a Banach space for the norm $\|f\|_{C^0_{\mathrm{per}}} = \sup_{\mathbb{R}} |f|$. Let us introduce the subset $\mathcal{A}$ of functions $f \in C^0_{\mathrm{per}}(\mathbb{R})$ whose Fourier series is absolutely convergent:

$$\sum_{k \in \mathbb{Z}} |\widehat{f}(k)| < \infty \quad \text{where} \quad \widehat{f}(k) = \frac{1}{2\pi} \int_{-\pi}^{\pi} f(x) e^{-ikx} \, dx.$$

---

[3] Due to Newman [149].

[4] See Wiener's original article [196] and Kahane's text [89].

We want to show that if $f \in \mathcal{A}$ satisfies $f(x) \neq 0$ for all $x \in \mathbb{R}$, then $1/f \in \mathcal{A}$.

Before proving this result, we will study the set $\mathcal{A}$. In particular, we will see that we can endow $\mathcal{A}$ with the structure of a Banach algebra (by definition, a Banach algebra is an algebra endowed with a norm satisfying inequality (4.8) below and for which $\mathcal{A}$ is a Banach space).

Proposition 4.20

1. If $f \in \mathcal{A}$, then the functions $S_n(f)(x) = \sum_{k=-n}^{n} \widehat{f}(k)e^{ikx}$ converge to f in $C_{\text{per}}^0(\mathbb{R})$ when n tends to $+\infty$.
2. $\|f\|_{\mathcal{A}} = \sum_{k \in \mathbb{Z}} |\widehat{f}(k)|$ is a norm on $\mathcal{A}$ and $\|f\|_{C_{\text{per}}^0} \leq \|f\|_{\mathcal{A}}$.
3. $(\mathcal{A}, \|\cdot\|_{\mathcal{A}})$ is a Banach space and the trigonometric polynomials are dense in $\mathcal{A}$.
4. $\mathcal{A}$ is an algebra and moreover, for all $f \in \mathcal{A}$ and all $g \in \mathcal{A}$,

$$\|fg\|_{\mathcal{A}} \leq \|f\|_{\mathcal{A}} \|g\|_{\mathcal{A}}. \tag{4.8}$$

Proof 1. Let us introduce the functions $e_k : x \mapsto e^{ikx}$. The triangle inequality implies that, for all $f \in \mathcal{A}$,

$$\|S_n(f)\|_{C_{\text{per}}^0} \leq \sum_{|k| \leq n} \left\| \widehat{f}(k)e_k \right\|_{C_{\text{per}}^0} = \sum_{|k| \leq n} |\widehat{f}(k)| \leq \sum_{k \in \mathbb{Z}} |\widehat{f}(k)| < \infty. \tag{4.9}$$

From this, we deduce that the Fourier series converges normally, therefore uniformly on $\mathbb{R}$ and its sum g is well defined, continuous and 2π-periodic. Furthermore, as $C_{\text{per}}^0(\mathbb{R})$ is continuously embedded in $L_{\text{per}}^2(\mathbb{R})$, we know from Theorem 4.3 that the Fourier series of f converges to f in $L_{\text{per}}^2(\mathbb{R})$. By uniqueness of the limit, we deduce that $g = f$.

2. It is clear that $\|\cdot\|_{\mathcal{A}}$ is a semi-norm. As $S_n(f)$ converges to f in $C_{\text{per}}^0(\mathbb{R})$, we can pass to the limit in (4.9) to obtain the desired inequality $\|f\|_{C_{\text{per}}^0} \leq \|f\|_{\mathcal{A}}$. This inequality implies that $f = 0$ if $\|f\|_{\mathcal{A}} = 0$, which proves that $\|\cdot\|_{\mathcal{A}}$ is a norm.

3. It suffices to show that the Fourier transform $\mathcal{F} : \mathcal{A} \to \ell^1(\mathbb{Z})$ defined by $\mathcal{F}(f) = \left(\widehat{f}(k) \right)_{k \in \mathbb{Z}}$ is a bijection and an isometry. It is an injective function because if $f \in C_{\text{per}}^0(\mathbb{R})$ satisfies $\widehat{f} = 0$, then $f = 0$ (this is a consequence of Theorem 4.3). This transform is also surjective. Indeed, if $a = (a_k)_{k \in \mathbb{Z}} \in \ell^1(\mathbb{Z})$ then the function defined by $f(x) = \sum_{k \in \mathbb{Z}} a_k e^{ikx}$ belongs to $C_{\text{per}}^0(\mathbb{R})$ (by normal convergence, as in point (1)). Moreover, for all $\ell \in \mathbb{Z}$, by taking the scalar product of $L_{\text{per}}^2(\mathbb{R})$ with the function $e_\ell : x \mapsto e^{i\ell x}$, we verify that $\widehat{f}(\ell) = a_\ell$, which shows that $\mathcal{F}(f) = a$. Finally, $\mathcal{F}$ is an isometry by definition of the norm $\|\cdot\|_{\mathcal{A}}$.

4. If f and g are two trigonometric polynomials, then we verify that $\widehat{fg}(k) = \sum_{p \in \mathbb{Z}} \widehat{f}(k-p)\widehat{g}(p)$. This formula remains true on $\mathcal{A}$ by density. We deduce from Fubini's theorem for series of real positive numbers that

$$\sum_{k \in \mathbb{Z}} |\widehat{fg}(k)| \leq \sum_{k \in \mathbb{Z}} \sum_{p \in \mathbb{Z}} |\widehat{f}(k-p)\widehat{g}(p)| \leq \left(\sum_{\ell \in \mathbb{Z}} |\widehat{f}(\ell)| \right) \left(\sum_{p \in \mathbb{Z}} |\widehat{g}(p)| \right),$$

from which the inequality (4.8) follows. $\qquad \square$

We now prove Wiener's theorem.

***Proof (of theorem* 4.19)** Let $f \in \mathcal{A}$ satisfy $f(x) \neq 0$ for all $x \in \mathbb{R}$. Then f does not change sign and we can assume without loss of generality that $m = \inf_{x \in \mathbb{R}} f(x) > 0$. According to point (3) of Proposition 4.20, the trigonometric polynomials are dense in $\mathcal{A}$, which implies that there exists $g \in C^2_{\mathrm{per}}(\mathbb{R})$ such that

$$\|f - g\|_{\mathcal{A}} \leqslant m/3.$$

As $\|f - g\|_{C^0_{\mathrm{per}}} \leqslant \|f - g\|_{\mathcal{A}}$ (see point (2) of Proposition 4.20), we deduce from the triangle inequality that $|g(x)| \geqslant 2m/3$ for all $x \in \mathbb{R}$. It follows that $\|(f - g)/g\|_{C^0_{\mathrm{per}}} \leqslant 1/2$. Then, by convergence of a geometric series, we verify that the series $\sum((g - f)/g)^n$ converges in $C^0_{\mathrm{per}}(\mathbb{R})$ and, moreover,

$$\frac{1}{f} = \frac{1}{g} \sum_{n=0}^{\infty} \left(\frac{g - f}{g}\right)^n. \tag{4.10}$$

We want to show that $1/f$ belongs to $\mathcal{A}$. As $1/g \in C^2_{\mathrm{per}}(\mathbb{R})$, Proposition 4.14 implies that $1/g \in \mathcal{A}$. Thus, as $\mathcal{A}$ is a Banach algebra (*cf.* point (4) of Proposition 4.20), it suffices to show that the series $\sum((g - f)/g)^n$ converges normally in $\mathcal{A}$.

Using the inequality $\|uv\|_{\mathcal{A}} \leqslant \|u\|_{\mathcal{A}}\|v\|_{\mathcal{A}}$, we have

$$\left\|\left(\frac{g - f}{g}\right)^n\right\|_{\mathcal{A}} \leqslant \|(f - g)^n\|_{\mathcal{A}}\left\|\frac{1}{g^n}\right\|_{\mathcal{A}} \leqslant \|f - g\|^n_{\mathcal{A}}\left\|\frac{1}{g^n}\right\|_{\mathcal{A}}$$
$$\leqslant \left(\frac{m}{3}\right)^n\left\|\frac{1}{g^n}\right\|_{\mathcal{A}}. \tag{4.11}$$

Furthermore, according to (4.5), we have

$$\left\|\frac{1}{g^n}\right\|_{\mathcal{A}} \leqslant \frac{\pi^2}{3}\left\|\left(\frac{1}{g^n}\right)''\right\|_{L^\infty} + \left\|\frac{1}{g^n}\right\|_{L^\infty}.$$

As $|g| \geqslant 2m/3$, we have $\|1/g^n\|_{L^\infty} \leqslant (3/(2m))^n$. Moreover,

$$\left\|\left(\frac{1}{g^n}\right)''\right\|_{L^\infty}$$
$$\leqslant n(n + 1)\left\|\frac{1}{g^{n+2}}\right\|_{L^\infty}\|(g')^2\|_{L^\infty} + n\left\|\frac{1}{g^{n+1}}\right\|_{L^\infty}\|g''\|_{L^\infty} \tag{4.12}$$
$$\leqslant n(n + 1)\left(\frac{3}{2m}\right)^{n+2}\|(g')^2\|_{L^\infty} + n\left(\frac{3}{2m}\right)^{n+1}\|g''\|_{L^\infty}.$$

By combining (4.11) and (4.12), it follows that there exists a constant $C(g, m)$ depending on g and m, but not on n, such that

$$\left\|\left(\frac{g - f}{g}\right)^n\right\|_{\mathcal{A}} \leqslant C(g, m)(1 + n^2)2^{-n}.$$

We deduce that $\sum \|((g - f)/g)^n\|_{\mathcal{A}} < +\infty$, which completes the proof. $\square$

4.6 Exercises

Exercise (solved) 4.1 Let $u = u(t,x)$ be a function of class C^2 on $\mathbb{R} \times \mathbb{R}$, 2π-periodic in x, and a solution of the heat equation

$$\frac{\partial u}{\partial t} - \partial_x(\gamma(x)\partial_x u) = 0,$$

where γ is a C^∞, 2π-periodic function and bounded below by a positive constant. Suppose also that $\int_{-\pi}^{\pi} u(0,x)\,dx = 0$. We want to show that there exists a constant C such that, for all time $t \geqslant 0$,

$$\int_{-\pi}^{\pi} u(t,x)^2\,dx \leqslant e^{-tC} \int_{-\pi}^{\pi} u(0,x)^2\,dx.$$

1. Provide a simple proof of the result in the case $\gamma = 1$.
2. Consider the general case.
   (a) Show that, for all time $t \geqslant 0$,

   $$\int_{-\pi}^{\pi} u(t,x)\,dx = 0.$$

   (b) Show that

   $$\frac{1}{2}\frac{d}{dt}\int_{-\pi}^{\pi} u(t,x)^2\,dx + \int_{-\pi}^{\pi} \gamma(x)(\partial_x u(t,x))^2\,dx = 0.$$

   (c) Show that there exists a constant C such that

   $$\frac{C}{2}\int_{-\pi}^{\pi} u(t,x)^2\,dx \leqslant \int_{-\pi}^{\pi} \gamma(x)(\partial_x u(t,x))^2\,dx.$$

   (d) Conclude.

Exercise 4.2 (Stabilization of the Schrödinger equation) Let $\chi: \mathbb{R} \to [0,1]$ be a function of class C^∞ and 2π-periodic. Consider a function $u = u(t,x)$ defined for $(t,x) \in \mathbb{R} \times \mathbb{R}$, with complex values, 2π-periodic in x. We further assume that $u \in C^\infty(\mathbb{R} \times \mathbb{R}; \mathbb{C})$ is a solution of the damped Schrödinger equation:

$$i\partial_t u + \partial_x^2 u + i\chi(x)u = 0.$$

Introduce the function $E: \mathbb{R} \to \mathbb{R}$, called the energy, defined by

$$E(t) = \int_0^{2\pi} |u(t,x)|^2\,dx. \qquad (4.13)$$

The aim of this exercise is to prove that the energy $E(t)$ decays exponentially to zero whenever χ is non-negative and non-zero. The proof is borrowed from an article by Machtyngier and Zuazua ([126]).

1. We assume in this question that χ is a positive constant function $\chi(x) = \lambda > 0$. The function $u \in C^\infty(\mathbb{R} \times \mathbb{R}; \mathbb{C})$ is therefore a solution of

$$i\partial_t u + \partial_x^2 u + i\lambda u = 0.$$

Show that

$$E'(t) + 2\lambda E(t) = 0$$

and deduce the inequality $E(t) \leqslant e^{-2\lambda t} E(0)$.

2. In this question we are interested in the equation with $\chi = 0$. We therefore assume that $u \in C^\infty(\mathbb{R} \times \mathbb{R}; \mathbb{C})$ is a solution of

$$i\partial_t u + \partial_x^2 u = 0.$$

   (a) Let $u_0(x) = u(0, x)$ and denote by $(c_n)_{n \in \mathbb{Z}}$ the Fourier coefficients of the function u_0. Show that $u(t, x) = \sum_{\mathbb{Z}} c_n e^{inx - in^2 t}$. We denote $u(t) = e^{it\partial_x^2} u_0$.
   (b) Show that

$$u(t, x) = c_0 + \sum_{n=1}^{+\infty} \left(c_n e^{inx} + c_{-n} e^{-inx} \right) e^{-in^2 t}$$

and then deduce that

$$\frac{1}{2\pi} \int_0^{2\pi} |u(t, x)|^2 \, dt = |c_0|^2 + \sum_{n \in \mathbb{Z}} \left| c_n e^{inx} + c_{-n} e^{-inx} \right|^2.$$

   (c) Let a, b be such that $0 < a < b < 2\pi$. Show that there exists $c > 0$ such that

$$\int_a^b \int_0^{2\pi} |u(t, x)|^2 \, dt \, dx \geqslant c \int_0^{2\pi} |u_0(x)|^2 \, dx. \tag{4.14}$$

3. Let $u_0 : \mathbb{R} \to \mathbb{C}$ be a 2π-periodic C^∞ function. We assume in this question that χ is a function satisfying $0 \leqslant \chi \leqslant 1$ and $\chi(x) = 1$ for all $x \in [a, b]$ with $0 < a < b < 2\pi$. The function u is therefore a solution of

$$i\partial_t u + \partial_x^2 u + i\chi u = 0, \qquad u(0, x) = u_0(x).$$

   (a) Let $T > 0$. Show that

$$\frac{1}{2} \int_0^{2\pi} |u(T, x)|^2 \, dx + \int_0^T \int_0^{2\pi} \chi(x) |u(t, x)|^2 \, dx \, dt = \frac{1}{2} \int_0^{2\pi} |u_0(x)|^2 \, dx.$$

   (b) Assume that $T \geqslant 2\pi$. Deduce from (4.14) that there exists $C > 0$ such that

$$\int_0^{2\pi} |u(T, x)|^2 \, dx \leqslant C \int_0^T \int_0^{2\pi} \chi(x) \left| (e^{it\partial_x^2} u_0)(x) \right|^2 \, dx \, dt.$$

(c) Show that

$$u(t) = e^{it\partial_x^2}u_0 + z(t) \quad \text{with} \quad z(t) = -\int_0^t e^{i(t-s)\partial_x^2}\chi u(s)\,ds$$

and then infer that

$$\int_0^{2\pi} |u(T,x)|^2\,dx \leqslant 2C \int_0^T \int_0^{2\pi} \chi(x)\,|u(t,x)|^2\,dx\,dt$$

$$+ 2C \int_0^T \|z(\tau)\|_{L^2(0,2\pi)}^2\,d\tau.$$

(d) Using the definition of z, show that

$$\int_0^T \|z(\tau)\|_{L^2(0,2\pi)}^2\,d\tau \leqslant C_T \int_0^T \int_0^{2\pi} \chi(x)\,|u(t,x)|^2\,dx\,dt,$$

and then deduce that there exists $0 < c_T < 1$ such that

$$\int_0^T \int_0^{2\pi} \chi(x)\,|u(t,x)|^2\,dx\,dt \geqslant c_T \int_0^{2\pi} |u_0(x)|^2\,dx.$$

(e) Recall that $E = E(t)$ is defined by (4.13). Let $T \geqslant 2\pi$. Prove that there exists $0 < c_T < 1$ such that
$$E(T) \leqslant (1 - c_T)E(0).$$

(f) Conclude that there exist $A, B > 0$ such that, for all $t \geqslant 0$ we have
$$E(t) \leqslant Ae^{-Bt}E(0).$$

Exercise 4.3 Prove the following extension, due to Lévy, of Wiener's theorem: if F is analytic on the range of $f(x)$, then the Fourier series of the function $x \mapsto F(f(x))$ is absolutely convergent. (Hint: consider

$$\sum_{n=0}^{\infty} \frac{F^{(n)}(g)}{n!}(g - f)^n.$$

instead of (4.10)).

Chapter 5
Fourier Transform

5.1 Introduction

In this chapter, we will study a decomposition analogous to the decomposition in Fourier series, but without making any periodicity assumption. Here too, the goal is to write a function as a sum of oscillatory exponentials. Recall that an oscillatory exponential is by definition a function of the form $x \mapsto \exp(ix \cdot \xi)$ with $\xi \in \mathbb{R}^n$. The difference with Fourier series is that this sum will be an integral over $\mathbb{R}^n$ instead of a sum indexed by $k \in \mathbb{Z}^n$.

The decomposition in Fourier series of a periodic function is well understood: it is the decomposition of an element of a Hilbert space on a Hilbert basis. On the other hand, the decomposition of a non-periodic function as a sum (in the sense of integrals) of oscillatory exponentials is less intuitive. To understand how we obtain this decomposition, we will start from the decomposition in Fourier series for functions that are $2T$-periodic with respect to each variable, then let T tend to $+\infty$. The heuristic idea is to see a function defined on $\mathbb{R}^n$ as a periodic function with period $+\infty$ with respect to each variable.

Consider a function f that is of class C^∞ and has compact support. For T large enough, the support of f is included in $Q_T = (-T, T)^n$. We will calculate the Fourier decomposition of f in $L^2(Q_T)$ and let T tend to $+\infty$. For this, we introduce the scalar product

$$(f, g) = \int_{Q_T} f(x)\overline{g(x)} \, dx$$

on $L^2(Q_T)$ and set

$$e_k(x) = (2T)^{-n/2} \exp(i\pi k \cdot x/T)$$

where $k \in \mathbb{Z}^n$. These functions are $2T$-periodic with respect to each variable and we have $(e_k, e_\ell) = \delta_k^\ell$. The Fourier coefficients of f are given by

$$\widehat{f_k} = (u, e_k) = \frac{1}{(2T)^{n/2}} \int_{Q_T} f(x) \exp\left(-\frac{i\pi k \cdot x}{T}\right) dx.$$

© The Author(s), under exclusive license to Springer Nature Switzerland AG 2024
T. Alazard, *Analysis and Partial Differential Equations*, Universitext,
https://doi.org/10.1007/978-3-031-70909-8_5

Fix $x \in \mathbb{R}^n$. As $f \in C_0^\infty(Q_T)$, we have

$$f(x) = \sum_{k \in \mathbb{Z}^n} \widehat{f}_k e_k(x)$$

(with normal convergence and therefore pointwise convergence). We can therefore write that

$$f(x) = \sum_{k \in \mathbb{Z}^n} \widehat{f}_k e_k(x)$$

$$= \sum_{k \in \mathbb{Z}^n} \frac{1}{(2T)^n} \left(\int_{Q_T} f(y) \exp\left(-\frac{i\pi k \cdot y}{T}\right) dy \right) \exp\left(\frac{i\pi k \cdot x}{T}\right).$$

As the support of f is included in Q_T, we observe that

$$\frac{1}{2^n} \left(\int_{Q_T} f(y) \exp\left(-\frac{i\pi k \cdot y}{T}\right) dy \right) \exp\left(\frac{i\pi k \cdot x}{T}\right) = F(k/T)$$

with

$$F(\xi) := \frac{1}{2^n} \exp(i\pi \xi \cdot x) \int_{\mathbb{R}^n} f(y) \exp(-i\pi \xi \cdot y) \, dy.$$

If we set $h = 1/T$, then $f(x)$ is equal to $\sum_{k \in \mathbb{Z}^n} h^n F(kh)$. When T tends to $+\infty$, the step h tends to 0 and this sum is a Riemann sum, which formally converges[1] towards $\int_{\mathbb{R}^n} F(\xi) \, d\xi$. We find that

$$f(x) = \frac{1}{2^n} \int_{\mathbb{R}^n} e^{i\pi x \cdot \xi} \left(\int_{\mathbb{R}^n} e^{-i\pi y \cdot \xi} f(y) \, dy \right) d\xi.$$

We prefer to write the previous relation in the form

$$f(x) = \frac{1}{(2\pi)^n} \int_{\mathbb{R}^n} e^{ix \cdot \xi} \left(\int_{\mathbb{R}^n} e^{-iy \cdot \xi} f(y) \, dy \right) d\xi. \tag{5.1}$$

This formula corresponds to a frequency description of the function f. In the physical literature, ξ is called the wave vector and $|\xi| = \sqrt{\xi_1^2 + \cdots + \xi_n^2}$ is the frequency.

Definition 5.1 Let $f \in L^1(\mathbb{R}^n)$. We call the Fourier transform of f the function, denoted $\widehat{f}$ or $\mathcal{F}(f)$, defined for all $\xi \in \mathbb{R}^n$ by

$$\widehat{f}(\xi) = \int_{\mathbb{R}^n} e^{-ix \cdot \xi} f(x) \, dx. \tag{5.2}$$

---

[1] As we will see later, it is easy to show that if f is of class C^∞ with compact support on $\mathbb{R}^n$, then the function F is integrable over $\mathbb{R}^n$. This allows us to justify the limit when T tends to $+\infty$.

The assumption that f is an integrable function is the minimal assumption for the formula (5.2) to make sense for the Lebesgue integral. That is why we start by defining the Fourier transform on $L^1(\mathbb{R}^n)$. But we will see that it is natural to work with other function spaces. There are several reasons for this. The first is that we want to give meaning to the formula (5.1), called the Fourier inversion formula,[2] which reads as

$$f(x) = \frac{1}{(2\pi)^n} \int_{\mathbb{R}^n} e^{ix\cdot\xi}\, \widehat{f}(\xi)\, \mathrm{d}\xi.$$

In another direction, we will seek to extend the Fourier transform to spaces different from $L^1(\mathbb{R}^n)$. We will see two results in this direction. As with Fourier series, an essential result is that the Fourier transform preserves the L^2 norm (up to a constant depending on π). This allows us to extend the definition of the Fourier transform from $L^1(\mathbb{R}^n) \cap L^2(\mathbb{R}^n)$ to $L^2(\mathbb{R}^n)$.

We will also see how to define the Fourier transform on a much larger space, the space of tempered distributions, which contains all the Lebesgue spaces $L^p(\mathbb{R}^n)$ as well as the Lebesgue spaces $L^p_{\text{per}}(\mathbb{R}^n)$ of periodic functions, and this for any p such that $1 \leqslant p \leqslant \infty$. In particular, this extended Fourier transform to the space of tempered distributions also contains the theory of Fourier series.

Let us add that the space of tempered distributions also contains many other useful spaces in the theory of partial differential equations. We will study the case of Hölder spaces, and introduce the Littlewood–Paley decomposition to provide a characterization of these spaces, which are very useful.

5.2 The Schwartz space

To construct a Fourier transform on a space that is as large as possible, we will use a duality argument.[3] The idea is to find a space, as small as possible, such that the Fourier transform is an isomorphism of this space into itself. This principle is very simple but we will see that its implementation is subtle. Already, concerning the choice of a smallest possible function space, the following proposition shows that we cannot use the space we spontaneously think of (i.e., the space $C_0^\infty(\mathbb{R}^n)$ of functions C^∞ with compact support).

Proposition 5.2 *There does not exist a non-zero function $f \in L^1(\mathbb{R})$ with compact support whose Fourier transform also has compact support.*

---

[2] Note that Cauchy and Poisson would have rediscovered Fourier's integral formula (which appears in Fourier's memoir of 1822, which earned him a prize for his answer, given in 1812, to a question posed by the Academy in 1811). This is explained in the article by Annaratone [5].

[3] The principle is as follows: if T is a continuous linear map from E to E then T^* is continuous from E' to E'. Moreover, if $E \subset L^1(\mathbb{R}^n)'$, then $L^1(\mathbb{R}^n) \subset E'$. Therefore, to extend the definition of the Fourier transform to a space larger than $L^1(\mathbb{R}^n)$, we will seek to define it as the adjoint of an isomorphism of a space E included in $L^1(\mathbb{R}^n)$. Caution: this only corresponds very approximately to what we are going to do. Indeed, we will not be able to limit ourselves to working within the framework of Banach spaces. We will have to work within the framework of Fréchet spaces.

Proof Let $f \in L^1(\mathbb{R})$ have compact support. Then we can define $F : \mathbb{C} \to \mathbb{C}$ by

$$F(z) = \int_{\mathbb{R}} e^{-ixz} f(x) \, dx.$$

Note that $F(\xi) = \widehat{f}(\xi)$ for all ξ in $\mathbb{R}$. In particular, F vanishes on an interval. Since F is an entire function (holomorphic on $\mathbb{C}$), we obtain that $F = 0$ because a non-zero entire function can only vanish on a discrete set.   □

Instead of working with functions of class C^∞ with compact support, we will work with functions of class C^∞ that decay rapidly at infinity, in the sense of the definition below, due to Laurent Schwartz.[4]

Definition 5.3

1. We say that a function f decays rapidly if the product of f by any polynomial is a bounded function.
2. We say that a function $f \in C^\infty(\mathbb{R}^n)$ belongs to the Schwartz space $\mathcal{S}(\mathbb{R}^n)$ if f and all its derivatives decay rapidly. It is equivalent to say that the following quantities

$$\mathcal{N}_p(f) = \sum_{|\alpha| \leqslant p, |\beta| \leqslant p} \left\| x^\alpha \partial_x^\beta f \right\|_{L^\infty}$$

are finite for all p.

Remark 5.4 Note that

$$C_0^\infty(\mathbb{R}^n) \subset \mathcal{S}(\mathbb{R}^n).$$

The fundamental example of a function from $\mathcal{S}(\mathbb{R}^n)$ that is not an element of $C_0^\infty(\mathbb{R}^n)$ is the Gaussian $x \mapsto \exp(-|x|^2)$. This function plays a special role in the study of the Fourier transform. More generally, for any complex number z with real part $\mathrm{Re}\, z > 0$, the function $\exp(-z\,|x|^2)$ belongs to $\mathcal{S}(\mathbb{R}^n)$.

Let us note that $\mathcal{N}_p$ is a norm on $\mathcal{S}(\mathbb{R}^n)$ for every integer p. However, if we consider $\mathcal{S}(\mathbb{R}^n)$ as a normed space for this norm, then we do not get a Banach space (a Cauchy sequence for this norm does not generally converge to a C^∞ element). The correct topological notion is that of a topological vector space equipped with a family of semi-norms. In this case, as we have seen in Chapter 1 we obtain a metrizable topology. It is easily verified that $\mathcal{S}(\mathbb{R}^n)$ is complete and we can then state the following result.

Proposition 5.5 *The Schwartz space is a Fréchet space for the topology induced by the family of semi-norms* $\{\mathcal{N}_p\}_{p \in \mathbb{N}}$.

The following proposition contains several simple, very useful properties.

---

[4] The first Frenchman to receive the Fields Medal, in 1950, for introducing the theory of distributions [170], which generalize functions and measures. This theory makes it possible (in a certain sense) to differentiate functions that, in the usual sense, are not differentiable. We will study a part of this theory in Chapter 7.

Proposition 5.6 *Suppose that f belongs to the Schwartz space $\mathcal{S}(\mathbb{R}^n)$. Then,*

1. *for all multi-indices α and β in $\mathbb{N}^n$, we have $x^\alpha \partial_x^\beta f \in \mathcal{S}(\mathbb{R}^n)$ (and the map $f \mapsto x^\alpha \partial_x^\beta f$ is continuous from $\mathcal{S}(\mathbb{R}^n)$ to $\mathcal{S}(\mathbb{R}^n)$);*
2. *the product of two elements of $\mathcal{S}(\mathbb{R}^n)$ belongs to $\mathcal{S}(\mathbb{R}^n)$ (and the product map is continuous from $\mathcal{S}(\mathbb{R}^n) \times \mathcal{S}(\mathbb{R}^n)$ to $\mathcal{S}(\mathbb{R}^n)$);*
3. *for all $p \in [1, +\infty]$, we have $f \in L^p(\mathbb{R}^n)$ (and the embedding from $\mathcal{S}(\mathbb{R}^n)$ to $L^p(\mathbb{R}^n)$ is continuous);*
4. *$C_0^\infty(\mathbb{R}^n)$ is dense in $\mathcal{S}(\mathbb{R}^n)$;*
5. *the Fourier transform $\widehat{f}$ belongs to $C^1(\mathbb{R}^n)$ and, for all j such that $1 \leqslant j \leqslant n$ and all $\xi \in \mathbb{R}^n$,*

$$\partial_{\xi_j} \widehat{f}(\xi) = \mathcal{F}\left((-ix_j)f\right);$$

6. *for all j such that $1 \leqslant j \leqslant n$ and all ξ in $\mathbb{R}^n$,*

$$\xi_j \widehat{f}(\xi) = -i\mathcal{F}\left(\partial_{x_j} f\right)(\xi).$$

Proof The first two points are immediate consequences of the definition of $\mathcal{S}(\mathbb{R}^n)$. To show (3), we start by observing that

$$
\begin{aligned}
\|f\|_{L^1} &= \int |f(x)|\, dx \\
&\leqslant \sup\left\{(1+|x|)^{n+1}|f(x)|\right\} \int \frac{dx}{(1+|x|)^{n+1}} \leqslant C\mathcal{N}_{n+1}(f).
\end{aligned}
\tag{5.3}
$$

Next, we observe that $f \in L^\infty(\mathbb{R}^n)$ (direct) and we conclude that $f \in L^p(\mathbb{R}^n)$ for all $p \in [1, +\infty]$ because $L^1(\mathbb{R}^n) \cap L^\infty(\mathbb{R}^n)$ is included in $L^p(\mathbb{R}^n)$.

To prove point (4), we consider a function $\chi \in C_0^\infty(\mathbb{R}^n)$ and show that, for any function f from $\mathcal{S}(\mathbb{R}^n)$, the sequence $\chi(\cdot/k)f$ converges to f in $\mathcal{S}(\mathbb{R}^n)$ as k tends to $+\infty$. For this, it suffices to verify that, for all $p \in \mathbb{N}$, the semi-norms $\mathcal{N}_p(f - \chi(\cdot/k)f)$ tend to 0. This calculation is left as an exercise.

To prove point (5), it suffices to observe that the hypotheses of the Lebesgue differentiation theorem are satisfied.

Finally, point (6) is obtained by writing that

$$\xi_j e^{-ix\cdot\xi} = i\partial_{x_j} e^{-ix\cdot\xi},$$

and then integrating by parts:

$$
\begin{aligned}
\xi_j \widehat{f}(\xi) &= \int \left(i\partial_{x_j} e^{-ix\cdot\xi}\right) f(x)\, dx = -i \int e^{-ix\cdot\xi} \partial_{x_j} f(x)\, dx \\
&= -i\mathcal{F}\left(\partial_{x_j} f\right)(\xi).
\end{aligned}
$$

This manipulation is justified because f is rapidly decreasing (we can then perform integration by parts on a ball $B(0, R)$ and then let R tend to $+\infty$). $\qquad\square$

Proposition 5.7 *The Schwartz space $S(\mathbb{R}^n)$ is dense in $L^p(\mathbb{R}^n)$ for all $p \in [1, +\infty)$.*

Proof This is a consequence of the inclusion $C_0^\infty(\mathbb{R}^n) \subset S(\mathbb{R}^n)$ and Corollary 6.9, which will be proved in the chapter on convolution. $\square$

The following proposition shows why $S(\mathbb{R}^n)$ is the right space to study the Fourier transform.

Proposition 5.8 *The Fourier transform maps $S(\mathbb{R}^n)$ to itself, and there exist constants C_p such that, for all f in $S(\mathbb{R}^n)$,*

$$N_p(\widehat{f}) \leqslant C_p N_{p+n+1}(f). \tag{5.4}$$

In particular, the Fourier transform is continuous from $S(\mathbb{R}^n)$ to $S(\mathbb{R}^n)$.

Proof Let $f \in S(\mathbb{R}^n)$. Then we can use the previous proposition to obtain

$$\left|\xi^\alpha \partial_\xi^\beta \widehat{f}(\xi)\right| = \left|\mathcal{F}\{\partial_x^\alpha (x^\beta f(x))\}\right|.$$

Suppose that $|\alpha| \leqslant p$ and $|\beta| \leqslant p$. Using the inequality $\|\widehat{u}\|_{L^\infty} \leqslant \|u\|_{L^1}$ and Leibniz's formula, we get

$$\left|\xi^\alpha \partial_\xi^\beta \widehat{f}(\xi)\right| \leqslant \left\|\partial_x^\alpha (x^\beta f)\right\|_{L^1} \leqslant K \sum_{|\alpha'| \leqslant p, |\beta'| \leqslant p} \left\|x^{\beta'} \partial_x^{\alpha'} f\right\|_{L^1}.$$

We deduce the desired inequality by applying (5.3). $\square$

We have said that Gaussian functions play an important role in the study of the Fourier transform. This is due to the following result, which states that the Fourier transform of a Gaussian function is a Gaussian function.

Proposition 5.9 *For all $a > 0$ and any dimension $n \geqslant 1$,*

$$\mathcal{F}(e^{-a|x|^2}) = \left(\frac{\pi}{a}\right)^{n/2} e^{-|\xi|^2/4a}.$$

Proof Let us start with the one-dimensional case, with $a = 1$. Let $f(x) = e^{-|x|^2}$. The Fourier transform of f, denoted $\mathcal{F}(f)(\xi)$, is a smooth function that satisfies

$$(\mathcal{F}f)'(\xi) = \int_{\mathbb{R}} (-ix)e^{-ix\xi}e^{-x^2}\,dx = \frac{i}{2}\int_{\mathbb{R}} e^{-ix\xi}\partial_x e^{-x^2}\,dx$$

$$= \frac{-i}{2}\int_{\mathbb{R}} (-i\xi)e^{-ix\xi}e^{-x^2}\,dx,$$

so $(\mathcal{F}f)'(\xi) = -\frac{1}{2}\xi(\mathcal{F}f)(\xi)$. Using $\int_{\mathbb{R}} e^{-x^2}\,dx = \sqrt{\pi}$, it follows that

$$(\mathcal{F}f)(\xi) = e^{-\xi^2/4}(\mathcal{F}f)(0) = \sqrt{\pi}\,e^{-\xi^2/4}.$$

The general result is then deduced from simple manipulations: if $f \in L^1(\mathbb{R}^n)$ then the Fourier transform of $f(x/\lambda)$ is $|\lambda|^n \widehat{f}(\lambda\xi)$. Moreover, the Fourier transform of $f_1(x_1) \cdots f_n(x_n)$ is $\widehat{f_1}(\xi_1) \cdots \widehat{f_n}(\xi_n)$. $\qquad\qquad\square$

We are then able to prove the following fundamental result.

Theorem 5.10 (Fourier Inversion) *If u belongs to $S(\mathbb{R}^n)$, then, for all x in $\mathbb{R}^n$,*

$$u(x) = \frac{1}{(2\pi)^n} \int e^{ix \cdot \xi} \widehat{u}(\xi)\, d\xi. \qquad (5.5)$$

Remark 5.11 We have seen that if $u \in S(\mathbb{R}^n)$ then $\widehat{u} \in S(\mathbb{R}^n)$. We have also seen that $S(\mathbb{R}^n) \subset L^1(\mathbb{R}^n)$ and therefore the function $\xi \mapsto e^{ix \cdot \xi} \widehat{u}(\xi)$ is indeed integrable. The previous formula makes sense for all $x \in \mathbb{R}^n$.

Proof Given $\varepsilon > 0$, we introduce

$$u_\varepsilon(x) = \frac{1}{(2\pi)^n} \int e^{ix \cdot \xi} \widehat{u}(\xi) e^{-\frac{1}{2}\varepsilon^2 |\xi|^2}\, d\xi.$$

Using the previous lemma we calculate (only manipulating convergent integrals)

$$\begin{aligned}
u_\varepsilon(x) &= \frac{1}{(2\pi)^n} \iint e^{i(x-y) \cdot \xi} u(y) e^{-\frac{1}{2}\varepsilon^2 |\xi|^2}\, dy\, d\xi \\
&= \frac{1}{(2\pi)^{n/2}} \int u(y) e^{-\frac{1}{2\varepsilon^2} |x-y|^2} \varepsilon^{-n}\, dy \\
&= \frac{1}{(2\pi)^{n/2}} \int \big(u(x + \varepsilon y) - u(x)\big) e^{-\frac{1}{2}|y|^2}\, dy + u(x).
\end{aligned}$$

Since

$$|u(x + \varepsilon y) - u(x)| \leqslant \varepsilon\, |y|\, \|\nabla u\|_{L^\infty},$$

we obtain the desired result by letting ε tend to 0. $\qquad\qquad\square$

Corollary 5.12 *The Fourier transform $\mathcal{F}$ is an isomorphism of $S(\mathbb{R}^n)$ onto itself, and*

$$\mathcal{F}^{-1} f = \frac{1}{(2\pi)^n} \overline{\mathcal{F}(\bar{f})}. \qquad (5.6)$$

Proof Let $f \in S(\mathbb{R}^n)$. The formula (5.5) implies that

$$f(x) = \frac{1}{(2\pi)^n} \int e^{ix \cdot \xi} \widehat{f}(\xi)\, d\xi = \frac{1}{(2\pi)^n} \overline{\int e^{-ix \cdot \xi} \overline{\widehat{f}(\xi)}\, d\xi},$$

equivalent to the desired result (5.6).

Since we have already seen in Proposition 5.8 that the Fourier transform is continuous from $S(\mathbb{R}^n)$ to $S(\mathbb{R}^n)$, the identity (5.5) implies that the inverse $\mathcal{F}^{-1}$ is continuous from $S(\mathbb{R}^n)$ to $S(\mathbb{R}^n)$ and hence the Fourier transform is an isomorphism of $S(\mathbb{R}^n)$. $\qquad\qquad\square$

Theorem 5.13 (Plancherel Identity) *If f and g belong to $S(\mathbb{R}^n)$, then*

$$\int_{\mathbb{R}^n} f(x)\overline{g}(x)\,dx = \frac{1}{(2\pi)^n} \int_{\mathbb{R}^n} \widehat{f}(\xi)\overline{\widehat{g}(\xi)}\,d\xi. \tag{5.7}$$

In particular, for any element f of $S(\mathbb{R}^n)$, we have

$$\|f\|_{L^2}^2 = \frac{1}{(2\pi)^n}\|\widehat{f}\|_{L^2}^2. \tag{5.8}$$

Proof We will start by showing that, if $\varphi, \psi \in S(\mathbb{R}^n)$, then

$$\int \widehat{\varphi}(x)\psi(x)\,dx = \int \varphi(y)\widehat{\psi}(y)\,dy. \tag{5.9}$$

As φ and ψ are rapidly decreasing, we can apply Fubini's theorem to obtain

$$\int \widehat{\varphi}\psi\,dx = \int \left(\int e^{-iy\cdot x}\varphi(y)\,dy\right)\psi(x)\,dx$$

$$= \int \left(\int e^{-iy\cdot x}\psi(x)\,dx\right)\varphi(y)\,dy = \int \varphi\widehat{\psi}\,dy.$$

Now consider two functions f and g from the Schwartz space $S(\mathbb{R}^n)$. We apply the identity (5.9) with $\varphi = f$ and $\psi = \mathcal{F}^{-1}(\overline{g})$. Then

$$\int f\overline{g} = \int \varphi\widehat{\psi} = \int \widehat{\varphi}\psi = \int \widehat{f}\mathcal{F}^{-1}\overline{g}.$$

Then we obtain (5.7) by verifying (using the Fourier inversion formula (5.6)) that

$$(\mathcal{F}^{-1}\overline{g})(\xi) = (2\pi)^{-n}\int e^{iy\xi}\overline{g}(y)\,dy = (2\pi)^{-n}\overline{\int e^{-iy\xi}g(y)\,dy} = (2\pi)^{-n}\overline{\widehat{g}(\xi)}.$$

The Plancherel identity (5.8) is then an obvious corollary of (5.7).   □

Corollary 5.14 *Let $f \in L^1(\mathbb{R}^n)$. If $\widehat{f} = 0$, then $f = 0$.*

Proof The proof uses the previous theorem and the fact that the Schwartz space is dense in $L^1(\mathbb{R}^n)$ (see Proposition 5.7). Let $g \in S(\mathbb{R}^n)$ and let $(f_n)_{n\in\mathbb{N}}$ be a sequence of functions from the Schwartz space that converges to f in $L^1(\mathbb{R}^n)$. We have directly

$$\left|\int (f_n - f)(x)\overline{g}(x)\,dx\right| \leq \|f_n - f\|_{L^1}\|g\|_{L^\infty} \xrightarrow[n\to+\infty]{} 0.$$

Furthermore, as $\widehat{f} = 0$, we have

$$\left| \int \widehat{f_n}(\xi)\overline{\widehat{g}(\xi)} \, \mathrm{d}\xi \right| = \left| \int (\widehat{f_n} - \widehat{f})(\xi)\overline{\widehat{g}(\xi)} \, \mathrm{d}\xi \right|$$

$$\leqslant \left\| \widehat{f_n} - \widehat{f} \right\|_{L^\infty} \left\| \widehat{g} \right\|_{L^1} \leqslant \left\| f_n - f \right\|_{L^1} \left\| \widehat{g} \right\|_{L^1} \xrightarrow[n \to +\infty]{} 0,$$

where we used the fact that $\widehat{g} \in S(\mathbb{R}^n) \subset L^1(\mathbb{R}^n)$ for all $g \in S(\mathbb{R}^n)$ (see Proposition 5.8). It follows from Theorem 5.13 that $\int f(x)\overline{g}(x) \, \mathrm{d}x = 0$ for all $g \in S(\mathbb{R}^n)$, hence $f = 0$ (this is a classical result, see Corollary 6.10).   □

5.3 Tempered Distributions

5.3.1 Definition of Tempered Distributions

Definition 5.15 The space of tempered distributions, denoted $S'(\mathbb{R}^n)$, is the topological dual of $S(\mathbb{R}^n)$.

Notation 5.16 Let $u \in S'(\mathbb{R}^n)$ and $f \in S(\mathbb{R}^n)$. We will denote by $\langle u, f \rangle_{S' \times S}$ (rather than $u(f)$) the complex number obtained by applying the linear form u to f.

It follows that a linear map $T : S(\mathbb{R}^n) \to \mathbb{C}$ belongs to $S'(\mathbb{R}^n)$ if and only if there exist $p \in \mathbb{N}$ and $C > 0$ such that

$$\forall f \in S(\mathbb{R}^n), \quad |\langle T, f \rangle_{S' \times S}| \leqslant C N_p(f) = C \sum_{|\alpha| \leqslant p, |\beta| \leqslant p} \left\| x^\alpha \partial_x^\beta f \right\|_{L^\infty}.$$

Let us show as a first example that every function $u \in L^\infty(\mathbb{R}^n)$ can define a tempered distribution. We define a linear form $U : S(\mathbb{R}^n) \to \mathbb{C}$ by

$$\langle U, v \rangle_{S' \times S} = \int_{\mathbb{R}^n} u(x)v(x) \, \mathrm{d}x. \tag{5.10}$$

We verify that U is indeed continuous from $S(\mathbb{R}^n)$ into $\mathbb{C}$ according to the following estimate

$$|\langle U, v \rangle_{S' \times S}| \leqslant \|u\|_{L^\infty} \|v\|_{L^1}$$

$$\leqslant \|u\|_{L^\infty} \left(\int \frac{\mathrm{d}x}{(1 + |x|)^{n+1}} \right) \sup_{x \in \mathbb{R}^n} |(1 + |x|)^{n+1} v(x)|,$$

which implies

$$|\langle U, v \rangle_{S' \times S}| \leqslant C \|u\|_{L^\infty} N_{n+1}(v).$$

By reasoning similarly, we show that the formula (5.10) defines a tempered distribution for every function $u \in L^p(\mathbb{R}^n)$ with $p \in [1, +\infty]$. This process allows us to embed Lebesgue spaces into $S'(\mathbb{R}^n)$. In fact, we can embed many other spaces, and we will see later the fundamental example of Sobolev spaces.

Definition 5.17 Given a function space X and a tempered distribution $U \in S'(\mathbb{R}^n)$, we will say that U belongs to X if there exists a $u \in X$ such that

$$\forall v \in S(\mathbb{R}^n), \quad \langle U, v \rangle_{S' \times S} = \int_{\mathbb{R}^n} u(x)v(x)\,dx.$$

5.3.2 Extension to Tempered Distributions

We have seen in the previous section that we can embed all Lebesgue spaces into $S'(\mathbb{R}^n)$. We can also embed Hölder spaces and Sobolev spaces (both will be defined later). We should think of the space of tempered distributions as the largest space in which we want to work.[5] It is then natural to want to extend the definition of important operators in analysis to $S'(\mathbb{R}^n)$. We will see that this can be done very simply.

Definition 5.18 Consider a linear map $A \colon S(\mathbb{R}^n) \to S(\mathbb{R}^n)$ which is assumed to be continuous. We will say that A admits a continuous adjoint on $S(\mathbb{R}^n)$ if there exists a continuous linear map $A^* \colon S(\mathbb{R}^n) \to S(\mathbb{R}^n)$ such that, for all $(u, v) \in S(\mathbb{R}^n) \times S(\mathbb{R}^n)$,

$$(Au, v) = (u, A^*v) \quad \text{where} \quad (f, g) = \int f(x)\overline{g(x)}\,dx.$$

Example 5.19

1. Consider the case of the Fourier transform $A = \mathcal{F}$ (which is indeed a continuous map from $S(\mathbb{R}^n)$ to $S(\mathbb{R}^n)$ according to Proposition 5.8). We claim that in this case, the map A^* is given by $A^*g = \mathcal{F}(\overline{g})$. This will imply that $\mathcal{F}$ admits a continuous adjoint on $S(\mathbb{R}^n)$.
   To see this, we recall the identity (5.9), which states that, for all $\varphi, \psi \in S(\mathbb{R}^n)$, we have
   $$\int \widehat{\varphi}(x)\psi(x)\,dx = \int \varphi(y)\widehat{\psi}(y)\,dy.$$
   By applying this with $\varphi = f$ and $\psi = \overline{g}$, we deduce that
   $$(Af, g) = (\widehat{f}, g) = \int \widehat{f}(x)\overline{g(x)}\,dx = \int f(y)\overline{\widehat{g}}(y)\,dy = (f, A^*g).$$

2. Let j be such that $1 \leq j \leq n$. If $A = \partial_{x_j}$, then A is indeed continuous from $S(\mathbb{R}^n)$ to $S(\mathbb{R}^n)$ because $N_p(Au) \leq N_{p+1}(u)$ and we have, by integrating by parts, $(Au, v) = (u, A^*v)$ with $A^* = -\partial_{x_j}$.

---

[5] There are larger spaces, like the space of distributions, but we will not study these spaces in this book.

3. Let us denote by $C_b^\infty(\mathbb{R}^n)$ the space of functions f of class C^∞ such that f and all its derivatives are bounded. If $c \in C_b^\infty(\mathbb{R}^n)$, then the operator A_c defined by $A_c(f)(x) = c(x)f(x)$ satisfies this property. Then $(A_c)^* = A_{\overline{c}}$.
4. If A and B satisfy this property then so does $A \circ B$, with $(A \circ B)^* = B^* \circ A^*$. We deduce from the two previous points that any differential operator A of the form $A(f)(x) = \sum_{|\alpha| \leqslant m} c_\alpha(x)\partial_x^\alpha f(x)$ with $c_\alpha \in C_b^\infty(\mathbb{R}^n)$ satisfies this property.
5. We will see another example later that generalizes the notion of differential operator (see the part dedicated to pseudo-differential operators).

Consider a continuous linear map $A \colon S(\mathbb{R}^n) \to S(\mathbb{R}^n)$ that admits a continuous adjoint A^* on $S(\mathbb{R}^n)$. Our goal is to show that there exists a continuous linear $\widetilde{A} \colon S'(\mathbb{R}^n) \to S'(\mathbb{R}^n)$ that extends the definition of A. For this we define

$$\forall u \in S'(\mathbb{R}^n), \ \forall v \in S(\mathbb{R}^n), \quad \langle \widetilde{A}u, v \rangle_{S' \times S} = \langle u, \overline{A^* \overline{v}} \rangle_{S' \times S}.$$

Let us prove that the operator thus constructed extends the definition of A.

Proposition 5.20 *Consider the map* $\mathcal{T} \colon u \in S(\mathbb{R}^n) \mapsto \mathcal{T}_u \in S'(\mathbb{R}^n)$ *defined by*

$$\mathcal{T}_u(v) = (u, \overline{v}) = \int u(x)v(x)\,\mathrm{d}x.$$

Then this map is well defined, linear, continuous and injective and moreover

$$\forall u \in S(\mathbb{R}^n), \quad \widetilde{A}\mathcal{T}_u = \mathcal{T}_{Au}.$$

Remark 5.21 The first part of the result means that $\mathcal{T}$ is injective from $S(\mathbb{R}^n)$ into $S'(\mathbb{R}^n)$; the second part means that $\widetilde{A}$ coincides with A on $S(\mathbb{R}^n)$.

Proof For all $u, v \in S(\mathbb{R}^n)$ we have already seen that

$$|\langle \mathcal{T}_u, v \rangle_{S' \times S}| \leqslant \|u\|_{L^\infty} \|v\|_{L^1} \leqslant C N_0(u) N_{n+1}(v),$$

which shows that $\mathcal{T}_u$ belongs to $S'(\mathbb{R}^n)$ and that $u \mapsto \mathcal{T}_u$ is continuous from $S(\mathbb{R}^n)$ into $S'(\mathbb{R}^n)$. Moreover, $\mathcal{T}$ is injective because $\mathcal{T}_{u_1} = \mathcal{T}_{u_2}$ implies $\mathcal{T}_{u_1-u_2}(\overline{u_1 - u_2}) = 0$, therefore $\|u_1 - u_2\|_{L^2} = 0$, from which $u_1 = u_2$.

With the previous definitions, for all u, v in $S(\mathbb{R}^n)$, we have

$$\langle \widetilde{A}\mathcal{T}_u, v \rangle_{S' \times S} = \langle \mathcal{T}_u, \overline{A^* \overline{v}} \rangle_{S' \times S} = (u, A^* \overline{v}) = (Au, \overline{v}) = \langle \mathcal{T}_{Au}, v \rangle_{S' \times S}.$$

This proves that $\widetilde{A}\mathcal{T}_u = \mathcal{T}_{Au}$. $\qquad\qquad\square$

In the following, we will simply denote by A instead of $\widetilde{A}$ the operator extended to $S'(\mathbb{R}^n)$. Using this construction, we can thus define the partial derivative ∂_{x_j} of any tempered distribution! By definition, we have

$$\forall u \in S'(\mathbb{R}^n), \ \forall v \in S(\mathbb{R}^n), \quad \langle \partial_{x_j} u, v \rangle_{S' \times S} = -\langle u, \partial_{x_j} v \rangle_{S' \times S}.$$

We deduce that we can define $\partial_x^\alpha u$ for all $\alpha \in \mathbb{N}^n$ and all $u \in S'(\mathbb{R}^n)$. We can therefore differentiate any distribution to any order (which is of course false for functions).

We will now apply the previous construction with the Fourier transform. Recall that (*cf.* (5.9)), for all φ, ψ in $S(\mathbb{R}^n)$, we have

$$\int \widehat{\varphi}(x)\psi(x)\,dx = \int \varphi(y)\widehat{\psi}(y)\,dy.$$

Consider a tempered distribution $u \in S'(\mathbb{R}^n)$. We can then apply the previous principle to define its Fourier transform, denoted $\mathcal{F}(u)$, by

$$\langle \mathcal{F}(u), v\rangle_{S'\times S} = \langle u, \widehat{v}\rangle_{S'\times S}.$$

We also denote by $\widehat{u}$ the Fourier transform of a tempered distribution.

Proposition 5.22 *The Fourier transform $\mathcal{F}$ is an isomorphism of $S'(\mathbb{R}^n)$ into itself (a continuous linear map with continuous inverse). Moreover, we have*

$$\mathcal{F}^{-1}u = (2\pi)^{-n}\overline{\mathcal{F}(\overline{u})}.$$

Proof This is a direct consequence of Proposition 5.8 and Corollary 5.12. □

We have already noticed that we can embed Lebesgue spaces into $S'(\mathbb{R}^n)$. In particular, we can consider the Fourier transform of a function belonging to $L^p(\mathbb{R}^n)$. The most important case in practice is that of $L^2(\mathbb{R}^n)$. In this case, we have the following result.

Proposition 5.23 *If $u \in L^2(\mathbb{R}^n)$, then $\mathcal{F}(u)$ belongs to $L^2(\mathbb{R}^n)$ and*

$$\|u\|_{L^2}^2 = \frac{1}{(2\pi)^n}\|\mathcal{F}(u)\|_{L^2}^2.$$

Remark 5.24 With the previous conventions, the fact that $\mathcal{F}(u)$ belongs to $L^2(\mathbb{R}^n)$ means that there exists a function $h \in L^2(\mathbb{R}^n)$ such that $\langle \mathcal{F}(u), v\rangle_{S'\times S} = \int_{\mathbb{R}^n} h(x)v(x)\,dx$ for all $v \in S(\mathbb{R}^n)$. Then we have $\|u\|_{L^2}^2 = \frac{1}{(2\pi)^n}\|h\|_{L^2}^2$.

Proof We use the definition of $\mathcal{F}$ on $S'(\mathbb{R}^n)$ and Theorem 5.13 to write

$$|\langle \mathcal{F}(u), v\rangle_{S'\times S}| = |\langle u, \widehat{v}\rangle_{S'\times S}| = \left|\int u\widehat{v}\,dx\right|$$

$$\leq \|u\|_{L^2}\|\widehat{v}\|_{L^2} = (2\pi)^{n/2}\|u\|_{L^2}\|v\|_{L^2}.$$

We deduce that the map $S(\mathbb{R}^n) \ni v \mapsto \langle \mathcal{F}(u), v\rangle_{S'\times S} \in \mathbb{C}$ extends to a continuous linear form on the Hilbert space $L^2(\mathbb{R}^n)$. According to the Riesz–Fréchet theorem 3.10, there exists $h \in L^2(\mathbb{R}^n)$ such that $\langle \mathcal{F}(u), v\rangle_{S'\times S} = \int h(x)v(x)\,dx$, which gives the desired result. □

Definition 5.25 We say that a function m belonging to $C^\infty(\mathbb{R}^n)$ is slowly growing if m and all its derivatives grow at most polynomially: for every multi-index $\alpha \in \mathbb{N}^n$, there exist $C_\alpha > 0$ and $N_\alpha \in \mathbb{N}$ such that $\left|\partial_\xi^\alpha m(\xi)\right| \leqslant C_\alpha(1 + |\xi|)^{N_\alpha}$ for all $\xi \in \mathbb{R}^n$.

If m is a slowly growing function and if $v \in S(\mathbb{R}^n)$, then mv also belongs to the Schwartz space $S(\mathbb{R}^n)$. We therefore verify directly that we can define an operator, denoted $m(D_x)$, on $S'(\mathbb{R}^n)$ in the following way:

$$\mathcal{F}(m(D_x)u) = m\mathcal{F}u.$$

We say that $m(D_x)$ is a Fourier multiplier.

5.4 The Littlewood–Paley Decomposition

A word of caution: this section is more difficult and requires some familiarity with the notion of convolution, which will be studied in the next chapter.

5.4.1 Dyadic Decomposition of Unity

We are going to introduce a dyadic decomposition of unity. This decomposition makes it possible to introduce a parameter (large or small) into a problem that has none. This is a simple and extremely fruitful idea.[6]

Lemma 5.26 Let $n \geqslant 1$. There exist $\psi \in C_0^\infty(\mathbb{R}^n)$ and $\varphi \in C_0^\infty(\mathbb{R}^n)$ such that the following properties are satisfied:

1. we have $0 \leqslant \psi \leqslant 1$, $0 \leqslant \varphi \leqslant 1$ and

$$\operatorname{supp}\psi \subset \{|\xi| \leqslant 1\}, \qquad \operatorname{supp}\varphi \subset \left\{\frac{3}{4} \leqslant |\xi| \leqslant 2\right\};$$

2. for all $\xi \in \mathbb{R}^n$,

$$1 = \psi(\xi) + \sum_{p=0}^{\infty} \varphi(2^{-p}\xi); \tag{5.11}$$

3. (almost orthogonality) for all $\xi \in \mathbb{R}^n$,

$$\frac{1}{3} \leqslant \psi^2(\xi) + \sum_{p=0}^{+\infty} \varphi^2(2^{-p}\xi) \leqslant 1. \tag{5.12}$$

---

[6] For an introduction to the use of this tool in recent research, we refer the reader to Bahouri [8] or Danchin [43]. There are many books that develop a systematic study of the Littlewood–Paley decomposition: Alinhac and Gérard [3], Bahouri, Danchin and Chemin [9], Coifman and Meyer [36, 138], Métivier [134], Muscalu and Schlag [145], Stein [176, 177], Tao [181] or Taylor [188, 189].

Proof Consider a function $\psi \in C_0^\infty(\mathbb{R}^n; \mathbb{R})$ satisfying $\psi(\xi) = 1$ for $|\xi| \leqslant 3/4$, and $\psi(\xi) = 0$ for $|\xi| \geqslant 1$ (the fact that such a function exists is proved in the appendix, see Proposition 17.34). Next, we define $\varphi(\xi) = \psi(\xi/2) - \psi(\xi)$ and we note that φ is supported in the annulus $\{3/4 \leqslant |\xi| \leqslant 2\}$. For all integer $N \in \mathbb{N}$ and all $\xi \in \mathbb{R}^n$, we have

$$\psi(\xi) + \sum_{p=0}^{N} \varphi(2^{-p}\xi) = \psi(2^{-N-1}\xi),$$

which immediately implies (5.11) by letting N tend to $+\infty$.

It remains to prove (5.12). For all integer $N \in \mathbb{N}$, we have

$$\psi^2(\xi) + \sum_{p=0}^{N} \varphi^2(2^{-p}\xi) \leqslant \left(\psi(\xi) + \sum_{p=0}^{N} \varphi(2^{-p}\xi)\right)^2.$$

On the other hand, we note that, for all $\xi \in \mathbb{R}^n$, there are never more than three non-zero terms in the set $\{\psi(\xi), \varphi(\xi), \dots, \varphi(2^{-p}\xi), \dots\}$. Therefore, using the elementary inequality $(a + b + c)^2 \leqslant 3(a^2 + b^2 + c^2)$, we obtain

$$\left(\psi(\xi) + \sum_{p=0}^{N} \varphi(2^{-p}\xi)\right)^2 \leqslant 3\left(\psi^2(\xi) + \sum_{p=0}^{N} \varphi^2(2^{-p}\xi)\right).$$

We then obtain (5.12) by letting N tend to $+\infty$ in the previous inequalities. $\qquad\square$

Let us define, given integers $p \geqslant -1$ and $N \geqslant 0$, the Fourier multipliers Δ_p and S_N as follows:

$$\Delta_{-1} := \psi(D_x) \quad \text{and} \quad \Delta_p := \varphi(2^{-p}D_x) \quad (p \geqslant 0),$$

$$S_N := \psi(2^{-N}D_x) = \sum_{p=-1}^{N-1} \Delta_p.$$

Then we obtain a decomposition of the identity.

Proposition 5.27 *We have* $I = \sum_{p \geqslant -1} \Delta_p$, *in the sense of distributions: for all* $u \in S'(\mathbb{R}^n)$, $\sum_p \Delta_p u$ *converges to* u *in* $S'(\mathbb{R}^n)$, *which means that* $\sum_p \langle \Delta_p u, \varphi \rangle_{S' \times S}$ *converges to* $\langle u, \varphi \rangle_{S' \times S}$ *for all* $\varphi \in S(\mathbb{R}^n)$.

Proof Let $u \in S'(\mathbb{R}^n)$ and $\theta \in S(\mathbb{R}^n)$. Let us set $u_p = \Delta_p u$. The partial sums $S_N u = \sum_{p=-1}^{N-1} u_p$ are well defined and

$$\langle \mathcal{F}(S_N u), \theta \rangle = \langle \psi(2^{-N} \cdot) \mathcal{F}(u), \theta \rangle = \langle \mathcal{F}(u), \psi(2^{-N} \cdot)\theta \rangle.$$

Since $\lim_{N \to +\infty} \psi(2^{-N} \cdot)\theta = \theta$ in $S(\mathbb{R}^n)$, we see that $\mathcal{F}(S_N u)$ converges to $\mathcal{F}(u)$ in $S'(\mathbb{R}^n)$ as N goes to $+\infty$. By continuity of $\mathcal{F}^{-1} : S'(\mathbb{R}^n) \to S'(\mathbb{R}^n)$ we indeed have $u = \sum_{p \geqslant -1} u_p$. $\qquad\square$

5.4.2 Characterization of Hölder Spaces

In this section, we will show that we can describe Hölder spaces using the Fourier transform.

Let us recall the definition of Hölder spaces.

Definition 5.28

1. Let $r \in (0, 1)$. We denote by $C^{0,r}(\mathbb{R}^n)$ the space of bounded functions u on $\mathbb{R}^n$ satisfying

$$\exists C > 0 / \forall x, y \in \mathbb{R}^n, \quad |u(x) - u(y)| \leqslant C |x - y|^r .$$

2. Let $k \in \mathbb{N}$ and $\alpha \in (0, 1]$. We denote by $C^{k,\alpha}(\mathbb{R}^n)$ the space of functions of class $C^k(\mathbb{R}^n)$ whose derivatives up to order k all belong to $C^{0,\alpha}(\mathbb{R}^n)$.
3. Let $r \in \mathbb{R}^+ \setminus \mathbb{N}$ such that $r = k + \alpha$ with $k \in \mathbb{N}$ and $\alpha \in (0, 1]$. Then we simply denote by $C^r(\mathbb{R}^n)$ the space $C^{k,\alpha}(\mathbb{R}^n)$. In particular, $C^r(\mathbb{R}^n) = C^{0,r}(\mathbb{R}^n)$ for $r \in (0, 1)$.

For $r \in (0, 1)$, the space $C^r(\mathbb{R}^n)$ is equipped with a Banach space structure by the norm

$$\|u\|_{C^r} = \|u\|_{L^\infty} + \sup_{x \neq y} \frac{|u(x) - u(y)|}{|x - y|^r}.$$

We provide below an equivalent norm.

Proposition 5.29 *Let $r \in (0, 1)$. There exists a constant $A_r > 0$ such that the following two properties hold:*

1. *If $u \in C^r(\mathbb{R}^n)$ then, for all $p \geqslant -1$,*

$$\left\| \Delta_p u \right\|_{L^\infty} \leqslant A_r \|u\|_{C^r} 2^{-pr}.$$

2. *Conversely, if, for all $p \geqslant -1$,*

$$\left\| \Delta_p u \right\|_{L^\infty} \leqslant C 2^{-pr},$$

then $u \in C^r(\mathbb{R}^n)$ and $\|u\|_{C^r} \leqslant A_r C$.

Proof 1. Consider $p \geqslant 0$. To prove the first point, we start by writing $\Delta_p u$ in integral form,

$$\Delta_p u(x) = 2^{pn} \int \mathcal{F}^{-1}(\varphi)(2^p(x - y))u(y) \, dy. \tag{5.13}$$

As φ vanishes in a neighborhood of 0, it follows that

$$\int \mathcal{F}^{-1}(z)z^k \, dz = 0,$$

for all $k \in \mathbb{N}$.

By using this identity with $k = 0$, we obtain

$$\Delta_p u(x) = 2^{pn} \int \mathcal{F}^{-1}(\varphi)(2^p(x-y))\,(u(y) - u(x))\,dy \quad \text{for} \quad p \geqslant 0.$$

Consequently, we infer that

$$\left|\Delta_p u(x)\right| \leqslant 2^{pn} \|u\|_{C^r} \int \left|\mathcal{F}^{-1}(\varphi)(2^p(x-y))\right| |y - x|^r \, dy$$

$$= 2^{-pr} \|u\|_{C^r} \int \left|\mathcal{F}^{-1}(\varphi)(z)z\right| dz.$$

The case $p = -1$ is handled similarly. We write

$$\Delta_{-1} u(x) = \int \mathcal{F}^{-1}(\psi)(x-y) u(y)\,dy,$$

then we use that $\mathcal{F}^{-1}(\psi) \in \mathcal{S}(\mathbb{R}^n)$ and therefore $\mathcal{F}^{-1}(\psi)$ belongs to $L^1(\mathbb{R}^n)$. We deduce that the norm L^∞ of $\Delta_{-1} u$ is controlled by the L^∞ norm of u.

2. Let us show the converse. We verify that $u \in L^\infty$ because $\Delta_p u \in L^\infty$ and $\sum \Delta_p u$ converges normally if $\|\Delta_p u\|_{L^\infty} \leqslant C 2^{-pr}$. It remains to estimate

$$\sup_{x \neq y,\ |x-y| \leqslant 1} \frac{|u(x) - u(y)|}{|x - y|^r}.$$

(Note that we can obviously restrict ourselves to $|x - y| \leqslant 1$.)
For this, we set, for an integer p to be determined,

$$u = S_p u + R_p u, \quad S_p u = \sum_{q=-1}^{p-1} \Delta_q u \quad \text{and} \quad R_p u = \sum_{q=p}^{+\infty} \Delta_q u.$$

By hypothesis, we have

$$\|R_p u\|_{L^\infty} \leqslant \sum_{q \geqslant p} \|\Delta_q u\|_{L^\infty} \leqslant \sum_{q \geqslant p} C 2^{-qr} = \frac{C}{1 - 2^{-r}} 2^{-pr},$$

from which we obviously deduce that $\left|R_p u(y) - R_p u(x)\right| \leqslant \frac{2C}{1-2^{-r}} 2^{-pr}$. On the other hand

$$\left|S_p u(x) - S_p u(y)\right| \leqslant |x - y| \sum_{q=-1}^{p-1} \|\nabla \Delta_q u\|_{L^\infty}.$$

Remembering the formula used at the beginning of the proof, we obtain

$$\|\nabla \Delta_q u\|_{L^\infty} \leqslant C' 2^q \|\Delta_q u\|_{L^\infty} \leqslant C' C 2^{q-qr}.$$

For $q = -1$, we prove similarly that $\|\nabla u_{-1}\|_{L^\infty} \leqslant C'' C$.

As $r < 1$, we have $1 - r > 0$ and therefore

$$\sum_{q=0}^{p-1} 2^{q-qr} \leqslant \frac{1}{2^{1-r} - 1} 2^{p(1-r)}.$$

Let us group the two estimates: there exist two constants K_1 and K_2 that depend only on r such that

$$|u(y) - u(x)| \leqslant K_1 C |x - y| 2^{p-pr} + K_2 2^{-pr}.$$

We choose p such that $2^{-1} \leqslant 2^p |x - y| \leqslant 1$ (which is possible because we assume $|x - y| \leqslant 1$). Then

$$|u(y) - u(x)| \leqslant K_3 C |x - y|^r,$$

which completes the proof.   □

5.4.3 Zygmund Spaces

We have shown (see Proposition 5.29) that if $r \in (0, 1)$, then

$$u \in C^r(\mathbb{R}^n) \iff \sup_{p \geqslant -1} 2^{pr} \|\Delta_p u\|_{L^\infty} < +\infty.$$

In fact, we have more generally

$$\forall r \in \mathbb{R}^+ \setminus \mathbb{N}, \quad u \in C^r(\mathbb{R}^n) \iff \sup_{p \geqslant -1} 2^{pr} \|\Delta_p u\|_{L^\infty} < +\infty. \tag{5.14}$$

Definition 5.30 Let r be a real number. We denote by $C_*^r(\mathbb{R}^n)$ the subspace of tempered distributions defined by

$$u \in C_*^r(\mathbb{R}^n) \iff \sup_{p \geqslant -1} 2^{pr} \|\Delta_p u\|_{L^\infty} < +\infty.$$

Remark 5.31 Note that we define these spaces for all $r \in \mathbb{R}$ and not just $r \geqslant 0$.

Thus, the previous result directly gives

$$r \in \mathbb{R}^+ \setminus \mathbb{N} \implies C_*^r(\mathbb{R}^n) = C^r(\mathbb{R}^n).$$

Moreover, for $k \in \mathbb{N}$, we can easily show that

$$\forall k \in \mathbb{N}, \quad C^k(\mathbb{R}^n) \subset W^{k,\infty}(\mathbb{R}^n) \subset C_*^k(\mathbb{R}^n).$$

However, we can show that

$$r \in \mathbb{N} \implies C_*^r(\mathbb{R}^n) \neq C^r(\mathbb{R}^n).$$

Let us give for example an elementary characterization of $C_*^1(\mathbb{R}^n)$.

Proposition 5.32 *The space $C_*^1(\mathbb{R}^n)$ is the space of bounded functions u such that*

$$\exists C > 0/\ \forall x, y \in \mathbb{R}^n, \quad |u(x + y) + u(x - y) - 2u(x)| \leqslant C\,|y|\,.$$

Proof Suppose that $u \in C_*^1(\mathbb{R}^n)$. Consider then $y \in B(0, 1)$ with $y \neq 0$. Using the dyadic decomposition of the space of frequencies and the Taylor inequality of order 2 between y and 0, we get

$$|u(x + y) + u(x - y) - 2u(x)| \leqslant C\,\|u\|_{C_*^1}\left(|y|^2 \sum_{q \leqslant N} 2^q + 4 \sum_{q > N} 2^{-q}\right),$$

where N is any integer. Choosing N as the integer part of $\left|\log_2(|y|^{-1})\right|$ (so that $2^N \approx |y|^{-1}$), we obtain

$$|u(x + y) + u(x - y) - 2u(x)| \leqslant C\,\|u\|_{C_*^1}\,|y|\,.$$

Conversely, choose a function u such that, for all y in $\mathbb{R}^n$, we have

$$|u(x + y) + u(x - y) - 2u(x)| \leqslant C\,|y|\,.$$

The fact that the function φ is radial, therefore even, implies that

$$\Delta_q u(x) = 2^{qn}\mathcal{F}^{-1}\varphi(2^q\cdot) * u(x) = 2^{qn}\int \mathcal{F}^{-1}\varphi(2^q y)u(x - y)\,dy$$

$$= 2^{qn}\int \mathcal{F}^{-1}\varphi(2^q y)u(x + y)\,dy.$$

Let us introduce $h(z) = \mathcal{F}^{-1}\varphi(z)$. As φ vanishes near the origin we deduce that the integral of h is zero, therefore

$$2^{qn}\mathcal{F}^{-1}\varphi(2^q\cdot) * u(x) = 2^{qn-1}\int \mathcal{F}^{-1}\varphi(2^q y)(u(x + y) + u(x - y) - 2u(x))\,dy.$$

As the function $z \mapsto |z|\,h(z)$ is integrable, we have

$$\left\|\Delta_q u\right\|_{L^\infty} \leqslant C2^{-q}\sup_{y \in \mathbb{R}^n}\frac{|u(x + y) + u(x - y) - 2u(x)|}{|y|},$$

which proves that u belongs to $C_*^1(\mathbb{R}^n)$. $\square$

5.5 Exercises

Exercise 5.1 Consider the functions $f(x) = e^x$ and $g(x) = e^x \cos(e^x)$. Show that $f \notin S'(\mathbb{R})$ and $g \in S'(\mathbb{R})$.

Exercise 5.2

1. Show that there exists a constant C such that, for all $\varepsilon \in (0, 1)$ and any function f belonging to the Hölder space $C^\varepsilon(\mathbb{R}^n)$, we have

$$\|f\|_{L^\infty} \leqslant \frac{C}{\varepsilon} \|f\|_{C_*^0} \log\left(e + \frac{\|f\|_{C^\varepsilon}}{\|f\|_{C_*^0}}\right).$$

   Hint: use the Littlewood–Paley decomposition and, for $N \in \mathbb{N}$ to be chosen, write

$$\|f\|_{L^\infty} \leqslant \sum_{q \leqslant N-1} \|\Delta_q f\|_{L^\infty} + \sum_{q \geqslant N} \|\Delta_q f\|_{L^\infty}.$$

2. Consider the tempered distribution

$$u = \sum_{q=0}^{\infty} e^{i 2^q x}.$$

   Show that u belongs to C_*^0 but not to L^∞.

Exercise 5.3 (Nash Inequality) Consider an integer $n \geqslant 1$ and a function u in the Schwartz space $S(\mathbb{R}^n)$.

1. Calculate the Fourier transform of ∇u and show that there exists a constant $C_1 > 0$ such that, for all $\rho > 0$, we have

$$\int_{|\xi| \geqslant \rho} |\widehat{u}(\xi)|^2 \, d\xi \leqslant \frac{C_1}{\rho^2} \int_{\mathbb{R}^n} |\widehat{\nabla u}(\xi)|^2 \, d\xi.$$

2. Show that there exists a constant $C_2 > 0$ such that, for all $\rho > 0$, we have

$$\int_{|\xi| \leqslant \rho} |\widehat{u}(\xi)|^2 \, d\xi \leqslant C_2 \rho^n \|u\|_{L^1}^2.$$

3. Deduce that there exists a constant $C > 0$ such that, for all $u \in S(\mathbb{R}^n)$, we have

$$\|u\|_{L^2(\mathbb{R}^n)}^{1+2/n} \leqslant C \|u\|_{L^1(\mathbb{R}^n)}^{2/n} \|\nabla u\|_{L^2(\mathbb{R}^n)}.$$

Exercise 5.4 (The div-curl lemma of Murat and Tartar) The following exercise requires knowledge of Sobolev spaces, which will be studied in Chapter 7. For the solution, we refer to the original articles by Murat [144] and Tartar [184] as well as the lecture notes by Prange [157].

Notation: Let Ω be an open set of $\mathbb{R}^n$. We denote by $C^\infty(\overline{\Omega})$ the space of functions $u: \Omega \to \mathbb{R}$ that are the restriction to Ω of functions of class C^∞ on $\mathbb{R}^n$. We denote by $C_0^\infty(\Omega)$ the functions of class C^∞ with compact support in Ω.

Consider a bounded open set $\Omega \subset \mathbb{R}^2$ and two sequences of vector fields, $E_n: \Omega \to \mathbb{R}^2$ and $B_n: \Omega \to \mathbb{R}^2$. We denote by (E_n^1, E_n^2) and (B_n^1, B_n^2) the coordinates of E_n and B_n. We assume that:

1. E_n and B_n belong to $C^\infty(\overline{\Omega})^2$ for every integer n.
2. We have

$$\sup_{n \in \mathbb{N}}\left(\|E_n\|_{L^2(\Omega)^2} + \|B_n\|_{L^2(\Omega)^2} + \|\text{div } E_n\|_{L^2(\Omega)} + \|\text{curl } B_n\|_{L^2(\Omega)} \right) < +\infty,$$

where $\|E_n\|_{L^2(\Omega)^2} = \sqrt{\left\|E_n^1\right\|_{L^2(\Omega)}^2 + \left\|E_n^2\right\|_{L^2(\Omega)}^2}$, and where div E_n and curl B_n are functions with values in $\mathbb{R}$ defined by

$$\text{div } E_n = \partial_{x_1} E_n^1 + \partial_{x_2} E_n^2, \qquad \text{curl } B_n = \partial_{x_2} B_n^1 - \partial_{x_1} B_n^2.$$

3. There exist $E \in L^2(\Omega)^2$ and $B \in L^2(\Omega)^2$ such that $E_n \rightharpoonup E$ and $B_n \rightharpoonup B$ in $L^2(\Omega)^2$, which means that each coordinate converges weakly ($E_n^j \rightharpoonup E^j$ for $j = 1, 2$ and similarly for B_n).

The goal of this exercise is to show a result, due to François Murat [144] and Luc Tartar [184], which states that, for all $\varphi \in C_0^\infty(\Omega)$,

$$\int_\Omega \varphi(x) E_n(x) \cdot B_n(x) \, dx \xrightarrow[n \to +\infty]{} \int_\Omega \varphi(x) E(x) \cdot B(x) \, dx, \qquad (*)$$

where $y \cdot y'$ is the scalar product of $\mathbb{R}^2$.

1. Let us fix a function $\varphi \in C_0^\infty(\Omega)$ and consider $\chi \in C_0^\infty(\Omega)$ equal to 1 on the support of φ, so that $\chi\varphi = \varphi$. We introduce

$$v_n = \varphi E_n, \quad w_n = \chi B_n, \quad v = \varphi E, \quad w = \chi B.$$

We extend these functions by 0 on $\mathbb{R}^2 \setminus \Omega$ (and we still denote them by v_n, w_n, v, w). Show that v_n and w_n belong to $H^1(\mathbb{R}^2)^2$. Show that $(v_n)_{n \in \mathbb{N}}$ converges weakly to v in $L^2(\mathbb{R}^2)^2$ and similarly $(w_n)_{n \in \mathbb{N}}$ converges weakly to w in $L^2(\mathbb{R}^2)^2$.
2. If $f = (f^1, f^2)$ is a function with values in $\mathbb{R}^2$, we denote by $\widehat{f} = (\widehat{f^1}, \widehat{f^2})$ its Fourier transform. Show that $(\widehat{v}_n)_{n \in \mathbb{N}}$ and $(\widehat{w}_n)_{n \in \mathbb{N}}$ are bounded in $L^2(\mathbb{R}^2)^2$ and in $L^\infty(\mathbb{R}^2)^2$.
3. Show that $(*)$ is equivalent to

$$\int_{\mathbb{R}^2} \widehat{v}_n(\xi) \cdot \overline{\widehat{w}_n(\xi)} \, d\xi \xrightarrow[n \to +\infty]{} \int_{\mathbb{R}^2} \widehat{v}(\xi) \cdot \overline{\widehat{w}(\xi)} \, d\xi. \qquad (**)$$

4. Show that, for all $\xi \in \mathbb{R}^2$, the sequences $(\widehat{v}_n(\xi))_{n \in \mathbb{N}}$ and $(\widehat{w}_n(\xi))_{n \in \mathbb{N}}$ converge to $\widehat{v}(\xi)$ and $\widehat{w}(\xi)$, respectively.

5. Let $R > 0$. Let $B(0, R)$ denote the ball of center 0 and radius R in $\mathbb{R}^2$. Show that

$$\int_{B(0,R)} \widehat{v}_n(\xi) \cdot \overline{\widehat{w}_n(\xi)} \, d\xi \xrightarrow[n \to +\infty]{} \int_{B(0,R)} \widehat{v}(\xi) \cdot \overline{\widehat{w}(\xi)} \, d\xi.$$

6. Let $\xi = (\xi_1, \xi_2) \in \mathbb{R}^2$ be non-zero. Let us set $\xi^\perp = (\xi_2, -\xi_1)$. Show that, for all $X \in \mathbb{R}^2$, we have

$$X = \left(X \cdot \frac{\xi}{|\xi|}\right) \frac{\xi}{|\xi|} + \left(X \cdot \frac{\xi^\perp}{|\xi|}\right) \frac{\xi^\perp}{|\xi|}.$$

Next, show that for all X, Y in $\mathbb{R}^2$, we have

$$|X \cdot Y| \leqslant \frac{1}{|\xi|} \left(|Y| \, |X \cdot \xi| + |X| \, |Y \cdot \xi^\perp|\right).$$

7. Show that the sequences of functions $\xi \mapsto \xi \cdot \widehat{v}_n(\xi)$ and $\xi \mapsto \xi^\perp \cdot \widehat{w}_n(\xi)$ are bounded in $L^2(\mathbb{R}^2)$.

8. Conclude that, for all $R > 0$, we have

$$\sup_{n \in \mathbb{N}} \int_{|\xi| > R} \widehat{v}_n(\xi) \cdot \overline{\widehat{w}_n(\xi)} \, d\xi \xrightarrow[R \to +\infty]{} 0,$$

and conclude the proof of $(*)$.

Chapter 6
Convolution

Convolution is found everywhere in analysis, but as a concept it is one of the last to be born: the term convolution only became established in the second half of the twentieth century. At least Wiener had made wonderful use of it under the name of Faltung, which served as a model in several respects for Schwartz's theory of distributions.

Jean-Pierre Kahane [92]

The aim of this chapter is to present a systematic study of the convolution. We will actually introduce two very useful tools in practice for studying the local averages of a function: the convolution product $f * g$ and the maximal function $M(f)$. These two tools will allow us to study the approximations of the identity introduced by Friedrichs and to prove fundamental inequalities due to Hardy, Littlewood and Sobolev. In the last two sections, we will study several results related to the concept of convolution and which were at the heart of the concerns of harmonic analysis in the twentieth century: the Calderón's reconstruction formula, the wavelet decomposition of Grossmann and Morlet as well as Wiener's Tauberian theorem.

6.1 Definition of the Convolution Product

The convolution product of two functions $f : \mathbb{R} \to \mathbb{R}$ and $g : \mathbb{R} \to \mathbb{R}$ is the function $h : \mathbb{R} \to \mathbb{R}$ formally defined by

$$h(x) = \int_{-\infty}^{\infty} f(x - y)g(y) \, dy.$$

The convolution product is a very useful tool that comes into play whenever we are not interested in a function but in its local averages.

© The Author(s), under exclusive license to Springer Nature Switzerland AG 2024   135
T. Alazard, *Analysis and Partial Differential Equations*, Universitext,
https://doi.org/10.1007/978-3-031-70909-8_6

For example, if f is the indicator of $[-1/2, 1/2]$, then

$$h(x) = \int_{x-1/2}^{x+1/2} g(y)\,dy,$$

so $h(x)$ is the average of g over $[x - 1/2, x + 1/2]$. Similarly, if $f_\varepsilon = (2\varepsilon)^{-1}\mathbf{1}_{[-\varepsilon,\varepsilon]}$, then $h_\varepsilon(x) = f_\varepsilon * g(x)$ is the average of g over $[x - \varepsilon, x + \varepsilon]$,

$$h_\varepsilon(x) = \frac{1}{2\varepsilon}\int_{x-\varepsilon}^{x+\varepsilon} g(y)\,dy.$$

We expect that $h_\varepsilon(x)$ will converge to $g(x)$ as ε tends to 0. This situation occurs very naturally in applications (because in Physics, for example, we do not know a function pointwise: we only have access to its local averages). It also occurs in theoretical questions. For example, to approximate a function by regular functions, it is natural to replace the value at a point by an average calculated over a neighborhood of that point. We will study the main results concerning the convolution product. The first result concerns the definition of $f * g$ under general assumptions.

Theorem 6.1 (Young) *Consider p, q, r such that*

$$1 \leqslant p, q, r \leqslant \infty, \quad \frac{1}{p} + \frac{1}{q} = 1 + \frac{1}{r}. \tag{6.1}$$

Then, for every f in $L^p(\mathbb{R}^n)$ and every g in $L^q(\mathbb{R}^n)$, the integral

$$f * g(x) = \int_{\mathbb{R}^n} f(x - y)g(y)\,dy$$

converges for almost every $x \in \mathbb{R}^n$. Moreover, $f * g$ belongs to $L^r(\mathbb{R}^n)$ and

$$\|f * g\|_{L^r(\mathbb{R}^n)} \leqslant \|f\|_{L^p(\mathbb{R}^n)}\|g\|_{L^q(\mathbb{R}^n)}.$$

We call $f * g$ the convolution product of f and g and we say that the binary operation $*$ is the convolution product.

Remark 6.2 Consider the particular case where f belongs to $L^1(\mathbb{R}^n)$, as in the example of local averages given in the introduction. Then the previous theorem implies that the operator $g \mapsto f * g$ is bounded from $L^q(\mathbb{R}^n)$ to $L^q(\mathbb{R}^n)$ for every $q \in [1, +\infty]$.

Proof Let us start by verifying that we can assume that f and g are functions with positive real values. Indeed, suppose that the result is true in this case. Then we can apply it to the functions $|f|$ and $|g|$. We deduce that the integral $\int |f(x - y)|\,|g(y)|\,dy$ is finite for almost every x, which implies from the triangle inequality that $f * g(x)$ is well defined for almost every x. Moreover, the triangle inequality also implies that $|f * g| \leqslant |f| * |g|$, so

$$\|f * g\|_{L^r(\mathbb{R}^n)} = \||f * g|\|_{L^r(\mathbb{R}^n)}$$
$$\leq \||f| * |g|\|_{L^r(\mathbb{R}^n)}$$
$$\leq \||f|\|_{L^p(\mathbb{R}^n)} \||g|\|_{L^q(\mathbb{R}^n)}$$
$$\leq \|f\|_{L^p(\mathbb{R}^n)} \|g\|_{L^q(\mathbb{R}^n)}.$$

This shows that it is sufficient to assume that f and g are functions with positive values. In this case, the function $h = f * g$ is well defined pointwise because the integral of a positive function is always well defined in $[0, +\infty]$.

Note also that the result is obvious if f or g are equal to 0. We can therefore also assume that $\|f\|_{L^p(\mathbb{R}^n)} = 1 = \|g\|_{L^q(\mathbb{R}^n)}$. It is now a matter of showing that the function $h = f * g$ belongs to $L^r(\mathbb{R}^n)$ and that its norm is less than 1.

If $r = \infty$ then p and q are two conjugate exponents ($1/p + 1/q = 1$) and the result is a consequence of Hölder's inequality. If $r = 1$, then necessarily $p = 1$ and $q = 1$. In this case, the desired result is a consequence of Fubini's theorem, noting that

$$\iint |f(x - y)| \, |g(y)| \, dx \, dy = \|f\|_{L^1(\mathbb{R}^n)} \|g\|_{L^1(\mathbb{R}^n)}.$$

We still have to deal with the case $1 < r < \infty$. In this case, p and q are necessarily finite and we can write that $|f(x - y)g(y)| = \alpha(x, y)\beta(x, y)$ with

$$\alpha(x, y) = |f(x - y)|^{p/r} |g(y)|^{q/r},$$
$$\beta(x, y) = |f(x - y)|^{1-p/r} |g(y)|^{1-q/r}.$$

Using Fubini's theorem, we directly have that

$$\iint \alpha(x, y)^r \, dx \, dy = \|f\|_{L^p(\mathbb{R}^n)}^p \|g\|_{L^q(\mathbb{R}^n)}^q = 1. \tag{6.2}$$

We will show that, for almost all x,

$$\int \beta(x, y)^{r'} \, dy \leq 1 \quad \text{where} \quad \frac{1}{r'} = 1 - \frac{1}{r}. \tag{6.3}$$

For this, we note that

$$\beta(x, y)^{r'} = |f(x - y)|^{p/s} |g(y)|^{q/t}$$

with

$$\frac{1}{s} = r'\left(\frac{1}{p} - \frac{1}{r}\right), \quad \frac{1}{t} = r'\left(\frac{1}{q} - \frac{1}{r}\right).$$

We then use the assumption (6.1) to verify that

$$\frac{1}{s} + \frac{1}{t} = r'\left(\frac{1}{p} + \frac{1}{q} - \frac{2}{r}\right) = r'\left(1 - \frac{1}{r}\right) = \frac{r}{r-1} \cdot \frac{r-1}{r} = 1.$$

So s and t are two conjugate real numbers and the inequality (6.3) is easily obtained by applying Hölder's inequality.

Next, we use Hölder's inequality again with the conjugate exponents r and r' to write that

$$|h(x)| \leqslant \left(\int \alpha(x,y)^r \, dy \right)^{1/r} \left(\int \beta(x,y)^{r'} \, dy \right)^{1/r'},$$

so, taking into account (6.3) and then (6.2),

$$\int |h(x)|^r \, dx \leqslant \iint \alpha(x,y)^r \, dy \, dx \leqslant 1.$$

This is the desired inequality.   □

We will now study certain properties of the convolution product that we will frequently need to use, starting with a property that relates the convolution product to the Fourier transform.

Proposition 6.3 *Let f and g be two functions from $L^1(\mathbb{R}^n)$. Then $f * g$ belongs to $L^1(\mathbb{R}^n)$ and for all $\xi \in \mathbb{R}^n$ we have*

$$\widehat{f * g}(\xi) = \widehat{f}(\xi)\widehat{g}(\xi).$$

Proof Young's theorem ensures that the convolution product $f * g$ belongs to $L^1(\mathbb{R}^n)$, which allows us to use Fubini's theorem to obtain that

$$\int e^{-ix \cdot \xi} f * g(x) \, dx = \int \left(\int e^{-i(x-y) \cdot \xi} f(x-y) \, dx \right) e^{-iy \cdot \xi} g(y) \, dy,$$

which directly implies the desired result.   □

We can easily deduce that the convolution product is commutative, associative and distributive with respect to addition.

Corollary 6.4 *For any triplet of functions f, g, h in $L^1(\mathbb{R}^n)$, we have*

$$f * g = g * f,$$
$$f * (g * h) = (f * g) * h,$$
$$f * (g + h) = (f * g) + (f * h).$$

Proposition 6.5 *The convolution product of two elements of the Schwartz space $S(\mathbb{R}^n)$ belongs to $S(\mathbb{R}^n)$. Furthermore, the convolution product is a bilinear map continuous from $S(\mathbb{R}^n) \times S(\mathbb{R}^n)$ to $S(\mathbb{R}^n)$.*

Proof Consider f and g in $S(\mathbb{R}^n)$. Then the convolution product $f * g$ is of class C^∞ on $\mathbb{R}^n$ and, for every multi-index β in $\mathbb{N}^n$, we have

$$\partial_x^\beta (f * g) = (\partial_x^\beta f) * g.$$

Furthermore, for every $m \in \mathbb{N}$, we have

$$|x|^m \leqslant (|x-y|+|y|)^m \leqslant (2\max\{|x-y|,|y|\})^m \leqslant 2^m(|x-y|^m+|y|^m).$$

Therefore, $\left|x^\alpha \partial_x^\beta (f * g)(x)\right|$ is bounded by a multiple of

$$\int \left(|x-y|^{|\alpha|}\left|\partial_x^\beta f(x-y)\right||g(y)| + \left|\partial_x^\beta f(x-y)\right||y|^{|\alpha|}|g(y)|\right) dy.$$

We then use the obvious inequality $\|F * G\|_{L^\infty} \leqslant \|F\|_{L^\infty}\|G\|_{L^1}$ and points (1) and (3) of Proposition 5.6 to bound the L^1 norms of g and $y^\alpha g$ by semi-norms of g in $\mathcal{S}(\mathbb{R}^n)$. □

Let us recall that the support of a continuous function f is defined as the closure of the set of points where it does not vanish:

$$\operatorname{supp} f = \overline{\{x \in \mathbb{R}^n : f(x) \neq 0\}}.$$

Let φ be a C^∞ function with compact support. Then, for any function f in $L^p(\mathbb{R}^n)$, the convolution product $\varphi * f$ is well defined (indeed we have $\varphi \in L^1(\mathbb{R}^n)$ so Theorem 6.1 guarantees that $\varphi * f$ belongs to $L^p(\mathbb{R}^n)$). An essential property of the convolution product is that this function is regular.

Proposition 6.6

1. Let $\varphi \in C_0^\infty(\mathbb{R}^n)$. For every $p \in [1, +\infty]$ and every $f \in L^p(\mathbb{R}^n)$, the function $\varphi * f$ is of class C^∞ on $\mathbb{R}^n$.
2. If in addition f has compact support, then $\varphi * f$ has compact support and

$$\operatorname{supp}(\varphi * f) \subset \operatorname{supp} \varphi + \operatorname{supp} f.$$

Proof 1. The first point is a direct corollary of the theorem on differentiation under the integral sign.
2. Assume that $\varphi * f(x) \neq 0$. Then there exists $y \in \mathbb{R}^n$ such that $\varphi(x-y)f(y) \neq 0$. We deduce that

$$x = (x-y) + y \in \operatorname{supp} \varphi + \operatorname{supp} f.$$

This proves that

$$\{x \in \mathbb{R}^n : \varphi * f(x) \neq 0\} \subset \operatorname{supp} \varphi + \operatorname{supp} f.$$

Since $\operatorname{supp} \varphi$ and $\operatorname{supp} f$ are compact, the sum of these two sets is also compact therefore closed. It follows that

$$\operatorname{supp}(\varphi * f) = \overline{\{x \in \mathbb{R}^n : \varphi * f(x) \neq 0\}} \subset \operatorname{supp} \varphi + \operatorname{supp} f,$$

which concludes the proof. □

6.2 Approximations of the Identity

The goal of this section is to construct a family of linear operators $(I_t)_{t\in(0,1]}$ that have the following properties:

1. if f belongs to $L^p(\mathbb{R}^n)$ for some $1 \leqslant p \leqslant \infty$, then $I_t(f)$ belongs to $L^p(\mathbb{R}^n)$ and moreover $\|I_t(f)\|_{L^p} \leqslant \|f\|_{L^p}$;
2. $I_t(f)$ is a function of class C^∞;
3. $I_t(f)$ converges to f as t tends to 0 (in a sense that will be specified).

These operators therefore allow us to approximate a function by regular functions. To construct them, we will consider the convolution product by regular functions with compact support.

Lemma 6.7 *There exists a function* $\Phi: \mathbb{R}^n \to [0, +\infty)$ *with compact support such that*

$$\operatorname{supp}\Phi \subset B(0,1) \quad and \quad \int_{\mathbb{R}^n} \Phi(x)\, dx = 1,$$

and such that Φ *is:*

- *radial:* $\Phi(x) = \Phi(y)$ *if* $|x| = |y|$;
- *decreasing:* $\Phi(x) \leqslant \Phi(y)$ *if* $|x| \geqslant |y|$.

Proof Consider the function $\phi: \mathbb{R} \to \mathbb{R}_+$ defined by

$$\text{for } y > 0, \quad \phi(y) = \exp(-1/y); \qquad \text{for } y \leqslant 0, \quad \phi(y) = 0.$$

This function is of class C^∞ on $\mathbb{R}$ (see the proof of Proposition 17.33). It is then sufficient to set

$$\Phi(x) = A\phi(1 - |x|^2), \tag{6.4}$$

where A is chosen so that $\int \Phi\, dx = 1$. $\square$

Given $t \in (0, 1]$, let us set

$$\Phi_t(x) = \frac{1}{t^n} \Phi(x/t).$$

We propose to study the operator[1] $I_t: f \mapsto f * \Phi_t$ of convolution by Φ_t. The family $\{I_t\}_{t\in(0,1]}$ is called an approximation of the identity.

---

[1] These operators were introduced by Friedrichs [53] and are now systematically used to study partial differential equations. We refer in particular to Stein [176], who developed a thorough study of the convergence of approximations of the identity in his famous book on the differentiability properties of functions.

Theorem 6.8

1. *If f is uniformly continuous and bounded on $\mathbb{R}^n$, then $f * \Phi_t$ converges to f uniformly as t tends to 0:*

$$\lim_{t \to 0} \sup_{x \in \mathbb{R}^n} |f * \Phi_t(x) - f(x)| = 0.$$

2. *If $p \in [1, +\infty)$ then $f * \Phi_t$ converges to f in $L^p(\mathbb{R}^n)$ as t tends to 0:*

$$\lim_{t \to 0} \|f * \Phi_t - f\|_{L^p} = 0.$$

Proof 1. Fix $x \in \mathbb{R}^n$ and $\varepsilon > 0$. Using

$$\int_{\mathbb{R}^n} \Phi_t(y) \, dy = \int_{\mathbb{R}^n} \Phi(y) \, dy = 1, \tag{6.5}$$

we can write

$$f(x) = \int_{\mathbb{R}^n} f(x) \Phi_t(y) \, dy,$$

from which

$$f * \Phi_t(x) - f(x) = \int_{\mathbb{R}^n} \Phi_t(y) \big(f(x - y) - f(x) \big) \, dy.$$

By uniform continuity, there exists $\delta > 0$ such that $|f(x - y) - f(x)| < \varepsilon$ if $|y| \leq \delta$. We write $f * \Phi_t(x) - f(x)$ as the sum of A_t and B_t with

$$A_t = \int_{|y| \leq \delta} \Phi_t(y) \big(f(x - y) - f(x) \big) \, dy, \quad B_t = \int_{|y| > \delta} \Phi_t(y) \big(f(x - y) - f(x) \big) \, dy.$$

We then directly have that A_t is bounded by ε and that B_t can be bounded by

$$B_t \leq 2\|f\|_{L^\infty} \int_{|y| > \delta} \Phi_t(y) \, dy = 2\|f\|_{L^\infty} \int_{|y| > \delta/t} \Phi(y) \, dy,$$

and we verify that the right-hand side tends to 0 when t tends to 0 (it vanishes for t sufficiently large). This concludes the proof of the first point.

2. Consider a continuous function g with compact support. We decompose $f * \Phi_t - f$ in the form

$$f * \Phi_t - f = g * \Phi_t - g + (f - g) * \Phi_t + g - f.$$

Then, using Young's inequality (with $p = 1$ in (6.1)), we find that

$$\|(f - g) * \Phi_t\|_{L^p(\mathbb{R}^n)} \leq \|f - g\|_{L^p(\mathbb{R}^n)},$$

so

$$\|f * \Phi_t - f\|_{L^p(\mathbb{R}^n)} \leq \|g * \Phi_t - g\|_{L^p(\mathbb{R}^n)} + 2\|f - g\|_{L^p(\mathbb{R}^n)}.$$

The second term on the right-hand side can be made arbitrarily small by the density of the space of continuous functions with compact support in $L^p(\mathbb{R}^n)$ (since p is finite). It remains only to show that the first term tends to 0 for any continuous function g with compact support. Let g be such a function and let K be the compact set $K = \operatorname{supp} g + \overline{B(0,1)}$. We directly verify that $g * \Phi_t$ is a function with compact support and that its support is included in K. We then have

$$\|g * \Phi_t - g\|_{L^p(\mathbb{R}^n)} = \|g * \Phi_t - g\|_{L^p(K)} \leqslant |K|^{1/p} \|g * \Phi_t - g\|_{L^\infty(\mathbb{R}^n)},$$

where $|K|$ is the Lebesgue measure of K. As a result, we can apply the result of point (1) to obtain that $\|g * \Phi_t - g\|_{L^p(K)}$ converges to 0. This completes the proof. $\qquad\square$

We deduce two very useful results.

Corollary 6.9 *The set $C_0^\infty(\mathbb{R}^n)$ of functions of class C^∞ with compact support is dense in $L^p(\mathbb{R}^n)$ for all $n \geqslant 1$ and for all $p \in [1, +\infty)$.*

Proof Consider a function $f \in L^p(\mathbb{R}^n)$ and a family $\{\Phi_t\}_{t\in(0,1]}$ as above. Then the functions $\Phi_t * f$ are C^∞ according to Proposition 6.6. Consider also a function θ which is of class C^∞, such that $0 \leqslant \theta \leqslant 1$, with compact support and which equals 1 over the unit ball (such a function is constructed in Proposition 17.34). Let θ_k be the function $\theta_k(x) = \theta(x/k)$. Then

$$\|f - \theta_k(\Phi_t * f)\|_{L^p} \leqslant \|f - \theta_k f\|_{L^p} + \|\theta_k(f - \Phi_t * f)\|_{L^p}$$
$$\leqslant \|f - \theta_k f\|_{L^p} + \|f - \Phi_t * f\|_{L^p}.$$

The first term on the right-hand side tends to 0 when k tends to $+\infty$ by dominated convergence and the second term tends to 0 when t tends to 0 according to Theorem 6.8. $\qquad\square$

Corollary 6.10 *Let $f \in L^1_{\mathrm{loc}}(\mathbb{R}^n)$ be such that*

$$\forall \varphi \in C_0^\infty(\mathbb{R}^n), \quad \int_{\mathbb{R}^n} f(y)\varphi(y)\,\mathrm{d}y = 0. \tag{6.6}$$

Then $f = 0$.

Proof We follow Friedrichs' proof of this classical result. To show that $f = 0$, it is enough to show that $\chi f = 0$ for any function χ with compact support. But $\chi f \in L^1(\mathbb{R}^n)$ so the functions $\Phi_t * (\chi f)$ converge to χf in $L^1(\mathbb{R}^n)$ when t tends to 0 according to Theorem 6.8. By definition of the convolution product, we have

$$(\Phi_t * (\chi f))(x) = \int_{\mathbb{R}^n} \chi(y)f(y)\Phi_t(x - y)\,\mathrm{d}y.$$

However, the hypothesis (6.6) applied with $\varphi(y) = \chi(y)\Phi_t(x - y)$ implies that this last integral is equal to 0 for all $t > 0$ and for all $x \in \mathbb{R}^n$. Consequently, $\Phi_t * (\chi f) = 0$ and by taking the limit as t tends to 0, we conclude that $\chi f = 0$. $\qquad\square$

6.3 The Hardy–Littlewood Maximal Function

We will study the Hardy–Littlewood maximal function.[2] It is a very useful tool that will allow us to extend the previous results on the convolution product in two different directions: (i) the study of the pointwise convergence of an approximate identity and (ii) the study of the convolution product by a singular function of the form $1/|x|^{n-\alpha}$, with $0 < \alpha < n$ (which are singular in the sense that they belong to no Lebesgue space $L^p(\mathbb{R}^n)$).

Consider a locally integrable function $f : \mathbb{R}^n \to \mathbb{C}$. By definition, the Hardy–Littlewood maximal function is the function $Mf : \mathbb{R}^n \to [0, +\infty]$ defined by

$$(Mf)(x) = \sup_{r>0} \left\{ \frac{1}{|B(x,r)|} \int_{B(x,r)} |f(y)|\, dy \right\}. \tag{6.7}$$

This function naturally comes into play when proving certain inequalities or convergence results.

Notice it may be useful to consider variants, such as the off-center maximal function defined by

$$(\widetilde{M}f)(x) = \sup_{x \in B} \frac{1}{|B|} \int_B |f(y)|\, dy,$$

where the supremum is taken over all balls containing x (and not only those which are centered at x). We directly have

$$(Mf)(x) \leqslant (\widetilde{M}f)(x).$$

Conversely, for any ball $B(z, \delta)$ containing x, we have $B(z, \delta) \subset B(x, 2\delta)$ so, using the monotonicity of the integral, we verify that

$$\frac{1}{|B(z,\delta)|} \int_{B(z,\delta)} |f(y)|\, dy \leqslant \frac{|B(x,2\delta)|}{|B(z,\delta)|} \frac{1}{|B(x,2\delta)|} \int_{B(x,2\delta)} |f(y)|\, dy$$
$$\leqslant 2^n (Mf)(x),$$

from which we get the inequality

$$(\widetilde{M}f)(x) \leqslant 2^n (Mf)(x).$$

For what follows, we could therefore use either one of these maximal functions indifferently. In the following, we only consider the maximal function Mf given by (6.7).

---

[2] Hardy and Littlewood were two English mathematicians, professors at Cambridge University in the first half of the 20th century, whose collaboration is at the origin of numerous methods of analysis. They notably introduced the maximal function in 1930 [71]. For a detailed exposition of the properties of the maximal function, we refer to the books by Stein [176, 177] and Tao [182].

We will be interested in the continuity of the operator M on the Lebesgue spaces. To begin, let us note that it is not even obvious that Mf is measurable (because we take a supremum over a non-countable set). Nevertheless, we will be able to directly show a stronger result.

Proposition 6.11 *For any locally integrable function f, the maximal function Mf is lower semi-continuous and therefore measurable.*

Proof We need to show that $U_\lambda := \{x \in \mathbb{R}^n : (Mf)(x) > \lambda\}$ is open for all $\lambda > 0$. Let $x \in U_\lambda$. Then by definition of Mf, there exists $r > 0$ such that

$$\frac{1}{|B(x,r)|} \int_{B(x,r)} |f(y)|\,dy > \lambda.$$

By an immediate continuity argument, there exists $r' > r$ such that

$$\frac{1}{|B(x,r')|} \int_{B(x,r)} |f(y)|\,dy > \lambda.$$

Let $z \in \mathbb{R}^n$ with $\varepsilon = |z - x|$ small, so that $B(x,r) \subset B(z, r + \varepsilon) \subset B(x, r')$. By monotonicity of the integral, we deduce that

$$(Mf)(z) \geqslant \frac{1}{|B(z, r + \varepsilon)|} \int_{B(z, r+\varepsilon)} |f(y)|\,dy$$

$$\geqslant \frac{1}{|B(x, r')|} \int_{B(x,r)} |f(y)|\,dy > \lambda,$$

which proves that $z \in U_\lambda$ and therefore that U_λ is an open set. $\square$

The central result is the following theorem.

Theorem 6.12 (Hardy–Littlewood)

1. *For all f in $L^1(\mathbb{R}^n)$, we have*

$$\|Mf\|_{L^1_w} \leqslant 2 \cdot 3^n \|f\|_{L^1}.$$

2. *For all $p \in (1, +\infty]$, there exists a constant C_p (also depending on the dimension n) such that*

$$\|Mf\|_{L^p} \leqslant C_p \|f\|_{L^p}.$$

Remark 6.13 It is important to note that the L^1 norm of Mf is finite if and only if $f = 0$. Indeed, suppose that f is non-zero. By making a translation and a dilation, we can assume that the integral of $|f|$ over $B(0, 1)$ is positive. Then, for all x with norm greater than 1, we can write

$$(Mf)(x) \geqslant \frac{1}{B(x, 2|x|)} \int_{B(x,2|x|)} |f(y)|\,dy \geqslant \frac{1}{B(x, 2|x|)} \int_{B(0,1)} |f(y)|\,dy.$$

We deduce that

$$(Mf)(x) \geqslant \frac{C}{|x|^n},$$

which proves that Mf cannot belong to $L^1(\mathbb{R}^n)$.

Proof The proof of the first point relies on the following covering lemma.

Lemma 6.14 (Vitali) *Let $E \subset \mathbb{R}^n$ be a measurable set of finite measure. Suppose that E is included in the union of a family $(B_a)_{a \in A}$ of open balls of $\mathbb{R}^n$. Then there exists a finite family $(B_a)_{a \in J}$, $J \subset A$ finite, of pairwise disjoint balls satisfying*

$$\left| \bigcup_{a \in J} B_a \right| \geqslant 2^{-1} 3^{-n} |E|.$$

Proof Measure theory assures us that there exists a compact $K \subset E$ whose measure $|K|$ is greater than half that of E. By compactness, there exists a finite set I_1 and a subfamily $\{B_a\}_{a \in I_1}$ of balls that cover K. Let us choose a ball, denoted B_1, with maximum radius among the B_a with $a \in I_1$. Then consider the subfamily $\{B_a\}_{a \in I_2}$ of balls B_a with $a \in I_1$ that do not intersect B_1. Suppose this subfamily is not empty. Let B_2 be a ball of maximum radius among the balls B_a with $a \in I_2$. We proceed like this by induction until the algorithm ends, and this provides us with a finite family B_i, $1 \leqslant i \leqslant N$, of balls. Now consider any ball B from the first family $\{B_a\}_{a \in I_1}$. Either this ball is part of the collection $\{B_i : 1 \leqslant i \leqslant N\}$, or it is not and then we can consider the smallest index i_0 such that $B \cap B_{i_0}$ is not empty. Then, by construction, the radius of B_{i_0} is larger than that of B. We deduce that B is included in the ball, denoted $3B_{i_0}$, which is the ball with the same center as B_{i_0} and radius equal to 3 times that of B_{i_0}. We then have

$$\frac{1}{2}|E| \leqslant |K| \leqslant \left| \bigcup_{a \in I_1} B_a \right| \leqslant \left| \bigcup_{1 \leqslant i \leqslant N} 3B_i \right| \leqslant \sum_{i=1}^N |3B_i| \leqslant 3^n \sum_{i=1}^N |B_i|.$$

This concludes the proof.   □

Now let us fix $\lambda > 0$ and consider a measurable set E of finite measure contained in $\{Mf > \lambda\}$. For every x belonging to E, there exists a ball $B_x = B(x, r_x)$ containing x such that

$$\frac{1}{|B_x|} \int_{B_x} |f(y)| \, dy > \lambda.$$

Consider the family $\{B_x\}_{x \in E}$. According to the previous lemma, there exists a finite subfamily $\{B_{x_i} : 1 \leqslant i \leqslant N\}$ such that

$$|E| \leqslant 2 \cdot 3^n \sum_{i=1}^N |B_{x_i}|.$$

This implies

$$|E| \leqslant 2 \cdot 3^n \sum_{i=1}^{N} \frac{1}{\lambda} \int_{B_{x_i}} |f(y)| \, dy.$$

As the balls B_{x_i} are pairwise disjoint, we can bound the sum in the right-hand side by $\lambda^{-1} \|f\|_{L^1}$. Then, by taking the supremum over all measurable sets E of finite measure included in $\{|Mf| > \lambda\}$, we deduce that

$$\lambda \, |\{|Mf| > \lambda\}| \leqslant 2 \cdot 3^n \|f\|_{L^1},$$

which concludes the proof of point (1).

It remains to prove statement (2). Note that the desired result is trivial in the case $p = +\infty$ because we directly verify that $\|Mf\|_{L^\infty} \leqslant \|f\|_{L^\infty}$. The case $p \in (1, +\infty)$ follows thanks to the Marcinkiewicz interpolation theorem (see Theorem 18.12 in Appendix 18). Notice that the function $f \mapsto Mf$ is not linear, but we can still apply the Marcinkiewicz theorem because, as explained in its proof, this interpolation result is true for sub-additive functions, which is the case for M. $\qquad\square$

6.4 Pointwise Convergence of an Approximation of the Identity

We studied approximations of the identity in Section 6.2. We will now change our point of view and study the question of pointwise convergence. Let us recall that we cannot deduce pointwise convergence from convergence in L^p norm (we can only deduce that a subsequence converges almost everywhere); see Exercise 18.1.

Let us recall the notations introduced in Section 6.2. We fix an integrable and normalized function $\Phi \colon \mathbb{R}^n \to [0, +\infty)$, so that

$$\int_{\mathbb{R}^n} \Phi(x) \, dx = 1.$$

We also assume that Φ is:

- radial: $\Phi(x) = \Phi(y)$ if $|x| = |y|$;
- decreasing: $\Phi(x) \leqslant \Phi(y)$ if $|x| \geqslant |y|$.

We then introduce

$$\Phi_t(x) = \frac{1}{t^n} \Phi(x/t) \quad \text{where} \quad t \in (0, 1].$$

The main result of this part states that we have convergence of $f * \Phi_t(x)$ towards $f(x)$ for almost all x as t tends to 0.

Theorem 6.15 *Let* Φ *be as above and let* $f \in L^p(\mathbb{R}^n)$ *for some* $p \in [1, +\infty]$. *Then, for almost all* $x \in \mathbb{R}^n$, *we have*

$$\lim_{t \to 0} f * \Phi_t(x) = f(x).$$

Proof The key observation is provided by the following lemma.

Lemma 6.16 *Let* Φ *be as above and let* $f \in L^p(\mathbb{R}^n)$. *For almost all* $x \in \mathbb{R}^n$, *we have*

$$\sup_{t>0} |f * \Phi_t(x)| \leq (Mf)(x). \tag{6.8}$$

Proof As $|f * \Phi_t| \leq |f| * \Phi_t$ and as the maximal function Mf is equal to $M|f|$, we can assume without loss of generality that $f \geq 0$. Recall that, by hypothesis, Φ is a radial decreasing function. We will then proceed by approximation and start by assuming that there exist radii ρ_p and positive numbers a_p, with $1 \leq p \leq N$, such that $\Phi = \sum_{p=1}^{N} a_p \mathbf{1}_{B(0,\rho_p)}$. Then

$$f * \Phi_t(x) = \int \frac{1}{t^n} \Phi\left(\frac{x-y}{t}\right) f(y) \, dy = \frac{1}{t^n} \sum_p a_p \int_{B(x,t\rho_p)} f(y) \, dy,$$

from which, by definition of the maximal function,

$$f * \Phi_t(x) \leq \frac{1}{t^n} (Mf)(x) \sum_p a_p |B(x,t\rho_p)| = \|\Phi\|_{L^1} (Mf)(x).$$

For any function Φ belonging to L^1, radial and decreasing, we can find a sequence of functions of the previous form that converges to Φ. We then use the monotone convergence theorem to pass to the limit in the previous inequality and obtain the desired result. □

To prove the theorem, we will use a density argument using the fact that if f is a uniformly continuous and bounded function, then it is a consequence of the first point of Theorem 6.8. We introduce

$$\theta(f)(x) = \limsup_{t \to 0} |f * \Phi_t(x) - f(x)|.$$

We want to show that $\theta(f)$ is zero almost everywhere. For this, we will show that the measure of the set $\{\theta(f) > \varepsilon\}$ is zero for all $\varepsilon > 0$. Note that the first point of Theorem 6.8 implies that $\theta(g) = 0$ for any uniformly continuous and bounded function g on $\mathbb{R}^n$. We deduce that $\theta(f) = \theta(f - g)$. According to the triangle inequality and Lemma 6.16, we have

$$\theta(f - g) \leq |f - g| + M(f - g).$$

Then

$$|\{\theta(f-g) > \varepsilon\}| \leqslant |\{|f-g| > \varepsilon/2\}| + |\{M(f-g) > \varepsilon/2\}|.$$

Suppose that $p = 1$. Then the Chebyshev inequality recalled in Lemma 18.9 implies that

$$|\{|f-g| > \varepsilon/2\}| \leqslant \frac{2}{\varepsilon}\|f-g\|_{L^1}$$

and the Hardy–Littlewood theorem 6.12 on the maximal function implies that

$$|\{M(f-g) > \varepsilon/2\}| \leqslant \frac{4 \cdot 3^n}{\varepsilon}\|f-g\|_{L^1}.$$

So, for any integrable, continuous, and bounded function g, we have

$$|\{\theta(f) > \varepsilon\}| = |\{\theta(f-g) > \varepsilon\}| \leqslant \frac{C}{\varepsilon}\|f-g\|_{L^1}.$$

However, the set of uniformly continuous and bounded functions is dense in L^1 (since this set contains that of continuous functions with compact support). This implies that $|\{\theta(f) > \varepsilon\}| = 0$. If $1 < p < \infty$, we proceed in the same way using that

$$|\{|f-g| > \varepsilon/2\}| \leqslant \frac{2^p}{\varepsilon^p}\|f-g\|_{L^p}^p,$$

and

$$|\{M(f-g) > \varepsilon/2\}| \leqslant \frac{2^p}{\varepsilon^p}\|M(f-g)\|_{L^p}^p \leqslant C_p^p \frac{2^p}{\varepsilon^p}\|f-g\|_{L^p}^p.$$

Finally, for $p = \infty$, we reduce to the previous case by multiplying f by the indicator function of $B(0,n)$. We then show that the set $\{x \in B(0,n/2) : \theta(f)(x) > 0\}$ is of measure zero for all n, which implies the desired result. $\qquad \square$

The previous theorem admits a very well-known corollary which states that the local averages of f on balls centered at x converge for almost every x to $f(x)$.

Corollary 6.17 (Lebesgue Differentiation Theorem) *Let $f \in L^1(\mathbb{R}^n)$. Then, for almost every $x \in \mathbb{R}^n$, we have*

$$f(x) = \lim_{t \to 0} \frac{1}{|B(x,t)|} \int_{B(x,t)} f(y) \, dy.$$

Proof For

$$\Phi = \frac{1}{|B(0,1)|} \mathbb{1}_{B(0,1)},$$

we have

$$f * \Phi_t(x) = \frac{1}{|B(x,t)|} \int_{B(x,t)} f(y) \, dy.$$

The Lebesgue differentiation theorem is therefore a consequence of the previous theorem. $\qquad \square$

As an example of application, note that by setting

$$\Phi(x) = \frac{1}{(4\pi)^{n/2}} \exp(-|x|^2/4),$$

we obtain the following result.

Corollary 6.18 *Let $f \in L^1(\mathbb{R}^n)$ and let t be a positive real number. We define the function u_t by*

$$u_t(x) = \frac{1}{(4\pi t)^{n/2}} \int_{\mathbb{R}^n} \exp\left(-\frac{|x-y|^2}{4t}\right) f(y)\, dx. \tag{6.9}$$

Then

$$\lim_{t \to 0^+} u_t(x) = f(x),$$

for almost every x in $\mathbb{R}^n$.

Remark 6.19

1. To interpret this result, let us start by noting that

$$u_t(x) = \frac{1}{(2\pi)^n} \int_{\mathbb{R}^n} e^{-t|\xi|^2} e^{ix\cdot\xi} \widehat{f}(\xi)\, d\xi. \tag{6.10}$$

We have already studied the question of the convergence of u_t towards f in the proof of the Fourier inversion theorem. We showed that u_t converges to f in $L^1(\mathbb{R}^n)$ when t tends to 0. However, as we have pointed out, this does not imply almost everywhere convergence.

2. The function $u(t, x) = u_t(x)$ defined by (6.9) is (formally) a solution of the heat equation

$$\partial_t u - \Delta u = 0.$$

This result gives meaning to the fact that u is the initial data solution $u|_{t=0} = f$.

6.5 The Hardy–Littlewood–Sobolev Inequality

We call *Riesz potentials* the operators I_α defined for $\alpha > 0$ by

$$I_\alpha(f)(x) = \int_{\mathbb{R}^n} \frac{f(y)}{|x-y|^{n-\alpha}}\, dy.$$

Note that $I_\alpha f(x)$ is well defined for all $\alpha > 0$ and any function f with compact support. To see this, we decompose the integration domain into two parts: the ball centered at x with radius 1, $B(x, 1)$, and its complement. Since f is bounded, the fact that the integral over $B(x, 1)$ is well defined comes from the Riemann criterion and the assumption $\alpha > 0$. On the complement, there is no integration problem because the integrand is bounded and has compact support.

Theorem 6.20 (Hardy–Littlewood–Sobolev) *Let $n \in \mathbb{N}^*$. Consider three positive real numbers (p, q, α) such that*

$$\frac{1}{p} - \frac{1}{q} = \frac{\alpha}{n}, \quad 1 < p < \frac{n}{\alpha}.$$

Then there exists a constant $C = C(p, q, n, \alpha)$ such that, for any function f in $C_0^1(\mathbb{R}^n)$,

$$\|I_\alpha f\|_{L^q} \leqslant C \|f\|_{L^p}.$$

Remark 6.21 This inequality was introduced by Hardy and Littlewood [70] when n equals 1 and generalized by Sobolev [174] to $n \geqslant 1$. Using rearrangement inequalities, Lieb [122] proved the existence of optimizers and determined the best constant in some cases.

Proof We can assume without loss of generality that $\|f\|_{L^p} = 1$.

We will use a decomposition similar to the one used to show that the integral is well defined. But here, we will introduce a parameter $R > 0$, which we will fix later, to decompose the integral into two parts:

$$I_\alpha(f)(x) = I_{\alpha,R}(f)(x) + I_\alpha^R(f)(x),$$

with

$$I_{\alpha,R}(f)(x) = \int_{B(x,R)} \frac{f(y)}{|x-y|^{n-\alpha}} \, dy, \quad I_\alpha^R(f)(x) = \int_{\mathbb{R}^n \setminus B(x,R)} \frac{f(y)}{|x-y|^{n-\alpha}} \, dy.$$

We can write $I_{\alpha,R}(f)(x)$ as the convolution product of f and the function

$$\Psi(x) = \mathbf{1}_{B(0,R)} |x|^{\alpha-n} .$$

Let $\gamma(R)$ be the L^1 norm of Ψ and let $\Phi = \Psi/\gamma(R)$. Then we can apply the inequality (6.8) with $t = 1$, to deduce that

$$\left| I_{\alpha,R}(f)(x) \right| \leqslant \gamma(R) M(f)(x).$$

Using polar coordinates, we calculate that

$$\gamma(R) = \int_{\mathbb{S}^{n-1}} \int_0^R \frac{r^{n-1} \, dr}{r^{n-\alpha}} \, d\theta = \frac{|\mathbb{S}^{n-1}|}{\alpha} R^\alpha.$$

To estimate $I_\alpha^R(f)(x)$, we proceed differently. We directly use the Hölder inequality to write

$$\left| I_\alpha^R(f)(x) \right| \leqslant C \|f\|_{L^p} \left(\int_R^{+\infty} r^{(\alpha-n)p'+n-1} \, dr \right)^{1/p'} \leqslant C \|f\|_{L^p} R^{\alpha-n/p},$$

where we used the hypothesis $p < \alpha/n$. Recalling that $\|f\|_{L^p} = 1$ and combining the above, we arrive at

$$|I_\alpha(f)(x)| \leqslant CR^\alpha M(f)(x) + CR^{\alpha-n/p}.$$

We then choose R such that $R^\alpha M(f)(x) = R^{\alpha-n/p}$, which leads us to the conclusion

$$|I_\alpha(f)(x)| \leqslant C(M(f)(x))^{1-\alpha p/n}.$$

Then, by definition of q, it follows that

$$|I_\alpha(f)(x)|^q \leqslant C(M(f)(x))^p.$$

As $p > 1$ by hypothesis, we can apply the Hardy–Littlewood theorem 6.12 and the hypothesis $\|f\|_{L^p} = 1$ to deduce that

$$\|I_\alpha(f)\|_{L^q}^q \leqslant C'\|f\|_{L^p}^p = C'.$$

This proves the desired result. $\square$

6.6 Sobolev Inequalities

A fundamental corollary of the Hardy–Littlewood–Sobolev theorem 6.20 states that we can estimate L^q norms of a function by L^p norms of its derivatives.

Theorem 6.22 (Sobolev Inequality) *Let $n \geqslant 2$ and let $p \in (1, n)$. Define p^* by*

$$\frac{1}{p^*} = \frac{1}{p} - \frac{1}{n}.$$

There exists a constant C such that, for every function $f \in C_0^\infty(\mathbb{R}^n)$,

$$\|f\|_{L^{p^*}(\mathbb{R}^n)} \leqslant C\|\nabla f\|_{L^p(\mathbb{R}^n)}.$$

Proof We will see in Chapter 8 (*cf.* (8.6)) that, for all $f \in C_0^\infty(\mathbb{R}^n)$,

$$f(x) = -\frac{1}{|\mathbb{S}^{n-1}|}\int_{\mathbb{R}^n} \frac{(x-y)\cdot\nabla f(y)}{|x-y|^n}\,dy,$$

so

$$|f| \leqslant \frac{1}{|\mathbb{S}^{n-1}|}I_1(|\nabla f|).$$

(A direct and elementary proof of this inequality is proposed in Exercise 7.9.) The desired result is then a consequence of the Hardy–Littlewood–Sobolev theorem 6.20.$\square$

We conclude this chapter with a result that extends Theorem 6.22 to the case of fractional derivatives.[3]

Theorem 6.23 (Fractional Sobolev Inequalities) *Consider an integer $n \geqslant 1$ and two real numbers $s \in (0, 1)$ and $p \in (0, n/s)$, then set*

$$p^* = \frac{np}{n - sp}.$$

There exists a constant C such that, for any function f in the Schwartz space $\mathcal{S}(\mathbb{R}^n)$ and any x in $\mathbb{R}^n$, we have

$$|f(x)|^{p^*} \leqslant C\|f\|_{L^{p^*}}^{p^*-p} \int_{\mathbb{R}^n} \frac{|f(x) - f(y)|^p}{|x - y|^{n+ps}} \, dy.$$

It follows that

$$\|f\|_{L^{p^*}} \leqslant C^{1/p} \left(\iint_{\mathbb{R}^n \times \mathbb{R}^n} \frac{|f(x) - f(y)|^p}{|x - y|^{n+ps}} \, dy \, dx \right)^{1/p}.$$

Proof We will denote by C several different constants that depend only on n and p and whose value may change from one line to another.

We start by verifying that the integral

$$\int_{\mathbb{R}^n} \frac{|f(x) - f(y)|^p}{|x - y|^{n+ps}} \, dy$$

is well defined for all f in $\mathcal{S}(\mathbb{R}^n)$ and any x in $\mathbb{R}^n$. For this, it is sufficient to truncate the integral into two parts: the integral over $B(x, 1)$ and that over $\mathbb{R}^n \setminus B(x, 1)$. On $B(x, 1)$, we use the estimate

$$|f(x) - f(y)| \leqslant K|x - y| \quad \text{with} \quad K = \sup_{m \in \mathbb{R}^n} |\nabla f(m)|,$$

and on $\mathbb{R}^n \setminus B(x, 1)$ we write $|f(x) - f(y)| \leqslant 2 \sup |f|$.

Let us now fix $x \in \mathbb{R}^n$ and a real number $t > 0$. Let C_t be the annulus

$$C_t = B(x, 2t) - B(x, t) = \{y \in \mathbb{R}^n : t \leqslant |y - x| < 2t\},$$

and $|C_t|$ its Lebesgue measure. Then

$$|f(x)|^p = \frac{1}{|C_t|} \int_{C_t} |f(x)|^p \, dy$$

$$\leqslant \frac{1}{|C_t|} \int_{C_t} \left(|f(y) - f(x)| + |f(y)| \right)^p \, dy.$$

---

[3] We will not define the notion of fractional derivative but refer to the book by Ponce [155], where one can find in particular the proof of Theorem 6.23 with a remark that credits Brezis for the result and the proof. A related result is proved by Brué and Nguyen in [18] (see also [19]).

Since

$$|a + b|^p \leqslant (2\max\{|a|, |b|\})^p \leqslant 2^p (|a|^p + |b|^p),$$

we deduce that

$$|f(x)|^p \leqslant \frac{2^p}{|C_t|} \int_{C_t} |f(y) - f(x)|^p \, dy + \frac{2^p}{|C_t|} \int_{C_t} |f(y)|^p \, dy.$$

Note that $p^* > p$ by definition of p^*. We can then apply Hölder's inequality $\|uv\|_{L^1} \leqslant \|u\|_{L^r} \|v\|_{L^{r'}}$ with $r = p^*/p$, to get

$$\frac{1}{|C_t|} \int_{C_t} |f(y)|^p \, dy \leqslant \frac{1}{|C_t|} \left(\int_{C_t} |f(y)|^{p^*} \, dy \right)^{p/p^*} \left(\int_{C_t} dy \right)^{1 - p/p^*}.$$

Note that $|C_t| \sim t^n$. We deduce that there exists a constant C such that

$$\frac{1}{|C_t|} \int_{C_t} |f(y)|^p \, dy \leqslant C \|f\|_{L^{p^*}}^p \, t^{sp-n}.$$

Therefore, there exists C such that

$$|f(x)|^p \leqslant Ct^{sp-n} \|f\|_{L^{p^*}}^p + \frac{C}{|C_t|} \int_{C_t} |f(y) - f(x)|^p \, dy.$$

Using again $|C_t| \sim t^n$ and the fact that on C_t we have $t \sim |y - x|$, it follows that

$$|f(x)|^p \leqslant Ct^{sp-n} \|f\|_{L^{p^*}}^p + Ct^{sp} \int_{\mathbb{R}^n} \frac{|f(y) - f(x)|^p}{|y - x|^{n+ps}} \, dy.$$

We then choose t such that

$$\|f\|_{L^{p^*}}^p \, t^{sp-n} = t^{sp} \int_{\mathbb{R}^n} \frac{|f(y) - f(x)|^p}{|y - x|^{n+ps}} \, dy$$

and we deduce the desired inequality.   □

6.7 Calderón's Reconstruction Formula and Wavelets

Consider a signal $f = f(t)$, that is, a function $f \colon \mathbb{R} \to \mathbb{C}$ depending on time t. The aim of Signal Theory is to extract the relevant information, the nature and origin of which can be very diverse depending on the phenomena under consideration. Fourier analysis allows an analysis of the frequencies contained in this signal. For example, if the signal f corresponds to the recording of a piece of music, we can find the notes it contains by looking at its Fourier transform. However, Fourier decomposition does not allow a signal to be analyzed simultaneously in frequency and time. This means that we cannot say which note was played first. Several other decompositions of a signal have been introduced to solve this problem, notably by Alfred Haar [65] and Denis Gabor [54] (we will see in Chapter 9 a use of the Gabor transform, also known as the wave packet transform). The wavelet decomposition is relatively recent: it was introduced independently by Alberto Calderón [23] in 1964 and by Alex Grossmann and Jean Morlet [64] in 1984. The link between Calderón's theoretical work and that of Grossmann and Morlet was made by Yves Meyer, who received the Abel Prize in 2017 "for his pivotal role in the development of the mathematical theory of wavelets" according to the jury's citation (see the text by Stéphane Jaffard [84]). For simplicity, we will limit ourselves to presenting the continuous wavelet transform (as opposed to a discrete transform), for regular functions (to ensure that the integrals are well defined) and in one dimension (to simplify the notation).

Definition 6.24

1. A wavelet is a real-valued function $\psi \in S(\mathbb{R})$ such that $\int_{\mathbb{R}} \psi(t)\, dt = 0$.
2. Given a wavelet $\psi \in S(\mathbb{R})$ and $s > 0$, we introduce the functions:

$$\psi_s(t) = \frac{1}{\sqrt{s}} \psi(t/s), \quad \widetilde{\psi_s}(t) = \psi_s(-t).$$

3. Let $f \in S(\mathbb{R})$. We denote by $Wf(u, s)$ the coefficients of the wavelet transform defined by

$$Wf(u, s) = f * \widetilde{\psi_s}(u) = \int_{\mathbb{R}} f(t) \frac{1}{\sqrt{s}} \psi((t-u)/s)\, dt \quad \text{with} \quad s > 0 \text{ and } u \in \mathbb{R}.$$

Remark 6.25

1. For all $s > 0$, we have

$$\|\psi_s\|_{L^2} = \|\psi\|_{L^2}, \quad \widehat{\psi_s}(\xi) = \sqrt{s}\widehat{\psi}(s\xi).$$

Furthermore, using the fact that ψ_s is real-valued, we directly verify that

$$\widehat{\widetilde{\psi_s}}(\xi) = \sqrt{s}\overline{\widehat{\psi}(s\xi)}. \tag{6.11}$$

2. The name wavelet comes from the fact that the graph of a regular, rapidly decreasing function with zero mean resembles a wave.

Wavelet decomposition is particularly suitable for studying the regularity of a signal. As an illustration, we will show how to study the regularity of the Weierstrass function. This function was the first example found of a function that is continuous everywhere, but differentiable nowhere. It is actually a family of functions $f_{a,b} : \mathbb{R} \to \mathbb{R}$ depending on two parameters $a \in (0, 1)$ and $b \geq 0$, defined by:

$$f_{a,b}(t) = \sum_{k=1}^{+\infty} a^k \cos(b^k t).$$

The function $f_{a,b}$ is well defined and continuous for all $a \in (0,1)$ and all $b \geqslant 0$ (by normal convergence). Weierstrass showed that it is not differentiable anywhere as soon as b is an odd integer satisfying $ab > 1 + 3/2\pi$. Hardy [69] later demonstrated that it is sufficient to assume that $ab \geqslant 1$, with a rather difficult proof. We will give a direct proof of Hardy's result, due to Pierre-Gilles Lemarié-Rieusset [114], which uses only the concept of wavelet transform.

Theorem 6.26 (Hardy) *Consider two positive real numbers a,b such that $a \in (0,1)$ and $ab \geqslant 1$. Then the Weierstrass function $f_{a,b}(t) = \sum_{k=1}^{+\infty} a^k \cos(b^k t)$ is continuous at every point, but differentiable nowhere.*

Proof We follow the proof of[4] [114]. Let $g \in L^1(\mathbb{R})$. By definition of the Fourier transform, we have

$$\int_{\mathbb{R}} \cos(b^k t) g(t)\, dx = \frac{1}{2}\int_{\mathbb{R}} \left(e^{ib^k t} + e^{-ib^k t}\right) g(t)\, dt = \frac{1}{2}\left(\widehat{g}(b^k) + \widehat{g}(-b^k)\right),$$

so, by uniform convergence,

$$\int_{\mathbb{R}} f(t) g(t)\, dt = \frac{1}{2}\sum_{k=1}^{+\infty} a^k \left(\widehat{g}(b^k) + \widehat{g}(-b^k)\right).$$

We will apply this identity to the functions $g_{s,u}(t) = \psi_s(t-u)$, whose Fourier transforms are given by

$$\widehat{g_{s,u}}(\xi) = \widehat{\psi_s}(\xi)e^{-i\xi u} = \sqrt{s}\,\widehat{\psi}(s\xi)e^{-i\xi u}.$$

As a result,

$$Wf(u,s) = \int_{\mathbb{R}} f(t) g_{s,u}(t)\, dt = \frac{1}{2}\sum_{k=1}^{+\infty} a^k \sqrt{s}\left(\widehat{\psi}(sb^k)e^{-ib^k u} + \widehat{\psi}(-sb^k)e^{ib^k u}\right).$$

Then, by choosing the scale $s = b^{-p}$ with $p \in \mathbb{N}^*$, we have

$$Wf(u,b^{-p}) = \frac{b^{-p/2}}{2}\sum_{k=1}^{+\infty} a^k \left(\widehat{\psi}(b^{k-p})e^{-ib^k u} + \widehat{\psi}(-b^{k-p})e^{ib^k u}\right).$$

Now, we will apply this formula for a well-chosen function ψ, such that its Fourier transform $\Psi = \widehat{\psi}$ is supported in an annulus of the form $\{\xi \in \mathbb{R} : \alpha \leqslant |\xi| \leqslant \beta\}$ where the numbers α and β are such that

$$0 < \alpha < 1 < \beta < \alpha b.$$

Note that the conditions $ab \geqslant 1$ and $a \in (0,1)$ imply that $b > 1$, which guarantees the existence of a pair (α,β) satisfying the previous conditions. Moreover, we have

---

[4] The proof in [114] uses conventions that make it even simpler and more natural. We prefer here to keep the notations used in Definition 6.24 to avoid creating confusion.

$\beta < \alpha b < b^\ell$ for all $\ell \geqslant 1$. Consequently, $\Psi(b^\ell) = 0$ as soon as $\ell \neq 0$ and we conclude that

$$W f(u, b^{-P}) = \frac{1}{2} a^P b^{-P/2} \left(\Psi(1) e^{-ib^P u} + \Psi(-1) e^{ib^P u} \right). \tag{6.12}$$

We now argue by contradiction. Suppose that f is differentiable at point t_0, so that we can write $f(t_0 + h) = f(t_0) + f'(t_0)h + h\varepsilon(h)$ where $\varepsilon(h)$ tends to 0 when h goes to 0 (and moreover ε is bounded on $\mathbb{R}$ since f is bounded). By hypothesis on the support of Ψ, we have

$$\int_{\mathbb{R}} \psi(t) \, dt = \Psi(0) = 0, \quad \int_{\mathbb{R}} t\psi(t) \, dt = i\Psi'(0) = 0,$$

which leads to

$$W f(t_0, b^{-P}) = \int_{\mathbb{R}} f(t) \frac{1}{\sqrt{b^{-P}}} \psi \left((t - t_0)/b^{-P} \right) dt$$

$$= b^{P/2} \int_{\mathbb{R}} f(t) \psi \left(b^P t - b^P t_0 \right) dt$$

$$= b^{-P/2} \int_{\mathbb{R}} f(t_0 + b^{-P} t') \psi(t') \, dt'$$

$$= b^{-3P/2} \int_{\mathbb{R}} \varepsilon(b^{-P} t) \psi(t) \, dt.$$

The dominated convergence theorem then implies that

$$W f(t_0, b^{-P}) = o(b^{-3P/2}). \tag{6.13}$$

By combining (6.12) applied with $u = t_0$ and (6.13), we obtain $\lim_{P \to +\infty} a^P b^P = 0$, which contradicts the hypothesis $ab \geqslant 1$. This concludes the proof. $\qquad\square$

As the previous proof illustrated so well, wavelet decomposition is a very powerful tool, both for theoretical harmonic analysis and for applications.[5] We will only prove two elementary results that are at the origin of this theory.[6]

Proposition 6.27 *Consider a wavelet* $\psi \in \mathcal{S}(\mathbb{R})$. *Then* $C_\psi = \int_0^{+\infty} \frac{|\hat{\psi}(\xi)|^2}{\xi} \, d\xi < +\infty$ *and, for all* $f \in \mathcal{S}(\mathbb{R})$,

$$\int_{\mathbb{R}} |f(t)|^2 \, dt = \frac{1}{C_\psi} \int_0^{+\infty} \int_{\mathbb{R}} |W f(u, s)|^2 \, du \, \frac{ds}{s^2}.$$

---

[5] We refer to the very beautiful article by Yves Meyer [137], and especially to section 11 which contains another application of wavelets to the study of the differentiability of a series introduced by Riemann, a question that inspired Weierstrass.

[6] We refer the reader to the books by Kahane and Lemarié-Rieusset [94], Mallat [128] and Meyer [135] for the general theory.

Proof As $\psi \in S(\mathbb{R})$, we have $\widehat{\psi} \in S(\mathbb{R})$ (see Proposition 5.8). We deduce that $\widehat{\psi} \in L^2(\mathbb{R})$, so $\int_1^{+\infty} |\widehat{\psi}(\xi)|^2 / \xi \, d\xi < +\infty$. Moreover, as $\widehat{\psi}(0) = \int_{\mathbb{R}} \psi(t) \, dt = 0$ and as $\widehat{\psi}$ is of class C^1, we have $|\widehat{\psi}(\xi)| \leq K\xi$ for $\xi \in [0, 1]$, which obviously implies that $\int_0^1 |\widehat{\psi}(\xi)|^2 / \xi \, d\xi < +\infty$. We deduce that $C_\psi < +\infty$.

Let us denote by $Wf(\cdot, s)$ the function $Wf(\cdot, s) \colon u \mapsto Wf(u, s)$. For all $s > 0$, the identity (6.11) implies that

$$\widehat{Wf(\cdot, s)}(\xi) = \widehat{f}(\xi) \widehat{\overline{\psi_s}}(\xi) = \widehat{f}(\xi) \sqrt{s} \, \overline{\widehat{\psi}(s\xi)}.$$

Then, Plancherel's identity (5.8) implies that

$$\|Wf(\cdot, s)\|_{L^2}^2 = \frac{1}{2\pi} \int_{\mathbb{R}} |\widehat{f}(\xi) \sqrt{s} \, \widehat{\psi}(s\xi)|^2 \, d\xi.$$

Observing that

$$\int_0^{+\infty} |\sqrt{s} \, \widehat{\psi}(s\xi)|^2 \, \frac{ds}{s^2} = \int_0^{+\infty} |\widehat{\psi}(s\xi)|^2 \, \frac{ds}{s} = \int_0^{+\infty} |\widehat{\psi}(\theta)|^2 \, \frac{d\theta}{\theta} = C_\psi,$$

and then using Fubini's theorem (for positive-valued functions) and Plancherel's identity (5.8) again, we obtain

$$\int_0^{+\infty} \int_{\mathbb{R}} |Wf(u, s)|^2 \, du \, \frac{ds}{s^2} = \int_0^{+\infty} \|Wf(\cdot, s)\|_{L^2}^2 \, \frac{ds}{s^2}$$

$$= \frac{C_\psi}{2\pi} \int_{\mathbb{R}} |\widehat{f}(\xi)|^2 \, d\xi = C_\psi \|f\|_{L^2}^2,$$

which concludes the proof. ☐

We then deduce a formula for reconstructing f from its wavelet transform.[7]

---

[7] The fact that such a formula must exist is well understood from a formal point of view. Indeed, we have just seen that the map $f \mapsto Wf$ is an isometry in the sense that

$$\|f\|_{L^2(\mathbb{R};dt)} = \|Wf\|_{L^2(\mathbb{R}\times\mathbb{R}_+;du\,ds/s^2)}.$$

Using a polarization identity, we deduce that

$$\langle g, f \rangle_{L^2(\mathbb{R};dt)} = \langle Wg, Wf \rangle_{L^2(\mathbb{R}\times\mathbb{R}_+;du\,ds/s^2)} = \langle g, W^*Wf \rangle_{L^2(\mathbb{R};dt)},$$

and therefore (formally) $f = W^*Wf$. Note that the same principle applies for all decompositions that are isometries between two spaces of type $L^2(X; d\mu)$. In particular, this observation applies to the Fourier transform and allows the Fourier inversion formula to be obtained from Plancherel's identity (5.8) (note that in Chapter 5 we did the opposite: we proved Plancherel's identity after having proved the Fourier inversion formula). This principle also applies to the wave packet transform that we will introduce later in Lemma 9.9.

Theorem 6.28 (Calderón's reconstruction formula) *Consider a wavelet* $\psi \in S(\mathbb{R})$. *Then, for all* $f \in S(\mathbb{R})$,

$$f = \frac{1}{C_\psi} \int_0^{+\infty} f * \widetilde{\psi_s} * \psi_s(t) \frac{ds}{s^2} = \frac{1}{C_\psi} \int_0^{+\infty} \int_{\mathbb{R}} Wf(u,s) \frac{1}{\sqrt{s}} \psi\left((t-u)/s\right) du \frac{ds}{s^2}$$

in the sense where the functions $\{S_{\varepsilon,\rho}\}_{0<\varepsilon<\rho}$ *defined by*

$$S_{\varepsilon,\rho}(t) = \int_\varepsilon^\rho f * \widetilde{\psi_s} * \psi_s(t) \frac{ds}{s^2}$$

converge towards f *in* $L^2(\mathbb{R})$ *when* ε *tends towards* 0 *and* ρ *towards* $+\infty$.

Proof For all $t > 0$, the function $s \mapsto f * \widetilde{\psi_s} * \psi_s(t)$ is continuous and therefore $S_{\varepsilon,\rho}(t)$ is well defined. Moreover, it follows from Young's theorem that, for all $s > 0$, the function $f * \widetilde{\psi_s} * \psi_s$ belongs to $L^2(\mathbb{R})$. Furthermore, using identities already seen in the proof of the previous proposition, we have successively

$$\mathcal{F}(f * \widetilde{\psi_s} * \psi_s)(\xi) = \widehat{f}(\xi)\widetilde{\widehat{\psi_s}}(\xi)\widehat{\psi_s}(\xi) = \widehat{f}(\xi)s|\widehat{\psi}(s\xi)|^2$$

and

$$\widehat{S_{\varepsilon,\rho}}(\xi) = \frac{1}{C_\psi} \int_\varepsilon^\rho \widehat{f}(\xi)|\widehat{\psi}(s\xi)|^2 \frac{ds}{s} = \widehat{f}(\xi)\left(\frac{1}{C_\psi}\int_{s\varepsilon}^{s\rho}|\widehat{\psi}(\zeta)|^2 \frac{d\zeta}{\zeta}\right),$$

which implies the desired convergence result, according to the dominated convergence theorem and Plancherel's identity (5.8). □

6.8 Wiener's Tauberian Theorem

Consider two functions f and h in $L^1(\mathbb{R}^n)$ with $n \geqslant 1$ arbitrary. We are interested in the functional equation $f * g = h$ with unknown $g \in L^1(\mathbb{R}^n)$. This equation generally does not have an exact solution. However, an important theorem by Wiener [196] demonstrates that it admits approximate solutions as long as the Fourier transform of f does not vanish. Recall that we denote by $\widehat{f} = \mathcal{F}f$ the Fourier transform of a function $f \in L^1(\mathbb{R}^n)$, defined by $\widehat{f}(\xi) = \int_{\mathbb{R}^n} e^{-ix\cdot\xi} f(x)\, dx$.

Theorem 6.29 (Wiener) *Let* $f \in L^1(\mathbb{R}^n)$ *such that* $\widehat{f}(\xi) \neq 0$ *for all* $\xi \in \mathbb{R}^n$. *For all* $h \in L^1(\mathbb{R}^n)$ *and for all* $\varepsilon > 0$, *there exists* $g \in L^1(\mathbb{R}^n)$ *such that*

$$\|f * g - h\|_{L^1} < \varepsilon. \tag{6.14}$$

Proof Introduce the subspace $\mathcal{A}$ of $C_b^0(\mathbb{R}^n)$ defined by $\mathcal{A} = \mathcal{F}(L^1(\mathbb{R}^n))$. By virtue of the injectivity of the Fourier transform (see Corollary 5.14), for all $U \in \mathcal{A}$, there exists a unique $u \in L^1(\mathbb{R}^n)$ such that $U = \widehat{u}$. We can therefore define a norm on $\mathcal{A}$

by setting

$$\|U\|_{\mathcal{A}} = \|u\|_{L^1}.$$

As the Fourier transform $\mathcal{F}$ is an isometry and a bijection from $L^1(\mathbb{R}^n)$ to $\mathcal{A}$, we obtain that $(\mathcal{A}, \|\cdot\|_{\mathcal{A}})$ is a Banach space.

Let us verify next that $\mathcal{A}$ is an algebra, that is, let us show that $\mathcal{A}$ is stable under multiplication. Consider two functions $U \in \mathcal{A}$ and $V \in \mathcal{A}$. Then there exist two integrable functions u, v in $L^1(\mathbb{R}^n)$ such that $U = \widehat{u}$ and $V = \widehat{v}$. Proposition 6.3 implies that the convolution product $u * v$ belongs to $L^1(\mathbb{R}^n)$, satisfies $\|u * v\|_{L^1} \leqslant \|u\|_{L^1} \|v\|_{L^1}$ and, moreover, $UV = \widehat{uv} = \mathcal{F}(u * v)$. We deduce that the product UV belongs to $\mathcal{A}$ with the additional estimate

$$\|UV\|_{\mathcal{A}} \leqslant \|U\|_{\mathcal{A}} \|V\|_{\mathcal{A}}. \tag{6.15}$$

Now consider two functions $f \in L^1(\mathbb{R}^n)$ and $h \in L^1(\mathbb{R}^n)$. We want to show that for all $\varepsilon > 0$, there exists $g \in L^1(\mathbb{R}^n)$ such that

$$\|f * g - h\|_{L^1} < \varepsilon. \tag{6.16}$$

The preceding discussion allows us to reformulate this in the following equivalent way: prove that there exists $G \in \mathcal{A}$ such that

$$\|FG - H\|_{\mathcal{A}} < \varepsilon \quad \text{where} \quad F = \widehat{f}, \quad H = \widehat{h}.$$

We will then obtain (6.16) with the unique function $g \in L^1(\mathbb{R}^n)$ such that $G = \widehat{g}$.

We have reduced the proof of Wiener's theorem to showing that the set $F\mathcal{A}$ is dense in $\mathcal{A}$, as soon as $F \in \mathcal{A}$ is a non-vanishing function. This is the subject of the following lemma.

Lemma 6.30 *Let $f \in L^1(\mathbb{R}^n)$ such that $\widehat{f}(\xi) \neq 0$ for all $\xi \in \mathbb{R}^n$. Let $F = \widehat{f}$. Then,*

1. $C_0^\infty(\mathbb{R}^n) \subset \mathcal{A}$ and $C_0^\infty(\mathbb{R}^n)$ is dense in $\mathcal{A}$;
2. $C_0^\infty(\mathbb{R}^n) \subset F\mathcal{A}$.

Proof **Step 1: $C_0^\infty(\mathbb{R}^n) \subset \mathcal{A}$**

Let $F \in C_0^\infty(\mathbb{R}^n)$. Then F belongs to the Schwartz space $S(\mathbb{R}^n)$. But the Fourier transform is an isomorphism from $S(\mathbb{R}^n)$ to $S(\mathbb{R}^n)$ (see Corollary 5.12), so there exists $f \in S(\mathbb{R}^n)$ such that $F = \widehat{f}$. As $S(\mathbb{R}^n) \subset L^1(\mathbb{R}^n)$, this proves $F \in \mathcal{A}$.

Step 2: density of $C_0^\infty(\mathbb{R}^n)$ in $\mathcal{A}$

Now consider a continuous linear form $\mu: \mathcal{A} \to \mathbb{C}$. As $\mathcal{F}: L^1(\mathbb{R}^n) \to \mathcal{A}$ is continuous (by definition of the norm $\|\cdot\|_{\mathcal{A}}$), it follows that $\mu \circ \mathcal{F}: L^1(\mathbb{R}^n) \to \mathbb{C}$ is a continuous linear form. According to a classical result on the duality of Lebesgue spaces (see Theorem 3.28), this implies that there exists a function $g \in L^\infty(\mathbb{R}^n)$ such that

$$\forall \varphi \in L^1(\mathbb{R}^n), \quad \mu \circ \mathcal{F}(\varphi) = \int_{\mathbb{R}^n} g\varphi \, dx.$$

It follows that

$$\forall F \in \mathcal{A}, \quad \mu(F) = \int_{\mathbb{R}^n} g\mathcal{F}^{-1}(F) \, dx.$$

Suppose that μ vanishes on $C_0^\infty(\mathbb{R}^n)$. Consider a non-zero function $\Phi \in C_0^\infty(\mathbb{R}^n)$ and let $\phi = \mathcal{F}^{-1}(\Phi)$ (then $\phi \neq 0$). Consider the function $\Phi_{a,b}(\xi) = e^{ia\cdot\xi}\Phi(\xi + b)$. Then $\Phi_{a,b} \in C_0^\infty(\mathbb{R}^n)$ and $\mathcal{F}^{-1}(\Phi_{a,b})(x) = e^{-ia\cdot b}e^{-ib\cdot x}\phi(a+x)$. Therefore, writing that $\mu(\Phi_{a,b}) = 0$, we obtain

$$\forall(a,b) \in \mathbb{R}^n \times \mathbb{R}^n, \quad \int_{\mathbb{R}^n} g(x)e^{-ib\cdot x}\phi(a+x) \, dx = 0.$$

Note that the function $x \mapsto g(x)\phi(a+x)$ belongs to $L^1(\mathbb{R}^n)$. The previous identity implies that the Fourier transform of this function is zero, and therefore this function is zero according to the injectivity of the Fourier transform (see Corollary 5.14). This means that for almost all $x \in \mathbb{R}^n$, and for all $a \in \mathbb{R}^n$, we have $g(x)\phi(a+x) = 0$. But $\phi \in S(\mathbb{R}^n)$ is non-zero, therefore, obviously, for all $x \in \mathbb{R}^n$, there exists $a \in \mathbb{R}^n$ such that $\phi(a+x) \neq 0$ and we deduce that $g(x) = 0$, hence $\mu = 0$. This proves that $C_0^\infty(\mathbb{R}^n)$ is dense in $\mathcal{A}$ by Corollary 3.24 of the Hahn–Banach theorem.

Step 3: a technical lemma

Consider two functions $F \in \mathcal{A}$ and $G \in C_0^\infty(\mathbb{R}^n)$ and a real number $\varepsilon \in (0, 1]$. Let us show that the functions F_ε defined by

$$F_\varepsilon(\xi) = (F(\xi) - F(0))G(\xi/\varepsilon)$$

converge to 0 in $\mathcal{A}$ when ε tends to 0.

Let $g \in L^1(\mathbb{R}^n)$ such that $G = \mathcal{F}(g)$. Then $\xi \mapsto G(\xi/\varepsilon)$ is the Fourier transform of $x \mapsto \varepsilon^n g(\varepsilon x)$. Therefore $F_\varepsilon = \mathcal{F}(h_\varepsilon)$ where

$$h_\varepsilon(x) = f * g_\varepsilon(x) - \left(\int f \, dy\right)g_\varepsilon(x)$$

$$= \int g_\varepsilon(x - y)f(y) \, dy - \left(\int f \, dy\right)g_\varepsilon(x)$$

$$= \int \varepsilon^n (g(\varepsilon(x - y)) - g(\varepsilon x))f(y) \, dy.$$

Then

$$\int |h_\varepsilon(x)| \, dx \leqslant \iint |g(u - \varepsilon y) - g(u)| \, |f(y)| \, dy du$$

which tends to 0 by the dominated convergence theorem.

Step 4: the case of functions localized in balls of sufficiently small radii

Let $\varepsilon \in (0, 1]$. Now consider a function $\Psi \in C_0^\infty(\mathbb{R}^n)$ equal to 1 on the ball $B(0, 1)$ and zero outside of $B(0, 2)$. We will show that, if ε is small enough, then $\Psi(\cdot/\varepsilon) \in F\mathcal{A}$.

Note that if $\Psi(\xi/\varepsilon) \neq 0$, then $\Psi(\xi/2\varepsilon) = 1$. Consequently, we verify that

$$\begin{cases} \dfrac{\Psi(\xi/\varepsilon)}{F(\xi)} = \dfrac{\Psi(\xi/\varepsilon)}{F(0)\left(1 + \frac{\Psi(\xi/2\varepsilon)(F(\xi)-F(0))}{F(0)}\right)} = \dfrac{\Psi(\xi/\varepsilon)}{F(0)} \dfrac{1}{1 + R_\varepsilon(\xi)}, \quad \text{where} \\[2em] R_\varepsilon(\xi) = \dfrac{\Psi(\xi/2\varepsilon)(F(\xi) - F(0))}{F(0)}. \end{cases}$$

Moreover, the technical result proved in the previous step implies that the function R_ε tends to 0 in $\mathcal{A}$ when ε tends to 0. In particular, for $\varepsilon \leqslant \varepsilon_0$ small enough, we have $\|R_\varepsilon\|_{\mathcal{A}} < 1$. As $\mathcal{A}$ is a Banach algebra, we deduce that the series $\sum_{k=0}^{+\infty}(-1)^k R_\varepsilon^k$ converges normally (and thus converges) towards the function $1/(1 + R_\varepsilon)$ and that this latter function belongs to $\mathcal{A}$. As $\Psi(\cdot/\varepsilon) \in C_0^\infty(\mathbb{R}^n) \subset \mathcal{A}$ (from the first step), using the fact that $\mathcal{A}$ is an algebra, we have therefore shown that

$$\frac{\Psi(\cdot/\varepsilon)}{F} = \frac{\Psi(\cdot/\varepsilon)}{F(0)} \frac{1}{1 + R_\varepsilon} \in \mathcal{A} \cdot \mathcal{A} \subset \mathcal{A}. \tag{6.17}$$

In particular, by multiplying the two members of (6.17) by F, we deduce that

$$\Psi(\cdot/\varepsilon) \in F\mathcal{A}. \tag{6.18}$$

Step 5: conclusion

Now consider a function $H \in C_0^\infty(\mathbb{R}^n)$ and let $K = \operatorname{supp} H$, which is compact by hypothesis. For all $\varepsilon > 0$, there exist $N \in \mathbb{N}$ and a collection $\{B_\ell\}_{1 \leqslant \ell \leqslant N}$ of balls of radius ε such that $K \subset \bigcup_{1 \leqslant \ell \leqslant N} B_\ell$. Proposition 17.35 implies that there exists a partition of unity subordinate to this cover, that is, N functions ζ_ℓ, $1 \leqslant \ell \leqslant N$, of class C^∞ and such that $\operatorname{supp} \zeta_\ell \subset B_\ell$ with the additional property:

$$\forall \xi \in K, \quad \sum_{\ell=1}^N \zeta_\ell(\xi) = 1.$$

We can then decompose H into a sum $H = H_1 + \cdots + H_N$ with $H_\ell = \zeta_\ell H$. For all ℓ, the function ζ_ℓ belongs to $C_0^\infty(\mathbb{R}^n)$ and therefore to $\mathcal{A}$ according to the result shown in the first step. Then $H_\ell \in \mathcal{A}$ as a product of two elements of $\mathcal{A}$. Moreover, as the functions H_ℓ are supported in balls of radius ε, we have $H_\ell(\xi) = \Psi((\xi - \xi_\ell)/\varepsilon)H_\ell(\xi)$ for a certain $\xi_\ell \in \mathbb{R}^n$. Now, observe that $\Psi((\cdot - \xi_\ell)/\varepsilon) \in F\mathcal{A}$ (the previous result (6.18) is of course invariant under translation by ξ_ℓ). By combining the above, we conclude that

$$H_\ell \in F\mathcal{A} \cdot \mathcal{A} \subset F\mathcal{A}.$$

So the sum $H = H_1 + \cdots + H_N$ also belongs to $F\mathcal{A}$.   □

According to the reasoning explained before the statement of the previous lemma, this completes the proof of Wiener's theorem.   □

The Tauberian theorem of Wiener that we have just proved is related to Wiener's theorem proved in Section 4.5. For the links between these statements, we refer to the original article by Wiener [196] as well as Gårding [56, Section 5] and Kahane [89]. Another classical formulation (see Chapter 9 of Rudin's book [164]) concerns the density of subspaces generated by the translates of functions from L^1.

6.9 Exercises

Exercise 6.1

1. Let $f \in L^1(\mathbb{R})$ and $g \in L^\infty(\mathbb{R})$. Show that $f * g$ is a continuous and bounded function. We can admit that the space of continuous functions with compact support $C_0(\mathbb{R}^n)$ is dense in $L^1(\mathbb{R}^n)$.
2. Consider a set $A \subset [0, 1]$ of Lebesgue measure $|A| > 0$. Let f be the indicator function of A and let $g(x) = f(-x)$. Show that $f * g(0) > 0$ and deduce that the set $A - A$ contains an open interval.

Exercise 6.2 (Friedrichs' Lemma)

Consider a function $\rho: \mathbb{R} \to \mathbb{R}_+$, C^∞ with compact support and such that $\int_{-\infty}^\infty \rho(x)\,dx = 1$. For all $\varepsilon > 0$, we define

$$\rho_\varepsilon(x) = \frac{1}{\varepsilon}\rho(x/\varepsilon),$$

and we denote by J_ε the operator defined by $J_\varepsilon u = \rho_\varepsilon * u$.

1. Prove that, for all u in $L^2(\mathbb{R})$, we have $J_\varepsilon u \in C^\infty(\mathbb{R})$. Recall the general statement that implies that, for all u in $L^2(\mathbb{R})$, $J_\varepsilon u$ converges to u in $L^2(\mathbb{R})$ when ε tends to 0.
2. Consider a function $a: \mathbb{R} \to \mathbb{R}$ that is of class C^1, bounded and whose derivative is bounded. We denote by L the differential operator defined by $Lu = au'$ where u' is the derivative of u. Show that L is continuous from $H^1(\mathbb{R})$ to $L^2(\mathbb{R})$.

3. We denote by $[J_\varepsilon, L]$ the commutator defined by

$$[J_\varepsilon, L]u = J_\varepsilon(Lu) - L(J_\varepsilon u) = \rho_\varepsilon * (au') - a(\rho_\varepsilon * u)'.$$

Show that for all $u \in C_0^1(\mathbb{R})$, $[J_\varepsilon, L]u$ converges to 0 in $L^2(\mathbb{R})$ when $\varepsilon \to 0$.

4. Let ρ'_ε be the derivative of the function ρ_ε. Verify that for all $u \in C_0^1(\mathbb{R})$,

$$[J_\varepsilon, L]u(x) = \int \rho'_\varepsilon(x - y)(a(y) - a(x))u(y)\, dy - \int \rho_\varepsilon(x - y)a'(y)u(y)\, dy,$$

then show that

$$\left| \int \rho'_\varepsilon(x - y)(a(y) - a(x))u'(y)\, dy \right| \leqslant C\left(\left| x\rho'_\varepsilon \right| * |u| \right).$$

5. Deduce that there exists a constant C such that, for all $u \in C_0^1(\mathbb{R})$,

$$\|[J_\varepsilon, L]u\|_{L^2} \leqslant C\|u\|_{L^2}.$$

Then deduce that $[J_\varepsilon, L]$ extends to a continuous linear operator on $L^2(\mathbb{R})$ with values in $L^2(\mathbb{R})$ and that $[J_\varepsilon, L]u$ tends to 0 when $\varepsilon \to 0$ for all u in $L^2(\mathbb{R})$.

Chapter 7
Sobolev Spaces

We will focus on the function spaces introduced in the period 1935-1938 by Sobolev [174], and which have since played an absolutely central role in the study of partial differential equations.[1]

7.1 Introduction

Consider two functions $u\colon [a,b] \to \mathbb{R}$ and $\phi\colon [a,b] \to \mathbb{R}$ of class C^1. The integration by parts formula is written as

$$\int_a^b u'(x)\phi(x)\,\mathrm{d}x = u(b)\phi(b) - u(a)\phi(a) - \int_a^b u(x)\phi'(x)\,\mathrm{d}x.$$

If the support of ϕ is contained in a compact subset of the open interval (a,b), then $\phi(a) = 0$ and $\phi(b) = 0$, so that

$$\int_a^b u'(x)\phi(x)\,\mathrm{d}x = - \int_a^b u(x)\phi'(x)\,\mathrm{d}x.$$

---

[1] In the late thirties and early forties, Sergei Sobolev, Jean Leray, and Laurent Schwartz realized that it was necessary to generalize the notion of derivative, and to introduce spaces adapted to these new definitions, in order to solve partial differential equations. Sobolev was interested in hyperbolic equations and Leray, in his fundamental article [115], gave an implicit definition of the space H^1, used to study so-called turbulent solutions of the Navier–Stokes equation. For his theory of distributions, which generalizes the notion of derivative in the weak sense introduced by Sobolev and Leray [170], Schwartz was awarded the Fields Medal, and was the first Frenchman to receive it. The links between the work of Sobolev and that of Leray and Schwartz, and more generally the deep ties that united the Russian and French schools of mathematics in the interwar period, notably thanks to Jacques Hadamard, are explained by Jean-Michel Kantor (see [95]). We also refer to Chapter 7 of Burenkov [21] for numerous references about the theory and the history of Sobolev spaces.

© The Author(s), under exclusive license to Springer Nature Switzerland AG 2024
T. Alazard, *Analysis and Partial Differential Equations*, Universitext,
https://doi.org/10.1007/978-3-031-70909-8_7

We now recall the following result.

Lemma 7.1 *Let $p \in [1, +\infty]$ and let $I \subset \mathbb{R}$ be an open interval. If $f \in L^p(I)$ satisfies*

$$\forall \phi \in C_0^\infty(I), \quad \int_I f(x)\phi(x)\, dx = 0,$$

then $f = 0$.

Proof This is a direct consequence of Corollary 6.10.   □

From this lemma and the preceding calculation, we deduce that the derivative u' is the unique function $v : [a, b] \to \mathbb{R}$ such that

$$\forall \phi \in C_0^\infty((a, b)), \quad \int_a^b v(x)\phi(x)\, dx = - \int_a^b u(x)\phi'(x)\, dx.$$

Note then that both sides of this identity are well defined for all functions u and v that are integrable over $[a, b]$. This suggests considering a generalization of the notion of derivative.

Definition 7.2 Let $p \in [1, +\infty]$ and let $I \subset \mathbb{R}$ be an open interval. We say that a function $u \in L^p(I)$ has a weak derivative in $L^p(I)$ if there exists $v \in L^p(I)$ such that

$$\forall \phi \in C_0^\infty(I), \quad \int_I v(x)\phi(x)\, dx = - \int_I u(x)\phi'(x)\, dx.$$

We denote v by u' (or $\partial u / \partial x$ or $\partial_x u$).

Proposition 7.3 *Every function u of class C^1 on a compact interval $[a, b]$ has a weak derivative in $L^p(a, b)$ for all $p \in [1, +\infty]$.*

Proof This is a consequence of the preceding discussion.   □

Remark 7.4 Let us emphasize that we must assume that u is of class C^1 on a compact interval. The function $u(x) = 1/x$ for example is of class C^1 on $(0, 1)$ but is not in $L^p(0, 1)$, whatever p. Another interesting example is that of the function $x^{-1/3}$, which is of class C^1 on $(0, 1)$, which belongs to $L^p(0, 1)$ if $1 \leqslant p < 3$, but whose usual derivative does not belong to any space L^p (thus $x^{-1/3}$ is not differentiable in the weak sense in L^p on $(0, 1)$).

The following result guarantees the uniqueness of the derivative in the weak sense.

Proposition 7.5 *Let $p \in [1, +\infty]$ and let $I \subset \mathbb{R}$ be an open interval. If $u \in L^p(I)$ has a weak derivative in $L^p(I)$, then it is unique.*

Proof This is a consequence of Lemma 7.1.   □

There exist functions that are differentiable in the weak sense without being differentiable in the classical sense (see Exercise 7.1). There are L^1 functions that do not have a weak derivative in L^1. We have already seen the example of the function $x \mapsto x^{-1/3}$ on $(0, 1)$; see also Exercise 7.2.

We can generalize the previous definition to the case of any dimension. Suppose that Ω is any open set of $\mathbb{R}^n$ and consider a function $u \in C^1(\overline{\Omega})$, which means that there exists a function $v \in C^1(\mathbb{R}^n)$ such that $u = v|_{\overline{\Omega}}$. Consider a function $\phi \in C_0^\infty(\Omega)$. Using the fact that the support of ϕ is included in Ω, one can write

$$\int_\Omega u \frac{\partial \phi}{\partial x_j}\, dx = \int_\Omega v \frac{\partial \phi}{\partial x_j}\, dx = \int_{\mathbb{R}^n} v \frac{\partial \phi}{\partial x_j}\, dx = - \int_{\mathbb{R}^n} \phi \frac{\partial v}{\partial x_j}\, dx = - \int_\Omega \phi \frac{\partial u}{\partial x_j}\, dx,$$

where the integration by parts used to get the third equality is easily justified since ϕ has compact support. Pay attention to the fact that we cannot apply the divergence theorem (whose proof is given by the solved Exercise 17.3) since, to apply this result, we would need to make a regularity assumption on the boundary of Ω, whereas here we are considering an arbitrary open set.

Definition 7.6 Let $p \in [1, +\infty]$ and let Ω be an open set of $\mathbb{R}^n$. We say that a function $u \in L^p(\Omega)$ has a weak derivative in $L^p(\Omega)$ with respect to the variable x_j if there exists $v \in L^p(\Omega)$ such that

$$\forall \phi \in C_0^\infty(\Omega), \quad \int_\Omega v\phi\, dx = - \int_\Omega u \frac{\partial \phi}{\partial x_j}\, dx. \tag{7.1}$$

We denote v by $\partial u/\partial x_j$ or $\partial_{x_j} u$ or simply $\partial_j u$.

If u is complex-valued, we say that u admits a weak derivative if its real part and its imaginary part admit weak derivatives. We define in the same way the weak derivative of a function with values in $\mathbb{R}^m$ or $\mathbb{C}^m$.

By definition, the Sobolev space $W^{1,p}(\Omega)$ is the set of functions u (with values in $\mathbb{R}^m$ or $\mathbb{C}^m$ with $m \geqslant 1$) belonging to $L^p(\Omega)$ and which admit weak derivatives $\partial_j u$ in $L^p(\Omega)$ for all j such that $1 \leqslant j \leqslant n$.

Remark 7.7

1. If u admits a weak derivative in $L^p(\Omega)$ with respect to the variable x_j, then it is unique according to Corollary 6.10.
2. Suppose that Ω is a bounded open set and that $u \in C^1(\overline{\Omega})$. We have already seen that

$$\forall \phi \in C_0^\infty(\Omega), \quad \int_\Omega u \frac{\partial \phi}{\partial x_j}\, dx = - \int_\Omega \phi \frac{\partial u}{\partial x_j}\, dx.$$

Furthermore, as Ω is a bounded open set, the function $\partial_{x_j} u$, which is continuous on $\overline{\Omega}$, is bounded on $\overline{\Omega}$ and therefore $\partial_{x_j} u$ belongs to $L^p(\Omega)$ for all p in $[1, +\infty]$. We deduce that u is weakly differentiable with respect to x_j and that its weak derivative is given by $\partial_{x_j} u$. Therefore, for every bounded open set Ω and every p in $[1, +\infty]$, the space $C^1(\overline{\Omega})$ is included in $W^{1,p}(\Omega)$. For the same reasons, whatever the open set Ω included in $\mathbb{R}^n$, the space $C_0^1(\Omega)$ is included in $W^{1,p}(\Omega)$.

However, $C^1(\Omega)$ is not included in $W^{1,p}(\Omega)$. Indeed, $C^1(\Omega)$ is not included in $L^p(\Omega)$ as we have already seen (think for example of $\exp(x^2)$ on $\Omega = \mathbb{R}$ or $1/|x|$ on $\Omega = (0, 1)$).

In identity (7.2), we say that ϕ is a test function. Note that in the definition of the weak derivative we can assume that the test functions belong to $C_0^1(\Omega)$ instead of $C_0^\infty(\Omega)$.

Proposition 7.8 *Let* $p \in [1, +\infty]$. *Consider an integer* $1 \leqslant j \leqslant n$ *and two functions* u, v *in* $L^p(\Omega)$. *Then* v *is the weak derivative of* u *with respect to* x_j *if and only if*

$$\forall \phi \in C_0^1(\Omega), \quad \int_\Omega v\phi \, dx = - \int_\Omega u \frac{\partial \phi}{\partial x_j} \, dx. \tag{7.2}$$

Proof Note that (7.2) trivially implies (7.1).

Conversely, suppose that (7.1) is satisfied and consider a function $\phi \in C_0^1(\Omega)$. Introduce a function Φ which is of class C^∞, with compact support, radial, decreasing and of integral 1 (see (6.4)). Let $\Phi_t(x) = t^{-n}\Phi(x/t)$, where $t \in (0, 1]$. Then Proposition 6.6 implies that $\Phi_t * \phi$ is a C^∞ function with compact support and moreover its support is included in Ω for small enough t. We can use (7.1) to write

$$\int_\Omega v(\Phi_t * \phi) \, dx = - \int_\Omega u \frac{\partial (\Phi_t * \phi)}{\partial x_j} \, dx = - \int_\Omega u\left(\Phi_t * \frac{\partial \phi}{\partial x_j}\right) dx. \tag{7.3}$$

However, as $\phi \in C_0^1(\mathbb{R}^n)$, Theorem 6.8 implies that $\Phi_t * \phi$ converges to ϕ in $L^1(\mathbb{R}^n) \cap L^\infty(\mathbb{R}^n)$. We deduce that the integral $\int_\Omega v(\Phi_t * \phi) \, dx$ converges to $\int_\Omega v\phi \, dx$. We proceed in the same way to study the right-hand side of (7.3) and we deduce (7.2). $\qquad \square$

Notation 7.9 We denote the Sobolev space $W^{1,2}(\Omega)$ by $H^1(\Omega)$. It is the set of functions $u \in L^2(\Omega)$ that admit weak derivatives $\partial_j u$ in $L^2(\Omega)$ for all j such that $1 \leqslant j \leqslant n$.

Proposition 7.10 *The Sobolev space* $H^1(\Omega)$ *is a Hilbert space for the scalar product* $\langle \cdot, \cdot \rangle$ *defined by*

$$\langle u, v \rangle := \sum_{1 \leqslant i \leqslant n} \int_\Omega (\partial_i u)\overline{(\partial_i v)} \, dx + \int_\Omega u\overline{v} \, dx.$$

Therefore, we have $\langle u, v \rangle = (\nabla u, \nabla v) + (u, v)$ *where, by abuse of notation,* $(\cdot, \cdot)$ *denotes in the same way the usual scalar product on* $L^2(\mathbb{R}^n)$ *and on* $L^2(\mathbb{R}^n)^n$.

Proof It is clear that $\langle \cdot, \cdot \rangle$ is a scalar product. Let us show that $H^1(\Omega)$ is complete for the norm $\|u\| = \sqrt{\langle u, u \rangle}$. Let $(u_k)_{k \in \mathbb{N}}$ be a sequence in $H^1(\Omega)$ that is Cauchy for this norm. Then the sequence $(u_k, \partial_1 u_k, \ldots, \partial_n u_k)$ is Cauchy in $L^2(\Omega)^{n+1}$. By completeness of $L^2(\Omega)$, this sequence converges to $(u, v_1, \ldots, v_n)$ in $L^2(\Omega)^{n+1}$. Let us fix j such that $1 \leqslant j \leqslant n$ and consider a function ϕ in $C_0^1(\Omega)$. Then ϕ belongs to

$L^2(\Omega)$ and by continuity of the scalar product on $L^2(\Omega) \times L^2(\Omega)$, by passing to the limit in the identity

$$\int_\Omega (\partial_j u_k)\phi \, dx = - \int_\Omega u_k \frac{\partial \phi}{\partial x_j} \, dx,$$

we verify that u is weakly differentiable with respect to the variable x_j and that its derivative is given by v_j. Then $(u_k)_{k\in\mathbb{N}}$ converges to u in $L^2(\Omega)$ and $(\partial_{x_j} u_k)_{k\in\mathbb{N}}$ converges to $\partial_{x_j} u$ in $L^2(\Omega)$, which proves that $(u_k)_{k\in\mathbb{N}}$ converges to u in $H^1(\Omega)$. $\square$

Proposition 7.11 *Let $p \in [1, +\infty]$. The Sobolev space $W^{1,p}(\Omega)$ is a Banach space for the norm*

$$\|u\|_{W^{1,p}} = \|u\|_{L^p} + \sum_{1 \leqslant j \leqslant n} \|\partial_{x_j} u\|_{L^p}.$$

Proof The proof is analogous to that of Proposition 7.10. $\square$

In the notation $W^{1,p}(\Omega)$, the index 1 refers to the fact that u admits weak partial derivatives of order 1 in L^p. We can also define weak derivatives of higher order. For this, we use a definition similar to that of the spaces of functions of class C^k: a function of class C^k is a function of class C^1 whose derivatives are of class C^{k-1}.

Definition 7.12 (Higher order Sobolev spaces) Let $k \in \mathbb{N}^*$ and let $p \in [1, +\infty]$. We say that a function u belongs to the Sobolev space $W^{k,p}(\Omega)$ if u belongs to $W^{1,p}(\Omega)$ and if the weak L^p derivatives of u belong to $W^{k-1,p}(\Omega)$. The space $W^{\infty,p}(\Omega)$ is the intersection of all these spaces:

$$W^{\infty,p}(\Omega) = \bigcap_{k\in\mathbb{N}^*} W^{k,p}(\Omega).$$

Proposition 7.13 *Let $k \in \mathbb{N}^*$ and let $p \in [1, +\infty]$. The Sobolev space $W^{k,p}(\Omega)$ is a Banach space for the norm*

$$\|u\|_{W^{k,p}} = \sum_{|\alpha| \leqslant k} \left\| \partial_x^\alpha u \right\|_{L^p}.$$

Furthermore, $C_0^\infty(\mathbb{R}^n)$ is dense in $W^{k,p}(\mathbb{R}^n)$ for all $k \geqslant 1$ and for all $p \in [1, +\infty)$.

Proof The proof is analogous to that of Proposition 7.10. $\square$

The notion of derivative in the weak sense allows the generalization of the notion of derivative. Similarly, we can generalize the notion of solution by introducing a notion of weak solution thanks to the theory of distributions. One of the objectives of this chapter is precisely to introduce this theory. For this, we will define the notion of a weak solution of the equation $-\Delta u = f$. Recall that the Laplacian Δ is defined by

$$\Delta = \sum_{1 \leqslant j \leqslant n} \partial_{x_j}^2.$$

Definition 7.14 (Weak solution) Let Ω be an open set of $\mathbb{R}^n$ and let $f \in L^1_{loc}(\Omega)$. A weak solution of the equation $-\Delta u = f$ is a function $u \in H^1(\Omega)$ satisfying

$$\forall \phi \in C^1_0(\Omega), \quad \sum_{1 \leq i \leq n} \int_\Omega (\partial_i u)(\partial_i \phi) \, dx = \int_\Omega f \phi \, dx. \qquad (7.4)$$

Remark 7.15

1. Let Ω be a bounded regular open set so that we can apply the divergence theorem[2]:

$$\int_\Omega \mathrm{div}\, X \, dx = \int_{\partial\Omega} X \cdot \nu \, d\sigma,$$

for any vector field $X \colon \overline{\Omega} \to \mathbb{R}^n$ that is of regularity $C^1(\overline{\Omega})$, where ν denotes the outgoing unit normal. Consider $f \in C^0(\overline{\Omega})$ and suppose that $u \in C^2(\overline{\Omega})$ satisfies $-\Delta u = f$ in the classical sense. Then u is also a weak solution, as we will verify using the divergence formula. For this, we use that ϕ vanishes on the boundary of Ω to write

$$0 = \int_{\partial\Omega} \phi \partial_\nu u \, d\sigma = \int_\Omega \mathrm{div}(\phi \nabla u) \, dx.$$

Now, we use the formula[3]

$$\mathrm{div}(f X) = \nabla f \cdot X + f \, \mathrm{div}\, X,$$

to deduce that $\mathrm{div}(\phi \nabla u) = \nabla \phi \cdot \nabla u + \phi \Delta u$. We then conclude that

$$\int_\Omega (\nabla \phi \cdot \nabla u + \phi \Delta u) \, dx = 0.$$

Then using $-\Delta u = f$, we obtain equation (7.4). Notice that, by repeating the arguments used in the computations preceding Definition 7.6, we can justify the previous identity even when Ω is an arbitrary open set.
2. There exist weak solutions that are not C^2 and for which the equation $-\Delta u = f$ does not hold in the pointwise sense. For example, we verify (exercise) that the function $u(x) = \mathrm{sign}(x)x^2$ is a weak solution of $\partial_x^2 u = 2 \, \mathrm{sign}(x)$ but the previous equation does not make sense pointwise at $x = 0$.
3. We obtain an equivalent definition by replacing ϕ with $\overline{\phi}$ in (7.4).
4. The assumption that f is locally integrable is the minimal assumption to give meaning to $\int_\Omega f \phi \, dx$.

---

[2] See Exercise 17.3 and its solution for the notion of regular open set and for the divergence theorem.
[3] Indeed, $\mathrm{div}(f X) = \sum_{1 \leq j \leq n} \partial_{x_j}(f X_j) = \sum_{1 \leq j \leq n} (\partial_{x_j} f) X_j + \sum_{1 \leq j \leq n} f \partial_{x_j} X_j = \nabla f \cdot X + f \, \mathrm{div}\, X$.

7.2 The Poincaré Inequality and the Poisson Problem

7.2.1 The Poincaré Inequality

We studied in Section 4.4 of Chapter 4 the Poincaré–Wirtinger inequality. This inequality allows us to control the L^2 norm of a periodic function with zero mean by the L^2 norm of its gradient. Such an inequality is clearly false for constant functions. The assumption that u is periodic and has zero mean allows us to "filter" constant functions. Let us consider more generally a function $u: \Omega \to \mathbb{R}$ where Ω is an open set of $\mathbb{R}^n$ with n any integer. One way to filter these constant functions is to assume that u vanishes somewhere, for instance on the boundary (this is the most relevant choice for applications). One difficulty is that we need to specify in what sense the function vanishes on the boundary. For this, we will consider functions that are limits of functions that vanish in a neighborhood of the boundary.

Definition 7.16 We define the space $H_0^1(\Omega)$ as the closure of $C_0^\infty(\Omega)$ in $H^1(\Omega)$.

Remark 7.17

1. In particular, $C_0^\infty(\Omega)$ is trivially included in $H_0^1(\Omega)$. We can also show (by convolution) that $C_0^1(\Omega)$ is included in $H_0^1(\Omega)$.
2. The space $C_0^\infty(\mathbb{R}^n)$ is dense in $H^1(\mathbb{R}^n)$ (this is a particular case of the following proposition). In particular, we have

$$H^1(\mathbb{R}^n) = H_0^1(\mathbb{R}^n).$$

Proposition 7.18 The space $C_0^\infty(\mathbb{R}^n)$ is dense in $W^{1,p}(\mathbb{R}^n)$ for all $n \geqslant 1$ and for all $p \in [1, +\infty)$.

Proof The proof is parallel to that of Corollary 6.9 and we will use the notations introduced there. Let us set $Q_{k,t}(f) = f - \theta_k(\Phi_t * f)$. We saw in the proof of Corollary 6.9 that the functions $Q_{k,t}(f)$ are of class C^∞ and converge to 0 in L^p when k tends to $+\infty$ and t tends to 0. It is therefore sufficient to prove that the weak derivatives $\partial_{x_j} Q_{k,t}(f)$ also converge to 0. For this, we start by noticing that $\partial_{x_j}(\Phi_t * f) = \Phi_t * \partial_{x_j} f$. Moreover, the inequality $\|\partial_{x_j} \theta_k\|_{L^\infty} \leqslant C/k$ implies that

$$\left\|\partial_{x_j} Q_{k,t}(f)\right\|_{L^p} = \left\|\partial_{x_j}\left(f - \theta_k(\Phi_t * f)\right)\right\|_{L^p}$$
$$\leqslant \left\|\partial_{x_j} f - \theta_k(\Phi_t * \partial_{x_j} f)\right\|_{L^p} + \frac{C}{k}\left\|\Phi_t * f\right\|_{L^p}.$$

Note that the first term on the right-hand side is equal to the norm L^p of $Q_{k,f}(\partial_{x_j} f)$, which converges to 0 as seen before, because $\partial_{x_j} f$ belongs to L^p. To estimate the second term, we use Young's inequality $\|\Phi_t * f\|_{L^p} \leqslant \|\Phi_t\|_{L^1} \|f\|_{L^p}$ and the fact that $\|\Phi_t\|_{L^1} = 1$. It follows that $\left\|\partial_{x_j} Q_{k,t}(f)\right\|_{L^p}$ converges to 0 when k tends to $+\infty$ and t tends to 0.   □

Theorem 7.19 (Poincaré Inequality) *Assume that Ω is an open set of $\mathbb{R}^n$ included in the strip $\mathcal{R} = \{x = (x', x_n) \in \mathbb{R}^{n-1} \times \mathbb{R} : |x_n| \leqslant R\}$. Then, for all $u \in H_0^1(\Omega)$, we have*

$$\|u\|_{L^2(\Omega)} \leqslant 2R\|\nabla u\|_{L^2(\Omega)}. \tag{7.5}$$

Proof Since $C_0^\infty(\Omega)$ is dense in $H_0^1(\Omega)$ (by definition of $H_0^1(\Omega)$) and since both terms of inequality (7.5) are continuous for the norm $\|\cdot\|_{H^1}$, it is sufficient to prove (7.5) for $u \in C_0^\infty(\Omega)$.

Consider a function $u \in C_0^\infty(\Omega)$. We start by extending u by 0 on $\mathbb{R}^n$: set

$$\widetilde{u}(x) = \begin{cases} u(x) & \text{if } x \in \Omega, \\ 0 & \text{if } x \notin \Omega. \end{cases}$$

The function $\widetilde{u}$ belongs to $C_0^\infty(\mathbb{R}^n)$. Moreover, $\operatorname{supp}\widetilde{u} = \operatorname{supp} u \subset \Omega$ and since Ω is included in $\mathcal{R}$, we see that $\widetilde{u}$ vanishes on the boundary $\partial \mathcal{R}$ of this strip. We can then write that

$$\widetilde{u}(x', x_n) = \int_{-R}^{x_n} \frac{\partial \widetilde{u}}{\partial x_n}(x', t)\, dt \qquad (x' = (x_1, \ldots, x_{n-1})).$$

The Cauchy–Schwarz inequality implies that

$$|\widetilde{u}(x', x_n)|^2 \leqslant \left(\int_{-R}^{x_n} 1\, dt\right) \int_{-R}^{x_n} \left|\frac{\partial \widetilde{u}}{\partial x_n}(x', t)\right|^2 dt.$$

We trivially deduce that

$$|\widetilde{u}(x', x_n)|^2 \leqslant \left(\int_{-R}^{R} 1\, dt\right) \int_{-R}^{R} \left|\frac{\partial \widetilde{u}}{\partial x_n}(x', t)\right|^2 dt.$$

Then, by integrating with respect to the variable x' over $\mathbb{R}^{n-1}$,

$$\int_{\mathbb{R}^{n-1}} |\widetilde{u}(x', x_n)|^2\, dx' \leqslant 2R \iint_{\mathcal{R}} \left|\frac{\partial \widetilde{u}}{\partial x_n}(x', t)\right|^2 dx'\, dt \leqslant 2R\, \|\nabla \widetilde{u}\|_{L^2(\mathcal{R})}^2.$$

By integrating then with respect to the variable x_n, we get

$$\iint_{\mathcal{R}} |\widetilde{u}(x', x_n)|^2\, dx'\, dx_n \leqslant (2R)^2\, \|\nabla \widetilde{u}\|_{L^2(\mathcal{R})}^2.$$

This directly implies the desired inequality (7.5). $\square$

7.2.2 Application to the Poisson Problem

Let us fix an integer $n \geqslant 1$ and consider a bounded open set $\Omega \subset \mathbb{R}^n$. Consider a non-negative function $V \in L^\infty(\Omega)$ and a function $f : \Omega \to \mathbb{R}$. The Poisson problem consists in studying the existence and uniqueness of a solution $u : \Omega \to \mathbb{R}$ of the following equations:

$$-\Delta u + Vu = f \quad \text{in } \Omega, \qquad u|_{\partial\Omega} = 0.$$

We recall that the Laplacian Δu of a function u is defined by

$$\Delta u = \sum_{i=1}^{n} \frac{\partial^2 u}{\partial x_i^2}.$$

For u and v belonging to $H^1(\Omega)$, we introduce

$$B(u, v) = \int_{\Omega} (\nabla u \cdot \nabla v + Vuv)\, dx \quad \text{where} \quad \nabla u \cdot \nabla v = \sum_{i=1}^{n} \frac{\partial u}{\partial x_i} \frac{\partial v}{\partial x_i}.$$

This is a bilinear map. The quadratic form associated with B is called the energy of u, it is given by

$$E(u) = B(u, u) = \int_{\Omega} \left(|\nabla u(x)|^2 + Vu^2 \right) dx.$$

This is a non-negative quantity because $V \geqslant 0$ by hypothesis and we consider real-valued functions.

Lemma 7.20 *The form B is a scalar product on $H_0^1(\Omega)$ and the function $u \mapsto E(u)^{1/2}$ is a norm equivalent to the norm $\|\cdot\|_{H^1(\Omega)}$.*

Proof It is clear that $B(\cdot, \cdot)$ is a scalar product. As Ω is a bounded open set by hypothesis, we can apply the Poincaré inequality (7.5), which implies that, for all $u \in H_0^1(\Omega)$,

$$\frac{1}{1 + \|V\|_{L^\infty}} E(u) \leqslant \|u\|_{H^1(\Omega)}^2 \leqslant (1 + C(\Omega)^2)\|\nabla u\|_{L^2}^2 \leqslant (1 + C(\Omega)^2)E(u),$$

equivalent to the desired result. $\qquad\qquad\square$

Definition 7.21 Let $f \in L^2(\Omega)$. A function $u \in H^1(\Omega)$ is a weak solution of the homogeneous Dirichlet problem

$$-\Delta u + Vu = f, \quad u|_{\partial\Omega} = 0,$$

if the following two properties are satisfied:

1. u is a weak solution of $-\Delta u + Vu = f$ in the following sense:

$$\forall \phi \in C_0^1(\Omega), \quad \int_{\Omega} (\nabla u \cdot \nabla \phi + Vu\phi)\, dx = \int_{\Omega} f\phi\, dx.$$

2. $u \in H_0^1(\Omega)$ (this is the meaning we give to the fact that u vanishes on the boundary).

Proposition 7.22 *Let $V \in L^\infty(\Omega)$ be a non-negative function. For all $f \in L^2(\Omega)$, the Poisson problem*

$$-\Delta u + Vu = f, \quad u|_{\partial\Omega} = 0,$$

has a unique weak solution $u \in H_0^1(\Omega)$.

Proof The map

$$\phi \longmapsto (f, \phi)_{L^2}$$

is a continuous linear form on $L^2(\Omega)$ and therefore a continuous linear form on $(H_0^1(\Omega), \|\cdot\|_{H^1(\Omega)})$ since the norm $\|\cdot\|_{H^1(\Omega)}$ dominates the norm $\|\cdot\|_{L^2(\Omega)}$.

Let us set $\|u\|_{H_0^1(\Omega)} = \sqrt{B(u,u)}$. We have seen that it is a norm equivalent to the norm $\|\cdot\|_{H^1(\Omega)}$. We deduce two things from this. First, the map $\phi \mapsto (f, \phi)_{L^2}$ is a continuous linear form on $(H_0^1(\Omega), \|\cdot\|_{H_0^1(\Omega)})$. And secondly, the space $H_0^1(\Omega)$ is a Hilbert space for the scalar product $B(\cdot, \cdot)$. We can then apply the Riesz–Fréchet representation theorem 3.10 to deduce the existence and uniqueness of an element u in $H_0^1(\Omega)$ such that $B(u, v) = (f, v)_{L^2}$ for all $v \in H_0^1(\Omega)$. As $C_0^1(\Omega) \subset H_0^1(\Omega)$, this proves the existence and uniqueness of a weak solution. $\qquad\square$

7.3 Sobolev Spaces Defined on any Open Set

Consider any open set Ω of $\mathbb{R}^n$. We will see in this section some very useful technical results for studying the Sobolev space $H^1(\Omega)$. We will start by studying the Leibniz rule. We will then see a criterion that allows us to show that a function belongs to $H^1(\Omega)$ and apply this criterion to prove a change of variables result. We will also use this criterion to show how to extend an element of $H^1(\Omega)$ to an element of $H^1(\mathbb{R}^n)$.

Recall that, by definition, a function $u \in L^2(\Omega)$ admits a weak derivative in $L^2(\Omega)$ with respect to the variable x_j if there exists $v_j \in L^2(\Omega)$ such that

$$\forall \phi \in C_0^\infty(\Omega), \quad \int_\Omega v_j \phi \, dx = -\int_\Omega u \frac{\partial \phi}{\partial x_j} \, dx. \tag{7.6}$$

We denote v_j by $\partial u/\partial x_j$ or simply $\partial_j u$. Moreover, we have seen in Proposition 7.8 that it is sufficient to suppose that $\phi \in C_0^1(\Omega)$ in (7.6).

Proposition 7.23 *Let Ω be an open set of $\mathbb{R}^n$. Let $u \in H^1(\Omega)$ and $v \in C_b^1(\Omega)$ (the index b means that v and its derivatives are bounded functions on Ω). Then the product uv belongs to $H^1(\Omega)$, the weak derivatives satisfy*

$$\partial_j(uv) = u\partial_j v + (\partial_j u)v, \quad j = 1, \dots, n, \tag{7.7}$$

and moreover

$$\|uv\|_{H^1(\Omega)} \leqslant 2\|u\|_{H^1(\Omega)}\|v\|_{W^{1,\infty}(\Omega)}, \tag{7.8}$$

where we have used the notation $\|v\|_{W^{1,\infty}(\Omega)} = \sup_{x\in\Omega} |v(x)| + \sup_{x\in\Omega} |\nabla v(x)|$.

Proof As the product of a function f in $L^\infty(\Omega)$ and a function g in $L^2(\Omega)$ belongs to $L^2(\Omega)$, we obtain that uv belongs to $L^2(\Omega)$ and, moreover, if (7.7) is true, then the estimate (7.8) holds. It is therefore sufficient to show that uv has a weak derivative in $L^2(\Omega)$ and that the relation (7.7) is satisfied. Let $\varphi \in C_0^1(\Omega)$. Then $\phi = v\varphi$ belongs to $C_0^1(\Omega)$ and we can apply (7.2) to write

$$\int_\Omega (\partial_j u)v\varphi \, dx = -\int_\Omega u\partial_j(v\varphi) \, dx,$$

which implies

$$\int_\Omega \left(u\partial_j v + (\partial_j u)v\right)\varphi \, dx = -\int_\Omega uv\partial_j\varphi \, dx.$$

Furthermore, we verify that $u\partial_j v + (\partial_j u)v$ belongs to $L^2(\Omega)$ (trivially, because u and $\partial_j u$ belong to $L^2(\Omega)$ while v and $\partial_j v$ belong to $L^\infty(\Omega)$). Then, by definition of weak differentiation, we conclude that uv has a weak derivative in $L^2(\Omega)$ with respect to the variable x_j and that this derivative satisfies (7.7). $\square$

We now show a criterion for belonging to $H^1(\Omega)$ which is based on a duality argument.

Proposition 7.24 *Let $\Omega \subset \mathbb{R}^n$ be an open set and let $u \in L^2(\Omega)$. Then u has a weak derivative in the direction x_j if and only if there exists a constant $C > 0$ such that, for all $\phi \in C_0^\infty(\Omega)$,*

$$\left|\int_\Omega u\partial_j\phi \, dx\right| \leqslant C\|\phi\|_{L^2(\Omega)}. \tag{7.9}$$

Then $u \in H^1(\Omega)$ if and only if the previous result is true for all j such that $1 \leqslant j \leqslant n$.

Proof If u has a weak derivative in the direction x_j then it follows from the Cauchy–Schwarz inequality and (7.2) that the inequality (7.9) is satisfied with $C = \|\partial_j u\|_{L^2(\Omega)}$. The delicate point is to show the converse. For this, consider the linear map $\Theta: C_0^\infty(\Omega) \to \mathbb{C}$ defined by

$$\Theta(\phi) = \int_\Omega u\partial_j\phi \, dx.$$

This map is well defined (it is immediate) and the inequality (7.9) implies that $|\Theta(\phi)| \leqslant C\|\phi\|_{L^2(\Omega)}$. As $C_0^\infty(\Omega)$ is dense in $L^2(\Omega)$, this means that we can extend Θ by continuity to a continuous linear form on $L^2(\Omega)$. The Riesz–Fréchet representation theorem implies the existence of a function $w_j \in L^2(\Omega)$ such that

$$\Theta(\phi) = (\phi, w_j) = \int_\Omega \phi\overline{w_j} \, dx.$$

By setting $v_j = -\overline{w_j}$, we obtain

$$\forall \phi \in C_0^\infty(\Omega), \quad \int_\Omega v_j \phi \, dx = - \int_\Omega u \partial_j \phi \, dx,$$

which proves the desired result.   □

We are now in position to show that the notion of Sobolev space is invariant under diffeomorphism.

Proposition 7.25 *Consider two open sets $\Omega_1 \subset \mathbb{R}^n$ and $\Omega_2 \subset \mathbb{R}^n$ and a diffeomorphism $\theta \colon \Omega_1 \to \Omega_2$ which is of class C^2 ($\theta \in C^2(\Omega_1)$, $\theta^{-1} \in C^2(\Omega_2)$) and such that:*

- *$\nabla\theta$ is bounded on Ω_1;*
- *$\nabla(\theta^{-1})$ is bounded on Ω_2.*

If $u \in H^1(\Omega_2)$, then $u \circ \theta \in H^1(\Omega_1)$.

Remark 7.26

1. This result remains true if we only assume that θ is a C^1-diffeomorphism.
2. We can further show that the weak derivatives of $u \circ \theta$ are given by

$$\frac{\partial}{\partial x_j}(u \circ \theta) = \sum_{k=1}^n \left(\frac{\partial u}{\partial x_k} \circ \theta \right) \frac{\partial \theta_k}{\partial x_j},$$

where $1 \leqslant j \leqslant n$ and where we have denoted by θ_k the coordinates of $\theta = (\theta_1, \ldots, \theta_n)$.

Proof Let $u \in H^1(\Omega_2)$. The fact that $u \circ \theta$ belongs to $L^2(\Omega_1)$ comes from general results on integration. It follows from Proposition 7.24 that it suffices to show that there exists $C > 0$ such that, for all $\phi \in C_0^\infty(\Omega_1)$ and for all j such that $1 \leqslant j \leqslant n$, we have

$$\left| \int_{\Omega_1} u(\theta(x)) \partial_j \phi(x) \, dx \right| \leqslant C \|\phi\|_{L^2(\Omega_1)}. \tag{7.10}$$

We refer to the solution of the solved Exercise 7.4 page 398 for the case $n = 1$. We assume in this proof that $n \geqslant 2$.

Let us set $\kappa = \theta^{-1}$ and denote by J the Jacobian matrix of κ, given by $J(y) = \det(\nabla\kappa_1, \ldots, \nabla\kappa_n)$ where $\kappa_1, \ldots, \kappa_n$ are the coordinates of κ. Then, by changing variables (see Theorem 2.16), we have

$$\int_{\Omega_1} u(\theta(x)) \partial_1 \phi(x) \, dx = \int_{\Omega_2} u(y)(\partial_1 \phi)(\kappa(y)) \, |J(y)| \, dy.$$

Since $J(y)$ does not change sign on Ω_2, we have

$$\left| \int_{\Omega_2} u(y)(\partial_1 \phi)(\kappa(y)) \, |J(y)| \, dy \right| = \left| \int_{\Omega_2} u(y)(\partial_1 \phi)(\kappa(y)) J(y) \, dy \right|.$$

We then use computations (as well as notations) analogous to those used in the proof of Lax's Lemma 2.17. Firstly, since $\nabla(\phi \circ \kappa) = \partial_1\phi\nabla\kappa_1 + \cdots + \partial_n\phi\nabla\kappa_n$, we have

$$(\partial_1\phi)(\kappa(y))J(y) = \det\big(\nabla(\phi \circ \kappa), \nabla\kappa_2, \ldots, \nabla\kappa_n\big).$$

Now, we can then expand the determinant and write

$$\det\big(\nabla(\phi \circ \kappa), \nabla\kappa_2, \ldots, \nabla\kappa_n\big) = M_1\partial_1(\phi \circ \kappa) + \cdots + M_n\partial_n(\phi \circ \kappa).$$

Since κ is of class C^2 by hypothesis, we can write

$$M_1\partial_1(\phi \circ \kappa) + \cdots + M_n\partial_n(\phi \circ \kappa) = \partial_1(M_1\phi \circ \kappa) + \cdots + \partial_n(M_n\phi \circ \kappa)$$
$$- \big(\partial_1 M_1 + \cdots + \partial_n M_n\big)\phi \circ \kappa.$$

Let us recall the remarkable identity $\partial_1 M_1 + \cdots + \partial_n M_n = 0$ (cf. (2.11)). We deduce that

$$\int_{\Omega_2} u(y)(\partial_1\phi)(\kappa(y))J(y)\,dy = \sum_{k=1}^{n} \int_{\Omega_2} u(y)\partial_k\big(M_k\phi \circ \kappa\big)\,dy.$$

However, as $u \in H^1(\Omega_2)$, there exists a constant $C > 0$ such that

$$\forall\varphi \in C_0^1(\Omega_2), \quad \left|\int_{\Omega_2} u(y)\partial_j\varphi(y)\,dy\right| \leqslant C\|\varphi\|_{L^2(\Omega_2)}.$$

It is easily verified that, for all k, the function $M_k\phi \circ \kappa$ belongs to $C_0^1(\Omega_2)$. It follows that

$$\left|\int_{\Omega_2} u(y)(\partial_1\phi)(\kappa(y))\,|J(y)|\,dy\right| \leqslant C\sum_{k=1}^{n}\|M_k\phi \circ \kappa\|_{L^2(\Omega_2)}.$$

We conclude the proof of (7.10) by observing that, on the one hand, M_k is bounded because the derivatives of κ are bounded by hypothesis and that, on the other hand,

$$\|\phi \circ \kappa\|_{L^2(\Omega_2)} \leqslant C'\|\phi\|_{L^2(\Omega_1)},$$

as can be seen by using the change of variables formula for an integral once again and the hypothesis that the derivatives of θ are bounded. $\qquad\square$

Let us recall that the space $H_0^1(\Omega)$ was defined at the beginning of this chapter as being the closure of $C_0^\infty(\Omega)$ in $H^1(\Omega)$.

Proposition 7.27 *Let Ω be any open set of $\mathbb{R}^n$. Given a function u defined on Ω, let $\tilde{u}$ be the function defined by*

$$\tilde{u}(x) = \begin{cases} u(x) & \text{if } x \in \Omega, \\ 0 & \text{if } x \in \mathbb{R}^n \setminus \Omega. \end{cases}$$

Then, for all $u \in H_0^1(\Omega)$, the function $\tilde{u}$ belongs to $H^1(\mathbb{R}^n)$.

Proof It is obvious that $\widetilde{u}$ belongs to $L^2(\mathbb{R}^n)$. In light of the duality criterion given by Proposition 7.24, it is sufficient to prove that, if $u \in H_0^1(\Omega)$, then

$$\forall \phi \in C_0^\infty(\mathbb{R}^n), \quad \left| \int_{\mathbb{R}^n} \widetilde{u}(\partial_j \phi) \, dx \right| \leqslant C \|\phi\|_{L^2(\mathbb{R}^n)}. \tag{7.11}$$

To obtain this result, note that by definition of $H_0^1(\Omega)$, there exists a sequence $(u_k)_{k \in \mathbb{N}}$ of functions belonging to $C_0^\infty(\Omega)$ and converging to u in $H^1(\Omega)$. Denote by $\widetilde{u}_k \in C_0^\infty(\mathbb{R}^n)$ the function obtained by extending u_k by 0 outside Ω. Then, for any $\phi \in C_0^\infty(\mathbb{R}^n)$, we get that

$$\left| \int_{\mathbb{R}^n} \widetilde{u}_k \partial_j \phi \, dx \right| = \left| - \int_{\mathbb{R}^n} (\partial_j \widetilde{u}_k) \phi \, dx \right|$$

$$= \left| \int_{\Omega} (\partial_j u_k) \phi \, dx \right| \leqslant \|\partial_j u_k\|_{L^2(\Omega)} \|\phi\|_{L^2(\Omega)},$$

where we performed an integration by parts to obtain the first equality (which is easily justified since we are handling smooth functions with compact support). Since $(u_k)_{k \in \mathbb{N}}$ is bounded in $H^1(\Omega)$, and since $(\widetilde{u}_k)_{k \in \mathbb{N}}$ converges to $\widetilde{u}$ in $L^2(\mathbb{R}^n)$, by taking the limit as k tends to $+\infty$, we obtain (7.11). This concludes the proof. $\qquad \square$

Proposition 7.28 *Let Ω be a bounded open set. Suppose that $u \in H^1(\Omega)$ is a function with compact support in Ω and such that $\operatorname{supp} u$ is at a positive distance from $\partial\Omega$. Then u belongs to $H_0^1(\Omega)$.*

Proof Consider an open set Ω' such that

$$\operatorname{supp} u \subset \Omega' \subset \Omega \quad \text{with} \quad \operatorname{dist}(\partial\Omega', \partial\Omega) > 0.$$

Let χ be a C^∞ function on $\mathbb{R}^n$ that equals 1 on $\operatorname{supp} u$ and 0 outside of Ω' (the fact that such a function exists is proved in the appendix, see Proposition 17.34). Extend u by 0 on $\mathbb{R}^n \setminus \Omega$ and denote this extension by $\widetilde{u}$. Let us prove that $\widetilde{u}$ belongs to $H^1(\mathbb{R}^n)$ (notice that we cannot apply the previous proposition here since we do not yet that u belongs to $H_0^1(\Omega)$; which is indeed the wanted result). To do so, we will apply the duality criterion given by Proposition 7.24. Consider a function $\phi \in C_0^\infty(\mathbb{R}^n)$. Then

$$\int_{\mathbb{R}^n} \widetilde{u} \partial_j \phi \, dx = \int_{\Omega} u \partial_j \phi \, dx = \int_{\Omega} \chi u \partial_j \phi \, dx$$

$$= \int_{\Omega} u \partial_j (\chi \phi) \, dx - \int_{\Omega} u \phi \partial_j \chi \, dx.$$

Since $\chi \phi \in C_0^\infty(\Omega)$ we have, by definition of the weak derivative,

$$\int_{\Omega} u \partial_j (\chi \phi) \, dx = - \int_{\Omega} \chi \phi (\partial_j u) \, dx.$$

Therefore, by combining the two previous identities, it follows from the Cauchy–Schwarz inequality that

$$\left| \int_{\mathbb{R}^n} \widetilde{u} \partial_j \phi \, dx \right| \leqslant \|u\|_{H^1(\Omega)} \|\chi\|_{W^{1,\infty}(\Omega)} \|\phi\|_{L^2(\mathbb{R}^n)} .$$

So Proposition 7.24 implies that $\widetilde{u}$ belongs to $H^1(\mathbb{R}^n)$. Once this is granted, we can use the fact that $C_0^\infty(\mathbb{R}^n)$ is dense in $H^1(\mathbb{R}^n)$ (see Proposition 7.18), to introduce a sequence of functions $u_k \in C_0^\infty(\mathbb{R}^n)$ that converges to $\widetilde{u}$ in $H^1(\mathbb{R}^n)$. As multiplication by χ is a continuous linear operator from $H^1(\mathbb{R}^n)$ to $H^1(\mathbb{R}^n)$ (see Proposition 7.23), it follows that χu_k converges to $\chi \widetilde{u}$ in $H^1(\mathbb{R}^n)$. But $\chi \widetilde{u} = \widetilde{u}$ because χ equals 1 on the support of $\widetilde{u}$. Since the restriction operator $f \mapsto f|_\Omega$ is continuous from $H^1(\mathbb{R}^n)$ to $H^1(\Omega)$, it follows that $\chi u_k = (\chi u_k)|_\Omega$ converges to u in $H^1(\Omega)$. As $\chi u_k \in C_0^\infty(\Omega)$, this shows that u is the limit in $H^1(\Omega)$ of a sequence of functions of class C^∞ with support in Ω. Therefore, u belongs to $H_0^1(\Omega)$ by definition of $H_0^1(\Omega)$.   □

The following proposition provides another example of a situation where we can explicitly extend a function to $\mathbb{R}^n$ while preserving the property of belonging to a Sobolev space.

Proposition 7.29 *Let $\mathbb{R}_+^n$ be the half-space such that $x_n > 0$. Let $u \in H^1(\mathbb{R}_+^n)$. Let u_* be the function defined on $\mathbb{R}^n$ by*

$$u_*(x) = \begin{cases} u(x) & \text{if } x_n \geqslant 0, \\ u(x', -x_n) & \text{if } x_n < 0, \end{cases}$$

where $x' = (x_1, \ldots, x_{n-1})$. Then $u_* \in H^1(\mathbb{R}^n)$ and the map $u \mapsto u_*$ is continuous from $H^1(\mathbb{R}_+^n)$ to $H^1(\mathbb{R}^n)$ with an operator norm bounded by 2.

Remark 7.30 Let B be the ball centered at 0 with radius 1. Let $B_+ = B \cap \mathbb{R}_+^n$. We will use below the fact that the previous result is true when we replace $\mathbb{R}_+^n$ with B_+ and $\mathbb{R}^n$ with B. The proof is the same.

Proof We refer to the solved Exercise 7.5.   □

For certain open sets Ω, by combining the two previous results, we will be able to extend any function $u \in H^1(\Omega)$. For this, we need to make an assumption about Ω that will allow us to reduce to the two previously treated cases. Let us introduce the following notations:

$$B = B(0,1), \quad B_+ = B \cap \{(x', x_n) : x' \in \mathbb{R}^{n-1}, \, x_n > 0\},$$
$$B_0 = B \cap \{(x', 0) : x' \in \mathbb{R}^{n-1}\}. \tag{7.12}$$

We say that an open set $\Omega \subset \mathbb{R}^n$ is of class C^k if, for every $x \in \partial\Omega$, there exists a neighborhood U of x in $\mathbb{R}^n$ and a bijective function $\theta \colon B \to U$ such that

$$\theta \in C^k(\overline{B}), \quad \theta^{-1} \in C^k(\overline{U}), \quad \theta(B_+) = U \cap \Omega, \quad \theta(B_0) = U \cap \partial\Omega.$$

Theorem 7.31 *Suppose that* $\Omega \subset \mathbb{R}^n$ *is a bounded open set of class* C^2. *Then there exists a bounded linear extension operator* E *from* $H^1(\Omega)$ *to* $H^1(\mathbb{R}^n)$ *such that for every* $u \in H^1(\Omega)$, *the restriction of* Eu *to* Ω *is equal to* u *and such that* $\|Eu\|_{H^1(\mathbb{R}^n)} \leqslant C\|u\|_{H^1(\Omega)}$.

Remark 7.32 The study of such extension operators has a long history going back to the works of Lichtenstein [121], Seeley [171], Calderón [22] and many others. We refer to the book by Burenkov [21] for a thorough study of this topic. Let us mention that Stein [176] proved that there exists a bounded linear extension operator T defined on $W^{l,p}(\Omega)$ with values in $W^{l,p}(\mathbb{R}^n)$, such that $Tu|_\Omega = u$ for all u in $W^{l,p}(\Omega)$ and which is given by a formula independent of $l \in \mathbb{N}^*$ and $p \in [1, +\infty]$. We also refer to Jones [87] for a generalization to more general domains.

Proof By the regularity assumption on Ω and by compactness, we can find a partition of unity (see Proposition 17.35) of class C^2, that is a finite collection of open sets $\{U_i\}_{0 \leqslant i \leqslant N}$ and a finite collection of functions $\{\zeta_i\}_{0 \leqslant i \leqslant N}$ such that the two following properties hold:

(i) U_0 is contained in Ω and the boundary $\partial\Omega$ is covered by the U_i's for $i \geqslant 1$. This means that

$$U_0 \subset \Omega \subset O := \bigcup_{i=0}^{N} U_i,$$

and, for all $1 \leqslant i \leqslant N$, there is a diffeomorphism $\theta_i : \overline{B} \to \overline{U_i}$ of class C^2 such that

$$\theta_i(B_+) = U_i \cap \Omega, \quad \theta_i(B_0) = U_i \cap \partial\Omega,$$

where B_+ and B_0 are as defined in (7.12).

(ii) The ζ_i's are defined on O and adapted to this cover. This means that, for all $0 \leqslant i \leqslant N$,

$$0 \leqslant \zeta_i \leqslant 1, \quad \zeta_i \in C^\infty(O), \quad \operatorname{supp} \zeta_i \subset U_i,$$

and furthermore $\sum_{i=0}^{N} \zeta_i(x) = 1$ for all $x \in \Omega$.

Now, for i such that $0 \leqslant i \leqslant N$, introduce the function $u_i : \Omega \to \mathbb{R}$ defined by $u_i(x) = \zeta_i(x)u(x)$. We define $v_0 = \widetilde{u_0}$, that is the extension of u_0 by 0 on $\mathbb{R}^n \setminus \Omega$. Then Proposition 7.27 implies that $v_0 \in H^1(\mathbb{R}^n)$. For $1 \leqslant i \leqslant N$, we set $w_i = u_i \circ \theta_i : B_+ \to \mathbb{R}$. According to Proposition 7.25, this function belongs to $H^1(B_+)$. By using Proposition 7.29, we can extend it to a function $w_{i,*}$ belonging to $H^1(B)$. Then $W_i = w_{i,*} \circ \theta_i^{-1}$ belongs to $H^1(U_i)$ and the function $\zeta_i W_i$ belongs to $H_0^1(U_i)$ according to Proposition 7.28. We then set $v_i = \widetilde{\zeta_i W_i}$, that is, we extend $\zeta_i W_i$ by 0 on $\mathbb{R}^n \setminus U_i$. Using Proposition 7.27 again, we see that v_i belongs to $H^1(\mathbb{R}^n)$. Then the function v defined by $v = v_0 + \sum_{i=1}^{N} v_i$ satisfies $v \in H^1(\mathbb{R}^n)$ and $v|_\Omega = u$. Moreover, the operator $E : u \mapsto v$ is linear and continuous from $H^1(\Omega)$ into $H^1(\mathbb{R}^n)$. $\square$

Corollary 7.33 *If* $\Omega \subset \mathbb{R}^n$ *is a bounded open set of class* C^2, *then* $C^\infty(\overline{\Omega})$ *is dense in* $H^1(\Omega)$.

Proof Let $u \in H^1(\Omega)$. Then the extension $Eu \in H^1(\mathbb{R}^n)$ given by Theorem 7.31 is the limit in $H^1(\mathbb{R}^n)$ of a sequence of functions belonging to $C_0^{\infty}(\mathbb{R}^n)$ (see Proposition 7.18). By restricting these functions to Ω, we obtain a sequence of functions in $C^{\infty}(\overline{\Omega})$ that converge to u in $H^1(\Omega)$. $\qquad\square$

7.4 Fourier Analysis and Sobolev Spaces

We have studied the Sobolev spaces $H^k(\Omega)$ where k is a natural number and Ω an arbitrary open set of $\mathbb{R}^n$. We also recalled in another chapter that we can extend the definition of the Fourier transform to the space $L^2(\mathbb{R}^n)$ and even to the space $S'(\mathbb{R}^n)$ of tempered distributions, which is the topological dual of the Schwartz space $S(\mathbb{R}^n)$. In the particular case $\Omega = \mathbb{R}^n$, using this Fourier transform, we will define and study, in this section, the Sobolev spaces $H^s(\mathbb{R}^n)$ where the index s is any positive real number.

Notation 7.34 We will often use the notation

$$\langle \xi \rangle = (1 + |\xi|^2)^{1/2}.$$

$\langle \xi \rangle$ is commonly called the Japanese bracket of ξ.

Definition 7.35 Let $s \in \mathbb{R}$. We say that a tempered distribution $u \in S'(\mathbb{R}^n)$ belongs to the Sobolev space $H^s(\mathbb{R}^n)$ if $\langle \xi \rangle^s \hat{u}$ belongs to $L^2(\mathbb{R}^n)$.

Remark 7.36 We can avoid the use of distribution theory by limiting ourselves to consider the case $s \geqslant 0$, which is the most important case in practice.

If $\Omega = \mathbb{R}^n$, we thus have two possible definitions for the spaces $H^k(\Omega)$ with $k \in \mathbb{N}$. The equivalence between these two definitions is a result whose proof is the subject of the solved Exercise 7.6.

Proposition 7.37 *Let $s \in \mathbb{R}$. Equipped with the scalar product*

$$(u, v)_{H^s} = (2\pi)^{-n} \int (1 + |\xi|^2)^s \hat{u}(\xi) \, \overline{\hat{v}(\xi)} \, d\xi,$$

and thus the norm

$$\|u\|_{H^s} = (2\pi)^{-n/2} \left\| (1 + |\xi|^2)^{s/2} \hat{u} \right\|_{L^2},$$

the Sobolev space $H^s(\mathbb{R}^n)$ is a Hilbert space.

Proof The map $u \mapsto (2\pi)^{-n/2}(1 + |\xi|^2)^{s/2}\hat{u}$ is by definition an isometric bijection of $H^s(\mathbb{R}^n)$ onto $L^2(\mathbb{R}^n)$. Since the latter space is complete, the same is true for $H^s(\mathbb{R}^n)$ equipped with the norm defined above. $\qquad\square$

Let us recall that the Schwartz space $S(\mathbb{R}^n)$ was introduced in Definition 5.3.

Proposition 7.38 *The Schwartz space $S(\mathbb{R}^n)$ is dense in $H^s(\mathbb{R}^n)$ for all $s \in \mathbb{R}$.*

Proof Consider the isometry $u \mapsto (2\pi)^{-n/2}(1+|\xi|^2)^{s/2}\widehat{u}$ from $H^s(\mathbb{R}^n)$ onto $L^2(\mathbb{R}^n)$. The inverse isometry transforms the dense subspace $S(\mathbb{R}^n)$ of $L^2(\mathbb{R}^n)$ into a dense subspace of $H^s(\mathbb{R}^n)$. But this map is a bijection from $S(\mathbb{R}^n)$ onto itself. We deduce that $S(\mathbb{R}^n)$ is dense in $H^s(\mathbb{R}^n)$.   □

Proposition 7.39 *For any real number $s > n/2$,*

$$H^s(\mathbb{R}^n) \subset C^0(\mathbb{R}^n) \cap L^\infty(\mathbb{R}^n),$$

with continuous embedding.

Proof From the Cauchy–Schwarz inequality, for all $f \in S(\mathbb{R}^n)$,

$$\|f\|_{L^\infty} \leqslant \|\widehat{f}\|_{L^1} \leqslant \|\langle\xi\rangle^{-s}\|_{L^2}\|\langle\xi\rangle^s\widehat{f}\|_{L^2}, \tag{7.13}$$

and we deduce the result by density of $S(\mathbb{R}^n)$ in $H^s(\mathbb{R}^n)$.   □

Theorem 7.40 *For any real number $s > n/2$, the product of two elements of $H^s(\mathbb{R}^n)$ is still in $H^s(\mathbb{R}^n)$. Moreover, there exists a constant C such that, for all u, v in $H^s(\mathbb{R}^n)$,*

$$\|uv\|_{H^s} \leqslant C\|u\|_{H^s}\|v\|_{H^s}.$$

Proof The proof is based on the following inequality: for all ξ, η in $\mathbb{R}^n$, we have

$$\forall s \geqslant 0, \quad (1+|\xi|^2)^{s/2} \leqslant 2^s\Big((1+|\xi-\eta|^2)^{s/2} + (1+|\eta|^2)^{s/2}\Big),$$

an inequality that follows from the triangle inequality and the bound $(a+b)^r \leqslant 2^r(a^r + b^r)$ for any triplet (a, b, r) of positive numbers. Then, for all u, v in $S(\mathbb{R}^n)$, we have (verify the following formula as an exercise)

$$\widehat{uv}(\xi) = (2\pi)^{-n}\int \widehat{u}(\xi-\eta)\widehat{v}(\eta)\,d\eta.$$

By multiplying both sides by $\langle\xi\rangle^s$ and using the previous inequality, we find

$$\langle\xi\rangle^s |\widehat{uv}(\xi)| \leqslant C\int \langle\xi-\eta\rangle^s |\widehat{u}(\xi-\eta)| |\widehat{v}(\eta)|\,d\eta + C\int |\widehat{u}(\xi-\eta)| \langle\eta\rangle^s |\widehat{v}(\eta)|\,d\eta.$$

If $s > n/2$, then $\mathcal{F}(H^s(\mathbb{R}^n)) \subset L^1(\mathbb{R}^n)$ as we have already seen (cf. (7.13)). We then recognize above two convolution products between a function of $L^1(\mathbb{R}^n)$ and another of $L^2(\mathbb{R}^n)$, which belong to $L^2(\mathbb{R}^n)$ according to Theorem 6.1. This implies that $\langle\xi\rangle^s\widehat{uv} \in L^2(\mathbb{R}^n)$, hence the desired result $uv \in H^s(\mathbb{R}^n)$.   □

We have seen that, for any real number $s > n/2$, the product of two elements of $H^s(\mathbb{R}^n)$ is still in $H^s(\mathbb{R}^n)$. The following proposition shows that we can also define the product φu for all $\varphi \in S(\mathbb{R}^n)$ and all $u \in H^s(\mathbb{R}^n)$ with s any real number.

Proposition 7.41 *For all $s \in \mathbb{R}$, if $u \in H^s(\mathbb{R}^n)$ and $\varphi \in \mathcal{S}(\mathbb{R}^n)$ then $\varphi u \in H^s(\mathbb{R}^n)$.*

Proof Since $\mathcal{S}(\mathbb{R}^n)$ is dense in $H^s(\mathbb{R}^n)$, it suffices to prove that there exists a positive constant C such that, for all $u \in \mathcal{S}(\mathbb{R}^n)$,

$$\|\varphi u\|_{H^s} \leqslant C \|u\|_{H^s} . \tag{7.14}$$

The proof of the latter uses an inequality, called Peetre's inequality, which states that for all ξ, η in $\mathbb{R}^n$, we have

$$\forall s \in \mathbb{R}, \quad (1 + |\xi|^2)^s \leqslant 2^{|s|} (1 + |\eta|^2)^s (1 + |\xi - \eta|^2)^{|s|}.$$

Suppose $s \geqslant 0$. To obtain this inequality, it is enough to use the triangle inequality

$$\begin{aligned} 1 + |\xi|^2 &\leqslant 1 + (|\eta| + |\xi - \eta|)^2 \\ &\leqslant 1 + 2|\eta|^2 + 2|\xi - \eta|^2 \\ &\leqslant 2(1 + |\eta|^2)(1 + |\xi - \eta|^2), \end{aligned}$$

then raise both sides to the power $s \geqslant 0$. If $s < 0$, then $-s > 0$ and the previous inequality implies

$$(1 + |\eta|^2)^{-s} \leqslant 2^{-s} (1 + |\xi|^2)^{-s} (1 + |\xi - \eta|^2)^{-s}.$$

We obtain the desired result by dividing by $(1 + |\eta|^2)^{-s}(1 + |\xi|^2)^{-s}$.

We then proceed as in the proof of Theorem 7.40. Indeed, for $u \in H^s(\mathbb{R}^n)$ and $\varphi \in \mathcal{S}(\mathbb{R}^n)$, we can still write $\widehat{\varphi u}(\xi)$ in the form of a convolution product. As $\widehat{\varphi}(\zeta)$ is in the Schwartz space, the function $\langle \zeta \rangle^{|s|} \widehat{\varphi}(\zeta)$ belongs to the Schwartz space and therefore to the space $L^1(\mathbb{R}^n)$. The previous inequality allows us to form the convolution product of a function of L^1 and of $\langle \eta \rangle^s |\widehat{u}(\eta)|$ which is in L^2. This completes the proof of (7.14). □

Proposition 7.42 (Interpolation in Sobolev spaces) *Let $s_1 < s_2$ be two real numbers and $s \in (s_1, s_2)$. Let us write s in the form $s = \alpha s_1 + (1 - \alpha)s_2$ with $\alpha \in [0, 1]$. There exists a constant $C(s_1, s_2)$ such that for all $u \in H^{s_2}(\mathbb{R}^n)$,*

$$\|u\|_{H^s} \leqslant C(s_1, s_2)\|u\|_{H^{s_1}}^{\alpha}\|u\|_{H^{s_2}}^{1-\alpha}.$$

Proof We have

$$\begin{aligned} \|u\|_{H^s}^2 &= (2\pi)^{-n} \int \langle \xi \rangle^{2s} |\widehat{u}(\xi)|^2 \, d\xi \\ &= (2\pi)^{-n} \int \langle \xi \rangle^{2\alpha s_1} |\widehat{u}(\xi)|^{2\alpha} \langle \xi \rangle^{2(1-\alpha)s_2} |\widehat{u}(\xi)|^{2(1-\alpha)} \, d\xi, \end{aligned}$$

so that the desired inequality is a consequence of Hölder's inequality. □

7.5 Sobolev Embeddings

We will now study several results involving the embedding of Sobolev spaces into
Lebesgue spaces. This is a fundamental property, which we will use for example
later in the context of the De Giorgi–Nash–Moser theory.

We start by studying the case of Sobolev spaces defined on $\mathbb{R}^n$.

Theorem 7.43 *Let $n \geqslant 1$ and s be a real number such that $0 \leqslant s < n/2$. Then the
Sobolev space $H^s(\mathbb{R}^n)$ is continuously embedded in $L^p(\mathbb{R}^n)$ for all p such that*

$$2 \leqslant p \leqslant \frac{2n}{n-2s}.$$

Remark 7.44 Theorem 7.43 states that, for any real number s in $[0, n/2)$, we have

$$\|f\|_{L^{2n/(n-2s)}} \leqslant C_s \|f\|_{H^s}.$$

In fact, we will show a stronger result (*cf.* (7.15)):

$$\|f\|_{L^{2n/(n-2s)}} \leqslant C\|f\|_{\dot{H}^s} := \left(\int |\xi|^{2s} |\widehat{f}(\xi)|^2 \, d\xi \right)^{1/2}.$$

In particular, for $s = 1$, this implies that

$$q = \frac{2n}{n-2} \implies \|f\|_{L^q} \leqslant C\|\nabla f\|_{L^2}.$$

Proof We will show that there exists a constant C such that, for all $f \in \mathcal{S}(\mathbb{R}^n)$, we
have

$$p = \frac{2n}{n-2s} \implies \|f\|_{L^p} \leqslant C\|f\|_{\dot{H}^s} := C\left(\int |\xi|^{2s} |\widehat{f}(\xi)|^2 \, d\xi \right)^{1/2}. \qquad (7.15)$$

This is a stronger result than the one stated. Indeed, for any real $2 \leqslant p < 2n/(n-2s)$,
there exists $s' \in [0, s]$ such that $p = 2n/(n - 2s')$ and then $\|f\|_{L^p} \leqslant C\|f\|_{\dot{H}^{s'}} \leqslant
C\|f\|_{H^s}$ (be careful: we cannot bound $\|f\|_{\dot{H}^{s'}}$ by $\|f\|_{\dot{H}^s}$ because we do not have
$|\xi|^{2s'} \leqslant |\xi|^{2s}$ for $|\xi| \leqslant 1$).

We use the proof of Chemin and Xu (*cf.* [34]), which is based on the measure
of level sets. We will denote by $\{|f| > \lambda\}$ the set $\{x \in \mathbb{R}^n : |f(x)| > \lambda\}$ and by
$|\{|f| > \lambda\}|$ the Lebesgue measure of this set.

Consider a function $f \in \mathcal{S}(\mathbb{R}^n)$. We can assume without loss of generality that
$\|f\|_{\dot{H}^s} = 1$. We start from the classical identity (see Lemma 18.8),

$$\|f\|_{L^p}^p = p \int_0^{+\infty} \lambda^{p-1} |\{|f| > \lambda\}| \, d\lambda.$$

To bound $|\{|f| > \lambda\}|$, we will use a decomposition in terms of low and high
frequencies.

Namely, for all $\lambda > 0$, we will decompose f in the form

$$f = g_\lambda + h_\lambda,$$

where the functions g_λ and h_λ are defined by their Fourier transforms:

$$\widehat{g_\lambda}(\xi) = \widehat{f}(\xi) \quad \text{if } |\xi| \leqslant A_\lambda, \qquad \widehat{g_\lambda}(\xi) = 0 \qquad \text{if } |\xi| > A_\lambda$$

$$\widehat{h_\lambda}(\xi) = 0 \qquad \text{if } |\xi| \leqslant A_\lambda, \qquad \widehat{h_\lambda}(\xi) = \widehat{f}(\xi) \quad \text{if } |\xi| > A_\lambda,$$

for a certain constant A_λ to be determined. Then, according to the triangle inequality,

$$\{|f| > \lambda\} \subset \{|g_\lambda| > \lambda/2\} \cup \{|h_\lambda| > \lambda/2\}.$$

We will choose the constant A_λ so that $\{|g_\lambda| > \lambda/2\} = \emptyset$. Then we will have

$$|\{|f| > \lambda\}| \leqslant |\{|h_\lambda| > \lambda/2\}| \leqslant \frac{4}{\lambda^2}\|h_\lambda\|_{L^2}^2,$$

because

$$\|h_\lambda\|_{L^2}^2 \geqslant \int_{\{|h_\lambda|>\lambda/2\}} |h_\lambda|^2 \, dx \geqslant \frac{\lambda^2}{4} |\{|h_\lambda| > \lambda/2\}|.$$

By combining the previous observations, we will obtain

$$\|f\|_{L^p}^p \leqslant 4p \int_0^{+\infty} \lambda^{p-3} \|h_\lambda\|_{L^2}^2 \, d\lambda. \tag{7.16}$$

Choice of A_λ

According to the Fourier inversion theorem, we have

$$|g_\lambda(x)| = \left| \frac{1}{(2\pi)^n} \int e^{ix\cdot\xi} \widehat{g_\lambda}(\xi) \, d\xi \right| = \left| \frac{1}{(2\pi)^n} \int_{|\xi| \leqslant A_\lambda} e^{ix\cdot\xi} \widehat{f}(\xi) \, d\xi \right|.$$

Since $2s < n$, we can use the Cauchy–Schwarz inequality to obtain

$$|g_\lambda(x)| \leqslant \frac{1}{(2\pi)^n} \left(\int_{|\xi| \leqslant A_\lambda} |\xi|^{-2s} \, d\xi \right)^{1/2} \left(\int |\xi|^{2s} |\widehat{f}(\xi)|^2 \, d\xi \right)^{1/2}.$$

By switching to polar coordinates, we get

$$\int_{|\xi| \leqslant A_\lambda} |\xi|^{-2s} \, d\xi = \int_0^{A_\lambda} \int_{\mathbb{S}^{n-1}} r^{n-1-2s} \, d\theta \, dr = \frac{|\mathbb{S}^{n-1}| A_\lambda^{n-2s}}{n - 2s}.$$

Since $\|f\|_{\dot{H}^s} = 1$ by hypothesis, we finally get $\|g_\lambda\|_{L^\infty} \leqslant C_1(s,n) A_\lambda^{\frac{n}{2}-s}$.

We then define A_λ by

$$C_1(s,n)A_\lambda^{\frac{n}{2}-s} = \frac{\lambda}{2}.$$

Then $\|g_\lambda\|_{L^\infty} \leqslant \lambda/2$. Since moreover g_λ is a continuous function (it is the Fourier transform of an integrable function), we deduce that $\{|g_\lambda| > \lambda/2\} = \varnothing$, which is the desired result.

End of the proof

By definition of h_λ, using identity (7.16) and Plancherel's formula (5.8), we find

$$\|f\|_{L^p}^p \leqslant 4p(2\pi)^n \int_0^{+\infty} \int_{|\xi| \geqslant A_\lambda} \lambda^{p-3}|\widehat{f}(\xi)|^2 \, d\xi \, d\lambda.$$

By definition of A_λ, if $|\xi| \geqslant A_\lambda$ then

$$\lambda \leqslant \Lambda(\xi) := 2C_1(s,n)|\xi|^{(n/2)-s},$$

so, using Fubini's theorem, it comes

$$\|f\|_{L^p}^p \leqslant 4p(2\pi)^n \int_{\mathbb{R}^n} \left(\int_0^{\Lambda(\xi)} \lambda^{p-3} \, d\lambda \right) |\widehat{f}(\xi)|^2 \, d\xi,$$

from which

$$\|f\|_{L^p}^p \leqslant C_2(s,n) \int_{\mathbb{R}^n} \Lambda(\xi)^{p-2}|\widehat{f}(\xi)|^2 \, d\xi,$$

However, if $p = 2n/(n-2s)$, we verify that

$$\left(\frac{n}{2} - s\right)(p - 2) = 2s.$$

Using again that $\|f\|_{\dot{H}^s} = 1$, we finally obtain

$$\|f\|_{L^p}^p \leqslant C_3(s,n) \int_{\mathbb{R}^n} |\xi|^{2s}|\widehat{f}(\xi)|^2 \, d\xi = C_3(s,n),$$

equivalent to the desired result. $\square$

Corollary 7.45 (Gagliardo–Nirenberg inequalities) *Let n and p be such that*

$$n \geqslant 3, \quad 2 \leqslant p < \frac{2n}{n-2}.$$

Then, there exists a constant C such that, for all u belonging to $H^1(\mathbb{R}^n)$, we have

$$\|u\|_{L^p(\mathbb{R}^n)} \leqslant C\|u\|_{L^2(\mathbb{R}^n)}^{1-\sigma}\|\nabla u\|_{L^2(\mathbb{R}^n)}^\sigma \quad \text{with} \quad \sigma = \frac{n(p-2)}{2p}.$$

Proof For the value of σ given by the statement, it directly follows from (7.15) that we have the inequality $\|u\|_{L^p} \leq C\|u\|_{\dot{H}^\sigma}$. The result is then deduced by an interpolation argument. More precisely, we use the fact that the inequality of Proposition 7.42 remains true when we replace the norms $\|\cdot\|_{H^s}$ by the semi-norms $\|\cdot\|_{\dot{H}^s}$; which is demonstrated by replacing $\langle\xi\rangle$ by $|\xi|$ in the proof of Proposition 7.42. □

We have seen in Remark 7.44 that, for all $f \in H^1(\mathbb{R}^n)$ with $n \geq 3$, we have

$$\|f\|_{L^{2n/(n-2)}} \leq C\|\nabla f\|_{L^2}.$$

A direct consequence of the Hardy–Littlewood–Sobolev theorem 6.20 proved in Chapter 6 is that we can generalize this result in the following way.

Theorem 7.46 (Sobolev embeddings for $W^{1,p}$) *Let $n \geq 2$ and let $p \in (1, n)$. Define p^* by*

$$\frac{1}{p^*} = \frac{1}{p} - \frac{1}{n}.$$

Then there exists a constant C such that, for every function $f \in W^{1,p}(\mathbb{R}^n)$,

$$\|f\|_{L^{p^*}(\mathbb{R}^n)} \leq C\|\nabla f\|_{L^p(\mathbb{R}^n)}.$$

Proof We demonstrated in Theorem 6.22 that there exists a constant C such that

$$\|f\|_{L^{p^*}(\mathbb{R}^n)} \leq C\|\nabla f\|_{L^p(\mathbb{R}^n)},$$

for any function f that is of class C^∞ and has compact support. We deduce the desired result because $C_0^\infty(\mathbb{R}^n)$ is dense in $W^{1,p}(\mathbb{R}^n)$ for $p < +\infty$.

We will now see how to deduce from the previous result on Sobolev spaces defined on $\mathbb{R}^n$ a similar result on the space $H_0^1(\Omega)$, where Ω is any open set. In this context, assuming that Ω is bounded, we can also obtain that Sobolev spaces are compactly embedded in Lebesgue spaces.

Theorem 7.47 (Rellich–Kondrachov) *Let $n \geq 2$ and let Ω be a bounded open set of $\mathbb{R}^n$. If $2 \leq q < 2n/(n-2)$, then $H_0^1(\Omega)$ is compactly embedded in $L^q(\Omega)$.*

Remark 7.48 When $n = 2$, this means that the embedding holds for any $q < +\infty$.

Proof Consider a sequence $(u_m)_{m\in\mathbb{N}}$ that is bounded in $H_0^1(\Omega)$. Then this sequence has a subsequence $(u_{\theta(m)})_{m\in\mathbb{N}}$ converging weakly in $H^1(\Omega)$ because $H^1(\Omega)$ is a Hilbert space. Let $u \in H^1(\Omega)$ be the weak limit. Since $H_0^1(\Omega)$ is a closed subset of $H^1(\Omega)$ by definition, it is also weakly closed according to Proposition 3.42, and we deduce that $u \in H_0^1(\Omega)$. If we replace u_m by $u_{\theta(m)} - u$, we can assume without loss of generality that the sequence $(u_m)_{m\in\mathbb{N}}$ converges weakly to 0 in $H^1(\Omega)$. To prove the theorem, we must show that $(u_m)_{m\in\mathbb{N}}$ converges strongly to 0 in $L^q(\Omega)$ for $2 \leq q < 2n/(n-2)$.

Lemma 7.49 *Let $\tilde{u}_m$ be the function obtained by extending u_m by 0 on $\mathbb{R}^n \setminus \Omega$. Then $\tilde{u}_m$ belongs to $H^1(\mathbb{R}^n)$, the sequence $(\tilde{u}_m)_{m \in \mathbb{N}}$ is bounded in $H^1(\mathbb{R}^n)$ and converges weakly to 0 in $H^1(\mathbb{R}^n)$.*

Proof Proposition 7.27 implies that $(\tilde{u}_m)_{m \in \mathbb{N}}$ is bounded in $H^1(\mathbb{R}^n)$. Moreover, the proof of Proposition 7.27 shows that the weak derivative of $\tilde{u}_m$ satisfies $\partial_j \tilde{u}_m = \widetilde{\partial_j u_m}$. Then, denoting by $(\cdot, \cdot)_{H^1(\mathbb{R}^n)}$ the scalar product of $H^1(\mathbb{R}^n)$, for all $v \in H^1(\mathbb{R}^n)$, we have

$$(\tilde{u}_m, v)_{H^1(\mathbb{R}^n)} = \int_{\mathbb{R}^n} \tilde{u}_m v \, dx + \int_{\mathbb{R}^n} \nabla \tilde{u}_m \cdot \nabla v \, dx$$

$$= \int_{\mathbb{R}^n} \tilde{u}_m v \, dx + \int_{\mathbb{R}^n} \widetilde{\nabla u_m} \cdot \nabla v \, dx$$

$$= \int_{\Omega} u_m v \, dx + \int_{\Omega} \nabla u_m \cdot \nabla v \, dx = (u_m, v)_{H^1(\Omega)}.$$

But $(u_m, v)_{H^1(\Omega)}$ converges to 0 when m tends to $+\infty$. This shows that $\tilde{u}_m$ converges weakly to 0 in $H^1(\mathbb{R}^n)$, which concludes the proof of the lemma. $\square$

The proof of the Rellich–Kondrachov theorem then relies on the following lemma.

Lemma 7.50 *Let $(w_m)_{m \in \mathbb{N}}$ be a bounded sequence in $H^1(\mathbb{R}^n)$ which converges weakly to 0 in $L^2(\mathbb{R}^n)$. Then, for all q such that $2 \leqslant q < 2n/(n-2)$ and for any function $\chi \in C_0^\infty(\mathbb{R}^n)$, the sequence $(\chi w_m)_{m \in \mathbb{N}}$ converges strongly to 0 in $L^q(\mathbb{R}^n)$.*

Proof The proof is in two steps. We first show the result for $q = 2$ then we consider the general case.

Case $q = 2$

Let us set $v_m = \chi w_m$. As the multiplication by a function of the Schwartz space is continuous from $H^1(\mathbb{R}^n)$ to $H^1(\mathbb{R}^n)$, the sequence $(v_m)_{m \in \mathbb{N}}$ is also bounded in $H^1(\mathbb{R}^n)$. Moreover, $v_m = \chi w_m$ belongs to $L^1(\mathbb{R}^n)$ using the fact that v_m has compact support. We can then consider its Fourier transform:

$$\widehat{v_m}(\xi) = \int e^{-ix \cdot \xi} \chi(x) w_m(x) \, dx.$$

According to the Plancherel identity (5.8), to show that $(v_m)_{m \in \mathbb{N}}$ converges strongly to 0 in $L^2(\mathbb{R}^n)$, it is enough to show that $(\widehat{v_m})_{m \in \mathbb{N}}$ converges strongly to 0 in $L^2(\mathbb{R}^n)$. For this, we use the decomposition

$$\int |\widehat{v_m}(\xi)|^2 \, d\xi \leqslant \int_{|\xi| \leqslant R} |\widehat{v_m}(\xi)|^2 \, d\xi + \int_{|\xi| > R} |\widehat{v_m}(\xi)|^2 \, d\xi,$$

where R is an arbitrary parameter.

Since

$$\widehat{v_m}(\xi) = \langle w_m, \chi e_\xi \rangle \quad \text{where} \quad e_\xi : x \mapsto e^{ix \cdot \xi},$$

the weak convergence of $(w_m)_{m \in \mathbb{N}}$ to 0 in $L^2(\mathbb{R}^n)$ implies that, for all $\xi \in \mathbb{R}^n$, $\widehat{v_m}(\xi)$ converges to 0 as m tends to $+\infty$. Then, we can apply the dominated convergence theorem to show that $\int_{|\xi| \leqslant R} |\widehat{v_m}(\xi)|^2 \, d\xi$ tends to 0 as m tends to $+\infty$. Moreover, we have the obvious upper bound

$$\int_{|\xi| > R} |\widehat{v_m}(\xi)|^2 \, d\xi \leqslant \frac{1}{1 + R^2} \int_{|\xi| > R} (1 + |\xi|^2) \, |\widehat{v_m}(\xi)|^2 \, d\xi$$

$$\leqslant \frac{1}{1 + R^2} \int_{\mathbb{R}^n} (1 + |\xi|^2) \, |\widehat{v_m}(\xi)|^2 \, dx$$

$$= \frac{(2\pi)^n}{1 + R^2} \|v_m\|_{H^1}^2.$$

As $(v_m)_{m \in \mathbb{N}}$ is bounded in $H^1(\mathbb{R}^n)$, we can make $\int_{|\xi| > R} |\widehat{v_m}(\xi)|^2 \, d\xi$ arbitrarily small by taking R large enough. This completes the proof that $(v_m)_{m \in \mathbb{N}}$ converges to 0 in $L^2(\mathbb{R}^n)$.

General case

Let $2 \leqslant q < 2n/(n-2)$. Then there exist $C > 0$ and $\lambda \in [0, 1)$ such that, for all $u \in L^2(\mathbb{R}^n) \cap L^{2n/(n-2)}(\mathbb{R}^n)$, we have

$$\|u\|_{L^q} \leqslant C \|u\|_{L^2}^{1-\lambda} \|u\|_{L^{2n/(n-2)}}^\lambda.$$

We have seen that the sequence $(v_m)_{m \in \mathbb{N}}$ is bounded in $H^1(\mathbb{R}^n)$. The Sobolev embedding theorem implies that $v_m \in L^2(\mathbb{R}^n) \cap L^{2n/(n-2)}(\mathbb{R}^n)$ and that $(v_m)_{m \in \mathbb{N}}$ is bounded in $L^{2n/(n-2)}(\mathbb{R}^n)$. The desired result then follows from the preceding inequality and the previous lemma, which implies that $\|v_m\|_{L^2}$ converges to 0. □

We are now able to prove the theorem. Consider a function $\chi \in C_0^\infty(\mathbb{R}^n)$ such that $\chi(x) = 1$ for $x \in \Omega$. Such a function exists because Ω is bounded by hypothesis. Using the two previous lemmas, we obtain that the sequence $(\chi \widetilde{u}_m)_{m \in \mathbb{N}}$ converges to 0 strongly in $L^q(\mathbb{R}^n)$, which directly implies that $(u_m)_{m \in \mathbb{N}}$ converges to 0 strongly in $L^q(\Omega)$. □

The previous proof uses a property of the elements of $H_0^1(\Omega)$, namely that if they are extended by 0 outside of Ω, we obtain a function that belongs to $H^1(\mathbb{R}^n)$. If Ω is an open set of class C^2, Theorem 7.31 states that there is an extension operator. By proceeding as above, we can then deduce the following result.

Theorem 7.51 (Rellich–Kondrachov) *Let Ω be a bounded open set of $\mathbb{R}^n$ of class C^2 (or the product of n bounded open intervals).*

- *If $n \geqslant 2$ and $2 \leqslant q < 2n/(n-2)$, then $H^1(\Omega)$ is compactly embedded in $L^q(\Omega)$.*
- *If $n = 1$, then $H^1(\Omega)$ is compactly embedded in $C^0(\overline{\Omega})$.*

We saw in Section 4.4 the Poincaré–Wirtinger inequality (*cf.* Lemma 4.17). We will now deduce from the Rellich–Kondrachov theorem a generalization of this inequality in any dimension.

Theorem 7.52 (Poincaré–Sobolev Inequality) *Let Ω be a bounded open set of $\mathbb{R}^n$ of class C^2 (or the product of n bounded open intervals). Given a function $u \in H^1(\Omega)$, let u_Ω be its average over Ω. There exists a constant $C(\Omega)$ such that*

$$\forall u \in H^1(\Omega), \quad \|u - u_\Omega\|_{L^2(\Omega)} \leqslant C(\Omega)\|\nabla u\|_{L^2(\Omega)}.$$

Proof Assume by contradiction that this result is false. Then we can find a sequence $(u_n)_{n\in\mathbb{N}}$ of elements of $H^1(\Omega)$ such that

$$\int_\Omega u_n(x)\,dx = 0, \quad \int_\Omega |u_n(x)|^2\,dx = 1, \quad \int_\Omega |\nabla u_n(x)|^2\,dx \leqslant \frac{1}{n+1}.$$

As $(u_n)_{n\in\mathbb{N}}$ is bounded in $H^1(\Omega)$, Theorem 7.51 implies that we can extract a subsequence $(u_{\theta(n)})_{n\in\mathbb{N}}$ that converges strongly in $L^2(\Omega)$ towards a function u. As $(\nabla u_n)_{n\in\mathbb{N}}$ converges strongly to 0 in $L^2(\Omega)$, we deduce that $(u_{\theta(n)})_{n\in\mathbb{N}}$ is in fact a Cauchy sequence in $H^1(\Omega)$, which converges in $H^1(\Omega)$. We deduce that

$$\int_\Omega u(x)\,dx = 0, \quad \int_\Omega |u(x)|^2\,dx = 1, \quad \int_\Omega |\nabla u(x)|^2\,dx = 0.$$

Therefore, u is a non-zero function, constant and zero on average. Hence the contradiction.   □

Remark 7.53 The previous result is not quantitative in the sense that it does not say how the constant $C(\Omega)$ depends on Ω. A quantitative estimate is given in Exercise 7.3, where we reproduce the original, very elegant, proof by Poincaré.

The optimal inequality was obtained by Payne and Weinberger [152]. They proved the following result: if $\Omega \subset \mathbb{R}^n$ is an open and convex domain, then for any function $u \in H^1(\Omega)$, we have

$$\|u - u_\Omega\|_{L^2} \leqslant \frac{\operatorname{diam}\Omega}{\pi}\|\nabla u\|_{L^2}.$$

7.6 Dyadic Characterization of Sobolev Spaces

In this section, we use the Littlewood–Paley decomposition introduced in Chapter 5 (*cf.* Section 5.4) to characterize Sobolev spaces.

Proposition 7.54 1. *For all $u \in L^2(\mathbb{R}^n)$,*

$$\sum_{p \geqslant -1} \|\Delta_p u\|_{L^2}^2 \leqslant \|u\|_{L^2}^2 \leqslant 3 \sum_{p \geqslant -1} \|\Delta_p u\|_{L^2}^2. \tag{7.17}$$

2. *Consider $s \in \mathbb{R}$. A tempered distribution $u \in S'(\mathbb{R}^n)$ belongs to the Sobolev space $H^s(\mathbb{R}^n)$ if and only if*

(a) *$\Delta_{-1}u \in L^2(\mathbb{R}^n)$ and for all $p \geqslant 0$, $\Delta_p u \in L^2(\mathbb{R}^n)$;*
(b) *the sequence $\delta_p = 2^{ps}\|\Delta_p u\|_{L^2}$ belongs to $\ell^2(\mathbb{N} \cup \{-1\})$.*

Moreover, there exists a constant C such that

$$\frac{1}{C}\|u\|_{H^s} \leqslant \left(\sum_{p=-1}^{+\infty} \delta_p^2 \right)^{1/2} \leqslant C\|u\|_{H^s}. \tag{7.18}$$

Proof The first point immediately follows from (5.12) and the Plancherel identity (5.8).

Since $\|u\|_{H^s} = \|\langle D_x \rangle^s u\|_{L^2}$, by applying (7.17) with u replaced by $\langle D_x \rangle^s u$, we get

$$\sum_{p \geqslant -1} \|\Delta_p \langle D_x \rangle^s u\|_{L^2}^2 \leqslant \|u\|_{H^s}^2 \leqslant 3 \sum_{p \geqslant -1} \|\Delta_p \langle D_x \rangle^s u\|_{L^2}^2.$$

Consider $p \geqslant 0$ and write

$$\|\Delta_p \langle D_x \rangle^s u\|_{L^2}^2 = (2\pi)^{-n} \int_{\mathbb{R}^n} (1 + |\xi|^2)^s \varphi^2(2^{-p}\xi)|\hat{u}(\xi)|^2 \, d\xi.$$

Since $(1 + |\xi|^2)^s \varphi^2(2^{-p}\xi) \sim 2^{2ps} \varphi^2(2^{-p}\xi)$, we see that

$$\frac{1}{C} 2^{2ps} \|\Delta_p u\|_{L^2}^2 \leqslant \|\Delta_p \langle D_x \rangle^s u\|_{L^2}^2 \leqslant C2^{2ps} \|\Delta_p u\|_{L^2}^2, \tag{7.19}$$

for a certain constant C depending only on s. We have a similar estimate for $\Delta_{-1}u$ and the desired result follows easily. □

Proposition 7.55 1. *Consider $s \in \mathbb{R}$ and $R \geqslant 1$. Suppose that $(u_j)_{j \geqslant -1}$ is a sequence of functions in $L^2(\mathbb{R}^n)$ such that*

$$\operatorname{supp} \widehat{u_{-1}} \subset \{|\xi| \leqslant R\}, \qquad \operatorname{supp} \widehat{u_j} \subset \left\{ \frac{1}{R}2^j \leqslant |\xi| \leqslant R2^j \right\},$$

and, in addition,

$$\sum_{j \geqslant -1} 2^{2js} \|u_j\|_{L^2}^2 < +\infty. \tag{7.20}$$

Then the series $\sum u_j$ converges to a function $u \in H^s(\mathbb{R}^n)$ and moreover,

$$\|u\|_{H^s}^2 \leqslant C \sum_{j \geqslant -1} 2^{2js} \|u_j\|_{L^2}^2,$$

for a certain constant C depending only on s and R.
2. *If $s > 0$, then the previous result is valid under the weaker assumption that $\operatorname{supp} \widehat{u_j}$ is included in the ball $B(0, R2^j)$.*

Proof We begin by proving that the series $\sum u_j$ is normally convergent in $H^r(\mathbb{R}^n)$ for all $r < s$ assuming that supp $\widehat{u}_j$ is included in a ball $\{|\xi| \leqslant R2^j\}$. By an argument parallel to that of (7.19), we see that $\|u_j\|_{H^r} \leqslant C2^{jr}\|u_j\|_{L^2}$. Therefore, the Cauchy–Schwarz inequality implies that

$$\sum_{j \geqslant -1} \|u_j\|_{H^r} \leqslant \left(\sum_{j \geqslant -1} 2^{2js}\|u_j\|_{L^2}^2\right)^{1/2}\left(\sum_{j \geqslant -1} 2^{2j(r-s)}\right)^{1/2} < +\infty.$$

This shows that the series $\sum u_j$ is normally convergent and therefore convergent in $H^r(\mathbb{R}^n)$. Now, we can define $u = \sum_{j \geqslant -1} u_j$.

1. Now assume that supp $\widehat{u}_j$ is included in an annulus $\{\frac{1}{R}2^j \leqslant |\xi| \leqslant R2^j\}$. Our goal is then to prove that u belongs to $H^s(\mathbb{R}^n)$. Notice that there exists an integer N depending only on R such that $\Delta_p u_j = 0$ if $|j - p| > N$. Consequently,

$$\|\Delta_p u\|_{L^2} \leqslant \sum_{|j-p| \leqslant N} \|\Delta_p u_j\|_{L^2} \leqslant \sum_{|j-p| \leqslant N} \|u_j\|_{L^2},$$

and hence the result follows from Proposition 7.54.

2. If we only assume that supp $\widehat{u}_j$ is included in a ball $\{|\xi| \leqslant R2^j\}$, then we just have, for an integer N,

$$\Delta_p u = \sum_{j \geqslant p-N} \Delta_p u_j.$$

It follows from the triangle inequality that

$$2^{ps}\|\Delta_p u\|_{L^2} \leqslant \sum_{j \geqslant p-N} 2^{(p-j)s}2^{js}\|u_j\|_{L^2}.$$

Now, since $s > 0$, the sequence $(2^{(p-j)s})_{j \geqslant p-N}$ belongs to ℓ^1 and the convolution inequality $\ell^1 * \ell^2 \hookrightarrow \ell^2$ gives the result. $\square$

Theorem 7.56 *Consider* $n \in \mathbb{N}^*$, $s \in (0, +\infty)$, $k \in \mathbb{N}$ *and* $\alpha \in (0, 1]$ *such that*

$$s > \frac{n}{2} + k + \alpha.$$

Then $H^s(\mathbb{R}^n)$ *is continuously embedded in the Hölder space* $C^{k,\alpha}(\mathbb{R}^n)$ *given by definition 5.28.*

Proof We will prove a more general result that deals with the embeddings of Sobolev spaces into the Zygmund spaces defined in Section 5.4.3.

Lemma 7.57 *For every integer* $n \geqslant 1$ *and for every* $s \in \mathbb{R}$, *we have*

$$H^s(\mathbb{R}^n) \subset C_*^{s-n/2}(\mathbb{R}^n),$$

and the embedding is continuous. In particular, $H^\infty(\mathbb{R}^n) \subset C^\infty(\mathbb{R}^n)$.

Proof Let $u \in H^s(\mathbb{R}^n)$. We have seen that

$$u = \sum_{p \geqslant -1} u_p, \quad \text{where} \quad u_p = \Delta_p u.$$

Then

$$u_p(x) = (2\pi)^{-n} \int_{\mathbb{R}^n} e^{ix \cdot \xi} \widehat{u}_p(\xi) \, d\xi = (2\pi)^{-n} \int_{C_p} e^{ix \cdot \xi} \widehat{u}_p(\xi) \, d\xi;$$

where C_p is the annulus $\{\xi \in \mathbb{R}^n : 2^{p-1} \leqslant |\xi| \leqslant 2^{p+1}\}$. Therefore

$$\|u_p\|_{L^\infty} \leqslant (2\pi)^{-n} \int_{C_p} |\widehat{u}_p(\xi)| \, d\xi$$

$$\leqslant (2\pi)^{-n} \left(\int_{C_p} \left(\frac{1 + |\xi|^2}{1 + 2^{2(p-1)}} \right)^s |\widehat{u}_p(\xi)|^2 \, d\xi \right)^{1/2} \left(\int_{C_p} d\xi \right)^{1/2}$$

$$\leqslant K 2^{p(n/2-s)} \|u\|_{H^s},$$

which implies the result.   □

The theorem follows directly because it follows from (5.14) that

$$C_*^{s-n/2}(\mathbb{R}^n) \subset C^{k,\alpha}(\mathbb{R}^n) \quad \text{if} \quad s > n/2 + k + \alpha.$$ □

7.7 Exercises

Exercise (solved) 7.1 Show that the function $x \mapsto |x|$ has a weak derivative on any open bounded interval $I \subset \mathbb{R}$ and that its derivative is given by $H|_I$ where $H = \mathbf{1}_{[0,+\infty)} - \mathbf{1}_{(-\infty,0)}$ is the function equal to 1 on $[0, +\infty)$ and -1 on $(-\infty, 0)$, called the Heaviside function.

Exercise (solved) 7.2 Show that the function $\mathbf{1}_{(0,1)}$, the indicator of the interval $(0, 1)$, is not weakly differentiable in $L^p(\mathbb{R})$ for any $p \in (1, +\infty]$.

Exercise (solved) 7.3 The aim of this exercise is to provide a direct proof, due to Henri Poincaré, of the inequality that bears his name. Let $n \geqslant 1$ be an integer, $r > 0$ a real number and B a ball of radius r in $\mathbb{R}^n$.

1. Consider $u: B \to \mathbb{R}$ of class C^1 and of zero mean on B.
   (a) Show that

$$\frac{1}{|B|} \int_B u^2 \, dx = \frac{1}{2|B|^2} \iint_{B^2} (u(x) - u(y))^2 \, dx \, dy.$$

(b) Deduce that

$$\frac{1}{|B|} \int_B u^2 \, dx \leq \frac{(2r)^2}{2|B|^2} \iint_{B^2} \int_0^1 |\nabla u(tx + (1-t)y)|^2 \, dt \, dx \, dy.$$

(c) Show that for any σ in B, we have:

$$\int_B 1_{tB+(1-t)y}(\sigma) \, dy \leq \min\left(1, \frac{t^n}{(1-t)^n}\right)|B|.$$

(d) Deduce that there exists a constant C such that

$$\left(\frac{1}{|B|} \int_B u^2 \, dx\right)^{1/2} \leq Cr\left(\frac{1}{|B|} \int_B |\nabla u|^2 \, dx\right)^{1/2}.$$

2. Conclude that for any function $u \in H^1(B)$, with complex values, we have

$$\left(\frac{1}{|B|} \int_B |u - u_B|^2 \, dx\right)^{1/2} \leq Cr\left(\frac{1}{|B|} \int_B |\nabla u|^2 \, dx\right)^{1/2},$$

where u_B is the average of u over B.

Exercise (solved) 7.4 Prove (7.10) in the case $n = 1$.

Exercise (solved) 7.5 Let $p \in [1, +\infty]$, $\Omega = \{(x_1, \ldots, x_n) \in \mathbb{R}^n : x_1 > 0\}$ and consider a function $u \in W^{1,p}(\Omega)$. We extend u by parity to a function u_* defined on $\mathbb{R}^n$ as follows:

$$u_*(x_1, \ldots, x_n) = \begin{cases} u(x_1, x_2, \ldots, x_n) & \text{if } x_1 > 0, \\ u(-x_1, x_2, \ldots, x_n) & \text{if } x_1 < 0. \end{cases}$$

Show that $u_* \in W^{1,p}(\mathbb{R}^n)$. Calculate $\|u_*\|_p$ and, for all i, $\|\partial_i u_*\|_{L^p}$ in terms of $\|u\|_{L^p}$ and $\|\partial_i u\|_{L^p}$.

Exercise (solved) 7.6 (Sobolev spaces of fractional order) Let $s \in \mathbb{R}$. We define:

$$H^s(\mathbb{R}^n) = \{f \in \mathcal{S}'(\mathbb{R}^n) : (1 + |\xi|^2)^{s/2}\widehat{f} \in L^2(\mathbb{R}^n)\},$$

where $\widehat{f}$ is the Fourier transform of f in the sense of tempered distributions. We equip this space with the following norm:

$$\|f\|_{H^s} = \left\|(1 + |\xi|^2)^{s/2}\mathcal{F}f\right\|_{L^2}.$$

1. Show that if $s \in \mathbb{N}$, this definition is equivalent to the definition (Definition 7.12) of the Sobolev space $W^{s,2}(\mathbb{R}^n)$.
   You can use that, if $f \in W^{s,2}(\mathbb{R}^n)$ and if $\alpha \in \mathbb{N}^n$ is a multi-index of length $|\alpha| \leq s$, then the Fourier transform of $\partial_x^\alpha f$ is given by $(i\xi)^\alpha \widehat{f}$.

2. Show that, for $s \in (0, 1)$, this definition is equivalent to:

$$H^s(\mathbb{R}^n) = \left\{ f \in L^2(\mathbb{R}^n) : \iint \frac{|f(x) - f(y)|^2}{|x - y|^{n+2s}} \, dx \, dy < +\infty \right\}.$$

Exercise 7.7

1. Consider an integer $n \geqslant 3$ and a real $s > n/2$. Verify that

$$\int_{\mathbb{R}^n} \frac{d\xi}{|\xi|^2 (1 + |\xi|^2)^{s-1}} < +\infty.$$

From this, deduce that there exists a constant C such that, for all $u \in H^s(\mathbb{R}^n)$,

$$\|u\|_{L^\infty(\mathbb{R}^n)} \leqslant C \|\nabla u\|_{H^{s-1}(\mathbb{R}^n)}. \tag{7.21}$$

2. Show that, for all integers j, k such that $1 \leqslant j, k \leqslant n$, there exists a constant C such that, for all $u \in H^2(\mathbb{R}^n)$,

$$\left\| \partial_{x_j} \partial_{x_k} u \right\|_{L^2(\mathbb{R}^n)} \leqslant C \|\Delta u\|_{L^2(\mathbb{R}^n)}.$$

Infer that there exists a constant C such that, for all $u \in H^2(\mathbb{R}^3)$,

$$\|u\|_{L^\infty(\mathbb{R}^3)} \leqslant C \|\nabla u\|_{L^2(\mathbb{R}^3)} + C \|\Delta u\|_{L^2(\mathbb{R}^3)}.$$

By applying the previous result to the function v defined by $v(x) = u(\lambda x)$, for a well-chosen parameter λ, show that there exists a constant C such that, for all $u \in H^2(\mathbb{R}^3)$,

$$\|u\|_{L^\infty(\mathbb{R}^3)} \leqslant C \|\nabla u\|_{L^2(\mathbb{R}^3)}^{1/2} \|\Delta u\|_{L^2(\mathbb{R}^3)}^{1/2}.$$

3. Show that inequality (7.21) is false in dimension $n = 2$.

Exercise 7.8 Let $\phi \in S(\mathbb{R}^n)$ such that $\phi(0) = 1$. Define for all $h \in (0, 1]$ a linear operator $A_h : S(\mathbb{R}^n) \to S(\mathbb{R}^n)$ by

$$(A_h u)(x) = (2\pi)^{-n} \int e^{ix \cdot \xi} \phi(h\xi) \widehat{u}(\xi) \, d\xi.$$

Show that, for all $r \in [2, +\infty]$, all $u \in S(\mathbb{R}^n)$ and all $h \in (0, 1]$,

$$\|A_h u\|_{L^r} \leqslant C h^{-n(1/2 - 1/r)} \|u\|_{L^2},$$

for a constant C that will be specified.

Exercise 7.9 Consider an integer $n \geqslant 2$ and a real number $p \in [1, n)$. We have seen that the Sobolev space $W^{1,p}(\mathbb{R}^n)$ is continuously embedded in $L^q(\mathbb{R}^n)$ for all real q such that

$$p \leqslant q \leqslant p^* := \frac{np}{n - p}.$$

We will see another proof in the case $p \leqslant q < p^*$.

1. Let $f \in C_0^1(\mathbb{R}^n)$ be a continuously differentiable function with compact support. Show that, for all $v \in \mathbb{S}^{n-1}$,

$$|f(x)| \leqslant \int_0^{+\infty} |(\nabla f)(x+tv)| \, dt.$$

From this, deduce (cf. the reminder at the end of the exercise) that

$$|f(x)| \leqslant \frac{1}{|\mathbb{S}^{n-1}|} \int_{\mathbb{R}^n} \frac{1}{|y|^{n-1}} |\nabla f(x-y)| \, dy.$$

2. Let us define

$$F_1(x) = \frac{1}{|\mathbb{S}^{n-1}|} \int_{|y| \geqslant 1} \frac{1}{|y|^{n-1}} |\nabla f(x-y)| \, dy,$$

and

$$F_2(x) = \frac{1}{|\mathbb{S}^{n-1}|} \int_{|y| \leqslant 1} \frac{1}{|y|^{n-1}} |\nabla f(x-y)| \, dy.$$

Using the Hölder or Young inequalities, show that

$$\|F_1\|_{L^\infty} \leqslant C_1 \|\nabla f\|_{L^p}, \quad \|F_2\|_{L^q} \leqslant C_2 \|\nabla f\|_{L^p} \quad \text{for all} \quad p \leqslant q < p^*.$$

3. From this, deduce that there exists a constant C such that, for all $f \in C_0^1(\mathbb{R}^n)$,

$$\|f\|_{L^q} \leqslant C \|f\|_{L^p} + C \|\nabla f\|_{L^p}.$$

Infer that $W^{1,p}(\mathbb{R}^n) \subset L^q(\mathbb{R}^n)$ for all $q \in [p, p^*)$.
4. Using a homogeneity argument, show that the previous inclusion is false for $q > p^*$.

Exercise 7.10 Let $d \geqslant 1$ and let $\varphi \colon \mathbb{R}^n \to [0, +\infty)$ be a C^∞ function, radial, which is equal to 1 on the ball $\{|\xi| \leqslant 1\}$ and which is equal to 0 outside the ball $\{|\xi| \leqslant 2\}$. For all $j \in \mathbb{Z}$, we define

$$\widehat{\Delta_j u}(\xi) = \left(\varphi\left(2^{-j}\xi\right) - \varphi\left(2^{-j+1}\xi\right) \right) \widehat{u}(\xi).$$

1. Show that, for all p, q such that $1 \leqslant p \leqslant q \leqslant +\infty$, there exists a constant C such that, for all $j \in \mathbb{Z}$,

$$\left\| \Delta_j u \right\|_{L^q(\mathbb{R}^d)} \leqslant C 2^{j(n/p - n/q)} \left\| \Delta_j u \right\|_{L^p(\mathbb{R}^n)}.$$

2. Show that for all p such that $1 \leqslant p \leqslant +\infty$, there exists a constant C such that, for all $j \in \mathbb{Z}$,

$$\left\| \Delta_j |\nabla| u \right\|_{L^p(\mathbb{R}^n)} \leqslant C 2^j \left\| \Delta_j u \right\|_{L^p(\mathbb{R}^n)}.$$

3. Let $\mathbb{N} \ni n \geqslant 1$ and $p, q \in [1, +\infty]$ with $p < q$, such that

$$\theta := \left(\frac{n}{p} - \frac{n}{q} \right) \in (0, 1).$$

We want to show that there exists a constant C such that, for all $u \in W^{1,p}(\mathbb{R}^n)$,

$$\|u\|_{L^q(\mathbb{R}^n)} \leqslant C\|u\|_{L^p(\mathbb{R}^n)}^{1-\theta}\|\nabla u\|_{L^p(\mathbb{R}^n)}^{\theta}.$$

Verify that the inequality is invariant under homothety $(u(x) \mapsto \lambda u(x))$ and scale change $(u(x) \mapsto u(\lambda x))$. Infer that we can suppose that $\|u\|_{L^p(\mathbb{R}^n)} = 1 = \|\nabla u\|_{L^p(\mathbb{R}^n)}$. By writing u in the form $\sum_{j \in \mathbb{Z}} \Delta_j u$, conclude by using the inequalities of the previous question.

Chapter 8
Harmonic Functions

8.1 The mean value property

In this entire chapter, $n \geqslant 2$ denotes the dimension of space, $|x|$ the Euclidean norm of a vector $x = (x_1, \ldots, x_n)$ in $\mathbb{R}^n$, so that $|x|^2 = x_1^2 + \cdots + x_n^2$, and $B(x, r)$ the open ball centered at x with radius $r > 0$ for this norm. By assumption, Ω will always denote a bounded open set in $\mathbb{R}^n$.

Definition 8.1 Let $u \in C^2(\Omega)$ be a real-valued function. We say that u is harmonic if

$$\Delta u = \sum_{j=1}^{n} \partial_{x_j}^2 u = 0.$$

Harmonic functions are remarkable functions. In particular, they satisfy the mean value property, which means that for any $x \in \Omega$, $u(x)$ is equal to the average of u over any ball $B \subset \Omega$ that is centered at x.

Theorem 8.2 *Let $u \in C^2(\Omega)$. Then u is harmonic if and only if, for any $x \in \Omega$ and any $r > 0$ such that $B(x, r) \subset \Omega$, we have*

$$u(x) = \frac{1}{|B(x, r)|} \int_{B(x, r)} u(y) \, dy. \tag{8.1}$$

Proof Let $x \in \Omega$. Let $r_0 = \text{dist}(x, \partial\Omega) = \inf_{y \in \mathbb{R}^n \setminus \Omega} |x - y|$ and consider the function $\phi \colon [0, r_0) \to \mathbb{R}$ defined by

$$\phi(r) = \frac{1}{|B(x, r)|} \int_{B(x, r)} u(y) \, dy = \frac{1}{|B(0, 1)|} \int_{B(0, 1)} u(x + r\zeta) \, d\zeta.$$

By differentiating the integral, we verify that

© The Author(s), under exclusive license to Springer Nature Switzerland AG 2024
T. Alazard, *Analysis and Partial Differential Equations*, Universitext,
https://doi.org/10.1007/978-3-031-70909-8_8

$$\phi'(r) = \frac{1}{|B(0,1)|} \int_{B(0,1)} \zeta \cdot (\nabla u)(x + r\zeta) \, d\zeta$$

$$= \frac{1}{|B(0,1)|} \int_{B(0,1)} \frac{1}{2} \operatorname{div}_\zeta \left(|\zeta|^2 \, (\nabla u)(x + r\zeta) \right) d\zeta$$

$$- \frac{1}{|B(0,1)|} \int_{B(0,1)} \frac{r}{2} |\zeta|^2 \, (\Delta u)(x + r\zeta) \, d\zeta.$$

We then use the divergence theorem (see the solved Exercise 17.3), which implies that for a vector field $X \in C^1(\overline{B(0,1)}; \mathbb{R}^n)$, we have

$$\int_{B(0,1)} \operatorname{div} X(\zeta) \, d\zeta = \int_{\partial B(0,1)} X(\zeta) \cdot \zeta \, dS(\zeta).$$

Since $|\zeta| = 1$ on the boundary of the ball $B(0,1)$, this implies that

$$\int_{B(0,1)} \operatorname{div}_\zeta \left(|\zeta|^2 \, (\nabla u)(x + r\zeta) \right) d\zeta = \int_{\partial B(0,1)} (\zeta \cdot \nabla u)(x + r\zeta) \, dS(\zeta)$$

$$= \int_{\partial B(0,1)} \frac{1}{r} \zeta \cdot \nabla_\zeta (u(x + r\zeta)) \, dS(\zeta)$$

$$= \int_{B(0,1)} \frac{1}{r} \Delta_\zeta (u(x + r\zeta)) \, d\zeta$$

$$= \int_{B(0,1)} r(\Delta u)(x + r\zeta) \, d\zeta,$$

where we used the divergence theorem again to get the third equality. It follows that

$$\phi'(r) = \frac{r}{2\,|B(0,1)|} \int_{B(0,1)} (1 - |\zeta|^2)(\Delta u)(x + r\zeta) \, d\zeta. \tag{8.2}$$

We deduce that if $\Delta u = 0$, then the derivative ϕ' is zero. Since $\phi(r)$ converges to $u(x)$ when r tends to 0, this proves that if u is harmonic, then u satisfies the mean value property.

Conversely, if u satisfies the mean value property (8.1) for all $x \in \Omega$ and all $r > 0$ such that $B(x,r) \subset \Omega$, then we have $\phi'(r) = 0$ and (8.2) implies that

$$I(r) = \frac{r}{2\,|B(0,1)|} \int_{B(0,1)} (1 - |\zeta|^2)(\Delta u)(x + r\zeta) \, d\zeta = 0.$$

Suppose that $\Delta u(x)$ is non-zero for a certain $x \in \Omega$ and, in this case, we can assume without loss of generality that $\Delta u(x) > 0$. Then, for r small enough, we have $(\Delta u)(x + r\zeta) > 0$ for all $\zeta \in B(0,1)$ and therefore the integral $I(r)$ is necessarily positive, which is absurd. We deduce that

$$\Delta u = 0,$$

which concludes the proof.   □

We will see several consequences of the mean value property. Let us start with the maximum principle, which states that the maximum is reached at the boundary.

Theorem 8.3 *Let Ω be a bounded and connected open set. If $u \in C^0(\overline{\Omega}) \cap C^2(\Omega)$ is a harmonic function, then*

$$\max_{\overline{\Omega}} u = \max_{\partial\Omega} u. \tag{8.3}$$

Furthermore, if there exists $x_0 \in \Omega$ such that $u(x_0) = \max_{\overline{\Omega}} u$, then u is constant on Ω.

Proof The function u reaches its maximum on $\overline{\Omega}$ because it is continuous and because this set is compact by hypothesis. If the maximum is reached on the boundary, the result (8.3) is proved. Let us assume that the maximum is reached inside Ω and consider $x_0 \in \Omega$ such that $u(x_0) = M := \max_{\overline{\Omega}} u$. We then choose $r > 0$ such that the ball $B(x_0, r)$ is included in Ω.

Using the mean value property, we then obtain

$$M = u(x_0) = \frac{1}{|B(x_0,r)|} \int_{B(x_0,r)} u(y)\,dy.$$

Since $u \leqslant M$ by definition of M, we deduce that $u(y) = M$ for all $y \in B(x_0, r)$ (otherwise the average $\frac{1}{|B(x_0,r)|} \int_{B(x_0,r)} u(y)\,dy$ would be strictly smaller than M). Thus, the set $\{x \in \Omega : u(x) = M\}$ is open. Since it is also closed by continuity of u, it is equal to Ω by connectivity.   □

We deduce the following uniqueness result.

Corollary 8.4 *If $u \in C^0(\overline{\Omega}) \cap C^2(\Omega)$ satisfies*

$$\begin{cases} \Delta u = 0 & \text{in} \quad \Omega, \\ u = 0 & \text{on} \quad \partial\Omega, \end{cases}$$

then $u = 0$. In particular, if $u \in C_0^2(\mathbb{R}^n)$ is harmonic, then $u = 0$.

Proof The maximum principle implies that $\max u \leqslant 0$. Since $-u$ satisfies the same properties, we also have $\min u \geqslant 0$, therefore $u = 0$.   □

The following result is another consequence of the averaging effect: the supremum of a *positive* harmonic function can be controlled by its infimum. We will return on this result later in Section 14.4, in the much more difficult context of equations with variable coefficients.

Theorem 8.5 (Harnack's Inequality) *Let Ω be an open set and ω be a subset that is open, connected, bounded, and such that $\overline{\omega} \subset \Omega$. Then, there exists a constant C such that, for any harmonic function $u : \Omega \to [0, +\infty)$,*

$$\sup_{\omega} u \leqslant C \inf_{\omega} u.$$

Proof Let $r = \frac{1}{4}\text{dist}(\omega, \partial\Omega)$. Note that we can cover the compact set $\overline{\omega}$ with a collection of balls $(B_i)_{1 \leqslant i \leqslant N}$ included in Ω and of radius r. We will show that for all $x \in \omega$ and all $y \in \omega$ we have $u(x) \geqslant 2^{-nN} u(y)$; which implies that

$$\sup_{y \in \omega} u(y) \leqslant 2^{nN} \inf_{x \in \omega} u(x),$$

which is the desired result with $C = 2^{nN}$.

Let x, y be in ω. To prove the inequality $u(x) \geqslant 2^{-nN} u(y)$, let us first assume that $|x - y| \leqslant r$. Then, using the mean value property and the hypothesis $u \geqslant 0$, we find

$$u(x) = \frac{1}{|B(x,2r)|} \int_{B(x,2r)} u \, dz \geqslant \frac{1}{|B(x,2r)|} \int_{B(y,r)} u \, dz \quad (\text{since } u \geqslant 0)$$

$$\geqslant \frac{|B(y,r)|}{|B(x,2r)|} \frac{1}{|B(y,r)|} \int_{B(y,r)} u \, dz = \frac{1}{2^n} u(y).$$

We deduce that if x and y are in the same ball of radius r, then $u(x) \geqslant 2^{-n} u(y)$, which implies the desired result.

Now suppose that x and y are not in the same ball of radius r. Using the fact that ω is a connected open set, therefore arcwise connected, there exists a sequence of balls $(B'_j)_{1 \leqslant j \leqslant N'}$ with $2 \leqslant N' \leqslant N$ and $B'_j \in \{B_1, \ldots, B_N\}$ such that $x \in B'_1$, $y \in B'_{N'}$ and $B'_j \cap B'_{j+1} \neq \emptyset$ for $1 \leqslant j \leqslant N' - 1$. Denote $x_0 = x$, $x_{N'} = y$ and consider for $1 \leqslant j \leqslant N' - 1$ a point x_j in $B'_j \cap B'_{j+1}$. Using successively the fact that x_ℓ and $x_{\ell+1}$ are in the same ball of radius r, we deduce from the above that

$$u(x) \geqslant 2^{-n} u(x_1) \geqslant \cdots \geqslant 2^{-n(N'-1)} u(x_{N'-1}) \geqslant 2^{-nN'} u(y).$$

We deduce that $u(x) \geqslant 2^{-nN} u(y)$ for all x, y in ω, which concludes the proof. □

8.2 The Fundamental Solution of the Laplacian

We will now introduce the notion of the fundamental solution of the Laplacian on $\mathbb{R}^n$ and use it to express a representation formula of a function in terms of its gradient. This will serve us later to demonstrate Sobolev inequalities.

To find non-trivial examples of harmonic functions, one can start by looking for radial functions u of the form $u(x) = v(|x|)$, where v is defined on $(0, +\infty)$. By proceeding in this way, we obtain the following particular solutions

$$u(x) = \begin{cases} a \log(|x|) + b & \text{if } n = 2, \\ \dfrac{a}{|x|^{n-2}} + b & \text{if } n \geqslant 3. \end{cases}$$

Definition 8.6 We call the fundamental solution of the Laplacian the function

$$\varphi(x) = \begin{cases} -\dfrac{1}{2\pi}\log(|x|) & \text{if } n = 2, \\[2mm] \dfrac{1}{(n-2)\,|\mathbb{S}^{n-1}|}\,\dfrac{1}{|x|^{n-2}} & \text{if } n \geqslant 3, \end{cases}$$

where $|\mathbb{S}^{n-1}|$ is the measure of the sphere $\partial B(0, 1)$.

Then, with this choice, we obtain the following result.

Proposition 8.7 *Suppose that $f \in C_0^2(\mathbb{R}^n)$. Then the function u, defined by the convergent integral*

$$u(x) = \int_{\mathbb{R}^n} \varphi(x - y) f(y)\, dy, \tag{8.4}$$

is of class C^2 on $\mathbb{R}^n$ and moreover

$$-\Delta u = f.$$

Remark 8.8 We can extend this result to consider less regular functions f that do not have compact support. For our needs, what is important is only to have a representation formula for functions of class C^∞ with compact support.

Proof We consider the case $n \geqslant 3$ (the case $n = 2$ is similar). Fix x in $\mathbb{R}^n$. Then the function $y \mapsto \varphi(x - y)$ is locally integrable and we immediately verify that the integral (8.4) is well defined. The fact that u is of class C^2 is obtained by writing, by an elementary change of variables,

$$u(x) = \int_{\mathbb{R}^n} \varphi(y) f(x - y)\, dy,$$

then using the theorem on differentiation of integrals with parameters. We thus obtain that

$$\Delta u(x) = \int_{\mathbb{R}^n} \varphi(y)(\Delta f)(x - y)\, dy.$$

But be careful: we cannot use this result to calculate Δu from the formula (8.4) because the function $\Delta\varphi$ is not integrable (if we could differentiate under the integral we would find $\Delta u = 0$ because $\Delta\varphi = 0$, but we will prove that $\Delta u = -f$). To circumvent this difficulty, let us introduce the functions

$$\varphi_\varepsilon(y) = \frac{1}{(n-2)\,|\mathbb{S}^{n-1}|}(|y|^2 + \varepsilon^2)^{-(n-2)/2} \quad \text{and} \quad u_\varepsilon(x) = \int \varphi_\varepsilon(y) f(x - y)\, dy,$$

where $\varepsilon > 0$. Then $u_\varepsilon \in C^2$ and moreover

$$\Delta u_\varepsilon(x) = \int_{\mathbb{R}^n} \varphi_\varepsilon(y)(\Delta f)(x - y)\, dy.$$

Using the dominated convergence theorem, we find that $\Delta u_\varepsilon(x)$ converges to $\Delta u(x)$ for all $x \in \mathbb{R}^n$. Furthermore, as φ_ε is a C^1 function on $\mathbb{R}^n$ and f has compact support, we can integrate by parts to obtain that

$$\Delta u_\varepsilon(x) = -\int (\Delta \varphi_\varepsilon(y)) f(x - y) \, dy.$$

A direct calculation gives us that

$$\Delta \varphi_\varepsilon(y) = -\frac{n}{\left|\mathbb{S}^{n-1}\right|} \frac{\varepsilon^2}{(|y|^2 + \varepsilon^2)^{1+n/2}},$$

so we get, after an elementary change of variables,

$$\Delta u_\varepsilon(x) = -\frac{n}{\left|\mathbb{S}^{n-1}\right|} \int \frac{f(x - \varepsilon z)}{(1 + |z|^2)^{1+n/2}} \, dz.$$

To prove that $-\Delta u(x) = f(x)$, all that remains is to verify that

$$\int \frac{1}{(1 + |z|^2)^{1+n/2}} \, dz = \frac{\left|\mathbb{S}^{n-1}\right|}{n}. \tag{8.5}$$

To do this, we calculate the integral in polar coordinates then we make the change of variable $r = 1/s$, to obtain

$$\begin{aligned}
\int \frac{1}{(1 + |z|^2)^{1+n/2}} \, dz &= \left|\mathbb{S}^{n-1}\right| \int_0^{+\infty} \frac{r^{n-1}}{(1 + r^2)^{1+n/2}} \, dr \\
&= \left|\mathbb{S}^{n-1}\right| \int_0^{+\infty} \frac{s}{(1 + s^2)^{1+n/2}} \, ds \\
&= \left|\mathbb{S}^{n-1}\right| \left[-\frac{1}{n(1 + s^2)^{n/2}} \right]_0^{+\infty},
\end{aligned}$$

which implies (8.5) and concludes the proof.   □

We will deduce a representation formula for u from its gradient.

Corollary 8.9 *Assume that $n \geq 2$. For any function $u \in C_0^2(\mathbb{R}^n)$ we have*

$$u(x) = \frac{1}{\left|\mathbb{S}^{n-1}\right|} \int_{\mathbb{R}^n} \frac{(x - y) \cdot \nabla u(y)}{|x - y|^n} \, dy. \tag{8.6}$$

Proof This result can be shown by several direct calculations. We choose to deduce it from the previous proposition. Let us write $-\Delta u = f$ with $f = -\text{div}(\nabla u)$. Then the identity (8.4) implies that

$$u(x) = -\int_{\mathbb{R}^n} \varphi(x - y) \, \text{div}(\nabla u(y)) \, dy.$$

Formally,[1] the result (8.6) is obtained by integrating by parts:

$$u(x) = \int_{\mathbb{R}^n} \nabla(\varphi(x-y)) \cdot \nabla u(y) \, dy - \int_{\mathbb{R}^n} (\nabla\varphi)(x-y) \cdot \nabla u(y) \, dy,$$

because, for $n = 2$ and also for $n \geqslant 3$, we have

$$\nabla\varphi(z) = -\frac{1}{|\mathbb{S}^{n-1}|} \frac{z}{|z|^n}.$$

To justify this integration by parts, we must be careful because the function φ is singular at the origin. For this, we write

$$\varphi(x-y) \operatorname{div}(\nabla u(y)) = \operatorname{div}\big(\varphi(x-y)\nabla u(y)\big) + (\nabla\varphi)(x-y) \cdot \nabla u(y), \qquad (8.7)$$

then, given $\varepsilon, R > 0$, we integrate this relation on the annulus $B(x, R) \setminus B(x, \varepsilon)$ and we let R tend to $+\infty$ and ε to 0. It is enough to verify that the integral of $\operatorname{div}(\varphi(x-y)\nabla u(y))$ converges to 0. For this, let us use the divergence theorem (see the solved Exercise 17.3):

$$\int_{B(x,R)\setminus B(x,\varepsilon)} \operatorname{div}\big(\varphi(x-y)\nabla u(y)\big) \, dy = \int_{\partial B(x,R)} \varphi(x-y)\partial_\nu u(y) \, dS(y)$$
$$- \int_{\partial B(x,\varepsilon)} \varphi(x-y)\partial_\nu u(y) \, dS(y).$$

The first term cancels out for large enough R because u has compact support. For the second term, we use the upper bound

$$\left| \int_{\partial B(x,\varepsilon)} \varphi(x-y)\partial_\nu u(y) \, dS(y) \right| \leqslant |\partial B(0,\varepsilon)| \, \|\varphi\|_{L^\infty(\partial B(0,\varepsilon))} \|\nabla u\|_{L^\infty(\mathbb{R}^n)},$$

and we note that the L^∞ norm of φ on the sphere $\partial B(0,\varepsilon)$ is of size $O(\varepsilon^{2-n} |\log \varepsilon|)$ (in any dimension). Therefore, by multiplying this by the measure of the sphere $\partial B(0,\varepsilon)$, we obtain a negligible quantity when ε tends to 0, which concludes the proof. □

8.3 Regularity of Harmonic Functions

To conclude, we are interested in the regularity of weakly harmonic functions.

By definition, a real-valued function u belonging to the Sobolev space $H^1(\Omega)$ is weakly harmonic if

$$\forall \varphi \in C_0^\infty(\Omega), \quad \int_\Omega \nabla u \cdot \nabla \varphi \, dx = 0. \qquad (8.8)$$

---

[1] When we say that we calculate formally, we mean that we manipulate the expressions symbolically, as if the functions were regular.

Note that if $u \in C^2(\overline{\Omega})$ satisfies $\Delta u = 0$ pointwise then $u \in H^1(\Omega)$ and (8.8) is satisfied.

Theorem 8.10 (Weyl) *If $u \in H^1(\Omega)$ is weakly harmonic, then $u \in C^\infty(\Omega)$.*

Proof The following proof uses the Caccioppoli inequality (we will revisit this fundamental lemma in the study of the De Giorgi–Nash–Moser theory).

Lemma 8.11 (Caccioppoli Inequality) *Suppose that $u \in C^\infty(\Omega)$ is harmonic and consider two concentric balls*

$$B(r) = B(x_0, r), \quad B(R) = B(x_0, R) \quad with \quad B(r) \subset B(R) \subset \Omega$$

where $x_0 \in \Omega$ and $0 < r < R$. Then, for any $c \in \mathbb{R}$, we have

$$\int_{B(r)} |\nabla u|^2 \, dx \leqslant \frac{16}{(R-r)^2} \int_{B(R) \setminus B(r)} |u - c|^2 \, dx.$$

Proof Let us introduce a function $\eta \in C_0^\infty(B(R))$, $0 \leqslant \eta \leqslant 1$, such that

$$\eta|_{B(r)} = 1, \qquad |\nabla \eta| \leqslant \frac{2}{R-r}. \tag{8.9}$$

To construct such a function, we can proceed in three steps. We start by choosing a Lipschitz function, which equals 1 on $B(0, r + \varepsilon)$ with $\varepsilon > 0$, and which equals 0 outside the ball $B(0, R - \varepsilon)$. Then, we regularize this function by convolution with a radial function of the form $\Phi_t(x) = t^{-n}\Phi(\cdot/t)$, where Φ is given by (6.4). Since convolution is a local average, for $t > 0$ small enough, the convolution product equals 1 on $B(0, r)$ and 0 outside $B(0, R)$. Moreover, as the slope of the initial function is equal to $1/(R - r - 2\varepsilon)$, the function obtained by convolution has a slope bounded by $2/(R - r)$ for ε and t small enough. We then obtain the desired function η by translating by x.

Now let us set $\varphi = (u - c)\eta^2$ (which belongs to $C_0^\infty(\Omega)$). Since u is harmonic, we have $\Delta u = 0$ so $\int_\Omega \varphi \Delta u \, dx = 0$. Since φ has compact support we can integrate by parts. We obtain

$$\int_\Omega \nabla u \cdot \left(\eta^2 \nabla u + 2(u - c)\eta \nabla \eta \right) dx = 0.$$

From this, thanks to the Cauchy–Schwarz inequality, we deduce

$$\int_\Omega \eta^2 |\nabla u|^2 \, dx \leqslant 2 \int_\Omega |u - c| \, \eta \, |\nabla u| \, |\nabla \eta| \, dx$$

$$\leqslant 2 \left(\int_\Omega |u - c|^2 \, |\nabla \eta|^2 \, dx \right)^{1/2} \left(\int_\Omega |\nabla u|^2 \eta^2 \, dx \right)^{1/2},$$

which immediately implies the desired result by dividing both sides by $\left(\int_\Omega |\nabla u|^2 \eta^2 \, dx \right)^{1/2}$ and then using the properties of the function η. $\square$

Lemma 8.12 *Consider a ball $B(x_0, R) \subset \Omega$. For every $k \in \mathbb{N}^*$, there exists a constant $K(R, k)$ such that for every $u \in C^\infty(\Omega)$ satisfying $\Delta u = 0$ we have*

$$\int_{B(x_0, R/2)} |\nabla^k u|^2 \, dx \leqslant K(R, k) \int_{B(x_0, R)} u^2 \, dx.$$

Proof Let us set $B(\rho) = B(x_0, \rho)$. The previous lemma implies that for every $R' < R$,

$$\int_{B(R')} |\nabla u|^2 \, dx \leqslant C(R, R') \int_{B(R)} |u|^2 \, dx.$$

If $u \in C^\infty(\Omega)$ and $\Delta u = 0$, then the derivatives of u are harmonic. The previous inequality can be applied with the derivatives of u and we deduce that

$$\int_{B(R'')} |\nabla^2 u|^2 \, dx \leqslant n C(R', R'') \int_{B(R')} |\nabla u|^2 \, dx$$

$$\leqslant n C(R', R'') C(R, R') \int_{B(R)} |u|^2 \, dx.$$

We obtain the desired result by iterating this argument. $\qquad \square$

Let us introduce an approximation of the identity

$$\phi_\varepsilon(y) = \varepsilon^{-n} \phi(y/\varepsilon) \qquad (\varepsilon \in (0, 1]),$$

where $\phi \in C_0^\infty(\mathbb{R}^n)$ is supported in the unit ball and such that $\phi \geqslant 0$ and $\int \phi = 1$. Let us introduce the open set $\Omega_\varepsilon = \{x \in \Omega : \mathrm{dist}(x, \partial\Omega) > \varepsilon\}$ and define, for $x \in \Omega_\varepsilon$,

$$u_\varepsilon(x) = \int_\Omega u(y) \phi_\varepsilon(x - y) \, dy.$$

By the theorem on differentiation of integrals, we have $u_\varepsilon \in C^\infty(\Omega_\varepsilon)$ and

$$\nabla u_\varepsilon(x) = \int_\Omega u(y) \nabla_x \phi_\varepsilon(x - y) \, dy.$$

Note that, for all $x \in \Omega_\varepsilon$, the support of the function $y \mapsto \phi_\varepsilon(x - y)$ is included in Ω. Since $u \in H^1(\Omega)$, by the very definition of weak differentiation, we have:

$$\nabla u_\varepsilon(x) = -\int_\Omega u(y) \nabla_y(\phi_\varepsilon(x - y)) \, dy = \int_\Omega \nabla u(y) \phi_\varepsilon(x - y) \, dy$$

$$= \int_{B(0, \varepsilon)} (\nabla u)(x - z) \phi_\varepsilon(z) \, dz,$$

where we used the fact that ϕ_ε is supported in the ball $B(0, \varepsilon)$ to restrict the integration domain to this ball in the last integral.

Let us show that u_ε is weakly harmonic in Ω_ε. For this, we start by observing that, for any function $\varphi \in C_0^\infty(\Omega_\varepsilon)$,

$$\int_{\Omega_\varepsilon} \nabla u_\varepsilon \cdot \nabla \varphi \, dx = \int_{\Omega_\varepsilon} \left(\int_{B(0,\varepsilon)} (\nabla u)(x - z) \phi_\varepsilon(z) \, dz \right) \cdot \nabla \varphi(x) \, dx$$

$$= \int_{B(0,\varepsilon)} \left(\int_{\Omega_\varepsilon} \nabla u(x - z) \cdot \nabla \varphi(x) \, dx \right) \phi_\varepsilon(z) \, dz$$

$$= \int_{B(0,\varepsilon)} \left(\int_\Omega \nabla u(t) \cdot \nabla \varphi(t + z) \, dt \right) \phi_\varepsilon(z) \, dz = 0.$$

Note that if $z \in B(0, \varepsilon)$ then the function $t \mapsto \varphi(t + y)$ is supported in Ω because φ is supported in Ω_ε. Therefore, the assumption that u is weakly harmonic in Ω implies that the inner integral is zero, which leads to $\int_{\Omega_\varepsilon} \nabla u_\varepsilon \cdot \nabla \varphi \, dx = 0$, hence the fact that u_ε is weakly harmonic in Ω_ε.

We can then apply the inequality of the previous lemma to u_ε. Since u_ε converges to u in $L^2(B(R))$, we find that u_ε is a Cauchy sequence in $H^k(B(R/2))$ and by taking the limit we find that $u \in H^k(B(R/2))$. This proves that, for all $s \in \mathbb{N}$, there exists $R_s > 0$ such that $u \in H^s(B(R_s))$. Given a function $\varphi \in C_0^\infty(B(R_s))$, we have $\varphi u \in H^s(B(R_s))$ and, by extending by 0 outside of $B(R_s)$, we deduce that $\varphi u \in H^s(\mathbb{R}^n)$. Lemma 7.57 implies that, if $s > n/2$ and $s - n/2$ is not an integer (which we can assume without loss of generality), then φu is a function of class $C^{s-(n/2)}$. Since this is true for any function φ and for all s, we deduce that u is of class C^∞, which is the desired result. $\qquad\square$

We will see other applications of the Caccioppoli inequality.

Proposition 8.13 *If $u \in H^1(\Omega)$ is weakly harmonic, then there exists a constant $\theta < 1$ such that, for all $r \geqslant 1$ and for any ball $B(r)$ included in Ω,*

$$\int_{B(r)} |\nabla u|^2 \, dx \leqslant \theta \int_{B(2r)} |\nabla u|^2 \, dx.$$

Proof We have seen that for all c,

$$\int_{B(r)} |\nabla u|^2 \, dx \leqslant \frac{16}{r^2} \int_{B(2r)\setminus B(r)} |u - c|^2 \, dx. \qquad (8.10)$$

We will apply this inequality with $c = u_{B(2r)\setminus B(r)}$ (that is the average of u over $B(2r) \setminus B(r)$). We have seen in the proof of the Poincaré inequality that if Ω is included in a strip $\{(x', y) : |y| \leqslant R\}$ then the constant $C(\Omega)$ in the Poincaré inequality can be bounded by $2R$. In particular,

$$C(B(2r) \setminus B(r)) \leqslant C(n)r$$

and we deduce that

$$\int_{B(2r)\setminus B(r)} |u - c|^2 \, dx \leqslant C(n)r \int_{B(2r)\setminus B(r)} |\nabla u|^2 \, dx.$$

By combining this inequality with (8.10), we find that there exists $C'(n)$ such that for all $r \geqslant 1$,

$$\int_{B(r)} |\nabla u|^2 \, dx \leqslant C'(n) \int_{B(2r) \setminus B(r)} |\nabla u|^2 \, dx.$$

Now, we add $C'(n) \int_{B(r)} |\nabla u|^2 \, dx$ to both sides of this inequality to deduce the desired result with $\theta = C'(n)/(1 + C'(n)) < 1$. $\qquad\square$

Corollary 8.14 *If $u \in H^1(\mathbb{R}^n)$ is a harmonic function then $u = 0$.*

Proof By taking the limit as $r \to +\infty$ in inequality (8.10) applied with $c = 0$, we find that $\nabla u = 0$, which implies that u is constant. Since $u \in L^2(\mathbb{R}^n)$, we conclude that $u = 0$. $\qquad\square$

8.4 Exercises

Exercise 8.1 (Hadamard's three lines lemma)

1. Let $\Omega \subset \mathbb{C}$ be a bounded connected open set. Let f be a holomorphic function on Ω, continuous on $\overline{\Omega}$. Using the maximum principle for harmonic functions, show that $\sup_{\Omega} |f| = \sup_{\partial\Omega} |f|$ and that, moreover, the maximum of $|f|$ can only be reached at a point of $\partial\Omega$ unless f is constant in Ω.
2. Let $S = \{x + iy : x \in [0,1], y \in \mathbb{R}\} \subset \mathbb{C}$ be a strip of the complex plane, and let $\varphi: S \to \mathbb{C}$ be a continuous and bounded function which is also holomorphic inside S. We want to show that $\sup_S |\varphi| = \sup_{\partial S} |\varphi|$.
   (a) Prove this in the case where φ tends to 0 at infinity.
   (b) Let $\delta > 0$ and $z_0 \in S$ such that $|\varphi(z_0)| \geqslant (1 - \delta) \sup_S |\varphi|$. By applying the result of (a) to the function $\psi_\varepsilon(z) = e^{\varepsilon(z - z_0)^2} \varphi(z)$, show that, for all $\varepsilon > 0$,

$$\sup_{\partial S} |\varphi| \geqslant e^{-\varepsilon}(1 - \delta) \sup_S |\varphi|.$$

Deduce the desired result from this.
3. For $\theta \in [0,1]$, we want to estimate $M_\theta := \sup_{y \in \mathbb{R}} |\varphi(\theta + iy)|$ in terms of M_0 and M_1.
   By applying the previous result to the function $\phi(z) = e^{-\lambda z} \varphi(z)$ (λ is a real number to choose), show that, for all $\theta \in [0,1]$,

$$M_\theta \leqslant M_1^\theta M_0^{1-\theta}.$$

This result is called Hadamard's three lines lemma.

   The next two exercise are taken from Chapter 2 in Krylov's classical textbook on elliptic and parabolic equations [106], to which we refer for the solutions.

Exercise 8.2 (Bernstein's method) Consider a harmonic function $u \in C^3(\overline{B_1})$ where $B_1 \subset \mathbb{R}^d$ denotes the open ball centered at 0 with radius 1. Consider a function $\zeta \in C_0^\infty(\mathbb{R}^d)$ with support in B_1 such that $\zeta(0) = 1$ and consider the function

$$w := \zeta^2 |\nabla u|^2 + \lambda |u|^2,$$

where λ is a constant to determine and where $|\cdot|$ denotes the Euclidean norm.

1. Using the inequality $2a^2 + 8ab + 8b^2 \geqslant 0$ (to be verified), show that

$$\Delta w \geqslant |\nabla u|^2 \left(2\lambda + \Delta(\zeta^2) - 8|\nabla\zeta|^2\right).$$

Deduce that, for a well-chosen λ, we have $\Delta w \geqslant 0$.

2. Using the maximum principle, deduce that

$$|\nabla u(x_0)| \leqslant K \max_{B_1(x_0)} |u|$$

for a certain constant K that depends only on the dimension d.

3. Deduce that

$$|\nabla u(x_0)| \leqslant \frac{K'}{R} \max_{B_R(x_0)} |u|.$$

4. Recall that $u \in C^\infty(B_R(x_0))$. Deduce that for any index i such that $1 \leqslant i \leqslant d$, we have

$$|\nabla \partial_{x_i} u(x_0)| \leqslant \frac{K'}{R/2} \max_{B_{R/2}(x_0)} |\partial_{x_i} u|,$$

and that

$$|\nabla \partial_{x_i} u(x_0)| \leqslant \left(\frac{2K'}{R}\right)^2 \max_{B_R(x_0)} |u|.$$

5. By induction, prove the following result.

Proposition 8.15 *Consider a bounded connected domain $\Omega \subset \mathbb{R}^d$ and $u \in C(\overline{\Omega}) \cap C^\infty(\Omega)$ harmonic in Ω. Then, for any multi-index α and any $x \in \Omega$, we have*

$$|D^\alpha u(x)| \leqslant \left(\frac{K|\alpha|}{\mathrm{dist}(x, \partial\Omega)}\right)^{|\alpha|} \max_\Omega |u|$$

for some constant K depending only on d.

Exercise 8.3 Let P_n be the set of all polynomial functions of degree less than or equal to n. Define the operator

$$Tu = \Delta\left((1 - |x|^2)u\right).$$

1. Show that if $u \in P_n$ then $Tu \in P_n$.
2. Using the maximum principle, show that if $Tu = 0$ then $u = 0$. (Apply the maximum principle on a well-chosen domain.)
3. Deduce that if f, g are polynomials, there exists a unique polynomial function $u \in C^2(B_1)$ such that

$$\Delta u = f \quad \text{in } B_1, \qquad u = g \quad \text{on } \partial B_1.$$

4. We want to deduce the following proposition.

Proposition 8.16 *Let $g \in C^0(\mathbb{R}^d)$. There exists a unique solution $u \in C^2(B_1) \cap C^0(\overline{B_1})$ of*

$$\Delta u = 0 \quad in \ B_1, \qquad u = g \quad on \ \partial B_1.$$

For this, we will consider a sequence of polynomials g_n that converges uniformly to g on ∂B_1. Using the maximum principle, we will show that the solutions u_n (given by the previous question) form a Cauchy sequence. Using Exercise 8.2, we will deduce the convergence of $D^\alpha u_n$ for every multi-index α.

Part III
Microlocal Analysis

Chapter 9
Pseudo-Differential Operators

How to act on a function to study its regularity? How to act on a partial differential equation to conjugate it to a simpler equation? There are many answers to these questions. Those that we will study in this part come from microlocal analysis. This is a theory that has developed since the sixties, notably under the impulse of Lars Hörmander, Alberto Calderón, Robert Kohn, and Louis Nirenberg.

This is a vast subject.[1] We will limit ourselves to studying one of the main objects of this theory: pseudo-differential operators. One of our objectives will be to construct a pseudo-differential calculus, that is to say, a way to associate with a function $a = a(x, \xi)$ defined on $\mathbb{R}^n \times \mathbb{R}^n$ and having certain decay properties when it is differentiated with respect to the variable ξ, an operator $\mathrm{Op}(a)$ whose properties (adjoint, continuity on function spaces, composition...) can be understood simply by looking at the properties of a.

In this chapter, we will start by studying the definition of these operators.

9.1 Symbols

Consider a differential operator $P = \sum_{|\alpha| \le m} p_\alpha(x) \partial_x^\alpha$ where the coefficients p_α belong to the space $C_b^\infty(\mathbb{R}^n; \mathbb{C})$ of functions of class C^∞ that are bounded along with all their derivatives. The function

$$p : \mathbb{R}^n \times \mathbb{R}^n \longrightarrow \mathbb{C}, \qquad p(x, \xi) = \sum_{|\alpha| \le m} p_\alpha(x)(i\xi)^\alpha$$

is called the symbol of P. With this definition, we have $P(e^{ix \cdot \xi}) = p(x, \xi)e^{ix \cdot \xi}$.

---

[1] We refer to Hörmander [80] for the general theory as well as to Alinhac and Gérard [3], Chazarain and Piriou [33], Grigis and Sjöstrand [63], Lerner [119], Métivier [134], Saint-Raymond [166], Taylor [186], Trèves [191] and Zworski [202]. The links between microlocal analysis and harmonic analysis are studied in the books of Stein [177], Coifman and Meyer [36] and Meyer [135, 136].

© The Author(s), under exclusive license to Springer Nature Switzerland AG 2024
T. Alazard, *Analysis and Partial Differential Equations*, Universitext,
https://doi.org/10.1007/978-3-031-70909-8_9

Now consider a function u in the Schwartz space. Using the Fourier inversion formula

$$u(x) = \frac{1}{(2\pi)^n} \int_{\mathbb{R}^n} e^{ix \cdot \xi} \widehat{u}(\xi) \, d\xi,$$

and the theorems on differentiation of integrals depending on a parameter, we see that we can write $Pu(x)$ in the form

$$Pu(x) = \frac{1}{(2\pi)^n} \int_{\mathbb{R}^n} e^{ix \cdot \xi} p(x, \xi) \widehat{u}(\xi) \, d\xi.$$

In the sixties, this point of view was used[2] to introduce operators that generalize differential operators. These are operators of the previous form, but where the function $p(x, \xi)$ is not necessarily a polynomial function. We propose in this chapter to study the definition of these operators when $p(x, \xi)$ is a symbol, that is to say a function satisfying the properties of the following definition.

Notation 9.1 Let $d \geqslant 1$ and $\Omega \subset \mathbb{R}^d$ be an open set. We denote by $C_b^\infty(\Omega)$ the set of functions of class C^∞ on Ω that are bounded along with all their derivatives.

Definition 9.2 Let $m \in \mathbb{R}$ and $\rho \in [0, 1]$. The class of symbols of order m, denoted $S_{\rho,0}^m(\mathbb{R}^n)$, is the set of functions $a \in C^\infty(\mathbb{R}^n \times \mathbb{R}^n)$ with complex values such that, for all multi-indices α and β in $\mathbb{N}^n$, there exists a constant $C_{\alpha\beta}$ such that

$$\forall (x, \xi) \in \mathbb{R}^n \times \mathbb{R}^n, \quad \left| \partial_x^\alpha \partial_\xi^\beta a(x, \xi) \right| \leqslant C_{\alpha\beta} (1 + |\xi|)^{m - \rho|\beta|}.$$

Notation 9.3 We will only be interested in this book in two particular subclasses: the class $S_{1,0}^m$ and the class $S_{0,0}^0$. We will simply write

$$S^m(\mathbb{R}^n) = S_{1,0}^m(\mathbb{R}^n) \tag{9.1}$$

and

$$C_b^\infty(\mathbb{R}^{2n}) = S_{0,0}^0(\mathbb{R}^n).$$

The index 0 in the notation $S_{\rho,0}^m(\mathbb{R}^n)$ is superfluous of course; it has been maintained for consistency with the notations used in the literature (see Section 9.3).

Recall (see Notation 7.34) that we denote by $\langle \xi \rangle$ the Japanese bracket defined by

$$\langle \xi \rangle = (1 + |\xi|^2)^{1/2}.$$

Notation 9.4 We also introduce

$$S^{-\infty} := \bigcap_{m \in \mathbb{R}} S^m \quad \text{and} \quad S^{+\infty} = \bigcup_{m \in \mathbb{R}} S^m.$$

---

[2] Notably by Unterberger and Bokobza [193, 194], Kohn and Nirenberg [102] and Hörmander [75].

Definition 9.5 (Elliptic symbols) Let $m \in \mathbb{R}$. A symbol $a \in S^m(\mathbb{R}^n)$ is elliptic if there exist two positive constants R and C such that,

$$\forall (x, \xi) \in \mathbb{R}^{2n}, \quad |\xi| \geqslant R \implies |a(x, \xi)| \geqslant C\langle \xi \rangle^m.$$

The elementary rules of differential calculus imply the following proposition.

Proposition 9.6 *If* $a \in S^m$, $b \in S^{m'}$, $\alpha, \beta \in \mathbb{N}^n$ *then*

$$\partial_x^\alpha \partial_\xi^\beta a \in S^{m - |\beta|}, \quad ab \in S^{m + m'}.$$

Of course, we have $S^0(\mathbb{R}^n) \subset C_b^\infty(\mathbb{R}^{2n})$.

Examples

1. If p is a function of x only and $p \in C_b^\infty(\mathbb{R}^n)$ then $p \in S^0(\mathbb{R}^n)$.
2. If $p = p(x, \xi)$ belongs to $C_0^\infty(\mathbb{R}^{2n})$ (compact support in x and ξ) then $p \in S^{-\infty}$.
3. Suppose that $p(x, \xi)$ is a polynomial in ξ of order $m \in \mathbb{N}$ whose coefficients are functions in $C_b^\infty(\mathbb{R}^n)$,

$$p(x, \xi) = \sum_{|\alpha| \leqslant m} p_\alpha(x) \xi^\alpha \qquad (p_\alpha \in C_b^\infty(\mathbb{R}^n)).$$

Then $p \in S^m(\mathbb{R}^n)$.
4. For all $m \in \mathbb{R}$, the symbol $\langle \xi \rangle^m$ belongs to $S^m(\mathbb{R}^n)$. Indeed, the function $\mathbb{R} \times \mathbb{R}^n \ni (\tau, \xi) \mapsto (\tau^2 + |\xi|^2)^{m/2}$ is positively homogeneous of order m on $\mathbb{R}^{n+1} \setminus \{0\}$ and therefore $\partial_\xi^\alpha ((\tau^2 + |\xi|^2)^{m/2})$ is homogeneous of order $m - |\alpha|$, bounded by $C_\alpha (\tau^2 + |\xi|^2)^{(m - |\alpha|)/2}$. As the derivative in ξ and the restriction to $\tau = 1$ commute, we deduce the result.
5. The symbol $|\xi|$ is not in $S^1(\mathbb{R}^n)$ because it is not regular at 0.
6. Let $a = a(\xi) \in C^\infty(\mathbb{R}^n \setminus 0)$ be a homogeneous function of degree m satisfying

$$\forall \lambda > 0, \quad a(\lambda \xi) = \lambda^m a(\xi).$$

For any function $\chi \in C_b^\infty(\mathbb{R}_\xi^n)$ vanishing in a neighborhood of 0, the symbol $\chi(\xi)a(\xi)$ belongs to $S^m(\mathbb{R}^n)$.
7. Let $a = a(x, \xi)$ be an elliptic symbol of order m. Then there exists a $\chi \in C_0^\infty(\mathbb{R}^n)$ such that

$$\frac{1 - \chi(\xi)}{a(x, \xi)} \in S^{-m}(\mathbb{R}^n).$$

8. Let $f = f(x)$ in $C_b^\infty(\mathbb{R})$. The symbol $p(x, \xi) = f(x) \sin(\xi)$ belongs to $C_b^\infty(\mathbb{R}^2)$ but not to $S^0(\mathbb{R})$ because the derivative in ξ of order α does not decrease like $(1 + |\xi|)^{-\alpha}$.

9.1.1 Definition of a Pseudo-Differential Operator

Let $m \in \mathbb{R}$ and $\rho \in [0, 1]$. For all $a \in S^m_{\rho,0}(\mathbb{R}^n)$, all u in the Schwartz class $S(\mathbb{R}^n)$ and all $x \in \mathbb{R}^n$, the function $\xi \mapsto a(x, \xi)\widehat{u}(\xi)$ belongs to $S(\mathbb{R}^n_\xi)$. Consequently, it is integrable and we can define

$$\mathrm{Op}(a)u(x) = (2\pi)^{-n} \int e^{ix \cdot \xi} a(x, \xi)\widehat{u}(\xi) \, d\xi.$$

We say that $\mathrm{Op}(a)$ is a pseudo-differential operator and we call a its symbol.

Theorem 9.7 *Let $m \in \mathbb{R}$ and $\rho \in [0, 1]$. If $a \in S^m_{\rho,0}(\mathbb{R}^n)$ and $u \in S(\mathbb{R}^n)$, the previous formula defines a function $\mathrm{Op}(a)u$ which belongs to $S(\mathbb{R}^n)$. Moreover, $\mathrm{Op}(a)$ is continuous from $S(\mathbb{R}^n)$ to $S(\mathbb{R}^n)$.*

Proof Note that $a(x, \xi)$ is a C^∞ function on $\mathbb{R}^{2n}$ which, together with all its derivatives, is bounded by powers of $\langle \xi \rangle$. Furthermore, $\widehat{u} \in S(\mathbb{R}^n)$, so we can easily verify that $\mathrm{Op}(a)u \in C^\infty(\mathbb{R}^n)$.

Using $\|\langle \xi \rangle^{-m} a\|_{L^\infty} < +\infty$ and $\|\langle \xi \rangle^{m+2n}\widehat{u}\|_{L^\infty} < +\infty$, we obtain the inequality

$$|\mathrm{Op}(a)u(x)| \leqslant (2\pi)^{-n} \int \|\langle \xi \rangle^{-m} a\|_{L^\infty} \|\langle \xi \rangle^{m+2n}\widehat{u}\|_{L^\infty} \langle \xi \rangle^{-2n} \, d\xi,$$

which implies that $\mathrm{Op}(a)u$ is bounded with

$$\|\mathrm{Op}(a)u\|_{L^\infty} \leqslant C \mathcal{N}_{m+2n}(\widehat{u}),$$

where we have denoted by $\mathcal{N}_p(\varphi) = \sum_{|\alpha| \leqslant p, |\beta| \leqslant p} \|x^\alpha \partial^\beta_x \varphi\|_{L^\infty}$ the canonical semi-norms on the Schwartz space (see Definition 5.3); recall (see Proposition 5.8) that the Fourier transform is continuous from $S(\mathbb{R}^n)$ to $S(\mathbb{R}^n)$ and that

$$\mathcal{N}_{m+2n}(\widehat{u}) \leqslant C_{m+2n} \mathcal{N}_{m+3n+1}(u).$$

To estimate the other semi-norms in $S(\mathbb{R}^n)$ of $\mathrm{Op}(a)u$, we must now look at $x^\alpha \partial^\beta_x \mathrm{Op}(a)u$. For this, we reduce to the case already studied using the formulas (to be verified as an exercise)

$$\partial_{x_j} \mathrm{Op}(a)u = \mathrm{Op}(a)(\partial_{x_j}u) + \mathrm{Op}(\partial_{x_j}a)u,$$
$$x_j \mathrm{Op}(a)u = \mathrm{Op}(a)(x_j u) + i\,\mathrm{Op}(\partial_{\xi_j}a)u.$$

Thus, $x^\alpha \partial^\beta_x \mathrm{Op}(a)u$ can be rewritten as a linear combination of terms

$$\mathrm{Op}(\partial^\gamma_x \partial^\delta_\xi a)(x^{\alpha-\delta}\partial^{\beta-\gamma}_x u).$$

We have reduced to the previous case. This shows that $\mathrm{Op}(a)u$ belongs to $S(\mathbb{R}^n)$ and that we have estimates of the semi-norms of $\mathrm{Op}(a)u$ in terms of a sum of semi-norms of u. □

9.2 Continuity of Pseudo-Differential Operators

Theorem 9.8 *If* $a \in C_b^\infty(\mathbb{R}^{2n})$, *the operator* $\mathrm{Op}(a)$ *extends uniquely to a continuous operator* $\mathcal{L}(L^2(\mathbb{R}^n))$.

This result is due to Calderón and Vaillancourt [25]. We will follow the proof of Hwang (see [83] as well as the reference textbook by Lerner [119]). To simplify the notation, we assume that the space dimension n is less than or equal to 3 (otherwise it suffices to replace the polynomial $P(\zeta)$ below by $(1 + |\zeta|^2)^k$, where k is an integer such that $4k > n$). Let us introduce the polynomial

$$P(\zeta) = 1 + |\zeta|^2 \qquad (\zeta \in \mathbb{R}^n, \ n = 1, 2, 3).$$

Lemma 9.9 *Given* $u \in S(\mathbb{R}^n)$, *we introduce its wave packet transform, defined by*

$$Wu(x, \xi) = \int_{\mathbb{R}^n} e^{-iy \cdot \xi} P(x - y)^{-1} u(y) \, dy \qquad ((x, \xi) \in \mathbb{R}^{2n}).$$

1. *Then* Wu *is a function of class* $C_b^\infty(\mathbb{R}^{2n})$ *and moreover for all multi-indices* α, β, γ,

$$\sup_{\mathbb{R}^{2n}} P(x) |\xi|^\gamma \left| (\partial_x^\alpha \partial_\xi^\beta Wu)(x, \xi) \right| < +\infty.$$

2. *There exists a constant* A *such that*

$$\|Wu\|_{L^2(\mathbb{R}^{2n})} = A \|u\|_{L^2(\mathbb{R}^n)} \tag{9.2}$$

for all u *in* $S(\mathbb{R}^n)$.
3. *For all* $\gamma \in \mathbb{N}^n$, *there exists an* A_γ *such that*

$$\left\| \partial_x^\gamma Wu \right\|_{L^2(\mathbb{R}^{2n})} \leq A_\gamma \|u\|_{L^2(\mathbb{R}^n)}.$$

Proof 1. We calculate that

$$\xi^\gamma (\partial_x^\alpha \partial_\xi^\beta Wu)(x, \xi) = \int i^{|\gamma|} \partial_y^\gamma \left(e^{-iy \cdot \xi} \right) (-iy)^\beta \partial_x^\alpha (P(x - y)^{-1}) u(y) \, dy$$

and we integrate by parts

$$\xi^\gamma (\partial_x^\alpha \partial_\xi^\beta Wu)(x, \xi)$$
$$= \sum_{\gamma' + \gamma'' = \gamma} \frac{\gamma! (-i)^{|\gamma|}}{\gamma'! \gamma''!} \int \partial_y^{\gamma'} \left(u(y)(-iy)^\beta \right) (-1)^{|\gamma''|} \left(\partial^{\gamma'' + \alpha} \frac{1}{P} \right) (x - y) e^{-iy \cdot \xi} \, dy.$$

The fact (already seen) that $\langle \xi \rangle^{-2}$ is a symbol of order -2 leads to

$$\left| \partial_\zeta^\alpha \langle \zeta \rangle^{-2} \right| \leq C_\alpha \langle \zeta \rangle^{-2 - |\alpha|} \leq C_\alpha \langle \zeta \rangle^{-2},$$

from which we deduce that

$$|\partial^\alpha (1/P)(x-y)| \leqslant C_\alpha (1+|x-y|^2)^{-1} \leqslant 2C_\alpha (1+|x|^2)^{-1}(1+|y|^2),$$

where the last inequality comes from the fact that

$$1+|x|^2 = 1+|x-y+y|^2 \leqslant 1+2|x-y|^2+2|y|^2 \leqslant 2(1+|x-y|^2)(1+|y|^2).$$

2. For all $x \in \mathbb{R}^n$, $Wu(x,\cdot)$ is the Fourier transform of $y \mapsto u(y)P(x-y)^{-1}$. Therefore

$$\int |W(x,\xi)|^2 \, d\xi = (2\pi)^n \int |u(y)P(x-y)^{-1}|^2 \, dy$$

according to the Plancherel formula (5.8). Then,

$$\iint |W(x,\xi)|^2 \, d\xi \, dx = (2\pi)^n \iint |u(y)P(x-y)^{-1}|^2 \, dy \, dx = A^2 \|u\|^2_{L^2(\mathbb{R}^n)}.$$

3. The last point is obtained directly by combining the previous observations. □

Remark 9.10 Let us define for $x \in \mathbb{R}^n$, $(y,\eta) \in \mathbb{R}^n \times \mathbb{R}^n$,

$$\varphi_{y,\eta}(x) = e^{i(x-y)\cdot\eta} P(x-y)^{-1},$$

and

$$\widetilde{W}u(y,\eta) = \langle u, \varphi_{y,\eta}\rangle_{L^2(\mathbb{R}^n)} = e^{iy\cdot\eta} Wu(y,\eta).$$

We then have the reconstruction formula

$$u(x) = \frac{1}{A} \iint \widetilde{W}u(y,\eta)\varphi_{x,\eta}(y) \, dy \, d\eta,$$

where A is the constant defined by (9.2). Indeed, according to (9.2), the operator $\mathcal{W} := \widetilde{W}/\sqrt{A}$ is an isometry, so

$$\mathcal{W}^*\mathcal{W} = I,$$

and we deduce the desired result.

Lemma 9.11 *For all u,v in $S(\mathbb{R}^n)$, there holds*

$$\hat{u}(\xi) = e^{-ix\cdot\xi}(I-\Delta_\xi)\big(e^{ix\cdot\xi} Wu(x,\xi)\big)$$

and

$$\bar{v}(x) = \frac{1}{(2\pi)^n} e^{-ix\cdot\xi}(I-\Delta_x)\big(e^{ix\cdot\xi} W\bar{v}(\xi,x)\big).$$

Proof As $(I-\Delta_\xi)e^{iX\cdot\xi} = P(X)e^{iX\cdot\xi}$, we have

$$e^{ix\cdot\xi}\hat{u}(\xi) = \int e^{i(x-y)\cdot\xi} u(y) \, dy = (I-\Delta_\xi) \int e^{i(x-y)\cdot\xi} P(x-y)^{-1} u(y) \, dy.$$

In a dual way, using the inverse Fourier transform, we have

$$e^{ix\cdot\xi}\overline{v}(x) = \frac{1}{(2\pi)^n} \int e^{i(\xi-\eta)\cdot x}\overline{v}(\eta)\,d\eta$$

$$= \frac{1}{(2\pi)^n}(I-\Delta_x) \int e^{i(\xi-\eta)\cdot x}P(\xi-\eta)^{-1}\overline{v}(\eta)\,d\eta,$$

which implies the second identity. □

***Proof (of Theorem* 9.8)** Given the density of $S(\mathbb{R}^n)$ in $L^2(\mathbb{R}^n)$, it is sufficient to prove the inequality

$$\|\mathrm{Op}(a)u\|_{L^2} \leqslant C\|u\|_{L^2}$$

for all u in $S(\mathbb{R}^n)$. Consider two functions u, v in $S(\mathbb{R}^n)$ and let

$$I := \iint e^{ix\cdot\xi} a(x,\xi)\widehat{u}(\xi)\overline{v}(x)\,d\xi\,dx.$$

We want to show that $|I| \leqslant C\|u\|_{L^2}\|v\|_{L^2}$. For this, using the above lemma, we will rewrite I as a scalar product in $L^2(\mathbb{R}^{2n})$ of functions involving Wu and $W\overline{v}$.

Let us start by writing I in the form

$$I = \iint a(x,\xi)\Big[(I-\Delta_\xi)\big(e^{ix\cdot\xi}Wu(x,\xi)\big)\Big]\overline{v}(x)\,d\xi\,dx.$$

Lemma 9.9 implies that $(I-\Delta_\xi)\big(e^{ix\cdot\xi}Wu(x,\xi)\overline{v}(x)\big)$ belongs to $S(\mathbb{R}^{2n})$, so we can integrate by parts in ξ to deduce that

$$I = \iint \Big[(I-\Delta_\xi)a(x,\xi)\Big]Wu(x,\xi)e^{ix\cdot\xi}\overline{v}(x)\,d\xi\,dx.$$

Using the identity for v, we get

$$I = \iint \Big[(I-\Delta_\xi)a(x,\xi)\Big]Wu(x,\xi)(I-\Delta_x)\big(e^{ix\cdot\xi}W\overline{v}(\xi,x)\big)\,d\xi\,dx.$$

Then, by integrating by parts in x,

$$I = \iint (I-\Delta_x)\Big[\big((I-\Delta_\xi)a(x,\xi)\big)Wu(x,\xi)\Big]e^{ix\cdot\xi}W\overline{v}(\xi,x)\,d\xi\,dx,$$

so

$$I = \sum_{|\beta|\leqslant 2, |\alpha|+|\gamma|\leqslant 2} C_{\alpha\beta\gamma} \iint (\partial_x^\alpha\partial_\xi^\beta a(x,\xi))\partial_x^\gamma Wu(x,\xi)W\overline{v}(\xi,x)e^{ix\cdot\xi}\,dx\,d\xi.$$

We conclude the proof via the Cauchy–Schwarz inequality and the previous results:

$$\left\|\partial_x^\gamma W u\right\|_{L^2(\mathbb{R}^{2n})} \leqslant A_\gamma \|u\|_{L^2(\mathbb{R}^n)},$$

$$\left\|W\overline{\widehat{v}}(\xi,x)\right\|_{L^2(\mathbb{R}^{2n})} = A\left\|\overline{\widehat{v}}\right\|_{L^2} = A(2\pi)^{n/2}\|v\|_{L^2(\mathbb{R}^n)},$$

where we used the Plancherel formula (5.8) in the last inequality. $\square$

9.3 Generalizations

Let us introduce the Hörmander symbol classes.

Definition 9.12 For $m \in \mathbb{R}$ and $0 \leqslant \delta \leqslant \rho \leqslant 1$, the class of symbols $S_{\rho,\delta}^m(\mathbb{R}^n)$ is the space of functions $a \in C^\infty(\mathbb{R}^{2n};\mathbb{C})$ such that, for all multi-indices α in $\mathbb{N}^n$ and β in $\mathbb{N}^n$, there exists a constant $C_{\alpha\beta}$ such that

$$\left|\partial_x^\alpha \partial_\xi^\beta a(x,\xi)\right| \leqslant C_{\alpha\beta}(1+|\xi|)^{m+\delta|\alpha|-\rho|\beta|}.$$

We say that a is a symbol of order m and of type (ρ,δ).

Note that $C_b^\infty(\mathbb{R}^{2n};\mathbb{C}) = S_{0,0}^0(\mathbb{R}^n)$. For any real number $m \in \mathbb{R}$ and $0 \leqslant \delta \leqslant \rho \leqslant 1$, and for any symbol $a \in S_{\rho,\delta}^m(\mathbb{R}^n)$, using arguments similar to those used to prove Theorem 9.7, we can prove that $\mathrm{Op}(a)$ is a continuous operator from $\mathcal{S}(\mathbb{R}^n)$ to $\mathcal{S}(\mathbb{R}^n)$.

Let us state a generalization of Theorem 9.8 to the case of general symbols.

Theorem 9.13 (Calderón–Vaillancourt) Let a be in $S_{\rho,\delta}^0(\mathbb{R}^n)$ with $0 \leqslant \delta \leqslant \rho \leqslant 1$ and $\delta < 1$. Then $\mathrm{Op}(a)$ can be extended as a bounded operator from $L^2(\mathbb{R}^n)$ to itself. Moreover,

$$\|\mathrm{Op}(a)\|_{\mathcal{L}(L^2)} \leqslant C \sup_{|\alpha|\leqslant[\frac{n}{2}]+1} \sup_{|\beta|\leqslant[\frac{n}{2}]+1} \sup_{(x,\xi)\in\mathbb{R}^{2n}} \left|(1+|\xi|)^{\delta|\alpha|-\rho|\beta|}\partial_x^\alpha \partial_\xi^\beta a(x,\xi)\right|$$

for a certain absolute constant C depending only on n,ρ,δ.

We will not use this result and we refer to [26] for the proof. The precise bound in terms of the semi-norms of a is proven for example by Coifman and Meyer [36].

Remark 9.14 (continuity on L^p) A pseudo-differential operator of order 0 and of type (ρ,δ) is not generally bounded on the Lebesgue spaces $L^p(\mathbb{R}^n)$ with $p \neq 2$. Nevertheless, Fefferman proved in [51] that, for all real numbers δ and ρ such that $0 \leqslant \delta \leqslant \rho \leqslant 1$ with $\delta < 1$, and any symbol $a \in S_{\rho,\delta}^m(\mathbb{R}^n)$, the operator $\mathrm{Op}(a)$ belongs to $\mathcal{L}(L^p(\mathbb{R}^n))$ provided that

$$m \leqslant -n(1-\rho)\left|\frac{1}{2}-\frac{1}{p}\right|.$$

We also refer to David and Journé (see [44]) for the boundedness of pseudo-differential operators on $L^p(\mathbb{R}^n)$ when $\rho = 1 = \delta$.

The aim of one of the exercises given at the end of this chapter is to show that Theorem 9.13 is no longer true when $(\rho, \delta) = (1, 1)$. This means that an operator of order 0 and type $(1, 1)$ is not generally bounded from $L^2(\mathbb{R}^n)$ to $L^2(\mathbb{R}^n)$. However, the following result, due to Stein, asserts that such an operator is bounded from $H^s(\mathbb{R}^n)$ to $H^s(\mathbb{R}^n)$ for all $s > 0$.

Theorem 9.15 (Stein) *Suppose that* $a \in S^0_{1,1}(\mathbb{R}^n)$. *Then the operator* $\mathrm{Op}(a) \in \mathcal{L}(H^s(\mathbb{R}^n))$ *for all* $s > 0$ *and* $\mathrm{Op}(a) \in \mathcal{L}(C^{0,\alpha}(\mathbb{R}^n))$ *for all* $\alpha \in (0, 1)$.

We will not use this result and refer to the book by Métivier [134] for the proof.

9.4 Exercises

Exercise (solved) 9.1 (Semi-classical operators) Let $h \in (0, 1]$ and let $a = a(x, \xi)$ be a symbol that belongs to $C^\infty_b(\mathbb{R}^{2n})$. We define

$$\mathrm{Op}_h(a)u(x) = \frac{1}{(2\pi)^n} \int e^{ix \cdot \xi} a(x, h\xi) \widehat{u}(\xi) \, d\xi.$$

We want to show that

$$\|\mathrm{Op}_h(a)\|_{\mathcal{L}(L^2)} \leqslant C \sup_{\mathbb{R}^{2n}} |a| + O(h^{1/2}).$$

1. Show that

$$\mathrm{Op}_h(a)u(x) = \big(\mathrm{Op}(a_h)u_h\big)(h^{-1/2}x)$$

where

$$a_h(x, \xi) = a(h^{1/2}x, h^{1/2}\xi), \quad u_h(y) = u(h^{1/2}y).$$

2. Deduce that there exists a constant C and an integer M such that for all $a \in C^\infty_b(\mathbb{R}^{2n})$ and all $h \in (0, 1]$,

$$\|\mathrm{Op}_h(a)\|_{\mathcal{L}(L^2)}$$
$$\leqslant C \sup_{(x,\xi) \in \mathbb{R}^{2n}} |a(x, \xi)| + C \sup_{1 \leqslant |\alpha| + |\beta| \leqslant M} \sup_{(x,\xi) \in \mathbb{R}^{2n}} h^{1/2(|\alpha| + |\beta|)} |\partial_x^\alpha \partial_\xi^\beta a|.$$

We refer to the book by Zworski [202] for a systematic study of semi-classical operators.

Exercise 9.2 (Wave packet transform) Let $u : \mathbb{R} \to \mathbb{C}$ be in the Schwartz space $S(\mathbb{R})$. The wave packet transform of u is the function $Wu : \mathbb{R} \times \mathbb{R} \to \mathbb{C}$ defined by

$$Wu(x, \xi) = \int_{\mathbb{R}} e^{i(x-y)\xi - 1/2(x-y)^2} u(y) \, dy.$$

1. Show that $(x, \xi) \mapsto x Wu(x, \xi)$ and $(x, \xi) \mapsto \xi Wu(x, \xi)$ are bounded on $\mathbb{R}^2$. Show more generally that Wu belongs to the Schwartz space $S(\mathbb{R}^2)$.
2. Show that, for all $x \in \mathbb{R}$,

$$\int_{\mathbb{R}} |Wu(x, \xi)|^2 \, d\xi = 2\pi \int_{\mathbb{R}} e^{-(x-y)^2/2} |u(y)|^2 \, dy.$$

Deduce that there exists a constant $A > 0$ such that, for all u in $S(\mathbb{R})$, we have

$$\iint_{\mathbb{R}^2} |Wu(x, \xi)|^2 \, dx \, d\xi = A \int_{\mathbb{R}} |u(y)|^2 \, dy.$$

(It is not asked to calculate A.)
3. Show that for any function u in the Schwartz space $S(\mathbb{R})$,

$$Wu(x, \xi) = c \, e^{ix\xi} \, (\widehat{Wu})(\xi, -x),$$

for a certain constant c (it is not asked to calculate c).
4. Let $\varepsilon \in (0, 1]$ and u be in the Schwartz space $S(\mathbb{R}^2)$. We introduce

$$W^\varepsilon u(x, \xi) = \varepsilon^{-3/4} \int_{\mathbb{R}} e^{i(x-y)\cdot\xi/\varepsilon - (x-y)^2/2\varepsilon} u(y) \, dy.$$

Verify that $A^{-1/2}W^\varepsilon$ is an isometry then show that there exists K such that, for all $\varepsilon \in (0, 1]$ and all functions u and v in the Schwartz space $S(\mathbb{R})$,

$$\left\| vW^\varepsilon u - W^\varepsilon(vu) \right\|_{L^2(\mathbb{R}^2)} \leqslant K\varepsilon^{1/2} \|\partial_x v\|_{L^\infty(\mathbb{R})} \|u\|_{L^2(\mathbb{R})}.$$

5. Show that there exists K' such that, for all $\varepsilon \in (0, 1]$ and any function u in the Schwartz space $S(\mathbb{R})$,

$$\left\| i\xi W^\varepsilon u - W^\varepsilon(\varepsilon\partial_x u) \right\|_{L^2(\mathbb{R}^2)} \leqslant K\varepsilon^{1/2} \|u\|_{L^2(\mathbb{R})}.$$

We refer to the articles by Córdoba and Fefferman [37] and Lerner [118] for the study of the wave packet transform.

Exercise 9.3 (An unbounded operator on L^2) Let $\chi \in C_0^\infty(\mathbb{R})$ such that

$$\operatorname{supp} \chi \subset \{\xi \in \mathbb{R}, \ 2^{-1/2} \leqslant |\xi| \leqslant 2^{1/2}\}, \quad \chi(\xi) = 1 \quad \text{if } 2^{-1/4} \leqslant |\xi| \leqslant 2^{1/4}.$$

We set

$$a(x, \xi) = \sum_{j=1}^{+\infty} \exp(-i2^j x)\chi(2^{-j}\xi).$$

1. Show that $a \in C^\infty(\mathbb{R}^2)$ satisfies

$$|\partial_x^\alpha \partial_\xi^\beta a(x, \xi)| \leqslant C_{\alpha,\beta}(1 + |\xi|)^{|\alpha|-|\beta|} \quad \forall \alpha, \beta \in \mathbb{N}^2, \ \forall(x, \xi) \in \mathbb{R}^2.$$

Does a belong to S^0 or to $C_b^\infty(\mathbb{R}^2)$?

2. Let f_0 be a function of the Schwartz space whose Fourier transform $\widehat{f_0}$ is supported in the interval $[-1/2, 1/2]$. For $N \in \mathbb{N}$, we set

$$f_N(x) = \sum_{j=2}^{N} \frac{1}{j} \exp(i2^j x) f_0(x).$$

Using the Plancherel formula, show that

$$\|f_N\|_{L^2}^2 = \left(\sum_{j=2}^{N} j^{-2}\right) \|f_0\|_{L^2}^2 \leqslant c.$$

3. Show that

$$\mathrm{Op}(a) f_N = \left(\sum_{j=2}^{N} j^{-1}\right) f_0.$$

4. Conclude.

Exercise 9.4 Fix $M > 0$. Let $q(x, \xi)$ be a C^∞ function on $\mathbb{R}^n \times \mathbb{R}^n$, with support in $\{(x, \xi) : |\xi| \leqslant 3\}$. We suppose that, for all $\beta \in \mathbb{N}^n$ such that $|\beta| \leqslant (n/2) + 2$, we have

$$\forall (x, \xi) \in \mathbb{R}^n \times \mathbb{R}^n, \quad \left|\partial_\xi^\beta q(x, \xi)\right| \leqslant M.$$

1. Introduce $Q(x, z) = (2\pi)^{-n} \int_{\mathbb{R}^n} e^{iz\cdot\xi} q(x, \xi) \, d\xi$. Using the relation $\partial_\xi e^{iz\cdot\xi} = i z e^{iz\cdot\xi}$, show that, for $\alpha \in \mathbb{N}^n$, the function $z \mapsto z^\alpha Q(x, z)$ can be written in the form of a Fourier transform of a function to be specified.

2. Deduce that, for all $|\alpha| \leqslant (n/2) + 2$, there exists a constant C_α such that

$$\int |z|^{2\alpha} |Q(x, z)|^2 \, dz \leqslant C_\alpha M^2.$$

Then deduce that $\int |Q(x, z)| \, dz \leqslant CM$.

3. Let $f \in S(\mathbb{R}^n)$. Show that $\mathrm{Op}(q) f(x) = \int Q(x, x - y) f(y) \, dy$ then deduce

$$\|\mathrm{Op}(q) f\|_{L^\infty} \leqslant CM \|f\|_{L^\infty}.$$

Exercise 9.5 (Continuity on Hölder spaces) The aim of this problem is to study the action of a pseudo-differential operator on Hölder spaces. We refer to the lecture notes of Guy Métivier [133, Chap. 10] or to his book [134] for the solution.

We will denote by $C, C_\alpha, C_{\alpha,\beta}, \ldots$ absolute constants (where C_α depends on the multi-index α...) which do not depend on the symbols, nor on the unknowns.

Consider a symbol $p = p(x, \xi)$ which is of class C^∞ on $\mathbb{R}^n \times \mathbb{R}^n$ and such that

$$\forall (x, \xi) \in \mathbb{R}^n \times \mathbb{R}^n, \quad \left|\partial_x^\alpha \partial_\xi^\beta p(x, \xi)\right| \leqslant C_{\alpha,\beta} (1 + |\xi|)^{-|\beta|}.$$

We recall that Hölder spaces can be studied using the Littlewood–Paley decomposition (see Section 5.4.2). To fix the notation, let us recall that there exist two functions χ_0 and χ, C^∞ on $\mathbb{R}^n$, with support respectively in the ball $\{|\xi| \leqslant 1\}$ and in the annulus $\{1/3 \leqslant |\xi| \leqslant 3\}$ and such that:

$$\forall \xi \in \mathbb{R}^n, \quad \chi_0(\xi) + \sum_{j=0}^{+\infty} \chi(2^{-j}\xi) = 1.$$

Let us introduce

$$p_{-1}(x,\xi) = p(x,\xi)\chi_0(\xi), \quad p_j(x,\xi) = p(x,\xi)\chi(2^{-j}\xi) \quad \text{for } j \in \mathbb{N}.$$

1. Show that there exists $M > 0$ such that, for all $\beta \in \mathbb{N}^n$ satisfying $|\beta| \leqslant (n/2) + 2$, we have

$$\left|\partial_\xi^\beta p_{-1}(x,\xi)\right| \leqslant M,$$

$$\left|\partial_\xi^\beta p_j(x,\xi)\right| \leqslant M 2^{-j|\beta|} \qquad (\forall j \in \mathbb{N}).$$

2. Show that

$$\|\mathrm{Op}(p_{-1})f\|_{L^\infty} \leqslant C\|f\|_{L^\infty}.$$

3. Let $a = a(x,\xi)$ be a symbol and $\lambda \in \mathbb{R}_*^+$. Let us set $b(x,\xi) = a(x/\lambda, \lambda\xi)$. Let H_λ be the map that associates to the function $u = u(x)$

$$(H_\lambda u)(x) = u(\lambda x).$$

Show that $\mathrm{Op}(a) = H_\lambda \circ \mathrm{Op}(b) \circ H_\lambda^{-1}$ and deduce that, if $\mathrm{Op}(b)$ is bounded from L^∞ to L^∞, then $\mathrm{Op}(a)$ is also and they then have the same norm.

4. For all $j \in \mathbb{N}$, show that we can choose λ_j so that $\tilde{p}_j(x,\xi) = p_j(\lambda_j^{-1}x, \lambda_j\xi)$ is supported in $\{(x,\xi) : |\xi| \leqslant 3\}$. Using Exercise 9.4, deduce that

$$\left\|\mathrm{Op}(p_j)f\right\|_{L^\infty} \leqslant C\|f\|_{L^\infty}.$$

5. Let us introduce $f_{-1} = \chi_0(D_x)f$ and $f_k = \chi(2^{-k}D_x)f$ for $k \geqslant 0$. Show that

$$\mathrm{Op}(p_j)f = \sum_{|j-k| \leqslant 3} \mathrm{Op}(p_j)f_k.$$

6. Let $r \in (0,1)$. Show that there exists $C > 0$ such that for all $j \in \mathbb{N}$ and all $f \in S(\mathbb{R}^n)$, we have

$$\left\|\mathrm{Op}(p_j)f\right\|_{L^\infty} \leqslant C\|f\|_{C^{0,r}} 2^{-jr}.$$

7. Let $\alpha \in \mathbb{N}^n$ such that $|\alpha| \leqslant 1$. Show that for all $j \in \mathbb{N} \cup \{-1\}$, we have

$$\partial_x^\alpha \mathrm{Op}(p_j) = \mathrm{Op}(q_j) \quad \text{where} \quad q_j(x,\xi) = \left((i\xi + \partial_x)^\alpha p_j\right)(x,\xi).$$

By repeating the previous steps, show that

$$\left\|\partial_x^\alpha \operatorname{Op}(p_j)f\right\|_{L^\infty} \le C_\alpha 2^{j(|\alpha|-r)}\|f\|_{C^{0,r}}.$$

8. ($*$) Let (f_j) be a sequence of functions of class $C^1(\mathbb{R}^n)$ that satisfy

$$\left\|\partial_x^\alpha f_j\right\|_{L^\infty} \le M2^{j(|\alpha|-r)} \quad \text{for all } |\alpha| \le 1.$$

Show that $f = \sum f_j$ belongs to $C^{0,r}(\mathbb{R}^n)$ and that its norm is bounded by CM.

9. Conclude: $\operatorname{Op}(p)$ is bounded from $C^{0,r}(\mathbb{R}^n)$ to itself for all $r \in (0,1)$.

Chapter 10
Symbolic Calculus

Let us recall that we say a function $a = a(x, \xi)$ defined on $\mathbb{R}^n \times \mathbb{R}^n$ and with complex values belongs to the class $S^m(\mathbb{R}^n)$, for a certain real number m, if for all multi-indices α and β in $\mathbb{N}^n$, there exists a constant $C_{\alpha\beta}$ such that

$$\forall (x, \xi) \in \mathbb{R}^n \times \mathbb{R}^n, \quad \left| \partial_x^\alpha \partial_\xi^\beta a(x, \xi) \right| \leqslant C_{\alpha\beta} (1 + |\xi|)^{m - |\beta|} \qquad (\alpha, \beta \in \mathbb{N}^n).$$

Given $m \in \mathbb{R}$, a symbol $a \in S^m(\mathbb{R}^n)$ and a function u in the Schwartz space $\mathcal{S}(\mathbb{R}^n)$, we have seen that we can define $\mathrm{Op}(a)u \in \mathcal{S}(\mathbb{R}^n)$ by

$$\mathrm{Op}(a)u(x) = (2\pi)^{-n} \int e^{ix \cdot \xi} a(x, \xi) \widehat{u}(\xi) \, d\xi.$$

We have also seen that if $a \in S^0(\mathbb{R}^n)$, then $\mathrm{Op}(a)$ uniquely extends as a bounded operator from $L^2(\mathbb{R}^n)$ to $L^2(\mathbb{R}^n)$.

Consider two pseudo-differential operators $A = \mathrm{Op}(a)$ and $B = \mathrm{Op}(b)$ with symbols $a, b \in S^m(\mathbb{R}^n)$. Then $\lambda A + \mu B$ is a pseudo-differential operator with symbol $\lambda a + \mu b \in S^m(\mathbb{R}^n)$. The questions that will interest us in this chapter concern the operators $A \circ B$ and A^*. We will see that these are also pseudo-differential operators and that we can calculate their symbols. The symbolic calculus is precisely the process that allows us to manipulate operators by working at the level of symbols.

10.1 Introduction to Symbolic Calculus

In this first section, we will see three distinct situations in which we can easily study the products (composition) and the adjoints of pseudo-differential operators.

These situations correspond to the following cases:

1. the *Fourier multipliers* (symbols not depending on x);
2. the *differential operators* (polynomial symbols in ξ);
3. the *microlocalization operators* (symbols with compact support in $\mathbb{R}^{2n}$).

© The Author(s), under exclusive license to Springer Nature Switzerland AG 2024
T. Alazard, *Analysis and Partial Differential Equations*, Universitext,
https://doi.org/10.1007/978-3-031-70909-8_10

A. Fourier Multipliers

Let $A = \mathrm{Op}(a)$ with $a = a(\xi)$ independent of x. Then A is a particular case of a Fourier multiplier. Recall that a Fourier multiplier is a linear operator that acts on L^2 or $\mathcal{S}'$ by multiplying the Fourier transform of a function (or a tempered distribution) by a given function, called the symbol. Given a function $m = m(\xi)$ with complex values, the Fourier multiplier of symbol m is the operator, denoted $m(D_x)$, defined by

$$\widehat{m(D_x)f}(\xi) = m(\xi)\widehat{f}(\xi).$$

If $m \in L^\infty(\mathbb{R}^n)$ then $m(D_x)$ is well defined on $L^2(\mathbb{R}^n)$ and $m(D_x) \in \mathcal{L}(L^2)$. If $m \in C^\infty(\mathbb{R}^n)$ is slowly growing in the sense of definition 5.25, then $m(D_x)$ is continuous from $\mathcal{S}'(\mathbb{R}^n)$ to $\mathcal{S}'(\mathbb{R}^n)$. We directly verify that

$$m_1(D_x) \circ m_2(D_x) = m(D_x) \quad \text{with } m(\xi) = m_1(\xi)m_2(\xi),$$

$$m(D_x)^* = m^*(D_x) \quad \text{with } m^*(\xi) = \overline{m(\xi)}.$$

Examples of Fourier multipliers

- ∂_{x_j} is the Fourier multiplier of symbol $i\xi_j$.
- The Laplacian Δ is the Fourier multiplier of symbol $-|\xi|^2$.
- The Hilbert transform is the Fourier multiplier of symbol $-i\xi/|\xi|$ ($\xi \in \mathbb{R}$).
- The square root of $-\Delta$ is the Fourier multiplier of symbol $|\xi| = \sqrt{\xi_1^2 + \cdots + \xi_n^2}$.
- Let $s \in \mathbb{R}$. The operator that realizes the canonical isomorphism of H^s onto L^2 is the Fourier multiplier of symbol $\langle \xi \rangle^s$, where $\langle \xi \rangle = (1 + |\xi|^2)^{1/2}$.
- Consider the equation

$$\partial_t u + i\langle D_x \rangle^s u = 0, \quad u|_{t=0} = u_0.$$

This equation can be solved by the Hille–Yosida theorem (see [15]) or by the Fourier transform. The operator that sends the initial data u_0 to the solution at time t is the Fourier multiplier of symbol $\exp(-it\langle \xi \rangle^s)$.

B. Differential Operators

Consider two differential operators

$$A = \sum_{|\alpha| \leqslant m} a_\alpha(x)\partial_x^\alpha, \quad B = \sum_{|\alpha| \leqslant m'} b_\alpha(x)\partial_x^\alpha$$

where the coefficients a_α, b_α belong to $C_b^\infty(\mathbb{R}^n)$.

Introduce their symbols

$$a(x, \xi) = \sum_{|\alpha| \leqslant m} a_\alpha(x)(i\xi)^\alpha, \quad b(x, \xi) = \sum_{|\alpha| \leqslant m'} b_\alpha(x)(i\xi)^\alpha,$$

so that $A = \mathrm{Op}(a)$ and $B = \mathrm{Op}(b)$. Let e_ξ be the exponential function $x \mapsto e^{ix \cdot \xi}$. Then

$$(Ae_\xi)(x) = a(x, \xi)e_\xi(x), \quad (Be_\xi)(x) = b(x, \xi)e_\xi(x).$$

Moreover, for any regular function $b(x, \xi)$,

$$A(be_\xi)(x) = \sum_\alpha a_\alpha(x)\partial_x^\alpha\left(e^{ix \cdot \xi} b(x, \xi)\right)$$

$$= \sum_\alpha a_\alpha(x)\left((i\xi + \partial_x)^\alpha b(x, \xi)\right)e^{ix \cdot \xi}$$

$$= e^{ix \cdot \xi} a\left(x, \xi + \frac{1}{i}\partial_x\right)b(x, \xi)$$

$$= e^{ix \cdot \xi} \sum_{\beta \in \mathbb{N}^n} \frac{1}{i^{|\beta|}\beta!}\left(\partial_\xi^\beta a(x, \xi)\right)\left(\partial_x^\beta b(x, \xi)\right),$$

where we used the Taylor formula for a polynomial. We deduce the following result.

Proposition 10.1 *If A and B are differential operators, then $A \circ B$ is a differential operator of symbol*

$$a\#b(x, \xi) = \sum_{\alpha \in \mathbb{N}^n} \frac{1}{i^{|\alpha|}\alpha!}\left(\partial_\xi^\alpha a(x, \xi)\right)\left(\partial_x^\alpha b(x, \xi)\right).$$

Note that the sum is finite since $\partial_\xi^\alpha a = 0$ if $|\alpha| > m$.

Proof The operator $A \circ B$ is of course a differential operator and we have seen that $(A \circ B)e_\xi = (a\#b)e_\xi$.   □

Exercise Let A be a differential operator. Show that A^* is a differential operator of symbol

$$a^*(x, \xi) = \sum_{\alpha \in \mathbb{N}^n} \frac{1}{i^{|\alpha|}\alpha!}\partial_\xi^\alpha \partial_x^\alpha \overline{a}(x, \xi).$$

C. Microlocalization Operators

Localization operators, of the form $u \mapsto \varphi u$ where $\varphi \in C_0^\infty(\mathbb{R}^n)$, are essential in Analysis, as are frequency localization operators, which are Fourier multipliers $u \mapsto \varphi(D_x)u$ with $\varphi \in C_0^\infty(\mathbb{R}^n)$. Pseudo-differential operators allow for simultaneous localization in x and in ξ, by considering an operator $\mathrm{Op}(a)$ with $a \in C_0^\infty(\mathbb{R}^{2n})$. If $a \in C_0^\infty(\mathbb{R}^{2n})$, we say that $\mathrm{Op}(a)$ is a microlocalization operator.

We will see that the adjoint of a microlocalization operator is a pseudo-differential operator whose symbol does not necessarily belong to $C_0^\infty(\mathbb{R}^{2n})$ but belongs to all spaces $S^m(\mathbb{R}^n)$ for $m \leqslant 0$.

Proposition 10.2 *Let $a = a(x, \xi)$ be a symbol belonging to $C_0^\infty(\mathbb{R}^{2n})$. Then*

$$a^*(x,\xi) = (2\pi)^{-n} \int e^{-iy\cdot\eta}\overline{a}(x-y,\xi-\eta)\,dy\,d\eta$$

defines a symbol a^* belonging to $S^{-\infty}$ and

$$(\mathrm{Op}(a)u, v) = (u, \mathrm{Op}(a^*)v)$$

for all u, v in $S(\mathbb{R}^n)$.

Remark 10.3 One goal of this chapter will be to prove a result that extends the previous proposition to the case of a general symbol $a \in S^\infty$. We start by looking at the case $a \in S^{-\infty}$ because the analysis is then much easier. The reader will note in particular that the integrals that appear in the proof below make no sense if a is a general symbol.

Proof Let $u \in S(\mathbb{R}^n)$. As a has compact support, we can use Fubini's theorem to write

$$\mathrm{Op}(a)u(x) = (2\pi)^{-n} \int e^{ix\cdot\xi} a(x,\xi)\widehat{u}(\xi)\,d\xi$$

$$= (2\pi)^{-n} \int e^{ix\cdot\xi} a(x,\xi) \left(\int e^{-iy\cdot\xi} u(y)\,dy\right) d\xi$$

$$= (2\pi)^{-n} \iint e^{i(x-y)\cdot\xi} a(x,\xi)u(y)\,dy\,d\xi$$

$$= (2\pi)^{-n} \int \left(\int e^{i(x-y)\cdot\xi} a(x,\xi)\,d\xi\right) u(y)\,dy.$$

Therefore,

$$\mathrm{Op}(a)u(x) = \int K(x,y)u(y)\,dy$$

where $K = K(x, y)$ (called the *kernel* of $\mathrm{Op}(a)$) is given by

$$K(x,y) = (2\pi)^{-n} \int e^{i(x-y)\cdot\xi} a(x,\xi)\,d\xi \tag{10.1}$$

$$= (2\pi)^{-n}(\mathcal{F}_\xi a)(x, y-x),$$

where $\mathcal{F}_\xi a(x, \zeta) = \int e^{-i\xi\cdot\zeta} a(x,\xi)\,d\xi$ is the Fourier transform of a with respect to the second variable. We deduce that $K \in S(\mathbb{R}^{2n})$.

Now, if v is also in $\mathcal{S}(\mathbb{R}^n)$ then

$$(\mathrm{Op}(a)u, v) = \int \left(\int K(x, y)u(y)\, dy \right) \overline{v(x)}\, dx$$

$$= \int u(y) \left(\overline{\int \overline{K(x, y)}v(x)\, dx} \right) dy,$$

so $(\mathrm{Op}(a)u, v) = (u, (\mathrm{Op}(a))^* v)$ with

$$(\mathrm{Op}(a))^* v(x) := \int \overline{K(y, x)}v(y)\, dy.$$

This means that $\mathrm{Op}(a)^*$ is an operator with kernel

$$K^*(x, y) = \overline{K(y, x)} = (2\pi)^{-n} \int e^{i(x-y)\cdot\theta}\overline{a(y, \theta)}\, d\theta.$$

We want to write $K^*(x, y)$ in the form $K^*(x, y) = (2\pi)^{-n}(\mathcal{F}_\xi a^*)(x, y - x)$. Then we must have

$$a^*(x, \xi) = (2\pi)^{-n} \int e^{i\xi\cdot z}(\mathcal{F}_\xi a^*)(x, z)\, dz$$

$$= \int K^*(x, x + z)e^{iz\cdot\xi}\, dz$$

$$= \int K^*(x, x - y)e^{-iy\cdot\xi}\, dy$$

$$= (2\pi)^{-n} \int \left(\int e^{i(x-(x-y))\cdot\theta}\overline{a(x - y, \theta)}\, d\theta \right) e^{-iy\cdot\xi}\, dy$$

$$= (2\pi)^{-n} \iint e^{iy\cdot(\theta-\xi)}\overline{a}(x - y, \theta)\, dy\, d\theta$$

$$= (2\pi)^{-n} \iint e^{-iy\cdot\eta}\overline{a}(x - y, \xi - \eta)\, dy\, d\eta.$$

The calculations already done at the beginning of the proof imply that $\mathrm{Op}(a)^*$ is the pseudo-differential operator of symbol a^*. $\square$

10.2 Oscillatory Integrals

To extend the results of the previous section to general pseudo-differential operators, we will need to define certain integrals, called oscillatory integrals, which play a crucial role in microlocal analysis. They are of the form

$$\int e^{i\phi(x)}a(x)\, dx \qquad (x \in \mathbb{R}^N,\ N \geqslant 1).$$

We say that ϕ is a *phase* and that a is an *amplitude*. We will always assume that a is a function of class C^∞ from $\mathbb{R}^N$ to $\mathbb{C}$ and that ϕ is real-valued.

We use N to denote the dimension here because we will apply the results of this part with $N = 2n$. Indeed, these integrals naturally appear to define symbols. For example, let us note that for (x_0, ξ_0) fixed in $\mathbb{R}^n \times \mathbb{R}^n$,

$$a^*(x_0, \xi_0) = (2\pi)^{-n} \int e^{-iy \cdot \eta} \overline{a}(x_0 - y, \xi_0 - \eta) \, dy \, d\eta$$

is written in the form

$$a^*(x_0, \xi_0) = \int e^{i\phi(x)} u(x) \, dx$$

with $N = 2n$, $x = (y, \eta)$ and $\phi(x) = -y \cdot \eta$.

If a is the symbol of a differential operator, polynomial in x, the integral is obviously divergent in the classical sense. To give it meaning, the idea is that, under a strong oscillation assumption of the term $e^{i\phi(x)}$, we can offset the growth of a.

A. Principle of non-stationary phase

The analysis of oscillatory integrals is based on the *principle of non-stationary phase*, which expresses the decrease of an oscillatory integral as a function of a large parameter.

Lemma 10.4 (Principle of non-stationary phase) *Let $N \geqslant 1$, $\varphi \in C_b^\infty(\mathbb{R}^N)$ be a real-valued function and $f \in C_0^\infty(\mathbb{R}^N)$. Let V be a neighborhood of the support of f. Assume that*

$$\inf_{x \in V} |\nabla\varphi(x)| > 0.$$

Then, for all $k \in \mathbb{N}$ and for all $\lambda \geqslant 1$,

$$\left| \int e^{i\lambda\varphi(x)} f(x) \, dx \right| \leqslant C_k \lambda^{-k} \sup_{|\alpha| \leqslant k} \|\partial_x^\alpha f\|_{L^1(\mathbb{R}^N)},$$

where C_k is a constant independent of λ and f.

Proof Let us introduce the differential operator

$$L := -i \frac{\nabla\varphi \cdot \nabla}{|\nabla\varphi|^2}, \quad \text{where} \quad \nabla\varphi \cdot \nabla = \sum_{1 \leqslant j \leqslant N} \frac{\partial\varphi}{\partial x_j} \frac{\partial}{\partial x_j},$$

which is well defined by assumption on φ. Moreover, L satisfies, for all $\lambda \in \mathbb{R}$,

$$L(e^{i\lambda\varphi}) = \lambda e^{i\lambda\varphi},$$

and therefore $L^k(e^{i\lambda\varphi(x)}) = \lambda^k e^{i\lambda\varphi(x)}$ for all $k \in \mathbb{N}$.

Then, by performing successive *integrations by parts*, we deduce that

$$\lambda^k \int e^{i\lambda\varphi(x)} f(x)\,dx = \int e^{i\lambda\varphi(x)} ({}^tL)^k f(x)\,dx,$$

where

$$^tLf = i \sum_{1\leqslant j\leqslant N} \frac{\partial}{\partial x_j}\left(\frac{1}{|\nabla\varphi|^2}\left(f\frac{\partial\varphi}{\partial x_j}\right)\right).$$

Note that $({}^tL)^k$ is a differential operator of order k whose coefficients are C^∞ (and depend on φ). We therefore obtain the desired result by bounding the last integral by $\|({}^tL)^k f\|_{L^1}$.

We see that C_k depends only on k, $\inf |\nabla\varphi|^2$ and $\sup_{|\alpha|\leqslant k+1} \|\partial^\alpha\varphi\|_{L^\infty}$.   □

B. Definition of an oscillatory integral

Definition 10.5 Let $m \in [0,+\infty)$. The space A^m of amplitudes of order m is the space of functions $a \in C^\infty(\mathbb{R}^N;\mathbb{C})$ such that

$$\forall \alpha \in \mathbb{N}^N, \quad \sup_{x\in\mathbb{R}^N} \left|(1+|x|)^{-m}\partial_x^\alpha a(x)\right| < +\infty.$$

We introduce the norms

$$\|a\|_{m,k} := \max_{|\alpha|\leqslant k} \sup_{x\in\mathbb{R}^N} \left|(1+|x|)^{-m}\partial_x^\alpha a(x)\right|.$$

We will assume that ϕ is a non-degenerate quadratic form, that is

$$\phi(x) = \frac{1}{2}(Ax)\cdot x \qquad (x\in\mathbb{R}^N),$$

where $A \in M_N(\mathbb{R})$ is an invertible symmetric matrix. Then $\nabla\phi(x) = Ax$ and we can apply the principle of non-stationary phase.

Theorem 10.6 *Let $m \geqslant 0$, ϕ be a non-degenerate quadratic form on $\mathbb{R}^N$, $a \in A^m$ and $\psi \in S(\mathbb{R}^N)$ such that $\psi(0) = 1$. Then the integral*

$$I(\varepsilon) := \int e^{i\phi(x)} a(x)\psi(\varepsilon x)\,dx$$

converges when ε tends to 0 towards a limit independent of ψ, which is equal to $\int e^{i\phi(x)} a(x)\,dx$ if $a \in L^1$. When $a \notin L^1$, we continue to denote the limit by $\int e^{i\phi(x)} a(x)\,dx$ and we have

$$\left|\int e^{i\phi(x)} a(x)\,dx\right| \leqslant C_{\phi,m}\,\|a\|_{m,m+N+1}\,. \tag{10.2}$$

Proof We want to use the principle of non-stationary phase, which requires us to introduce a large parameter. For this, we will use a dyadic decomposition of unity. Let us recall how to obtain such a decomposition: we start by considering a function $\chi_0 \in C_0^\infty(\mathbb{R}^N; \mathbb{R})$ which satisfies $\chi_0(x) = 1$ for $|x| \leqslant 1/2$, and $\chi_0(x) = 0$ for $|x| \geqslant 1$. We then define $\chi(x) = \chi_0(x/2) - \chi_0(x)$. This function χ is supported in the annulus $\{1/2 \leqslant |x| \leqslant 2\}$. Moreover, for all $x \in \mathbb{R}^N$, we have the equality

$$1 = \chi_0(x) + \sum_{j=0}^\infty \chi(2^{-j}x).$$

Let us recall that the convergence of this series is not problematic because, for all $x \in \mathbb{R}^N$, we have $\chi(2^{-p}x) = 0$ for all integer p large enough. Let us define

$$S_p(x) = \chi_0(x) + \sum_{j=0}^p \chi(2^{-j}x) = \chi_0(2^{-p-1}x)$$

and introduce the (well-defined) integrals

$$I_p := \int e^{i\phi(x)} a(x) S_p(x)\,dx, \quad R_p(\varepsilon) := \int e^{i\phi(x)} a(x)(1 - \psi(\varepsilon x)) S_p(x)\,dx.$$

It follows from the dominated convergence theorem that

$$I(\varepsilon) = \int e^{i\phi(x)} a(x)\psi(\varepsilon x)\,dx = \lim_{p\to+\infty} \int e^{i\phi(x)} a(x)\psi(\varepsilon x) S_p(x)\,dx.$$

Also, by definition,

$$\int e^{i\phi(x)} a(x)\psi(\varepsilon x) S_p(x)\,dx = I_p - R_p(\varepsilon).$$

Therefore, it will be enough to show that the limit $\lim_{p\to+\infty} I_p$ exists and that $\lim_{p\to+\infty} R_p(\varepsilon) = O(\varepsilon)$. This will establish that the integral $I(\varepsilon)$ has a limit when ε tends to 0 and that this limit is independent of ψ.

After changing variables $z = 2^{-p}x$, we have

$$I_p - I_{p-1} = \int e^{i2^{2p}\phi(z)} a(2^p z)\chi(z) 2^{Np}\,dz,$$

where we used the fact that ϕ is quadratic to write

$$\phi(tz) = t^2\phi(z).$$

Furthermore, on the support of $\chi(z)$ we have $|z| \geqslant 1/2$, therefore

$$\inf_{z \in \mathrm{supp}\,\chi} |\nabla\phi(z)| \geqslant c_0 > 0.$$

Consequently, we are in position to apply the principle of non-stationary phase. Specifically, using Lemma 10.4 with $f(x) = a(2^p x)\chi(x)$ and $\lambda = 2^{2p}$, we obtain that, for all $k \in \mathbb{N}$,

$$\left| \int e^{i2^{2p}\phi(z)} a(2^p z)\chi(z) 2^{Np} \, dz \right| \leqslant C_k 2^{Np-2pk} \max_{|\alpha| \leqslant k} \int_{|x| \leqslant 2} \left| \partial_x^\alpha (a(2^p x)\chi(x)) \right| dx,$$

where we used the fact that supp χ is contained in the ball $B(0, 2)$. The assumption that a is an amplitude of order m implies that there exists a constant $K > 0$ such that for all $p \geqslant 1$,

$$\int_{|x| \leqslant 2} \left| \partial_x^\alpha (a(2^p x)\chi(x)) \right| dx \leqslant K 2^{p(|\alpha|+m)} \|a\|_{m, |\alpha|}.$$

We deduce that

$$\left| \int e^{i2^{2p}\phi(z)} a(2^p z)\chi(z) 2^{pN} \, dz \right| \leqslant KC_k 2^{p(N+k+m-2k)} \|a\|_{m,k}.$$

We choose $k = N + m + 1$ so that

$$\left| I_p - I_{p-1} \right| \leqslant KC_{N+m+1} 2^{-p} \|a\|_{m, N+m+1}.$$

Similarly, we obtain that $\left| R_p(\varepsilon) - R_{p-1}(\varepsilon) \right| \leqslant \varepsilon C 2^{-p}$. This concludes the proof. $\square$

C. A Hörmander inequality

Consider a family of operators T_h, depending on a small parameter h, of the form

$$(T_h f)(\xi) := \int e^{i\phi(x,\xi)/h} a(x, \xi) f(x) \, dx \qquad (x, \xi \in \mathbb{R}^n).$$

Suppose that the phase ϕ is real-valued and that the amplitude a has compact support in x and in ξ. Then, it is easily verified that, for all $h > 0$, T_h is a continuous linear map from $L^2(\mathbb{R}^n)$ to $L^2(\mathbb{R}^n)$. We will prove an estimate, due to Hörmander [76], which states that if the mixed second derivative $\phi''_{x\xi}$ is not singular on the support of the amplitude, then $h^{-n/2} T_h$ is uniformly bounded in $\mathcal{L}(L^2)$.

Theorem 10.7 *Let $a \in C_0^\infty(\mathbb{R}^n \times \mathbb{R}^n)$. If $\phi \in C^\infty(\mathbb{R}^n \times \mathbb{R}^n)$ is real-valued and satisfies*

$$(x, \xi) \in \text{supp } a \implies \det \left[\frac{\partial^2 \phi}{\partial x \partial \xi}(x, \xi) \right] \neq 0,$$

then there exists a constant C such that, for all $h \in (0, 1]$ and all $f \in L^2(\mathbb{R}^n)$,

$$\|T_h f\|_{L^2} \leqslant Ch^{n/2} \|f\|_{L^2}.$$

Remark 10.8 By combining this inequality with the fact that

$$\|T_h f\|_{L^\infty} \leqslant \|f\|_{L^1}$$

and the Riesz–Thorin convexity theorem 18.13, we find that, if $p \in [1,2]$ and $1/p + 1/p' = 1$, then

$$\|T_h f\|_{L^{p'}} \leqslant h^{n/p'} \|f\|_{L^p}, \quad f \in C_0^\infty(\mathbb{R}^n).$$

Proof The proof of this result will allow us to implement several very useful ideas in practice. We start by recalling classical results on bounded operators on $\mathcal{L}(L^2)$. First, the equality $\|T\|_{\mathcal{L}(L^2)} = \|T^*\|_{\mathcal{L}(L^2)}$. From this, we deduce that $\|TT^*\|_{\mathcal{L}(L^2)} \leqslant \|T^*\|^2_{\mathcal{L}(L^2)}$. As furthermore

$$\|T^* f\|^2_{L^2} = \langle T^* f, T^* f \rangle = \langle TT^* f, f \rangle \leqslant \|TT^*\|_{\mathcal{L}(L^2)} \|f\|^2_{L^2},$$

we verify that

$$\|T\|^2_{\mathcal{L}(L^2)} = \|T^*\|^2_{\mathcal{L}(L^2)} = \|TT^*\|_{\mathcal{L}(L^2)}.$$

Therefore, it is sufficient to prove that the operator norm[1] of $T_h T_h^*$ is bounded by Ch^n. Let us write

$$(T_h T_h^* f)(\xi) = \int K_h(\xi, \eta) f(\eta) \, d\eta$$

where

$$K_h(\xi, \eta) = \int e^{i(\Phi(x,\xi) - \Phi(x,\eta))/h} a(x, \xi) \overline{a}(x, \eta) \, dx.$$

We then use Schur's lemma (proven at the end of this proof) which states that an operator T defined by

$$(Tf)(x) = \int K(x, y) f(y) \, dy,$$

satisfies

$$2 \|T\|_{\mathcal{L}(L^2)} \leqslant \sup_y \int |K(x,y)| \, dx + \sup_x \int |K(x,y)| \, dy.$$

It remains to estimate the kernel K_h. By introducing a partition of unity (see Proposition 17.35), we can always assume that the support of a is included in a ball of small diameter δ. We can therefore limit ourselves to considering the case where ξ and η are close. Then

$$|\partial_x(\Phi(x,\xi) - \Phi(x,\eta))| = |\Phi''_{x\xi}(x,\eta)(\xi - \eta)| + O(|\xi - \eta|^2) \geqslant c|\xi - \eta|.$$

---

[1] It is often more convenient to estimate the operator norm of TT^* than that of T; we then say that we use "the TT^* argument".

As a result, we can use the principle of non-stationary phase (*cf.* Lemma 10.4) to obtain the upper bound

$$|K_h(\xi, \eta)| \leqslant C_N \left(\frac{|\xi - \eta|}{h} \right)^{-N},$$

for all $N \in \mathbb{N}$. As furthermore K_h is bounded, we deduce that

$$|K_h(\xi, \eta)| \leqslant C_N' (1 + |\xi - \eta| / h)^{-N},$$

from which

$$\sup_\eta \int |K_h(\xi, \eta)| \, d\xi \leqslant C h^n, \quad \sup_\xi \int |K_h(\xi, \eta)| \, d\eta \leqslant C h^n,$$

which concludes the proof. $\qquad\square$

Lemma 10.9 (Schur's Lemma) *Let $K(x, y)$ be a continuous function on $\mathbb{R}^n \times \mathbb{R}^n$ such that*

$$\sup_y \int |K(x, y)| \, dx \leqslant A_1, \quad \sup_x \int |K(x, y)| \, dy \leqslant A_2.$$

Then the operator P with kernel K, defined for $u \in C_0^0(\mathbb{R}^n)$ by

$$Pu(x) = \int K(x, y) u(y) \, dy,$$

extends uniquely to a continuous operator from $L^2(\mathbb{R}^n)$ to $L^2(\mathbb{R}^n)$ and

$$\|Pu\|_{L^2} \leqslant \sqrt{A_1 A_2} \, \|u\|_{L^2}.$$

Proof From the Cauchy–Schwarz inequality

$$|Pu(x)|^2 \leqslant \int |K(x, y)| \, |u(y)|^2 \, dy \int |K(x, y)| \, dy \leqslant A_2 \int |K(x, y)| \, |u(y)|^2 \, dy,$$

from which

$$\int |Pu(x)|^2 \, dx \leqslant A_2 \iint |K(x, y)| \, |u(y)|^2 \, dy \, dx$$

$$\leqslant A_2 \int |u(y)|^2 \left(\int |K(x, y)| \, dx \right) dy$$

$$\leqslant A_1 A_2 \int |u(y)|^2 \, dy,$$

which implies the desired inequality. $\qquad\square$

Remark 10.10 The Schur lemma implies that if $f \in L^1$ and $g \in L^2$, then there holds $\|f * g\|_{L^2} \leqslant \|f\|_{L^1} \|g\|_{L^2}$. To see this, it is enough to apply the above inequality with $K(x, y) = f(x - y)$.

10.3 Adjoint and Composition

Let $s \in \mathbb{R}$. Recall that the Sobolev space $H^s(\mathbb{R}^n)$ is the space of tempered distributions f such that $(1 + |\xi|^2)^{s/2} \widehat{f}(\xi)$ belongs to $L^2(\mathbb{R}^n)$, equipped with the norm

$$\|f\|_{H^s}^2 := \frac{1}{(2\pi)^n} \int (1 + |\xi|^2)^s |\widehat{f}(\xi)|^2 \, d\xi.$$

It will be convenient to use the following definition.

Definition 10.11 Let $m \in \mathbb{R}$. We say that an operator is of order m if it is bounded from $H^\mu(\mathbb{R}^n)$ to $H^{\mu-m}(\mathbb{R}^n)$ for all $\mu \in \mathbb{R}$.

Example 10.12 • The identity is an operator of order 0 and the Laplacian is an operator of order 2.
• A differential operator $P = \sum_{|\alpha| \leqslant k} p_\alpha(x) \partial_x^\alpha$ with $k \in \mathbb{N}$ and $p_\alpha \in C_b^\infty(\mathbb{R}^n)$ is an operator of order k.
• The convolution operator by a function in the Schwartz space is an operator of order $-\infty$ (which means it is of order $-n$ for all $n \in \mathbb{N}$ or that it sends $H^{-\infty}(\mathbb{R}^n) = \bigcup_{s \in \mathbb{R}} H^s(\mathbb{R}^n)$ to $H^\infty(\mathbb{R}^n) = \bigcap_{s \in \mathbb{R}} H^s(\mathbb{R}^n)$).

In this chapter, we will prove (and give meaning to) the following statement.

Theorem 10.13 1. *If $a \in S^m(\mathbb{R}^n)$ then $\mathrm{Op}(a)$ is of order m.*
2. *Suppose that $a \in S^m(\mathbb{R}^n)$ and $b \in S^{m'}(\mathbb{R}^n)$. Then $\mathrm{Op}(a) \circ \mathrm{Op}(b)$ is a pseudo-differential operator of symbol denoted $a\#b$ and defined by*

$$a\#b(x, \xi) = (2\pi)^{-n} \iint e^{i(x-y)\cdot(\xi-\eta)} a(x, \xi) b(y, \eta) \, d\xi \, dy.$$

Furthermore, $\mathrm{Op}(a) \circ \mathrm{Op}(b) = \mathrm{Op}(ab) + R$ where R is of order $m + m' - 1$ and more generally, the operator

$$\mathrm{Op}(a) \circ \mathrm{Op}(b) - \mathrm{Op}\left(\sum_{|\alpha| \leqslant k} \frac{1}{i^{|\alpha|} \alpha!} (\partial_\xi^\alpha a(x, \xi))(\partial_x^\alpha b(x, \xi)) \right)$$

is of order $m + m' - k - 1$, for every integer $k \in \mathbb{N}$.
3. *The adjoint $\mathrm{Op}(a)^*$ is a pseudo-differential operator of symbol a^* defined by*

$$a^*(x, \xi) = (2\pi)^{-n} \iint e^{-iy\cdot\eta} \overline{a}(x - y, \xi - \eta) \, dy \, d\eta.$$

Furthermore, $\mathrm{Op}(a)^* = \mathrm{Op}(\overline{a}) + R$ where R is of order $m - 1$ and more generally

$$\mathrm{Op}(a^*) - \mathrm{Op}\left(\sum_{|\alpha| \leqslant k} \frac{1}{i^{|\alpha|} \alpha!} \partial_\xi^\alpha \partial_x^\alpha \overline{a}(x, \xi) \right)$$

is of order $m - k - 1$ for every integer $k \in \mathbb{N}$.

Corollary 10.14 *Let $a \in S^m(\mathbb{R}^n)$ and $b \in S^{m'}(\mathbb{R}^n)$. We will denote by $\{a, b\}$ the Poisson bracket of a and b defined by*

$$\{a, b\} = \sum_{1 \leqslant j \leqslant n} \left(\frac{\partial a}{\partial \xi_j} \frac{\partial b}{\partial x_j} - \frac{\partial b}{\partial \xi_j} \frac{\partial a}{\partial x_j} \right).$$

Then the commutator

$$[\mathrm{Op}(a), \mathrm{Op}(b)] = \mathrm{Op}(a) \circ \mathrm{Op}(b) - \mathrm{Op}(b) \circ \mathrm{Op}(a)$$

is an operator of order $m + m' - 1$ whose symbol c can be written in the form

$$c = \frac{1}{i} \{a, b\} + c', \quad where \quad c' \in S^{m+m'-2}.$$

To prove Theorem 10.13, we start by studying the adjoint with the following proposition.

Proposition 10.15 *Let $m \in \mathbb{R}$. If $a \in S^m(\mathbb{R}^n)$ then the oscillatory integral*

$$a^*(x, \xi) = (2\pi)^{-n} \iint e^{-iy \cdot \eta} \overline{a}(x - y, \xi - \eta) \, dy \, d\eta$$

defines a symbol a^* that belongs to $S^m(\mathbb{R}^n)$.

Proof Let $\phi(y, \eta) = -y \cdot \eta$. Then ϕ is a non-degenerate quadratic form on $\mathbb{R}^{2n}$ (we have $\phi(X) = (AX) \cdot X$, where A is the invertible symmetric matrix $A = -\frac{1}{2} \left(\begin{smallmatrix} 0 & I \\ I & 0 \end{smallmatrix} \right)$). Given $(x, \xi) \in \mathbb{R}^n \times \mathbb{R}^n$, we define $b_{x,\xi}(y, \eta) = \overline{a}(x - y, \xi - \eta)$. To study $b_{x,\xi}$, we will use the following inequality, which has already been seen in the proof of Proposition 7.41.

Lemma 10.16 (Peetre's Lemma) *Let $n \geqslant 1$. For all $m \in \mathbb{R}$ and all ξ, η in $\mathbb{R}^n$, we have*

$$\langle \xi + \eta \rangle^m \leqslant 2^{|m|} \langle \xi \rangle^{|m|} \langle \eta \rangle^m,$$

where recall that $\langle \xi \rangle = (1 + |\xi|^2)^{1/2}$. □

The previous lemma implies that

$$\langle \xi - \eta \rangle^m \leqslant 2^{|m|} \langle \xi \rangle^m \langle \eta \rangle^{|m|} \qquad \forall \xi, \eta \in \mathbb{R}^n.$$

Then, the assumption that a is a symbol leads to

$$\left| \partial_y^\alpha \partial_\eta^\beta a(x - y, \xi - \eta) \right| \leqslant C_{\alpha\beta} \langle \xi - \eta \rangle^{m-|\beta|} \leqslant C_{\alpha\beta} \langle \xi - \eta \rangle^m$$
$$\leqslant C_{\alpha\beta} 2^{|m|} \langle \xi \rangle^m \langle \eta \rangle^{|m|}$$
$$\leqslant C_{\alpha\beta} 2^{|m|} \langle \xi \rangle^m (1 + |y|^2 + |\eta|^2)^{|m|/2}$$

for all α, β in $\mathbb{N}^n$.

By definition of the amplitude classes, it follows that

$$b_{x,\xi} \in A^{|m|}(\mathbb{R}^{2n})$$

and moreover

$$\|b_{x,\xi}\|_{|m|,|m|+2n+1} = \max_{|\alpha|+|\beta|\leqslant|m|+2n+1} \left|\langle(y,\eta)\rangle^{-|m|}\partial_y^{\alpha}\partial_{\eta}^{\beta}b_{x,y}(y,\eta)\right| \leqslant C\langle\xi\rangle^{m}.$$

Since $a^{*}(x,\xi)$ is an oscillatory integral:

$$a^{*}(x,\xi) = \iint e^{i\phi(y,\eta)}b_{x,\xi}(y,\eta)\, dy\, d\eta,$$

the previous estimate and inequality (10.2) imply that $\langle\xi\rangle^{-m}a^{*}$ is a bounded function. It remains to estimate the derivatives. For this, we will show that a^{*} is of class C^{∞} and that, for all multi-indices α, β, we have

$$\partial_x^{\alpha}\partial_{\xi}^{\beta}(a^{*}) = (\partial_x^{\alpha}\partial_{\xi}^{\beta}a)^{*}.$$

Let us admit this identity. Then the previous argument applied with the symbol $\partial_x^{\alpha}\partial_{\xi}^{\beta}a \in S^{m-|\beta|}(\mathbb{R}^n)$ instead of $a \in S^m(\mathbb{R}^n)$ implies that

$$\langle\xi\rangle^{-(m-|\beta|)}\partial_x^{\alpha}\partial_{\xi}^{\beta}(a^{*})$$

is bounded for all α, β in $\mathbb{N}^n$. This proves that a^{*} is in $S^m(\mathbb{R}^n)$.

It remains to show that $\partial_x^{\alpha}\partial_{\xi}^{\beta}(a^{*}) = (\partial_x^{\alpha}\partial_{\xi}^{\beta}a)^{*}$. For this, we will show that we can differentiate the oscillatory integral that defines a^{*} under the integral sign. Recall that, for any function $\psi \in C_0^{\infty}(\mathbb{R}^{2n})$ such that $\psi(0) = 1$, we have

$$a^{*}(x,\xi) = \lim_{\varepsilon\to 0} \iint e^{-iy\cdot\eta}\overline{a}(x-y,\xi-\eta)\psi(\varepsilon y,\varepsilon\eta)\, dy\, d\eta.$$

We will use once again an integration by parts argument that relies on the identity

$$(1+|y|^2)^{-k}(1+|\eta|^2)^{-k}(I-\Delta_y)^k(I-\Delta_\eta)^k e^{-iy\cdot\eta} = e^{-iy\cdot\eta}.$$

As the integrands are regular functions with compact support, we can integrate by parts and obtain that

$$\iint e^{-iy\cdot\eta}\overline{a}(x-y,\xi-\eta)\psi(\varepsilon y,\varepsilon\eta)\, dy\, d\eta$$

$$= \iint e^{-iy\cdot\eta}(I-\Delta_y)^k(I-\Delta_\eta)^k\left[\frac{\overline{a}(x-y,\xi-\eta)\psi(\varepsilon y,\varepsilon\eta)}{(1+|y|^2)^k(1+|\eta|^2)^k}\right]dy\, d\eta.$$

Recall that we have shown that

$$\left|\partial_y^{\gamma}\partial_{\eta}^{\delta}a(x-y,\xi-\eta)\right| \leqslant C_{\gamma\delta}2^{|m|}\langle\xi\rangle^{m}(1+|\eta|^2)^{|m|/2}.$$

Furthermore,

$$\left|\partial_y^\gamma (1+|y|^2)^{-k}\right| \leqslant C_{k,\gamma}(1+|y|^2)^{-k}, \quad \left|\partial_\eta^\delta (1+|\eta|^2)^{-k}\right| \leqslant C_{k,\delta}(1+|\eta|^2)^{-k}.$$

It is then easily verified that, if $k > (n + |m|)/2$, then we can use the dominated convergence theorem and deduce that

$$a^*(x,\xi) = \iint e^{-iy\cdot\eta}(I-\Delta_y)^k(I-\Delta_\eta)^k \left[\frac{\bar{a}(x-y,\xi-\eta)}{(1+|y|^2)^k(1+|\eta|^2)^k}\right] dy\, d\eta.$$

The key point is that we have written $a^*(x,\xi)$ in the form of a Lebesgue convergent integral, and whose integrand depends on x,ξ in a C^∞ way. We conclude that we can apply the theorems of differentiation under the integral sign for integrals converging in the usual sense of Lebesgue.   □

Proposition 10.17 *Let $m \in \mathbb{R}$ and $a \in S^m(\mathbb{R}^n)$. Then, for all u, v in $S(\mathbb{R}^n)$ we have*

$$(\mathrm{Op}(a)u, v) = (u, \mathrm{Op}(a^*)v),$$

where $(f, g) = \int_{\mathbb{R}^n} f(x)\overline{g(x)}\, dx$.

Proof To prove this result, we will make the additional assumption that a has compact support in x and refer the reader to the book by Saint-Raymond [166] for the general case.

The proof is based on continuity arguments. We start by equipping the symbol classes $S^m(\mathbb{R}^n)$ with the most natural topology, which is that of a Fréchet space. Recall that the Schwartz space $S(\mathbb{R}^n)$ is also a Fréchet space whose topology is induced by the following family of semi-norms, indexed by $p \in \mathbb{N}$,

$$M_p(\varphi) = \sum_{|\alpha|+|\beta|\leqslant p} \sup_{x\in\mathbb{R}^n} \left|x^\alpha \partial_x^\beta \varphi(x)\right|.$$

The convergence of a sequence $(\varphi_k)_{k\in\mathbb{N}}$ of $S(\mathbb{R}^n)$ towards a function $\varphi \in S(\mathbb{R}^n)$ therefore equates to

$$\forall p \in \mathbb{N}, \quad \lim_{k\to\infty} M_p(\varphi_k - \varphi) = 0.$$

Similarly, the topology on the symbol class $S^m(\mathbb{R}^n)$ is induced by the following family of semi-norms, indexed by $p \in \mathbb{N}$,

$$N_p^m(a) = \sum_{|\alpha|+|\beta|\leqslant p} \sup_{(x,\xi)\in\mathbb{R}^n\times\mathbb{R}^n} \left\{\langle\xi\rangle^{-(m-|\beta|)}\left|\partial_x^\alpha \partial_\xi^\beta a(x,\xi)\right|\right\}.$$

The convergence of a sequence of symbols equates to the convergence in the sense of semi-norms: for $m \in \mathbb{R}$, we say that a sequence (a_k) of symbols belonging to $S^m(\mathbb{R}^n)$ converges towards a in $S^m(\mathbb{R}^n)$ if and only if

$$\forall p \in \mathbb{N}, \quad \lim_{n\to+\infty} N_p^m(a_k - a) = 0.$$

Lemma 10.18 *Let $m \in \mathbb{R}$, $a \in S^m(\mathbb{R}^n)$ and (a_k) be a sequence of symbols belonging to $S^m(\mathbb{R}^n)$ and converging towards a in $S^m(\mathbb{R}^n)$.*

1. *For all $u \in S(\mathbb{R}^n)$, the sequence $(\mathrm{Op}(a_k)u)$ converges towards $\mathrm{Op}(a)u$ in $S(\mathbb{R}^n)$.*
2. *The sequence (a_k^*) converges towards a^* in $S^m(\mathbb{R}^n)$.*
3. *Let $\ell \in \mathbb{R}$, $b \in S^\ell(\mathbb{R}^n)$ and (b_ℓ) be a sequence of symbols belonging to $S^\ell(\mathbb{R}^n)$ and converging to b in $S^\ell(\mathbb{R}^n)$. Then $(a_k b_k)$ converges to ab in $S^{m+\ell}(\mathbb{R}^n)$.*

Proof We have already seen that if $a \in S^m(\mathbb{R}^n)$ and $u \in S(\mathbb{R}^n)$ then $\mathrm{Op}(a)u \in S(\mathbb{R}^n)$. The proof of this result directly leads to the result stated in point (1). Similarly, the previous proposition gives the continuity result stated in point (2). Finally, the result stated in point (3) is a direct consequence of Leibniz's rule. $\square$

Lemma 10.19 *Let $\chi \in C_0^\infty(\mathbb{R}^n)$ such that $\chi(0) = 1$. Introduce $r_\varepsilon(\xi) = \chi(\varepsilon\xi) - 1$. Then r_ε converges to 0 in $S^1(\mathbb{R}^n)$.*

Proof We will show that $\left|\partial_\xi^\alpha r_\varepsilon(\xi)\right| \leq C_\alpha \varepsilon \langle\xi\rangle^{1-|\alpha|}$ for every multi-index α in $\mathbb{N}^n$. For $\alpha = 0$, we write

$$r_\varepsilon(\xi) = \varepsilon \int_0^1 \chi'(t\varepsilon\xi) \cdot \xi \, dt$$

and deduce $\langle\xi\rangle^{-1} r_\varepsilon(\xi) = O(\varepsilon)$ because χ' is bounded. For $|\alpha| > 0$, we directly verify that

$$\left|\langle\xi\rangle^{|\alpha|-1}\partial_\xi^\alpha r_\varepsilon(\xi)\right| = \varepsilon\left|\varepsilon^{|\alpha|-1}\langle\xi\rangle^{|\alpha|-1}\partial_\xi^\alpha\chi(\varepsilon\xi)\right|$$

then we use the bound

$$\left|\varepsilon^{|\alpha|-1}\langle\xi\rangle^{|\alpha|-1}\partial_\xi^\alpha\chi(\varepsilon\xi)\right| \leq \left|\langle\varepsilon\xi\rangle^{|\alpha|-1}\partial_\xi^\alpha\chi(\varepsilon\xi)\right| \leq \sup_{\mathbb{R}^n}\left|\langle\zeta\rangle^{|\alpha|-1}\partial_\xi^\alpha\chi(\zeta)\right|$$

to obtain the desired result. $\square$

We are now ready to prove the proposition. Consider a symbol $a = a(x,\xi)$ in $S^m(\mathbb{R}^n)$. We fix $\chi \in C_0^\infty(\mathbb{R}^n)$ satisfying $\chi(0) = 1$ and we introduce, for all $k \in \mathbb{N}^*$,

$$a_k(x,\xi) = \chi(\xi/k)\, a(x,\xi).$$

As a_k has compact support in ξ and also in x (by the additional hypothesis on a), we can apply Proposition 10.2 to write that

$$(\mathrm{Op}(a_k)u, v) = (u, \mathrm{Op}(a_k^*)v). \tag{10.3}$$

To prove the theorem, we still need to show that we can take the limit in this equality. For this, we start by combining Lemma 10.19 with point (3) of Lemma 10.18 to obtain that (a_k) converges to a in $S^{m+1}(\mathbb{R}^n)$. Point (2) of Lemma 10.18 then implies that (a_k^*) converges to a^* in $S^{m+1}(\mathbb{R}^n)$. We can then apply point (1) of this lemma to obtain that $\mathrm{Op}(a_k)u$ converges to $\mathrm{Op}(a)u$ in $S(\mathbb{R}^n)$ and similarly, we obtain that $\mathrm{Op}(a_k^*)u$ converges to $\mathrm{Op}(a^*)u$ in $S(\mathbb{R}^n)$. We can then take the limit in identity (10.3), which concludes the proof. $\square$

We can now define the action of $\mathrm{Op}(a)$ on a tempered distribution. For this, let us recall the principle that we saw in the chapter on the Fourier transform. Let $a \in S^m(\mathbb{R}^n)$ with $m \in \mathbb{R}$. Then $\mathrm{Op}(a) \colon S(\mathbb{R}^n) \to S(\mathbb{R}^n)$ is a continuous linear map. We then define an operator A from $S'(\mathbb{R}^n)$ to $S'(\mathbb{R}^n)$ by

$$\forall (u, v) \in S(\mathbb{R}^n)^2, \qquad \langle Au, v \rangle_{S' \times S} = \langle u, \overline{\mathrm{Op}(a^*)\overline{v}} \rangle.$$

Then Proposition 5.20 shows that the operator A thus defined extends the definition of $\mathrm{Op}(a)$. We therefore still denote it $\mathrm{Op}(a)$.

We will conclude this section by considering the composition of pseudo-differential operators.

Let $A_1 = \mathrm{Op}(a_1)$ and $A_2 = \mathrm{Op}(a_2)$ be two pseudo-differential operators. Suppose that a_1 and a_2 belong to $C_0^\infty(\mathbb{R}^{2n})$ and consider $u \in S(\mathbb{R}^n)$. Then

$$A_1 A_2 u(x) = (2\pi)^{-n} \int e^{ix \cdot \xi} a_1(x, \xi) \widehat{A_2 u}(\xi) \, d\xi \quad \text{where}$$

$$\widehat{A_2 u}(\xi) = \int e^{-iy \cdot \xi} A_2 u(y) \, dy$$

$$= (2\pi)^{-n} \iint e^{-iy \cdot (\xi - \eta)} a_2(y, \eta) \widehat{u}(\eta) \, d\eta \, dy,$$

so

$$A_1 A_2 u(x) = (2\pi)^{-2n} \iiint e^{iy \cdot \eta + i\xi \cdot (x-y)} a_1(x, \xi) a_2(y, \eta) \widehat{u}(\eta) \, d\xi \, dy \, d\eta.$$

Thus, we obtain that $A_1 A_2 u(x)$ is equal to

$$(2\pi)^{-n} \int e^{ix \cdot \eta} \left((2\pi)^{-n} \iint e^{i(x-y) \cdot (\xi - \eta)} a_1(x, \xi) a_2(y, \eta) \, d\xi \, dy \right) \widehat{u}(\eta) \, d\eta.$$

Formally, $A_1 A_2 = \mathrm{Op}(b)$ where

$$b(x, \eta) = (2\pi)^{-n} \iint e^{i(x-y) \cdot (\xi - \eta)} a_1(x, \xi) a_2(y, \eta) \, d\xi \, dy. \qquad (10.4)$$

The formula that defines b is still a convolution.

Proposition 10.20 *If $a_1 \in S^{m_1}(\mathbb{R}^n)$ and $a_2 \in S^{m_2}(\mathbb{R}^n)$, then $\mathrm{Op}(a_1) \circ \mathrm{Op}(a_2) = \mathrm{Op}(b)$, where $b = a_1 \# a_2 \in S^{m_1 + m_2}(\mathbb{R}^n)$ is given by the oscillatory integral*

$$b(x, \eta) = (2\pi)^{-n} \iint e^{i(x-y) \cdot (\xi - \eta)} a_1(x, \xi) a_2(y, \eta) \, d\xi \, dy.$$

We will not give the proof, which is analogous to that concerning the adjoint (we refer the reader to [166]).

We have seen in this section that, for all $m \in \mathbb{R}$ and all $a \in S^m(\mathbb{R}^n)$, we can define $\mathrm{Op}(a)$ on the space of tempered distributions. In particular, we can define

Op$(a)u$ for all u in a Sobolev space $H^s(\mathbb{R}^n)$ with any $s \in \mathbb{R}$. Thanks to the previous proposition on the composition of pseudo-differential operators, we will now see that Op(a) is an operator of order m, as was asserted in point (1) of Theorem 10.13.

Proposition 10.21 *Let $m \in \mathbb{R}$ and $a \in S^m(\mathbb{R}^n)$. The operator* Op(a) *maps* $H^s(\mathbb{R}^n)$ *to* $H^{s-m}(\mathbb{R}^n)$ *for all* $s \in \mathbb{R}$.

Proof For $\mu \in \mathbb{R}$, let $(1 - \Delta)^{\mu/2}$ denote the Fourier multiplier of symbol $\langle \xi \rangle^\mu = (1 + |\xi|^2)^{\mu/2}$. Then $(1 - \Delta)^{\mu/2}$ is an isomorphism from $H^\mu(\mathbb{R}^n)$ to $L^2(\mathbb{R}^n)$. It is therefore sufficient to show that the operator

$$A_{s,m} := (I - \Delta)^{(s-m)/2} \circ \text{Op}(a) \circ (I - \Delta)^{-s/2}$$

is bounded from $L^2(\mathbb{R}^n)$ to $L^2(\mathbb{R}^n)$. Note that if $b = b(\xi)$, then

$$\text{Op}(a) \circ \text{Op}(b) = \text{Op}(ab)$$

so Op$(a) \circ (1 - \Delta)^{-s/2}$ is the operator of symbol $a(x, \xi)\langle \xi \rangle^{-s}$. Since $a \in S^m(\mathbb{R}^n)$ and $\langle \xi \rangle^{-s} \in S^{-s}(\mathbb{R}^n)$, the product of these two symbols belongs to $S^{m-s}(\mathbb{R}^n)$. On the other hand, to manipulate $(I - \Delta)^{(s-m)/2} \circ \text{Op}(a\langle \xi \rangle^{-s/2})$, we use the composition theorem, which implies that $A_{s,m}$ is a pseudo-differential operator whose symbol belongs to $S^0(\mathbb{R}^n)$. Therefore, it is a bounded operator on $L^2(\mathbb{R}^n)$ according to the continuity theorem proved in the previous chapter (*cf.* Theorem 9.8).   □

We conclude by proving the part concerning the symbolic calculation of pseudo-differential operators. For this, let us introduce the concept of an asymptotic sum of symbols. This concept allows us to give a rigorous sense to statements such as: a is the sum of a term (usually its so-called principal symbol) and a "better" remainder.

Definition 10.22 Let $a_j \in S^{m_j}(\mathbb{R}^n)$ be a sequence of symbols indexed by $j \in \mathbb{N}$, such that m_j decreases towards $-\infty$. We will say that $a \in S^{m_0}(\mathbb{R}^n)$ is the asymptotic sum of the a_j if

$$\forall k \in \mathbb{N}, \quad a - \sum_{j=0}^{k} a_j \in S^{m_{k+1}}(\mathbb{R}^n).$$

We then write $a \sim \sum a_j$.

Proposition 10.23

1. *Let $m \in \mathbb{R}$ and $a \in S^m(\mathbb{R}^n)$. Then*

$$a^* \sim \sum_j A_j \quad \text{with } A_j = \sum_{|\alpha|=j} \frac{1}{i^{|\alpha|}\alpha!} \partial_\xi^\alpha \partial_x^\alpha \overline{a}.$$

2. *Let $m_1, m_2 \in \mathbb{R}$. If $a_1 \in S^{m_1}(\mathbb{R}^n)$ and $a_2 \in S^{m_2}(\mathbb{R}^n)$, then*

$$a_1 \# a_2 \sim \sum_j A_j \quad \text{with } A_j = \sum_{|\alpha|=j} \frac{1}{i^{|\alpha|}\alpha!} (\partial_\xi^\alpha a_1)(\partial_x^\alpha a_2).$$

Remark 10.24 In practice, by abuse of notation, we simply write

$$a^* \sim \sum_{\alpha} \frac{1}{i^{|\alpha|}\alpha!} \, \partial_\xi^\alpha \partial_x^\alpha \overline{a},$$

and

$$a_1 \# a_2 \sim \sum_{\alpha} \frac{1}{i^{|\alpha|}\alpha!} \left(\partial_\xi^\alpha a_1\right)\left(\partial_x^\alpha a_2\right).$$

Proof We will limit ourselves to proving point (1). We use the Taylor formula (whose statement is recalled after this proof)

$$\overline{a}(x-y, \xi-\eta) = \sum_{|\alpha+\beta|<2k} \frac{(-y)^\alpha}{\alpha!} \frac{(-\eta)^\beta}{\beta!} \, \partial_x^\alpha \partial_\xi^\beta \overline{a}(x,\xi) + r_k(x,\xi,y,\eta)$$

with

$$r_k(x,\xi,y,\eta) = \sum_{|\alpha+\beta|=2k} 2k \frac{(-y)^\alpha}{\alpha!} \frac{(-\eta)^\beta}{\beta!} \, r_{\alpha\beta}(x,\xi,y,\eta)$$

and

$$r_{\alpha\beta}(x,\xi,y,\eta) = \int_0^1 (1-t)^{2k-1} \partial_x^\alpha \partial_\xi^\beta \overline{a}(x-ty, \xi-t\eta) \, dt.$$

We will use the following calculation (which is detailed in the solved Exercise 10.1): for all α and β in $\mathbb{N}^n$,

$$(2\pi)^{-n} \int e^{-iy\cdot x} \frac{y^\alpha}{\alpha!} \frac{x^\beta}{\beta!} \, dy \, dx = \begin{cases} 0 \text{ if } \alpha \neq \beta, \\ (-i)^{|\alpha|}/\alpha! \text{ if } \alpha = \beta. \end{cases}$$

This result implies that the sum over $|\alpha+\beta| < 2k$ corresponds to the asymptotic expansion for a^*. All that remains is to prove that

$$\int e^{-iy\cdot\eta} r_k(x,\xi,y,\eta) \, dy \, d\eta \in S^{m-k}.$$

To do so, observe that, simply denoting different harmless numerical constants by $*$, we obtain

$$\int e^{-iy\cdot\eta} y^\alpha \eta^\beta r_{\alpha\beta}(x,\xi,y,\eta) \, dy \, d\eta = * \int \partial_\eta^\alpha \left(e^{-iy\cdot\eta}\right) \eta^\beta r_{\alpha\beta}(x,\xi,y,\eta) \, dy \, d\eta.$$

We will then integrate by parts to treat $r_{\alpha\beta}$. Namely, we will use the fact that, if $a \in A^m$ and $b \in A^\ell$, for $\alpha \in \mathbb{N}^N$ we have,

$$\int e^{i\phi(x)} a(x) \partial^\alpha b(x) \, dx = \int b(x)(-\partial)^\alpha \left(e^{i\phi(x)} a(x)\right) dx.$$

This implies that

$$\int e^{-iy\cdot\eta} y^\alpha \eta^\beta r_{\alpha\beta}(x,\xi,y,\eta)\, dy\, d\eta$$

$$= *\int e^{-iy\cdot\eta} \sum_\gamma (\partial_\eta^\gamma \eta^\beta) \partial_\eta^{\alpha-\gamma} r_{\alpha\beta}(x,\xi,y,\eta)\, dy\, d\eta$$

$$= \sum_\gamma *\int e^{-iy\cdot\eta} \eta^{\beta-\gamma} \partial_\eta^{\alpha-\gamma} r_{\alpha\beta}(x,\xi,y,\eta)\, dy\, d\eta$$

$$= \sum_\gamma *\int e^{-iy\cdot\eta} \partial_y^{\beta-\gamma} \partial_\eta^{\alpha-\gamma} r_{\alpha\beta}(x,\xi,y,\eta)\, dy\, d\eta.$$

By definition of $r_{\alpha\beta}$, we have

$$\partial_y^{\beta-\gamma} \partial_\eta^{\alpha-\gamma} r_{\alpha\beta}(x,\xi,y,\eta)$$

$$= *\int_0^1 (1-t)^{2k-1} t^{2k-2|\gamma|} \partial_x^{\alpha+\beta-\gamma} \partial_\xi^{\alpha+\beta-\gamma} \overline{a}(x-ty,\xi-t\eta)\, dt.$$

Since $\gamma \leqslant \alpha$ and $\gamma \leqslant \beta$ we have $|\gamma| \leqslant k$ and $|\alpha + \beta - \gamma| \geqslant k$, therefore $\partial_x^{\alpha+\beta-\gamma} \partial_\xi^{\alpha+\beta-\gamma} \overline{a} \in S^{m-k}$. Then

$$\int e^{-iy\cdot\eta} r_k(x,\xi,y,\eta)\, dy\, d\eta = \int e^{-iy\cdot\eta} s_k(x,\xi,y,\eta)\, dy\, d\eta,$$

where s_k is an amplitude $s_k \in A^{|m-k|}$ with

$$\|s_k\|_{|m-k|,|m-k|+2n+1} \leqslant C_k \langle\xi\rangle^{m-k}.$$

We deduce that

$$\langle\xi\rangle^{k-m} \int e^{-iy\cdot\eta} r_k(x,\xi,y,\eta)\, dy\, d\eta$$

is bounded and then that $\int e^{-iy\cdot\eta} r_k(x,\xi,y,\eta)\, dy\, d\eta$ belongs to S^{m-k}. $\square$

To be complete, we prove the version of the Taylor formula that was used above.

Theorem 10.25 *Let u be a function of class C^k on $\mathbb{R}^n$. Then, we have*

$$u(x+y) = \sum_{|\alpha|<k} \frac{1}{\alpha!} y^\alpha \partial_x^\alpha u(x) + \sum_{|\alpha|=k} \frac{k}{\alpha!} y^\alpha \int_0^1 (1-t)^{k-1} (\partial_x^\alpha u)(x+ty)\, dt,$$

for all x and y in $\mathbb{R}^n$.

Proof We verify that

$$\frac{d}{dt}\left(\sum_{|\alpha|=k-1}\frac{1}{\alpha!}y^\alpha\partial_x^\alpha u(x+ty)\right) = \sum_{|\beta|=k}\left(\sum_{\alpha\leqslant\beta,|\alpha|=k-1}\frac{1}{\alpha!}\right)y^\beta(\partial_x^\beta u)(x+ty)$$

$$= \sum_{|\beta|=k}\left(\sum_{1\leqslant j\leqslant n}\frac{\beta_j}{\beta!}\right)y^\beta(\partial_x^\beta u)(x+ty)$$

$$= \sum_{|\beta|=k}\frac{k}{\beta!}y^\beta(\partial_x^\beta u)(x+ty).$$

So the function

$$v(t) = \sum_{|\alpha|<k}\frac{1}{\alpha!}(1-t)^{|\alpha|}y^\alpha(\partial_x^\alpha u)(x+ty)$$

satisfies $v(1) = u(x+y)$ and

$$v(0) = \sum_{|\alpha|<k}\frac{1}{\alpha!}y^\alpha\partial^\alpha u(x), \quad \partial_t v_k = \sum_{|\alpha|=k}\frac{k}{\alpha!}y^\alpha(1-t)^{k-1}(\partial_x^\alpha u)(x+ty),$$

so the Taylor formula is a consequence of the fundamental theorem of integral calculus. □

10.4 Applications of Symbolic Calculus

10.4.1 Action on Sobolev Spaces

We will give a tricky proof, due to Hörmander, of the following result that we have already seen in the previous chapter.

Proposition 10.26 If $a \in S^0(\mathbb{R}^n)$, then $\mathrm{Op}(a)$ is bounded on $L^2(\mathbb{R}^n)$.

Proof Let $A = \mathrm{Op}(a)$. The idea is as follows. As

$$\|Au\|_{L^2}^2 = (Au, Au) = (A^*Au, u),$$

to show the inequality $\|Au\|_{L^2}^2 \leqslant M\|u\|_{L^2}^2$ for some $M > 0$, it is enough to show that $(Bu, u) \geqslant 0$ where $B = M - A^*A$. Note that B is a self-adjoint operator.

To prove that B is positive for M large enough, we will show that we can write, approximately, B in the form of a square. More precisely, we will show that we can write B in the form

$$B = C^*C + R,$$

where $C = \mathrm{Op}(c)$ with $c \in S^0(\mathbb{R}^n)$ and $R = \mathrm{Op}(r)$, $r \in S^{-1}$. To do this, choose $M \geqslant 2\sup|a(x,\xi)|^2$, and take $c(x,\xi) = (M - |a(x,\xi)|^2)^{1/2}$.

It is quite easy to verify that c belongs to $S^0(\mathbb{R}^n)$. The theorem of composition of operators implies that $C^*C = M - A^*A + R$ where $R = \text{Op}(r)$ with $r \in S^{-1}$. Thus

$$\|Au\|_{L^2}^2 \leqslant M \|u\|_{L^2}^2 + (Ru, u).$$

Now we need to bound the error (Ru, u). As $\|Ru\|_{L^2}^2 = (Ru, Ru) = (R^*Ru, u)$, R will be continuous on L^2 if R^*R is, with $\|R\|_{\mathcal{L}(L^2)} \leqslant \|R^*R\|_{\mathcal{L}(L^2)}^{1/2}$.

Now notice that $r^*\#r \in S^{-2}$. Then, by iterating the argument, we see that it suffices to show that, for k large enough, any operator of symbol $r \in S^{-k}$ is continuous on L^2. We will prove this result using Schur's lemma (*cf.* Lemma 10.9) and the following remark: if $r \in S^{-n-1}$ then the kernel $K(x, y)$ of $\text{Op}(r)$ (as given by formula (10.1) with a replaced by r) is a continuous bounded function, because

$$|K(x, y)| \leqslant (2\pi)^{-n} \int |r(x, \xi)| \, d\xi \leqslant \frac{C_0}{(2\pi)^n} \int \frac{d\xi}{(1 + |\xi|)^{n+1}} \leqslant C.$$

Moreover, $(x_j - y_j)K(x, y)$ is the kernel of $\text{Op}(i\partial_{\xi_j} r) \in \text{Op}\, S^{-n-2} \subset \text{Op}\, S^{-n-1}$ so by iterating $(n+1)$ times, we finally find $(1 + |x - y|^{n+1})K(x, y) \leqslant C$. The decay of K at infinity implies in particular:

$$\int |K(x, y)| \, dx \leqslant A, \quad \int |K(x, y)| \, dy \leqslant A.$$

We conclude the proof with Schur's lemma.   □

10.4.2 Applications to Sub-elliptic Problems

Proposition 10.27 *Let $\mu \in \mathbb{R}$ and $e \in S^\mu$. Assume that there exists $c > 0$ such that $|e(x, \xi)| \geqslant c(1 + |\xi|)^\mu$ for all $(x, \xi) \in \mathbb{R}^{2n}$. Then e^{-1} belongs to $S^{-\mu}$. Moreover, for all $s \in \mathbb{R}$, there exist constants $K_0, K_1 > 0$ such that, for all $u \in H^s$,*

$$\|u\|_{H^s} \leqslant K_0 \|\text{Op}(e)u\|_{H^{s-\mu}} + K_1 \|u\|_{H^{s-1}}.$$

Proof The fact that $e^{-1} \in S^{-\mu}$ is directly demonstrated. Then $e\#e^{-1} = 1 + b$ with $b \in S^{-1}$ and we deduce

$$\text{Op}(e^{-1})\,\text{Op}(e)u = \text{Op}(1)u + \text{Op}(b)u = u + \text{Op}(b)u,$$

which leads to

$$\|u\|_{H^s} \leqslant \left\|\text{Op}(e^{-1})\right\|_{\mathcal{L}(H^{s-\mu};H^s)} \|\text{Op}(e)u\|_{H^{s-\mu}} + \|\text{Op}(b)\|_{\mathcal{L}(H^{s-1},H^s)} \|u\|_{H^{s-1}}.$$

The wanted result then follows by point (1) of Theorem 10.13 which implies that $K_0 = \|\text{Op}(e^{-1})\|_{\mathcal{L}(H^{s-\mu};H^s)} < +\infty$ and $K_1 = \|\text{Op}(b)\|_{\mathcal{L}(H^{s-1},H^s)} < +\infty$.   □

Proposition 10.28 (Gårding inequality) *Let $m \in \mathbb{R}$ and $a \in S^m(\mathbb{R}^n)$ be a symbol such that*

$$\exists c > 0 / \ \forall(x, \xi) \in \mathbb{R}^{2n}, \qquad \operatorname{Re} a(x, \xi) \geq c(1 + |\xi|)^m.$$

Then there exist constants $C_0, C_1 > 0$ such that, for all $u \in S(\mathbb{R}^n)$,

$$\operatorname{Re}(\operatorname{Op}(a)u, u) \geq C_0 \|u\|_{H^{m/2}}^2 - C_1 \|u\|_{H^{(m-1)/2}}^2.$$

Remark 10.29 The previous proposition remains true if a is a symbol with matrix values (in this case $\operatorname{Re} a = a + a^*$). The proof is reduced to showing that, for any symbol $a \in S^0$ such that $a(x, \xi)$ is uniformly hermitian positive definite (for $(x, \xi) \in \mathbb{R}^d \times \mathbb{R}^d$), there exists $b \in S^0$ such that $b(x, \xi)^* b(x, \xi) = a(x, \xi)$.

Proof To prove this inequality, which is a relation between the positivity of a symbol and that of the associated operator, we will use symbolic calculus to write $A = \operatorname{Op}(a)$ in the form of a square (that is, P^*P) plus an operator of order $m - 1$.

Let us set

$$B := \operatorname{Re} A = \frac{1}{2}(A + A^*),$$

so that $\operatorname{Re}(Au, u) = (Bu, u)$. As $A^* \in \operatorname{Op}(\bar{a}) + \operatorname{Op} S^{m-1}$, we have $B = \operatorname{Op}(b)$ with $b = \frac{1}{2}(a + a^*) = \operatorname{Re} a + d$, where $d \in S^{m-1}$.

We then denote the positive square root of $\operatorname{Re} a$ by e, which is a symbol belonging to $S^{m/2}$. Moreover, the composition of symbols is such that

$$f := e^* \# e - \operatorname{Re} a \in S^{m-1}.$$

We deduce that $b = e^* \# e + g$ where $g = d - f \in S^{m-1}$. We can then write

$$
\begin{aligned}
\operatorname{Re}(Au, u) &= (\operatorname{Op}(b)u, u) \\
&= (\operatorname{Op}(e)^* \operatorname{Op}(e)u, u) + (\operatorname{Op}(g)u, u) \\
&= \|\operatorname{Op}(e)u\|_{L^2}^2 + (\operatorname{Op}(g)u, u) \\
&\geq \|\operatorname{Op}(e)u\|_{L^2}^2 - \|\operatorname{Op}(g)u\|_{H^{(1-m)/2}} \|u\|_{H^{(m-1)/2}}.
\end{aligned}
$$

The previous proposition implies that

$$\|u\|_{H^{m/2}} \leq K_0 \|\operatorname{Op}(e)u\|_{L^2} + K_1 \|u\|_{H^{(m/2)-1}},$$

and the continuity theorem for pseudo-differential operators implies that

$$\|\operatorname{Op}(g)u\|_{H^{(1-m)/2}} \leq K_2 \|u\|_{H^{(m-1)/2}}.$$

By combining the previous inequalities, we obtain the desired result. □

Exercise Show the following improvement. Let $m \in \mathbb{R}_+$ and $a \in S^m(\mathbb{R}^n)$. Suppose there exist two constants c, R such that,

$$|\xi| \geq R \implies \operatorname{Re} a(x, \xi) \geq c|\xi|^m.$$

Then, for all N, there exists a constant C_N such that,

$$\mathrm{Re}(Au,u) \geqslant \frac{c}{2} \|u\|^2_{H^{m/2}} - C_N \|u\|^2_{H^{-N}},$$

for all $u \in \mathcal{S}(\mathbb{R}^n)$. [Use the following inequalities: (i) $2xy \leqslant \eta x^2 + (1/\eta)y^2$ and (ii) for all $\varepsilon > 0$ and all $N > 0$, there exists a $C_{\varepsilon,N} > 0$ such that

$$\|u\|_{H^{-1}} \leqslant \varepsilon\|u\|_{L^2} + C_{\varepsilon,N} \|u\|_{H^{-N}},$$

which results from the easy inequality $\langle \xi \rangle^{-2} \leqslant \varepsilon^2 + C_{\varepsilon,N} \langle \xi \rangle^{-2N}$.]

Let us recall the notation of the Poisson bracket,

$$\{a,b\} = \sum_{1 \leqslant j \leqslant n} \frac{\partial a}{\partial \xi_j} \frac{\partial b}{\partial x_j} - \frac{\partial b}{\partial \xi_j} \frac{\partial a}{\partial x_j}.$$

Theorem 10.30 *Let $P = \mathrm{Op}(p)$ be a pseudo-differential operator such that*

$$p = p_1 + p_0 \quad with \quad p_1 \in S^1(\mathbb{R}^n) \quad and \quad p_0 \in S^0(\mathbb{R}^n).$$

Suppose there exists a constant c such that,

$$i\{p_1, \overline{p_1}\} \geqslant c(1 + |\xi|).$$

Then there exists a constant C such that

$$\|u\|_{H^{1/2}} \leqslant C \|Pu\|_{L^2} + C\|u\|_{L^2}.$$

Remark 10.31 The preceding condition on $i\{p_1, \overline{p_1}\}$ is called Hörmander's hypoellipticity condition. It is easily verified that for all $p \in C^1(\mathbb{R}^n)$ with complex values, the Poisson bracket $i\{p, \overline{p}\}$ is a function with real values.

Proof Let us introduce the operator $Q = P^*P - PP^*$. Then

$$\|Pu\|^2_{L^2} = (P^*Pu, u) = (PP^*u, u) + ((P^*P - PP^*)u, u)$$
$$= \|P^*u\|^2_{L^2} + (Qu, u) \geqslant (Qu, u).$$

Thus, any positivity estimate of Q will give an estimate on $\|Pu\|^2_{L^2}$.

Let us first recall that if $A = \mathrm{Op}(a) \in \mathrm{Op}\, S^{m_1}$ and $B = \mathrm{Op}(b) \in \mathrm{Op}\, S^{m_2}$ are two pseudo-differential operators, then $A^* \in \mathrm{Op}\, S^{m_1}$ and $[A, B] \in \mathrm{Op}\, S^{m_1+m_2-1}$. Moreover,

$$A^* \in \mathrm{Op}(\overline{a}) + \mathrm{Op}\, S^{m_1-1}, \qquad [A, B] \in \mathrm{Op}\left(\frac{1}{i}\{a, b\}\right) + \mathrm{Op}\, S^{m_1+m_2-2}.$$

So $Q = \mathrm{Op}(q)$ with $q = q_1 + q_0$, where $q_1 \in S^1(\mathbb{R}^n)$, $q_0 \in S^0(\mathbb{R}^n)$ and

$$q_1 = \frac{1}{i}\{\overline{p_1}, p_1\}.$$

By hypothesis, we deduce that $\mathrm{Re}\, q_1 \geqslant c|\xi|$ if $|\xi| \geqslant R$. The Gårding inequality implies that

$$\mathrm{Re}(Qu, u) \geqslant \frac{1}{C} \|u\|_{H^{1/2}}^2 - C\|u\|_{L^2}^2 .$$

This concludes the proof. $\square$

10.5 Exercises

Exercise (solved) 10.1 (Calculations of oscillatory integrals)

1. Let $a \in A^m(\mathbb{R}^n \times \mathbb{R}^n)$. Show that, for all j such that $1 \leqslant j \leqslant n$,

$$\int e^{-iy \cdot x} x_j a(x, y)\, dx\, dy = -i \int e^{-iy \cdot x} \partial_{y_j} a(x, y)\, dx\, dy.$$

2. Let $a \in A^m(\mathbb{R}^n)$. Show that:

$$\frac{1}{(2\pi)^n} \int e^{-iy \cdot x} a(y)\, dy\, dx = \frac{1}{(2\pi)^n} \int e^{-iy \cdot x} a(x)\, dy\, dx = a(0).$$

3. Let $\alpha, \beta \in \mathbb{N}^n$. Show that:

$$\frac{1}{(2\pi)^n} \int e^{-iy \cdot x} \frac{y^\alpha x^\beta}{\alpha! \beta!}\, dy\, dx = \begin{cases} 0 & \text{if } \alpha \neq \beta, \\ (-i)^{|\alpha|}/\alpha! & \text{if } \alpha = \beta. \end{cases}$$

Exercise (solved) 10.2 (Asymptotic sum of symbols)

1. Let $(m_j)_{j\in\mathbb{N}}$ be a strictly decreasing sequence of integers such that $\lim_{j\to+\infty} m_j = -\infty$. Consider a sequence of symbols $(a_j)_{j\in\mathbb{N}}$ satisfying $a_j \in S^{m_j}$ as well as a function $\chi \in C_0^\infty(\mathbb{R}^n)$ such that $\chi = 1$ in the neighborhood of 0.
2. Show that there exists a sequence $\varepsilon_j \to 0$ such that, for all α, β, if we set $\tilde{a}_j(x, \xi) = (1 - \chi(\varepsilon_j \xi)) a_j(x, \xi)$, we have, for all sufficiently large j:

$$\forall (x, \xi) \in \mathbb{R}^n \times \mathbb{R}^n, \quad |\partial_x^\alpha \partial_\xi^\beta \tilde{a}_j(x, \xi)| \leqslant \frac{1}{2^j}(1 + |\xi|)^{1+m_j-|\beta|}.$$

3. We set $a = \sum_{j\in\mathbb{N}} \tilde{a}_j$. Show that this indeed defines a C^∞ function.
4. Show that, for all $k \in \mathbb{N}^*$, $a - \sum_{j<k} a_j \in S^{m_k}$.

Exercise (solved) 10.3 Let $a \in S^m(\mathbb{R}^n)$ with $m \in \mathbb{R}$.

1. Recall the expression of the kernel of $\mathrm{Op}(a)$.
2. Show that if $m = -\infty$, then the kernel is C^∞.
3. Let $x, y \in \mathbb{R}^n$ such that $x \neq y$. Let $\phi, \psi \in C_0^\infty(\mathbb{R}^n)$ such that:

    a. ϕ equals 1 in the neighborhood of x;
    b. ψ equals 1 in the neighborhood of y;
    c. $\mathrm{supp}(\phi) \cap \mathrm{supp}(\psi) = \emptyset$.

Show that $M_\phi \operatorname{Op}(a)M_\psi$ belongs to $\operatorname{Op}(S^{-\infty})$ where M_ϕ and M_ψ respectively denote the multiplications by ϕ and ψ.

4. Calculate the kernel of $M_\phi \operatorname{Op}(a)M_\psi$ in terms of the kernel of $\operatorname{Op}(a)$.

5. Show that the kernel of $\operatorname{Op}(a)$ is of class C^∞ in the neighborhood of (x, y).

Exercise (solved) 10.4 If $\Omega \subset \mathbb{R}^n$ is an open set, we call $S^m_{\mathrm{loc}}(\Omega)$ the set of functions $a : \Omega \times \mathbb{R}^n \to \mathbb{C}$ such that, for all $\phi \in C_0^\infty(\Omega)$, ϕa belongs to $S^m(\mathbb{R}^n)$.

1. For all $a \in S^m_{\mathrm{loc}}(\Omega)$, we define, for all "suitable" functions u:

$$\forall x \in \Omega, \quad \operatorname{Op}_\Omega(a)u(x) = \frac{1}{(2\pi)^n} \int_{\mathbb{R}^n} e^{ix\cdot\xi} a(x,\xi)\widehat{u}(\xi)\, d\xi.$$

(a) Show that this formula defines an operator $C_0^\infty(\Omega) \to C^\infty(\Omega)$.

(b) For all $\phi \in C_0^\infty(\Omega)$, we denote by $M_\phi : C^\infty(\Omega) \to C^\infty(\mathbb{R}^n)$ the operator that, to u, associates ϕu (extended by 0 outside of Ω).
Show that, for all $v \in C_0^\infty(\Omega)$, if $\phi, \widetilde{\phi} \in C_0^\infty(\Omega)$ equal 1 on the support of v, then:

$$(\operatorname{Op}(\phi a))^* v = (\operatorname{Op}(\widetilde{\phi} a))^* v.$$

(c) Deduce that we can extend $\operatorname{Op}_\Omega(a)$ to an operator from $\mathcal{E}'(\Omega)$ to $\mathcal{D}'(\Omega)$.

2. We suppose that $A : C_0^\infty(\Omega) \to C^\infty(\Omega)$ is a linear operator satisfying the following property: for all $\phi, \psi \in C_0^\infty(\Omega)$, $M_\phi A M_\psi$ belongs to $\operatorname{Op}(S^m)$.
We will show that there exists $a \in S^m_{\mathrm{loc}}(\Omega)$ such that $A = \operatorname{Op}_\Omega(a) + R$, for a certain operator R of the form:

$$Ru: x \in \Omega \mapsto \int_\Omega K(x, y)u(y)\, dy$$

with $K \in C^\infty(\Omega \times \Omega)$.
We recall (see Proposition 17.35) the existence of a *partition of unity* of Ω, that is, a sequence $(\psi_j)_{j\in\mathbb{N}}$ of elements of $C_0^\infty(\Omega)$ satisfying the following two properties:

- for all compact $K \subset \Omega$, $\{j : \operatorname{supp}(\psi_j) \cap K \neq \varnothing\}$ is finite;
- for all $x \in \Omega$, $\sum_j \psi_j(x) = 1$.

(a) Show that, for all $u \in C_0^\infty(\Omega)$, it makes sense to write:

$$Au = \sum_{j,k\in\mathbb{N}} A_{jk}u$$

where A_{jk} denotes $M_{\psi_j} A M_{\psi_k}$.

(b) Let $I = \{(j, k) : \operatorname{supp}(\psi_j) \cap \operatorname{supp}(\psi_k) \neq \varnothing\}$. Show that $\sum_{(j,k)\in I} A_{jk}$ is of the form $\operatorname{Op}_\Omega(a)$ with $a \in S^m_{\mathrm{loc}}(\Omega)$.

(c) Show that, for all $(j, k) \notin I$, A_{jk} is an operator with a C^∞ kernel. Show that the support of the kernel is included in $\operatorname{supp}(\psi_j) \times \operatorname{supp}(\psi_k)$.
(Hint: use Exercise 10.3.)

(d) Prove the desired result.

Exercise (solved) 10.5 (Parametrix of an elliptic problem) Let $P(\xi)$ be a polynomial in ξ, of degree m, with n variables, such that there exist $R, C > 0$ such that

$$|\xi| \geq R \implies |P(\xi)| \geq C|\xi|^m.$$

If $P(\xi)$ satisfies these properties, we say that the Fourier multiplier $P(D)$ is elliptic.

1. Let $\chi \in C_0^\infty(\mathbb{R}^n)$ such that $\chi = 1$ on the ball $B(0, R) = \{|\xi| \leq R\}$. Show that, for all $u \in C_0^\infty(\mathbb{R}^n)$, the following integral makes sense (in the sense of oscillatory integrals)

$$\frac{1}{(2\pi)^n} \iint e^{ix\cdot\xi} \frac{1 - \chi(\xi)}{P(\xi)} u(x) \, dx \, d\xi.$$

2. Let U be a bounded open set not containing 0. We consider the distribution T on $C_0^\infty(U)$ defined by:

$$T(u) = \frac{1}{(2\pi)^n} \iint e^{ix\cdot\xi} \frac{1 - \chi(\xi)}{P(\xi)} u(x) \, dx \, d\xi.$$

   Show that T identifies with a function of $L^2(U)$.
3. Similarly, show that, for all $\alpha \in \mathbb{N}^n$, $\partial^\alpha T$ identifies with a function of $L^2(U)$. Conclude that, on $C_0^\infty(\mathbb{R}^n \setminus \{0\})$, T identifies with a function of $C^\infty(\mathbb{R}^n \setminus \{0\})$.
4. Show that $P(D)T = \delta_0 + r$ with $r \in C^\infty(\mathbb{R}^n)$. The distribution T is called a *parametrix* of $P(D)$.
5. Show that for all $\varepsilon > 0$, we can find a parametrix of $P(D)$ with support in $B(0, \varepsilon)$.

Exercise 10.6 (Sums of squares of vector fields) The aim of this exercise (which is very difficult) is to prove a famous result of Lars Hörmander [74] on the hypoellipticity of certain sums of squares of vector fields. We will follow the proof by Kohn [101] and refer to the books of Helffer and Nier [72] and Trèves [191] for the solution.

Notation

We only consider real-valued functions defined on an open set of $\mathbb{R}^n$ with $n \geq 1$ an arbitrary integer. Given $s \in \mathbb{R}$, we denote by $H^s(\mathbb{R}^n)$ the Sobolev space of order s and by $\langle D_x \rangle^s$ the Fourier multiplier of symbol $\langle \xi \rangle^s = (1 + |\xi|^2)^{s/2}$. We denote by $\langle u, v \rangle$ the scalar product on $L^2(\mathbb{R}^n)$,

$$\langle u, v \rangle = \int_{\mathbb{R}^n} u(x)v(x) \, dx.$$

We will use:

- the Cauchy–Schwarz inequality;
- the fact that $\|u\|_{H^s} = (2\pi)^{-n/2} \|\langle D_x \rangle^s u\|_{L^2}$;
- the duality: $\langle Au, v \rangle = \langle u, A^* v \rangle$ and $|\langle u, v \rangle| \leq \|u\|_{H^s} \|v\|_{H^{-s}}$ for $s \in \mathbb{R}$.

Given two operators A and B, we denote by AB the composition $A \circ B$ (so $A^2 = A \circ A$) and by $[A, B] = AB - BA$ their commutator.

In this problem we are interested in the operator of order 2

$$L = \sum_{1 \leqslant j \leqslant m} X_j^2,$$

where $X_1, \dots, X_m$ are differential operators of order 1: for $1 \leqslant j \leqslant m$, X_j is defined by

$$(X_j u)(x) = \sum_{1 \leqslant i \leqslant n} a_{i,j}(x) \frac{\partial u}{\partial x_i}(x),$$

where $a_{i,j}$ is of class C^∞ on $\mathbb{R}^n$ with values in $\mathbb{R}$ for all i such that $1 \leqslant i \leqslant n$. Note that we only assume that the functions $a_{i,j}$ are C^∞ (and not C^∞ and bounded together with their derivatives). For example, we want to study the case $L = \partial_x^2 + x^2 \partial_y^2 = X_1^2 + X_2^2$ with $X_1 = \partial_x$ and $X_2 = x \partial_y$.

Preliminary Questions

1. Show that the adjoint X_j^* of X_j satisfies $X_j^* u = -X_j u + c_j u$ where $c_j \in C^\infty(\mathbb{R}^n)$ is a function that we will determine. That is to say, show that for all $u, v \in C_0^\infty(\mathbb{R}^n)$,

$$\int_{\mathbb{R}^n} (X_j u)(x) v(x) \, dx = \int_{\mathbb{R}^n} \left(-u(x)(X_j v)(x) + c_j(x) u(x) v(x) \right) dx.$$

2. Show that there exists a constant $C > 0$ such that, for all $u \in C_0^\infty(\mathbb{R}^n)$,

$$\sum_{1 \leqslant j \leqslant m} \|X_j u\|_{L^2}^2 \leqslant C \|Lu\|_{L^2}^2 + C \|u\|_{L^2}^2.$$

Study of a class of operators

Let us fix a bounded open set V. We denote by Ψ_V^0 the set of operators $P \in \mathcal{L}(L^2(\mathbb{R}^n))$ that can be written in the form

$$Pu = \varphi_1 \operatorname{Op}(a)(\varphi_2 u)$$

where

- $\varphi_1, \varphi_2 \in C_0^\infty(\mathbb{R}^n)$ and $\operatorname{supp} \varphi_k \subset V$ for $k = 1, 2$;
- a is a symbol with complex values belonging to S^0.

Let $\varepsilon \in (0, 1/2]$. We denote by $\mathcal{A}_\varepsilon$ the set of operators $P \in \Psi_V^0$ such that

$$\exists C > 0 / \ \forall u \in C_0^\infty(V), \quad \|Pu\|_{H^\varepsilon}^2 \leqslant C \|Lu\|_{L^2}^2 + C \|u\|_{L^2}^2.$$

1. Show that if P_1 and P_2 belong to Ψ^0_V, then $P_1 P_2 \in \Psi^0_V$ and $P^*_1 \in \Psi^0_V$.
2. Show that if $P \in \mathcal{A}_\varepsilon$ then $P^* \in \mathcal{A}_\varepsilon$.
3. Show that for every $\varepsilon \in (0, 1/2]$, $\mathcal{A}_\varepsilon$ is stable under left or right composition with a pseudo-differential operator from Ψ^0_V: if $P \in \mathcal{A}_\varepsilon$ and $Q \in \Psi^0_V$, then

$$QP \in \mathcal{A}_\varepsilon, \quad PQ \in \mathcal{A}_\varepsilon.$$

4. Let θ_1 and θ_2 be two functions of class C^∞ with compact support such that $\operatorname{supp} \theta_k \subset V$ for $k = 1, 2$ and $\theta_1 \equiv 1$ on the support of θ_2. Let us introduce the operator S defined by

$$Su = \theta_1 \langle D_x \rangle^{-1} (\theta_2 u).$$

Show that $X_j S \in \Psi^0_V$. Also show that, for every j such that $1 \leqslant j \leqslant n$ and every $\varepsilon \in [0, 1/2]$, we have $X_j S \in \mathcal{A}_\varepsilon$.
5. Let $\varepsilon, \delta \in (0, 1/2]$ with $\delta \leqslant \varepsilon/2$. Consider $P \in \mathcal{A}_\varepsilon$. We want to show in this question that

$$[X_j, P] \in \mathcal{A}_\delta$$

for every j such that $1 \leqslant j \leqslant n$.
(a) Write $\left\| [X_j, P]u \right\|^2_{H^\delta}$ in the form $\langle [X_j, P]u, Tu \rangle$, where $T = \operatorname{Op}(\tau)$ is a pseudo-differential operator with $\tau \in S^{2\delta}$.
(b) Show that there exists a constant $C > 0$ such that

$$\left| \langle PX_j u, Tu \rangle \right| \leqslant \left\| X_j u \right\|^2_{L^2} + \left\| TP^* u \right\|^2_{L^2} + C \|u\|^2_{H^{2\delta - 1}}.$$

(c) Obtain a similar estimate for $\left| \langle X_j Pu, Tu \rangle \right|$ and conclude.
6. We denote by $\mathcal{A}$ the set of operators $P \in \Psi^0_V$ such that $P \in \mathcal{A}_\varepsilon$ for a certain $\varepsilon \in (0, 1/2]$. That is, $P \in \mathcal{A}_\varepsilon$ if and only if

$$\exists \varepsilon \in (0, 1/2], \ \exists C > 0/ \ \forall u \in C^\infty_0(V), \quad \|Pu\|^2_{H^\varepsilon} \leqslant C \|Lu\|^2_{L^2} + C \|u\|^2_{L^2}.$$

Let i, j be such that $1 \leqslant i, j \leqslant m$. Show that the commutator

$$[X_i, X_j] = X_i X_j - X_j X_i$$

is a differential operator of order 1.
Show that for all j, k such that $1 \leqslant j, k \leqslant m$, we have

$$[X_j, X_k]S \in \mathcal{A}.$$

(Hint: observe that $[X_j, X_k S] \in \mathcal{A}$.)

The sub-Laplacian on the Heisenberg group

Consider the case of the space dimension $n = 3$. Consider the operator $L = X^2 + Y^2$ with $X = \partial_{x_2} + 2x_1 \partial_{x_3}$ and $Y = \partial_{x_1} - 2x_2 \partial_{x_3}$.

1. Let V be a bounded open set. Show that, for every k such that $1 \leqslant k \leqslant 3$,

$$\exists \varepsilon \in (0, 1/2], \ \exists C > 0/ \ \forall u \in C_0^\infty(V), \quad \left\| \partial_{x_k}(Su) \right\|_{H^\varepsilon}^2 \leqslant C \, \|Lu\|_{L^2}^2 + C\|u\|_{L^2}^2.$$

2. Deduce that, for any compact $K \subset \mathbb{R}^n$, there exist $\varepsilon > 0$ and a constant $C > 0$ such that, for any $u \in C_0^\infty(\mathbb{R}^n)$ with $\operatorname{supp} u \subset K$,

$$\|u\|_{H^\varepsilon}^2 \leqslant C \, \|Lu\|_{L^2}^2 + C\|u\|_{L^2}^2.$$

General case

We identify a differential operator $X = \sum_{1 \leqslant i \leqslant n} a_i \partial_{x_i}$ with the vector field $a = (a_1, \ldots, a_n): \mathbb{R}^n \to \mathbb{R}^n$. Given $x \in \mathbb{R}^n$, we denote by $X(x)$ the vector $a(x) \in \mathbb{R}^n$.

We again consider a general operator $L = \sum_{1 \leqslant j \leqslant m} X_j^2$ and we assume that there exists $r \in \mathbb{N}^*$ such that for any $x \in \mathbb{R}^n$,

$$\operatorname{vect}\left\{ [X_{i_1}, [X_{i_2}, \ldots, [X_{i_{p-1}}, X_{i_p}] \ldots]](x) : p \leqslant r, \ i_k \in \{1, \ldots, m\} \right\} = \mathbb{R}^n.$$

(a) Show that this condition is satisfied for the following two examples:

- $n = 2$, $X_1 = \partial_x$ and $X_2 = x\partial_y$;
- $n = 4$ (we denote by (x, y, z, t) the coordinates of a point in $\mathbb{R}^4$). We consider the two vector fields $X_1 = \partial_x$, $X_2 = \frac{1}{2}x^2\partial_t + x\partial_z + \partial_y$.

(b) Show that

$$[X_{i_1}, [X_{i_2}, \ldots [X_{i_{p-1}}, X_{i_p}] \ldots]]S \in \mathcal{A}$$

for any p-tuple of indices.

(c) Show that, for any compact $K \subset \mathbb{R}^n$, there exists $\varepsilon > 0$ and a constant $C > 0$ such that, for any $u \in C_0^\infty(\mathbb{R}^n)$ with $\operatorname{supp} u \subset K$,

$$\|u\|_{H^\varepsilon}^2 \leqslant C \, \|Lu\|_{L^2}^2 + C\|u\|_{L^2}^2.$$

Supplement

1. We now consider the operator

$$\mathcal{L} = \sum_{1 \leqslant j \leqslant m} X_j^2 + X_0,$$

where $X_0, X_1, \ldots, X_m$ are $m + 1$ vector fields as before.
Show that there exists a constant C such that, for any $u \in C_0^\infty(\mathbb{R}^n)$,

$$\sum_{1 \leqslant i \leqslant m} \|X_i u\|_{L^2}^2 \leqslant C \, \|\mathcal{L}u\|_{L^2}^2 + C\|u\|_{L^2}^2.$$

We say that $P \in \Psi_V^0$ belongs to $\mathcal{A}_L$ if and only if

$$\exists \varepsilon \in (0, 1/2],\ \exists C > 0 /\ \forall u \in C_0^\infty(V),\quad \|Pu\|_{H^\varepsilon}^2 \leqslant C \|\mathcal{L}u\|_{L^2}^2 + C\|u\|_{L^2}^2.$$

Show that, with S as before and $P \in \Psi_V^0$,

$$X_0 S \in \mathcal{A}_L, \quad [X_0, P] \in \mathcal{A}_L$$

and that, for all i, j such that $0 \leqslant i, j \leqslant m$, we have $[X_j, X_k] S \in \mathcal{A}_L$.
Deduce that $\mathcal{A}_L = \Psi_V^0$ if there exists $r \in \mathbb{N}$ such that

$$\text{vect}\left\{ [X_{i_1}, [X_{i_2}, \ldots, [X_{i_{p-1}}, X_{i_p}] \ldots]](x) : p \leqslant r,\ i_k \in \{0, \ldots, m\} \right\} = \mathbb{R}^n.$$

2. So, we have shown that, for any compact $K \subset \mathbb{R}^n$, there exist $\varepsilon > 0$ and $C > 0$ such that, for any $u \in C_0^\infty(\mathbb{R}^n)$ with $\operatorname{supp} u \subset K$,

$$\|u\|_{H^\varepsilon}^2 \leqslant C \|\mathcal{L}u\|_{L^2}^2 + C\|u\|_{L^2}^2.$$

We say that this is a sub-ellipticity estimate. Using this estimate and the structure of L, show that L is hypoelliptic, which means that if $u \in \mathcal{D}'(\Omega)$ satisfies $Lu \in C^\infty(\omega)$ with $\omega \subset \Omega$ then $u \in C^\infty(\omega)$.

Chapter 11
Hyperbolic Equations

We will focus on hyperbolic equations, which are partial differential equations fundamental for both theory and applications, as they govern propagation phenomena. In
the next chapter, we will use such equations to study the propagation of singularities.

11.1 Transport Equations

Let $n \geqslant 1$ and $v \in \mathbb{R}^n$. The transport equation is the prototype of a first-order
hyperbolic equation. It is the equation

$$\partial_t u + v \cdot \nabla u = 0$$

where the unknown $u \in C^1(\mathbb{R} \times \mathbb{R}^n)$ is a real-valued function.

Proposition 11.1 *Let $u_0 \in C^1(\mathbb{R}^n)$. There exists a unique function $u \in C^1(\mathbb{R} \times \mathbb{R}^n)$
that is a solution to the Cauchy problem*

$$\begin{cases} \partial_t u + v \cdot \nabla u = 0, \\ u_{|t=0} = u_0. \end{cases}$$

This solution is given by the formula $u(t,x) = u_0(x - tv)$.

Proof The idea is to introduce a function $t \mapsto X(t)$ such that, if u is a solution
of $\partial_t u + v \cdot \nabla u = 0$, then $u(t, X(t))$ is a constant function. Here, this amounts to
introducing $X(t) = x + vt$ with $x \in \mathbb{R}^n$ arbitrary. Indeed,

$$\frac{\mathrm{d}}{\mathrm{d}t} u(t, x + vt) = (\partial_t u + v \cdot \nabla u)(t, x + vt).$$

So, if u is a solution to the Cauchy problem then $u(t, x + vt) = u_0(x)$ from which
$u(t,x) = u_0(x - vt)$. Conversely, one directly verifies that $(t,x) \mapsto u_0(x - tv)$ is a
function of class C^1 that is a solution to the Cauchy problem. □

© The Author(s), under exclusive license to Springer Nature Switzerland AG 2024
T. Alazard, *Analysis and Partial Differential Equations*, Universitext,
https://doi.org/10.1007/978-3-031-70909-8_11

In the case where the constant vector v is replaced by a function with variable coefficients, we still have a representation formula for the solution based on the use of the characteristic curves of the vector field. As we will not use this point of view, we will limit ourselves in this section to defining these curves and stating[1] the main result. Let us start by recalling Gronwall's lemma. This very simple lemma plays a fundamental role in the study of evolution equations.

Lemma 11.2 (Gronwall's Lemma) *Let* $A, B \geqslant 0$ *and* $b, \phi \colon \mathbb{R}_+ \to \mathbb{R}_+$ *be two continuous functions such that*

$$\phi(t) \leqslant A + B \int_0^t \phi(s)\, ds + \int_0^t b(s)\, ds$$

for all $t \geqslant 0$*. Then, for all* $t \geqslant 0$*,*

$$\phi(t) \leqslant A e^{Bt} + \int_0^t b(s) e^{B(t-s)}\, ds.$$

Proof Let us introduce

$$w(t) = A + B \int_0^t \phi(s)\, ds + \int_0^t b(s)\, ds.$$

By hypothesis, this function is of class C^1 on $\mathbb{R}_+$ and $w'(t) = B\phi(t) + b(t) \leqslant Bw(t) + b(t)$. Therefore

$$\left(w(t) e^{-Bt}\right)' \leqslant b(t) e^{-Bt},$$

and we deduce the desired result by integrating this inequality.   □

Let $T > 0$ and $(t, x) \mapsto V(t, x) \in \mathbb{R}^n$ be a vector field defined for $(t, x) \in \mathbb{R} \times \mathbb{R}^n$, admitting partial derivatives of order 1 with respect to the variables x_j for $j = 1, \ldots, n$ and satisfying the following assumptions:

$$V \text{ and } \nabla_x V \text{ are continuous on } [0, T] \times \mathbb{R}^n, \tag{H1}$$

and there exists a constant $\kappa > 0$ such that, for all $(t, x) \in [0, T] \times \mathbb{R}^n$,

$$|V(t, x)| \leqslant \kappa(1 + |x|). \tag{H2}$$

Recall that we say that γ is an integral curve of the field V passing through x at time t if $\gamma \colon s \mapsto \gamma(s) \in \mathbb{R}^n$ satisfies

$$\frac{d}{ds}\gamma(s) = V(s, \gamma(s)), \qquad \gamma(t) = x.$$

The Cauchy–Lipschitz theorem implies the local existence of such an integral curve.

---

[1] We refer the reader to the books by Alinhac [2], Courant and Hilbert [38, 39] and John [86].

Using Gronwall's lemma, the assumption (H2) allows us to show that for all $(t,x) \in [0,T] \times \mathbb{R}^n$ the integral curve $s \mapsto \gamma(s)$ of V passing through x at time t is defined for all $s \in [0,T]$. In the following, we will denote this integral curve by $s \mapsto X(s,t,x)$, which is therefore by definition a solution of

$$\partial_s X(s,t,x) = V(s, X(s,t,x)), \qquad X(t,t,x) = x.$$

The main identity states that for all $t_1, t_2, t_3 \in [0,T]$, we have

$$X(t_3, t_2, X(t_2, t_1, x)) = X(t_3, t_1, x).$$

To obtain this identity, note that the functions

$$t_3 \longmapsto X(t_3, t_2, X(t_2, t_1, x)) \qquad \text{and} \qquad t_3 \longmapsto X(t_3, t_1, x)$$

are two integral curves of V passing through $X(t_2, t_1, x)$ for $t_3 = t_2$. By uniqueness in the Cauchy–Lipschitz theorem, these maps coincide on their maximum interval of definition.

We state the following result without proving it (as we will not use this point of view in the following).

Proposition 11.3 (Method of characteristics) *For all* $(s,t) \in [0,T] \times [0,T]$ *the function*

$$X(s,t,\cdot) : x \longmapsto X(s,t,x)$$

is a C^1-diffeomorphism from $\mathbb{R}^n$ onto itself. Moreover,

$$X \in C^1([0,T] \times [0,T] \times \mathbb{R}^n; \mathbb{R}^n)$$

and

$$\partial_t X(0,t,x) + \sum_{j=0}^{d} V_j(t,x)\partial_{x_j} X(0,t,x) = 0, \quad \text{for all } (t,x) \in [0,T] \times \mathbb{R}^n.$$

If $u_0 \in C^1(\mathbb{R}^n; \mathbb{R})$, the function defined by $u(t,x) = u_0(X(0,t,x))$ is C^1 on $[0,T] \times \mathbb{R}^n$ and satisfies

$$\partial_t u(t,x) + V(t,x) \cdot \nabla u(t,x) = 0, \quad u(0,x) = u_0(x).$$

If $\partial_t u + c \cdot \nabla u = 0$ and $u(0) = u_0$, we have seen that $u(t,x) = u_0(x - ct)$. From this, we deduce that all the L^p norms of u_0 are preserved by the equation. This means that $\|u(t)\|_{L^p} = \|u_0\|_{L^p}$ for all p such that $1 \leq p \leq +\infty$. In particular,

$$\|u(t)\|_{L^2} = \|u_0\|_{L^2}.$$

We will see how to prove a similar estimate for an equation with variable coefficients (without using the fact that we can integrate such an equation by the method of characteristics).

Now consider $V \in C_b^{\infty}(\mathbb{R} \times \mathbb{R}^n)$ with real values and a regular solution $u \in C^1(\mathbb{R}; H^1(\mathbb{R}^n))$ of the equation

$$\partial_t u + V(t, x) \cdot \nabla u = 0.$$

By multiplying the equation by u and integrating, we find that

$$\frac{d}{dt} \int_{\mathbb{R}^n} u(t, x)^2 \, dx = 2 \int_{\mathbb{R}^n} u \partial_t u \, dx = -2 \int_{\mathbb{R}^n} u(V \cdot \nabla u) \, dx$$

and by integrating by parts, we deduce that

$$\int_{\mathbb{R}^n} u(V \cdot \nabla u) = -\frac{1}{2} \int_{\mathbb{R}^n} (\operatorname{div} V) u^2 \, dx,$$

from which

$$\frac{d}{dt} \int_{\mathbb{R}^n} u(t, x)^2 \, dx \leqslant \|\operatorname{div} V\|_{L^{\infty}} \int_{\mathbb{R}^n} u^2 \, dx.$$

Gronwall's lemma then gives

$$\forall t \geqslant 0, \quad \|u(t)\|_{L^2(\mathbb{R}^n)}^2 \leqslant e^{t \|\operatorname{div} V\|_{L^{\infty}}} \|u_0\|_{L^2(\mathbb{R}^n)}^2 .$$

Note that if $\operatorname{div} V = 0$ then the norm $\|u(t)\|_{L^2(\mathbb{R}^n)}$ is preserved.

The goal of this chapter will be to generalize this estimate in many directions and to apply it to prove the existence of solutions to general hyperbolic equations.

11.2 Pseudo-Differential Hyperbolic Equations

We consider symbols $a_t(x, \xi)$ depending on a parameter $t \in \mathbb{R}$, with complex values. In the rest of this chapter, $x, \xi \in \mathbb{R}^n$, where $n \geqslant 1$ is a fixed integer.

Definition 11.4

1. Consider a symbol $a = a(x, \xi)$ belonging to $S^1(\mathbb{R}^n)$. We say that a is hyperbolic if a can be written in the form $a = a_1 + a_0$, where $a_1 \in S^1(\mathbb{R}^n)$ is a symbol with purely imaginary values and a_0 belongs to $S^0(\mathbb{R}^n)$. It is equivalent to say that a is a symbol of $S^1(\mathbb{R}^n)$ whose real part belongs to $S^0(\mathbb{R}^n)$.
2. Consider a family $(a_t)_{t \in \mathbb{R}}$ of symbols of $S^1(\mathbb{R}^n)$ such that $t \mapsto a_t$ is continuous and bounded from $\mathbb{R}$ to $S^1(\mathbb{R}^n)$. We say that $(a_t)_{t \in \mathbb{R}}$ is a time-dependent symbol of order 1. By definition, this symbol is hyperbolic if $\operatorname{Re}(a_t)$ is bounded in $S^0(\mathbb{R}^n)$.

Example 11.5 The simplest example of a hyperbolic symbol is the following: $a(x, \xi) = i\xi$ in dimension $n = 1$, then $\operatorname{Op}(a)u = \partial_x u$. In any dimension $n \geqslant 1$, note that the symbol $a_t(x, \xi) = iV(t, x) \cdot \xi$ is hyperbolic and in this case, we have $\operatorname{Op}(a)u = V \cdot \nabla u$.

Consider a hyperbolic symbol $(a_t)_{t\in\mathbb{R}}$ and a continuous function u from $\mathbb{R}$ to $S'(\mathbb{R}^n)$. By definition, $\mathrm{Op}(a)u$ is defined by $(\mathrm{Op}(a)u)(t) = \mathrm{Op}(a_t)u(t)$. If $u \in C^0(\mathbb{R}; H^s(\mathbb{R}^n))$ with $s \in \mathbb{R}$, then the continuity result of pseudo-differential operators on Sobolev spaces (see Proposition 10.21) implies $\mathrm{Op}(a)u \in C^0(\mathbb{R}; H^{s-1}(\mathbb{R}^n))$.

Let us also give:

- a time $T > 0$ and any real number s;
- a function $u_0 \in H^s(\mathbb{R}^n)$, called the initial data;
- a function $f \in C^0([0, T]; H^s(\mathbb{R}^n))$, called the source term.

We are interested in the following Cauchy problem

$$\begin{cases} \dfrac{\partial u}{\partial t} + \mathrm{Op}(a)u = f, \\ u_{|t=0} = u_0, \end{cases} \tag{11.1}$$

where the unknown is the function $u = u(t, x)$, the variable $t \in \mathbb{R}_+$ corresponds to time and the variable $x \in \mathbb{R}^n$ $(n \geqslant 1)$ corresponds to the space variable.

Theorem 11.6 *Let $T > 0$, $n \geqslant 1$ and $s \in \mathbb{R}$. For any initial data $u_0 \in H^s(\mathbb{R}^n)$ and any $f \in C^0([0, T]; H^s(\mathbb{R}^n))$, there exists a unique function*

$$u \in C^0([0, T]; H^s(\mathbb{R}^n)) \cap C^1([0, T]; H^{s-1}(\mathbb{R}^n))$$

that satisfies

$$\frac{\partial u}{\partial t} + \mathrm{Op}(a)u = f$$

and that is such that $u(0) = u_0$.

Proof The first ingredient to prove this theorem is an *a priori* estimate.

Lemma 11.7 *Let $s \in \mathbb{R}$, $T > 0$. There exists a constant C such that for all $u \in C^1([0, T]; H^s)$, all $f \in C^0([0, T]; H^s)$, all $u_0 \in H^s$ and all $t \in [0, T]$, if u is a solution of (11.1) then*

$$\|u(t)\|_{H^s} \leqslant e^{Ct}\|u_0\|_{H^s} + \int_0^t e^{C(t-t')}\|f(t')\|_{H^s}\,dt'. \tag{11.2}$$

Moreover, there exist two constants K and N that depend only on s such that, with $\alpha_t := (a_t^* - \bar{a}_t) + 2\,\mathrm{Re}\,a_t$ where a_t^* is the symbol of the adjoint of $\mathrm{Op}(a_t)$, we have:

$$C \leqslant K \sum_{|\alpha|+|\beta|\leqslant N} \sup_{t\in[0,T]} \sup_{x,\xi}\left|\langle\xi\rangle^{-|\beta|}\partial_x^\alpha \partial_\xi^\beta \alpha_t(x,\xi)\right|.$$

Proof We start with the case $s = 0$. As u is C^1 with values in L^2 we have

$$
\begin{aligned}
\frac{\mathrm{d}}{\mathrm{d}t}\, \|u(t)\|_{L^2}^2 &= \frac{\mathrm{d}}{\mathrm{d}t}\, (u(t), u(t)) \\
&= 2\,\mathrm{Re}\,(\partial_t u(t), u(t)) \\
&= -2\,\mathrm{Re}\,(\mathrm{Op}(a_t)u(t), u(t)) + 2\,\mathrm{Re}\big(f(t), u(t)\big).
\end{aligned}
\tag{11.3}
$$

We saw in the previous chapter that the adjoint $\mathrm{Op}(a_t)^*$ is a pseudo-differential operator whose symbol, denoted a_t^*, is such that $a_t^* = \overline{a}_t + b_t$ with $b_t \in S^0(\mathbb{R}^n)$. We can then write

$$
(\mathrm{Op}(a_t)u(t), u(t)) = (u(t), \mathrm{Op}(a_t)^* u(t)) = (u(t), \mathrm{Op}(\overline{a}_t)u(t) + \mathrm{Op}(b_t)u(t)).
$$

The hypothesis that (a_t) is hyperbolic means that $\overline{a}_t = -a_t + 2\,\mathrm{Re}\,a_t$ with $\mathrm{Re}\,a_t \in S^0(\mathbb{R}^n)$. So $a_t^* = -a_t + \alpha_t$, where α_t belongs to $S^0(\mathbb{R}^n)$ (uniformly in t). We deduce that

$$
(\mathrm{Op}(a_t)u(t), u(t)) = \big(u(t), -\mathrm{Op}(a_t)u(t) + \mathrm{Op}(\alpha_t)u(t)\big),
$$

from which we get that

$$
2\,\mathrm{Re}\,(\mathrm{Op}(a_t)u(t), u(t)) = \big(u(t), \mathrm{Op}(\alpha_t)u(t)\big).
$$

The Cauchy–Schwarz inequality and the continuity on L^2 of pseudo-differential operators of order 0 (see Theorem 9.8 or Proposition 10.21 with $m = 0$) imply that

$$
|(\mathrm{Op}(\alpha_t)u(t), u(t))| \leqslant K \sup_t \|\mathrm{Op}(\alpha_t)\|_{\mathcal{L}(L^2)} \|u(t)\|_{L^2}^2 \leqslant C_0 \|u(t)\|_{L^2}^2,
$$

where C_0 is a constant that does not depend on t. By substituting this inequality into (11.3), we conclude that

$$
\frac{\mathrm{d}}{\mathrm{d}t}\, \|u(t)\|_{L^2}^2 \leqslant C_0 \, \|u(t)\|_{L^2}^2 + 2\|f(t)\|_{L^2}\|u(t)\|_{L^2}.
\tag{11.4}
$$

We would like to write that $\frac{\mathrm{d}}{\mathrm{d}t}\|u(t)\|_{L^2}^2 = 2\,\|u(t)\|_{L^2}\frac{\mathrm{d}}{\mathrm{d}t}\|u(t)\|_{L^2}$ and simplify the inequality by dividing by $\|u(t)\|_{L^2}$. As we do not yet know that $\|u(t)\|_{L^2}$ cannot be zero (unless u is identically zero), we proceed as follows: given $\delta > 0$, we deduce from (11.4) that the function $y(t) = \sqrt{\|u(t)\|_{L^2}^2 + \delta}$ satisfies

$$
\frac{\mathrm{d}}{\mathrm{d}t} y(t)^2 \leqslant C_0 y(t)^2 + 2\|f(t)\|_{L^2} y(t),
$$

and since $\|u(t)\|_{L^2}^2 + \delta > 0$, the function $y(t)$ is of class C^1 and we can simplify

$$
2\frac{\mathrm{d}y(t)}{\mathrm{d}t} \leqslant C_0 y(t) + 2\|f(t)\|_{L^2}.
$$

Gronwall's lemma implies that

$$\|u(t)\|_{L^2} \leqslant y(t) \leqslant y(0)e^{C_0 t/2} + \int_0^t \|f(t')\|_{L^2}\, e^{C_0(t-t')/2}\, dt',$$

for all $\delta > 0$. By letting δ tend to 0, we obtain that

$$\|u(t)\|_{L^2} \leqslant \|u(0)\|_{L^2}\, e^{C_0 t/2} + \int_0^t \|f(t')\|_{L^2}\, e^{C_0(t-t')/2}\, dt',$$

which concludes the proof of the lemma in the case $s = 0$. Now consider an arbitrary Sobolev index of regularity $s \in \mathbb{R}$ and introduce $\Lambda_s = \langle D_x \rangle^s = (I - \Delta)^{s/2}$, which is the Fourier multiplier of symbol $\langle \xi \rangle^s = (1 + |\xi|^2)^{s/2}$. Then we commute $L = \partial_t + \mathrm{Op}(a)$ to the operator $\Lambda_s = \langle D_x \rangle^s = (I - \Delta)^{s/2}$, which gives

$$\Lambda_s L u = \widetilde{L}\Lambda_s u, \quad \widetilde{L} = \partial_t + \widetilde{A}, \quad \widetilde{A} = \Lambda_s \mathrm{Op}(a)\Lambda_{-s}.$$

Then, it follows from symbolic calculus (*cf.* point 2 in Theorem 10.13) that $\widetilde{A}$ is a pseudo-differential operator with a hyperbolic symbol. Since Λ_s is an isometry from H^s to L^2, we conclude the proof by applying the previous L^2 estimate to $\widetilde{L}$. □

Let us now prove Theorem 11.6 using the previous *a priori* estimate[2].

Let us verify uniqueness. If u_1 and u_2 are two different solutions, then the difference belongs to $C^1([0, T]; H^{s-1}(\mathbb{R}^n))$. We can use the energy estimate (11.2) applied with $u = u_1 - u_2$ and s replaced by $s - 1$, and we obtain that $u_1 = u_2$.

We now prove existence following a method introduced by Hörmander, relying on the Hahn–Banach theorem. We will start by proving the existence of a weak solution, in the sense of the following definition.

Definition 11.8 Let $s \in \mathbb{R}$. Consider $u \in C^0([0, T]; H^s(\mathbb{R}^n))$. We say that u is a solution in the sense of distributions of the Cauchy problem (11.1) if the following property is satisfied. For any function $\varphi = \varphi(t)$ belonging to $C_0^\infty((-\infty, T))$ and any function $\psi = \psi(x)$ belonging to the Schwartz space $\mathcal{S}(\mathbb{R}^n)$, we have

$$-\int_0^T \varphi'(t)\langle u(t), \psi \rangle\, dt + \int_0^T \varphi(t)\langle \mathrm{Op}(a_t)u(t), \psi \rangle\, dt$$

$$= \int_0^T \varphi(t)\langle f(t), \psi \rangle\, dt + \varphi(0)\langle u_0, \psi \rangle,$$

where $\langle v, \psi \rangle$ denotes the duality product between a tempered distribution v and a function ψ from the Schwartz space.

---

[2] There are several way to do this. Here, we proceed by using a duality argument due to Hörmander (see [80]). In Exercise 11.1, we propose another method, based on the regularization of the equations, following an idea introduced by Friedrichs [53] and developed in a systematic way by Taylor [187, 189] for linear or nonlinear problems.

Let $L = \partial_t + \mathrm{Op}(a)$ and $L^* = -\partial_t + \mathrm{Op}(a)^*$. We consider the space $\mathcal{T}$ of functions $v = v(t, x) \in \mathbb{R}$ that are C^∞ and of the form $v(t, x) = \varphi(t)\psi(x)$ where[3] $\mathrm{supp}\,\varphi \Subset (-\infty, T)$ and $\psi \in S(\mathbb{R}^n)$, and we denote by E the subspace $L^*(\mathcal{T})$.

Let us start by proving that there exists a constant C such that, for all $v \in \mathcal{T}$,

$$\sup_{t \in [0,T]} \|v(t)\|_{H^{-s}} \leqslant C \int_0^T \|L^* v(t)\|_{H^{-s}}\, dt. \tag{11.5}$$

For this, let us introduce $\widetilde{v}(t, x) = v(T - t, x)$ and note that

$$(\partial_t + \mathrm{Op}(a)^*)\widetilde{v} = (L^* v)(T - t, x), \quad \widetilde{v}(0, x) = 0.$$

We have seen in the previous chapter that $\mathrm{Op}(a)^*$ is a pseudo-differential operator whose symbol, denoted a^*, satisfies $a^* - \overline{a} \in S^0(\mathbb{R}^n)$. Then $\mathrm{Re}\,a^* - \mathrm{Re}\,a \in S^0(\mathbb{R}^n)$ and we see that a is hyperbolic if and only if a^* is. We can therefore apply the inequality (11.2) by replacing a with a^*. Moreover, as $\widetilde{v}(0, x) = 0$, the inequality (11.5) is a consequence of the a priori estimate (11.2) applied with s replaced by $-s$ (note that the assumptions that allow us to use the estimate (11.2) are satisfied because $\widetilde{v} \in C^1([0, T]; H^{-s}(\mathbb{R}^n))$).

Now, let y be an element of E. By definition, y can be written as $y = L^* v$ with $v \in \mathcal{T}$. Moreover, if $v' \in \mathcal{T}$ is such that $y = L^* v'$, then $L^*(v - v') = 0$. The estimate (11.5) implies $\widetilde{v} = \widetilde{v}'$ and therefore $v = v'$. We have shown that for every y belonging to E, there exists a unique $v \in \mathcal{T}$ such that $y = L^* v$.

Let $y = L^* v \in E$ with $v \in \mathcal{T}$. We define $\Psi(y)$ by

$$\Psi(y) = \int_0^T \langle f(t), v(t) \rangle\, dt + \langle u_0, v(0, \cdot) \rangle.$$

We have

$$|\Psi(y)| \leqslant \left(\|u_0\|_{H^s} + \int_0^T \|f(t)\|_{H^s}\, dt \right) \sup_{t \in [0,T]} \|v(t)\|_{H^{-s}}$$

and furthermore the estimate (11.5) implies that

$$\sup_{t \in [0,T]} \|v(t)\|_{H^{-s}} \leqslant C \int_0^T \|L^* v\|_{H^{-s}}\, dt = C \int_0^T \|y\|_{H^{-s}}\, dt,$$

therefore

$$|\Psi(y)| \leqslant C \left(\|u_0\|_{H^s} + \int_0^T \|f(t)\|_{H^s}\, dt \right) \int_0^T \|y\|_{H^{-s}}\, dt.$$

The map $\Psi \colon E \to \mathbb{R}$ is therefore a continuous linear form. With the Hahn–Banach theorem, we extend Ψ to a continuous linear form $\widetilde{\Psi}$ defined on $L^1([0, T]; H^{-s})$.

---

[3] The notation $A \Subset B$ means that $\overline{A} \subset B$.

The continuous linear forms on $L^1([0,T];H^{-s})$ are represented by functions of $L^\infty([0,T];H^s)$ and hence we have shown the existence of

$$u \in L^\infty([0,T];H^s)$$

such that

$$\forall v \in \mathcal{T}, \quad \int_0^T \langle u(t), L^* v(t) \rangle \, dt = \int_0^T \langle f(t), v(t) \rangle \, dt + \langle u_0, v(0) \rangle.$$

This proves the existence of a weak solution. The following lemma states that weak solutions are in fact classical solutions.

Lemma 11.9 *We have* $u \in C^1([0,T];H^{s-2}(\mathbb{R}^n))$, *moreover* $u(0) = u_0$ *and finally we have the following equality between functions of* $C^0([0,T];H^{s-2}(\mathbb{R}^n))$:

$$\frac{\partial u}{\partial t} + \mathrm{Op}(a)u = f.$$

Proof We will just sketch the proof. Let $\psi \in \mathcal{S}(\mathbb{R}^n)$ and let $\varphi \in C_0^\infty((0,T))$. The previous identity applied with $v = \varphi(t)\psi(x)$ implies that

$$-\int_0^T \varphi'(t)\langle u(t), \psi \rangle \, dt = \int_0^T \varphi(t) \left(-\langle \mathrm{Op}(a)u, \psi \rangle + \langle f(t), \psi \rangle \right) dt.$$

It follows that, in the sense of distributions, we have

$$\frac{\partial u}{\partial t} = -\mathrm{Op}(a)u + f.$$

We deduce that $u \in L^\infty([0,T];H^s)$ is such that its time derivative belongs to $L^\infty([0,T];H^s)$. We deduce that $u \in C^0([0,T];H^{s-1})$, from which we get $\mathrm{Op}(a)u \in C^0([0,T];H^{s-2})$. Using the equation again, we show that $u \in C^1([0,T];H^{s-2})$. $\square$

To conclude, we regularize f and u_0, construct a sequence of regular solutions and pass to the limit. Namely, we introduce two sequences $(f^n)_{n\in\mathbb{N}}$ and $(u_0^n)_{n\in\mathbb{N}}$ with $f^n \in C^0([0,T];H^{s+2})$ and $u_0^n \in H^{s+2}$ converging to f in $C^0([0,T];H^s)$ and to u in H^s, respectively. The previous work gives a sequence of solutions u^n belonging to $C^1([0,T];H^s)$. The energy estimate (11.2) then shows that the sequence of solutions to the approximated problems is Cauchy in $C^0([0,T];H^s)$, thus converges to $u \in C^0([0,T];H^s)$, which is the wanted solution. $\square$

11.3 Exercises

Exercise (solved) 11.1 (Resolution by regularization) The aim of this exercise is to show that the solution obtained in Section 11.2 is the limit of solutions of approximated problems. This result can be seen as another way to prove the existence of a solution once an *a priori* estimate has been demonstrated.

Let a_t be a time-dependent hyperbolic symbol. Let $T > 0$, $s \in \mathbb{R}$, and consider

$$u_0 \in H^s(\mathbb{R}^n) \quad \text{and} \quad f \in C^0([0,T]; H^s(\mathbb{R}^n)).$$

Let $\varepsilon \in (0,1)$. Consider the following Cauchy problem:

$$\partial_t u + \mathrm{Op}(a_t)J_\varepsilon u = f, \quad u(0) = u_0. \tag{11.6}$$

Here J_ε, called the Friedrichs mollifier, is defined by:

$$\widehat{J_\varepsilon v}(\xi) = \chi(\varepsilon\xi)\widehat{v}(\xi),$$

with $\chi \in C^\infty(\mathbb{R}^n, \mathbb{R})$, supported in $B(0,2)$ and equal to 1 on $B(0,1)$.

1. Show that $\mathrm{Op}(a_t)J_\varepsilon = \mathrm{Op}(a_t^\varepsilon)$, where:

$$a_t^\varepsilon(x,\xi) = a_t(x,\xi)\chi(\varepsilon\xi).$$

2. Show that there exists a unique solution $u_\varepsilon \in C^1([0,T]; H^s(\mathbb{R}^n))$ of (11.6).
3. Show that there exists a constant C such that, for all $\varepsilon > 0$, all $t \in [0;T]$ and all functions $v \in C^1([0,T]; H^s(\mathbb{R}^n))$,

$$\|v(t)\|_{H^s} \leqslant C\|v(0)\|_{H^s} + C\int_0^t \|(\partial_t v + \mathrm{Op}(a^\varepsilon)v)(\tau)\|_{H^s}\, d\tau.$$

Deduce that $(u_\varepsilon)_{\varepsilon \in (0,1)}$ is bounded in $C^0([0,T]; H^s(\mathbb{R}^n))$.
4. Show that $(u_\varepsilon)_{\varepsilon \in (0,1)}$ is Cauchy in $C^0([0,T]; H^{s-2}(\mathbb{R}^n))$.
5. Let $s_1 < s_2$ be two real numbers and $\sigma \in (s_1, s_2)$. Write σ in the form $\sigma = \alpha s_1 + (1-\alpha)s_2$ with $\alpha \in [0,1]$. Show that there exists a constant $C(s_1, s_2)$ such that, for all $u \in H^{s_2}(\mathbb{R}^n)$:

$$\|u\|_{H^\sigma} \leqslant C(s_1, s_2)\|u\|_{H^{s_1}}^\alpha \|u\|_{H^{s_2}}^{1-\alpha}. \tag{11.7}$$

Deduce that (u_ε) is Cauchy in $C^0([0,T]; H^\sigma(\mathbb{R}^n))$ for $s - 2 < \sigma < s$ and that u_ε converges in $C^0([0,T]; H^\sigma(\mathbb{R}^n)) \cap C^1([0,T]; H^{\sigma-1}(\mathbb{R}^n))$ to a limit denoted by u.
6. Show that $u \in C^0([0,T]; H^s) \cap C^1([0,T]; H^{s-1})$ and that u is a solution of the equation:

$$\partial_t u + \mathrm{Op}(a_t)u = f, \quad u(0) = u_0.$$

Chapter 12
Microlocal Singularities

In this chapter, we aim to provide an introduction to microlocal analysis, which is the study of singularities of functions of several real variables.

12.1 Local Properties

Let $u \in C_0^\infty(\mathbb{R}^n)$ and $a \in S^m(\mathbb{R}^n)$ with m any real number. The continuity theorem of pseudo-differential operators implies that $\mathrm{Op}(a)$ is continuous from $H^s(\mathbb{R}^n)$ to $H^{s-m}(\mathbb{R}^n)$ for all $s \in \mathbb{R}$. Since $C_0^\infty(\mathbb{R}^n)$ is included in $H^s(\mathbb{R}^n)$ for any $s \in \mathbb{R}$, we deduce that

$$\mathrm{Op}(a)\big(C_0^\infty(\mathbb{R}^n)\big) \subset H^\infty(\mathbb{R}^n) \subset C_b^\infty(\mathbb{R}^n),$$

where the second inclusion comes from the Sobolev embedding theorem. One may wonder if we can do better. For example, is it true that $\mathrm{Op}(a)u$ is a function with compact support? This result is true, trivially, if a is a polynomial in ξ (with coefficients depending on x). Indeed, in this case, $\mathrm{Op}(a)$ is a differential operator and $\mathrm{Op}(a)u$ is supported in $\mathrm{supp}\, u$. Conversely, a classical result of differential calculus states that local operators (which do not increase the support) are necessarily differential operators. Therefore, given a pseudo-differential operator, it is generally false that if u belongs to $C_0^\infty(\mathbb{R}^n)$, then $\mathrm{Op}(a)u \in C_0^\infty(\mathbb{R}^n)$. However, we do have several results concerning the local theory of pseudo-differential operators, and we will describe them. Among these results, the simplest is given by the following proposition.

Proposition 12.1 *Let $a \in S^m(\mathbb{R}^n)$ and $u \in L^2(\mathbb{R}^n)$ be a function with compact support. Consider a function $\varphi \in C_b^\infty(\mathbb{R}^n)$ that vanishes on a neighborhood of the support of u. Then $\varphi \, \mathrm{Op}(a)u$ is a function of class $C_b^\infty(\mathbb{R}^n)$.*

Proof Consider a function $\psi \in C_0^\infty(\mathbb{R}^n)$ that equals 1 on the support of u and whose support is included in $\varphi^{-1}(\{0\})$.

By definition of pseudo-differential operators, if $a = a(x)$ does not depend on ξ, and if $b = b(x, \xi)$ is any symbol, then $\mathrm{Op}(a) \circ \mathrm{Op}(b)u = a(\mathrm{Op}(b)u) = \mathrm{Op}(ab)u$. Here, this implies $\varphi \, \mathrm{Op}(a)u = \mathrm{Op}(\varphi a)u$.

T. Alazard, *Analysis and Partial Differential Equations*, Universitext,
https://doi.org/10.1007/978-3-031-70909-8_12

Moreover, we have $u = \psi u$. The composition theorem (see Theorem 10.13) implies that

$$\varphi \, \mathrm{Op}(a)u = \mathrm{Op}(\varphi a)\{\psi u\} = \mathrm{Op}((\varphi a)\#\psi)u,$$

together with

$$(\varphi a)\#\psi \sim \sum_\alpha \frac{1}{i^{|\alpha|}\alpha!}\varphi(\partial_\xi^\alpha a)(\partial_x^\alpha \psi).$$

By hypothesis on φ, ψ, we have $\varphi(\partial_x^\alpha \psi)=0$ for all $\alpha \in \mathbb{N}^n$, so $(\varphi a)\#\psi \sim 0$. We deduce that $\mathrm{Op}((\varphi a)\#\psi)$ is a regularizing operator, bounded from $H^{s_1}(\mathbb{R}^n)$ to $H^{s_2}(\mathbb{R}^n)$ for all real numbers s_1, s_2. This concludes the proof. $\square$

Recall that

$$S^\infty(\mathbb{R}^n) = \bigcup_{m\in\mathbb{R}} S^m(\mathbb{R}^n), \qquad S^{-\infty}(\mathbb{R}^n) = \bigcap_{m\in\mathbb{R}} S^m(\mathbb{R}^n).$$

Thus, $S^\infty(\mathbb{R}^n)$ is the space of all symbols while $S^{-\infty}(\mathbb{R}^n)$ is the space of regularizing symbols. Of course, $S^{-\infty}(\mathbb{R}^n) \subset S^\infty(\mathbb{R}^n)$.

We define

$$H^{-\infty}(\mathbb{R}^n) = \bigcup_{s\in\mathbb{R}} H^s(\mathbb{R}^n), \qquad H^\infty(\mathbb{R}^n) = \bigcap_{s\in\mathbb{R}} H^m(\mathbb{R}^n),$$

and this time we have $H^\infty(\mathbb{R}^n) \subset H^{-\infty}(\mathbb{R}^n)$.

We have seen that if $a \in S^m(\mathbb{R}^n)$ with $m \in \mathbb{R}$ and $u \in H^s(\mathbb{R}^n)$ with $s \in \mathbb{R}$ then $\mathrm{Op}(a)u \in H^{s-m}(\mathbb{R}^n)$. If $m \leqslant 0$ then $\mathrm{Op}(a)u$ is more regular than u. In particular

$$a \in S^{-\infty}(\mathbb{R}^n), \; u \in H^{-\infty}(\mathbb{R}^n) \implies \mathrm{Op}(a)u \in H^\infty(\mathbb{R}^n) \subset C_b^\infty(\mathbb{R}^n).$$

Proposition 12.2 *Consider a regularizing pseudo-differential operator* $\mathrm{Op}(a)$ *such that* $a \in S^{-\infty}(\mathbb{R}^n)$. *Then* $\mathrm{Op}(a)$ *is continuous from* $H^{-\infty}(\mathbb{R}^n)$ *to* $S(\mathbb{R}^n)$.

Proof Let $a \in S^{-\infty}(\mathbb{R}^n)$ and $u \in H^{-\infty}(\mathbb{R}^n)$. The above shows that $\mathrm{Op}(a)u$ belongs to $H^\infty(\mathbb{R}^n)$ and therefore to $C_b^\infty(\mathbb{R}^n)$. Then we apply an argument which we have already encountered (see the proof of Theorem 9.7) which tells us that, for all $\alpha \in \mathbb{N}^n$, $x^\alpha \, \mathrm{Op}(a)u$ is a linear combination of terms $\mathrm{Op}(\partial_\xi^\delta a)(x^{\alpha-\delta}u)$, which belong to $C_b^\infty(\mathbb{R}^n)$ for the same reasons ($H^{-\infty}(\mathbb{R}^n)$ is stable under derivation and multiplication by a smooth function). Thus, we conclude that $\mathrm{Op}(a)u \in S(\mathbb{R}^n)$. $\square$

We recall the definition of the singular support of a distribution.

Definition 12.3 We say that a distribution $f \in S'(\mathbb{R}^n)$ is of class C^∞ in the neighborhood of x_0 if there exists a neighborhood ω of x_0 such that, for any function $\varphi \in C_0^\infty(\omega)$, we have $\varphi f \in C^\infty(\mathbb{R}^n)$.

The singular support of f, denoted supp sing f, is the complement of the set of points in the neighborhood of which f is of class C^∞.

This concept allows us to generalize Proposition 12.1.

Proposition 12.4 *For all $a \in S^{\infty}(\mathbb{R}^n)$ and all $u \in H^{-\infty}(\mathbb{R}^n)$, we have*

$$\text{supp sing} \operatorname{Op}(a)u \subset \text{supp sing} \, u.$$

Proof Let $a \in S^{\infty}(\mathbb{R}^n)$, $u \in S'(\mathbb{R}^n)$ and $\Omega = \mathbb{R}^n \setminus \text{supp sing} \, u$. Thus, $\psi u \in C_0^{\infty}(\mathbb{R}^n)$ for all $\psi \in C_0^{\infty}(\Omega)$. Moreover, for all $\varphi \in C_0^{\infty}(\Omega)$, we can find $\psi \in C_0^{\infty}(\Omega)$ with $\psi = 1$ on the support of φ (since we have $\text{dist}(\text{supp}(\varphi), \partial\Omega) > 0$), and

$$\varphi \operatorname{Op}(a)u = \varphi \operatorname{Op}(a)(\psi u) + \varphi \operatorname{Op}(a)\big((1 - \psi)u\big).$$

The first term is in S since $\psi u \in C_0^{\infty}(\mathbb{R}^n) \subset S$, and the second term can be written as $\operatorname{Op}(b)u$ where $b = \varphi a \# (1 - \psi)$. As we have already seen, the symbol b satisfies $b \sim 0$ since $\text{supp}(\varphi) \cap \text{supp}(1 - \psi) = \varnothing$ by construction of ψ. Furthermore, if $u \in H^{-\infty}(\mathbb{R}^n)$, then we also have $(1 - \psi)u \in H^{-\infty}(\mathbb{R}^n)$. We deduce that

$$\varphi \operatorname{Op}(a)\big((1 - \psi)u\big) \in H^{\infty}(\mathbb{R}^n).$$

Therefore, for all $\varphi \in C_0^{\infty}(\Omega)$, we have $\varphi \operatorname{Op}(a)u \in C_0^{\infty}(\Omega)$. We deduce that $\operatorname{Op}(a)u \in C^{\infty}(\Omega)$ (regularity is a local notion), which is the desired property. $\qquad\square$

12.2 The Wave Front Set

The wave front set of a tempered distribution $f \in S'(\mathbb{R}^n)$, denoted $\text{WF}(f)$, is a subset of $\mathbb{R}^n \times (\mathbb{R}^n \setminus \{0\})$ which describes not only the points where f is singular, but also the co-directions in which it is singular. This set is defined by its complement.

Definition 12.5 Let $f \in S'(\mathbb{R}^n)$.

1. We say that f is *microlocally of class* C^{∞} at a point $(x_0, \xi_0) \in \mathbb{R}^n \times (\mathbb{R}^n \setminus \{0\})$ if there exists an open set $\omega \subset \mathbb{R}^n$ containing x_0 and an open cone Γ of $\mathbb{R}^n \setminus \{0\}$ containing ξ_0 such that we have

$$\forall \varphi \in C_0^{\infty}(\omega), \ \forall N \in \mathbb{N}, \ \exists C_N > 0, \ \forall \xi \in \Gamma, \ \big|\widehat{\varphi f}(\xi)\big| \leqslant C_N (1 + |\xi|)^{-N}. \quad (12.1)$$

2. The set of points (x_0, ξ_0) where f is not microlocally C^{∞} is called the wave front set of f and denoted $\text{WF}(f)$.

The wave front set is a conical subset of $\mathbb{R}^n \times (\mathbb{R}^n \setminus \{0\})$, which means that for all $t > 0$,

$$(x, \xi) \in \text{WF}(f) \iff (x, t\xi) \in \text{WF}(f).$$

The wave front set allows us to specify the notion of singular support. Indeed, we have the following proposition.

Proposition 12.6 *The projection on $\mathbb{R}^n$ of $\text{WF}(u)$ is $\text{supp sing}(u)$: that is*

$$\pi(\text{WF}(u)) = \text{supp sing}(u) \quad \text{where} \quad \pi(x, \xi) = x.$$

Proof Let $x_0 \in \mathbb{R}^n$ not belonging to supp sing(u). If $\varphi \in C_0^\infty(\mathbb{R}^n)$ is supported in a sufficiently small ball centered at x_0, then φu is a C^∞ function with compact support and therefore belongs to the Schwartz space. Since the Fourier transform of a function from $\mathcal{S}(\mathbb{R}^n)$ belongs to $\mathcal{S}(\mathbb{R}^n)$, we deduce that $\widehat{\varphi u}$ is rapidly decreasing in all directions. In particular, no (x_0, ξ_0) belongs to WF(u).

Conversely, suppose that x_0 is such that no (x_0, ξ_0) belongs to WF(u). For each ξ_0, we can find an open set ω containing x_0 and a cone Γ containing ξ_0 such that (12.1) holds. By compactness of the sphere, we can find a finite number of such pairs (ω_j, Γ_j) so that the Γ_j cover $\mathbb{R}^n \setminus \{0\}$. For $\varphi \in C_0^\infty(\mathbb{R}^n)$ whose support is contained in $\cap_j \omega_j$, we deduce that the function $\widehat{\varphi u}$ is rapidly decreasing, hence $x_0 \notin$ supp sing(u), which completes the proof.   □

Let P be a differential operator of order m whose coefficients p_α are real and C^∞,

$$P = \sum_{|\alpha| \le m} p_\alpha(x) \partial_x^\alpha.$$

An important question in PDEs is to determine the wave front of the distribution solutions of the equation $Pf = 0$. The basic results relate the geometry of the operator to the geometry of the singularities of its solutions. The two simplest geometric objects associated with the PDE $P(f) = 0$ are the following.

1. The *principal symbol*

$$p_m(x, \xi) = i^m \sum_{|\alpha|=m} p_\alpha(x) \xi^\alpha,$$

   which is a homogeneous polynomial of degree m in ξ.
2. The *characteristic variety* of P, denoted Car(P), which is the closed set (homogeneous in ξ) defined by

$$\mathrm{Car}(P) = \{(x, \xi) \in \mathbb{R}^n \times (\mathbb{R}^n \setminus \{0\}) : p_m(x, \xi) = 0\}.$$

The first important result of the theory is the following, which states that the microlocal singularities are contained in the characteristic variety.

Theorem 12.7 (Sato–Hörmander) *If P is a differential operator with coefficients belonging to $C_b^\infty(\mathbb{R}^n)$, then for all $u \in \mathcal{S}'(\mathbb{R}^n)$,*

$$P(u) = 0 \implies \mathrm{WF}(u) \subset \mathrm{Car}(P).$$

Proof We follow the proof given by Hörmander and start with a technical lemma. Given a differential operator Q and a function $\varphi \in C_0^\infty(\mathbb{R}^n)$, we seek $\psi \in C_0^\infty(\mathbb{R}^n)$ that satisfies, approximately, the equation

$$Q(\psi e^{ix \cdot \xi}) = \varphi e^{ix \cdot \xi}.$$

To solve approximately means that we will have an error and that this error will be measured in terms of the natural parameter, which is the frequency $|\xi|$ (here $|\xi|$

is large). Also note that $e^{-ix\cdot\xi}Q(fe^{ix\cdot\xi}) = q_m(x,\xi)f + \cdots$ where the dots hide a polynomial in ξ of degree less than $m-1$. Thus, as a first approximation, we seek $\psi_{\xi,N}$ as a perturbation of $\varphi/q_m(x,\xi)$.

Lemma 12.8 *Consider a differential operator $Q = \sum_{|\alpha|\leqslant m} a_\alpha(x)\partial_x^\alpha$ of order m and let $q_m(x,\xi) = \sum_{|\alpha|=m} a_\alpha(x)(i\xi)^\alpha$ be its principal symbol. Let ω be an open set of $\mathbb{R}^n$ and $V \subset \mathbb{R}^n \setminus \{0\}$ a cone such that*

$$\exists C > 0/\ \forall (x,\xi) \in \omega \times V, \quad |q_m(x,\xi)| \geqslant C|\xi|^m.$$

For any integer N, for any $\varphi \in C_0^\infty(\omega)$ and any $\xi \in V$, there exist $\psi_{\xi,N} \in C_0^\infty(\omega)$ and $r_{\xi,N} \in C_0^\infty(\omega)$ such that

$$Q\left(\frac{\psi_{\xi,N}(x)}{q_m(x,\xi)}e^{ix\cdot\xi}\right) = \varphi(x)e^{ix\cdot\xi} + r_{\xi,N}(x)e^{ix\cdot\xi}$$

with $\sup_{x\in\mathbb{R}^n}\langle\xi\rangle^N \left|\partial_x^\alpha r_{\xi,N}(x)\right| < \infty$ for all multi-indices $\alpha \in \mathbb{N}^n$.

Proof We introduce an operator R_ξ (which depends on ξ) by setting

$$Q\left(\frac{\psi}{q_m(x,\xi)}e^{ix\cdot\xi}\right) = (\psi + R_\xi(\psi))e^{ix\cdot\xi}.$$

Then, we must solve, approximately, the equation $\psi + R_\xi(\psi) = \varphi$. Let us start by giving an expression for $R_\xi(\psi)$. For this, we calculate

$$e^{-ix\cdot\xi}Q\left(\frac{\psi}{q_m(x,\xi)}e^{ix\cdot\xi}\right)$$

directly with Leibniz's rule, by separating the expression into several terms: the first term corresponds to the case where the derivatives of order $|\alpha| = m$ act on the oscillating factor $e^{ix\cdot\xi}$; the sum of the other terms corresponds to $R_\xi(\psi)$, it is the sum of the terms for which either $|\alpha| \leqslant m-1$ and all derivatives act on $e^{ix\cdot\xi}$, or at least one derivative acts on the factor $\psi/q_m(x,\xi)$. We find

$$e^{-ix\cdot\xi}Q\left(\frac{\psi}{q_m(x,\xi)}e^{-ix\cdot\xi}\right) = (I) + R_\xi(\psi),$$

where

$$(I) = e^{-ix\cdot\xi}\sum_{|\alpha|=m}\frac{\psi}{q_m(x,\xi)}a_\alpha(x)\partial_x^\alpha(e^{ix\cdot\xi}),$$

$$R_\xi(\psi) = e^{-ix\cdot\xi}\sum_{|\alpha|\leqslant m-1}\frac{\psi}{q_m(x,\xi)}a_\alpha(x)\partial_x^\alpha(e^{ix\cdot\xi})$$

$$+ e^{-ix\cdot\xi}\sum_{|\alpha|\leqslant m}\sum_{\substack{\beta+\gamma=\alpha\\|\beta|>0}}a_\alpha(x)\partial_x^\beta\left(\frac{\psi}{q_m(x,\xi)}\right)\partial_x^\gamma(e^{ix\cdot\xi}).$$

Then $(I) = \psi$ because

$$\sum_{|\alpha|=m} a_\alpha(x)\partial_x^\alpha(e^{ix\cdot\xi}) = q_m(x,\xi)e^{ix\cdot\xi},$$

by definition of q_m. Let us set

$$\psi_{\xi,N} := \sum_{n=0}^{N-1}(-R_\xi)^n(\varphi), \quad r_{\xi,N} = (-1)^{N+1}R_\xi^N(\varphi).$$

Then $\psi_{\xi,N} + R_\xi(\psi_{\xi,N}) = \varphi + r_{\xi,N}$ and we verify that $r_{\xi,N}$ satisfies the desired properties.   □

Let us now prove the theorem. Let $(x_0,\xi_0) \notin \mathrm{Car}(P)$. Consider an open set ω of $\mathbb{R}^n$ and a cone $\Gamma \subset \mathbb{R}^n \setminus \{0\}$ such that

$$\exists C > 0 / (x,\xi) \in \omega \times \Gamma \implies |p_m(x,\xi)| \geqslant C|\xi|^m.$$

Then, with $Q = {}^tP$ and $V = -\Gamma$, we have

$$\exists C > 0 / \forall(x,\xi) \in \omega \times V, \quad |q_m(x,-\xi)| \geqslant C|\xi|^m.$$

Let us fix a function $\varphi \in C_0^\infty(\omega)$. To show that $(x_0,\xi_0) \notin \mathrm{WF}(u)$, we will estimate $\widehat{\varphi u}(\xi)$. The previous lemma implies that, for any integer N and any $\xi \in V$, there exist $\psi_{\xi,N} \in C_0^\infty(\mathbb{R}^n)$ and $r_{\xi,N}$ such that

$$ {}^tP\left(\frac{\psi_{\xi,N}}{q_m}e^{-ix\cdot\xi}\right) = \varphi e^{-ix\cdot\xi} + r_{\xi,N}e^{-ix\cdot\xi} $$

with $\sup|\partial_x^\alpha r_{\xi,N}| = O(|\xi|^{-N})$.

Then we can write

$$\begin{aligned}
\widehat{\varphi u}(\xi) = \langle u, \varphi e^{-ix\cdot\xi}\rangle &= \langle u, {}^tP\left((\psi_{\xi,N}/q_m)e^{-ix\cdot\xi}\right) - r_{\xi,N}e^{-ix\cdot\xi}\rangle\\
&= \langle Pu, (\psi_{\xi,N}/q_m)e^{-ix\cdot\xi}\rangle - \langle u, r_{\xi,N}e^{-ix\cdot\xi}\rangle\\
&= -\langle u, r_{\xi,N}e^{-ix\cdot\xi}\rangle,
\end{aligned}$$

where we used that $Pu = 0$.

Let us recall that by definition of tempered distributions, there exists an integer p and a constant C such that

$$\forall\kappa \in S(\mathbb{R}^n), \quad |\langle u,\kappa\rangle| \leqslant C \sup_{x\in\mathbb{R}^n, |\alpha|+|\beta|\leqslant p} \left|x^\alpha\partial_x^\beta\kappa(x)\right|,$$

where C_K^∞ is the space of functions of class C^∞ with support in K.

As $r_{\xi,N}$ is a C^∞ function with support in ω, there exists an integer M that depends only on ω such that, for all $\xi \in \Gamma$,

$$\left| \langle u, r_{\xi,N} e^{-ix\cdot\xi} \rangle \right| \leqslant C \sum_{|\alpha| \leqslant M} \sup \left| \partial_x^\alpha \left(r_{\xi,N} e^{-ix\cdot\xi} \right) \right|$$

but $\sup \left| \partial_x^\alpha \left(r_{\xi,N} e^{-ix\cdot\xi} \right) \right| = O(|\xi|^{|\alpha|-N})$, therefore $\left| \langle u, re^{-ix\cdot\xi} \rangle \right| \leqslant C_N \langle \xi \rangle^{M-N}$. (The constant C_N depends on ω and φ, but this does not pose any problem.) By taking N large enough, we conclude the proof. □

12.3 The Theorem of Propagation of Singularities

The theorem of propagation of singularities states that not only is the wave front set of the function contained in the characteristic variety, but it is necessarily a union of trajectories for a natural dynamical system.[1]

Definition 12.9 Consider a function $b = b(x,\xi) \in C^2(\mathbb{R}^{2n})$ with real values. We denote by $H_b : \mathbb{R}^{2n} \to \mathbb{R}^{2n}$ the vector field defined by

$$H_b(x,\xi) = \left(\frac{\partial b}{\partial \xi_1}(x,\xi), \ldots, \frac{\partial b}{\partial \xi_n}(x,\xi), -\frac{\partial b}{\partial x_1}(x,\xi), \ldots, -\frac{\partial b}{\partial x_n}(x,\xi) \right). \quad (12.2)$$

We say that H_b is the Hamiltonian vector field of b. Its integral curves are called bicharacteristics. For $(x,\xi) \in \mathbb{R}^n \times \mathbb{R}^n$, we denote by $t \mapsto \Phi^t_{H_b}(x,\xi) = (x(t),\xi(t))$ the unique maximal solution of the system

$$\frac{dx}{dt} = \frac{\partial b}{\partial \xi}(x(t),\xi(t)), \quad \frac{d\xi}{dt} = -\frac{\partial b}{\partial x}(x(t),\xi(t)),$$
$$x(0) = x, \quad \xi(0) = \xi. \quad (12.3)$$

Remark 12.10 We also use the notation

$$H_b = \sum_{j=1}^{n} \left(\frac{\partial b}{\partial \xi_j} \frac{\partial}{\partial x_j} - \frac{\partial b}{\partial x_j} \frac{\partial}{\partial \xi_j} \right)$$

to denote the vector field given by (12.2).

---

[1] To explain why such a result must exist, let us quote an argument from Gilles Lebeau. The starting postulate of this argument is that an equation should reduce the number of independent variables. This means that if $f = f(x)$ with $x \in \mathbb{R}^n$ is a solution of a partial differential equation $P(f) = 0$, then f actually depends only on $n - 1$ variables. When we microlocalize, we double the number of variables to go from $n - 1$ to $2n - 2$ variables. When we say that the wave front set is contained in the characteristic variety, we go from $2n$ to $2n - 1$. To go from $2n - 1$ dimensions to $2n - 2$, we need to use the theorem of propagation of singularities.

Proposition 12.11 *Suppose that b is a real-valued symbol with $b \in S^1(\mathbb{R}^n)$. Then the flow $\Phi^t_{H_b} : \mathbb{R}^{2n} \rightarrow \mathbb{R}^{2n}$ is defined for $t \in \mathbb{R}^+$. Moreover, if $p \in S^0(\mathbb{R}^n)$, then $p(\Phi^t_{H_b}(x, \xi))$ defines a symbol that belongs to $S^0(\mathbb{R}^n)$ for all $t \in \mathbb{R}^+$.*

Proof Given $(x, \xi) \in \mathbb{R}^{2n}$, the Cauchy problem (12.3) can be written in the form

$$\begin{cases} m'(t) = H_b(m(t)) & \text{where } m(t) = (x(t), \xi(t)), \\ m(0) = (x, \xi). \end{cases}$$

Since b is a C^∞ function, the vector field H_b is C^1 and therefore the Cauchy–Lipschitz theorem implies that there exists a unique maximal solution $m \colon [0, T^*) \rightarrow \mathbb{R}^{2n}$. Let us prove that this solution is globally defined, which means that $T^* = +\infty$. Recall the following alternative: either $T^* = +\infty$, or $\lim\sup_{t \rightarrow T^*} |m(t)| = +\infty$. To prove that this last condition is impossible, we will estimate $y(t) = |m(t)|^2$. Since the symbol b belongs to S^1, there exists a constant $C > 0$ such that $|H_b(m)| \leqslant C + C|m|$ for all $m \in \mathbb{R}^{2n}$. It follows that

$$\frac{d}{dt} y(t) = 2m'(t) \cdot m(t) \leqslant 2C(1 + |m|)|m| \leqslant C^2 + 3Cy(t),$$

where we assumed, without loss of generality, that $C \geqslant 1$ and used the inequality $2C|m| \leqslant C^2 + |m|^2 = C^2 + y$. It then follows from Gronwall's lemma that

$$y(t) \leqslant y(0)e^{3Ct} + \frac{C}{3}(e^{3Ct} - 1),$$

from which we deduce that $T^* = +\infty$.

Moreover, the regular dependence of the solution of an ordinary differential equation on the initial data implies that, for all $t \geqslant 0$, the flow $(x, \xi) \mapsto \Phi^t_{H_b}(x, \xi)$ is of class C^∞. Let $\Phi^t(x, \xi) = (X^t(x, \xi), \Xi^t(x, \xi))$. We claim that, for all multi-indices α and β in $\mathbb{N}^n$, there exist constants $C_{\alpha\beta}$ and $C'_{\alpha\beta}$ such that

$$\forall (x, \xi) \in \mathbb{R}^{2n}, \quad \left| \partial_x^\alpha \partial_\xi^\beta X^t(x, \xi) \right| \leqslant C_{\alpha\beta} \langle \xi \rangle^{-|\beta|} \text{ if } |\alpha| + |\beta| > 0, \tag{12.4}$$

$$\forall (x, \xi) \in \mathbb{R}^{2n}, \quad \left| \partial_x^\alpha \partial_\xi^\beta \Xi^t(x, \xi) \right| \leqslant C'_{\alpha\beta} \langle \xi \rangle^{1-|\beta|} \text{ for all } \alpha, \beta \in \mathbb{N}^n. \tag{12.5}$$

We start by studying $\Xi^t(x, \xi)$. Since $\partial_x b$ is a symbol of order 1, as above, we have

$$\left| \frac{d\Xi^t}{dt}(x, \xi) \right| \leqslant C + C \left| \Xi^t(x, \xi) \right|, \tag{12.6}$$

for some constant $C \geqslant 1$. Since $\Xi^0(x, \xi) = \xi$, the same argument as above implies that

$$\left| \Xi^t(x, \xi) \right|^2 \leqslant |\xi|^2 e^{3Ct} + \frac{C}{3}(e^{3Ct} - 1).$$

This proves that (12.5) holds when $\alpha = \beta = 0$. Moreover, this implies that there exists t_1 small enough (that is for $Ce^{3Ct_1} < 4$), such that for all $t \in [0, t_1]$, we have $|\Xi^t(x, \xi)| \leqslant 2(1 + |\xi|)$. Now let us fix $t_0 \leqslant \min\{t_1, 1/(6C)\}$. Then, by substituting the estimate $|\Xi^t(x, \xi)| \leqslant 2(1 + |\xi|)$ into (12.6) it follows that, for all $t \in [0, t_0]$,

$$\left|\Xi^t(x, \xi) - \xi\right| \leqslant \int_0^t \left|\frac{\mathrm{d}}{\mathrm{d}s}\Xi^s(x, \xi)\right| \mathrm{d}s \leqslant \frac{1}{2}(1 + |\xi|).$$

Therefore, for all (x, ξ) in $\mathbb{R}^{2n}$ and any time t in $[0, t_0]$,

$$\frac{1}{2}|\xi| - \frac{1}{2} \leqslant \left|\Xi^t(x, \xi)\right| \leqslant \frac{1}{2} + \frac{3}{2}|\xi|. \tag{12.7}$$

Let us define

$$S_t(x, \xi) = \begin{pmatrix} \mathrm{d}_x X^t & \langle \xi \rangle \, \mathrm{d}_\xi X^t \\ \langle \xi \rangle^{-1} \, \mathrm{d}_x \Xi^t & \mathrm{d}_\xi \Xi^t \end{pmatrix},$$

where the differentials $\mathrm{d}_x X^t$, $\mathrm{d}_\xi X^t$, $\mathrm{d}_x \Xi^t$ and $\mathrm{d}_\xi \Xi^t$ are identified with matrices. We can form an evolution equation on S_t, namely

$$\frac{\partial}{\partial t} S_t(x, \xi) = A(t, x, \xi) S_t(x, \xi) \quad ; \quad S_0(x, \xi) = I_{\mathbb{R}^{2n}},$$

where

$$A = \begin{pmatrix} \mathrm{d}_x \nabla_\xi b \circ \Phi^t(x, \xi) & \langle \xi \rangle \, \mathrm{d}_\xi \nabla_\xi b \circ \Phi^t(x, \xi) \\ -\langle \xi \rangle^{-1} \, \mathrm{d}_x \nabla_x b \circ \Phi^t(x, \xi) & -\mathrm{d}_\xi \nabla_x b \circ \Phi^t(x, \xi) \end{pmatrix}.$$

Assume that (12.4) and (12.5) hold for all multi-indices α, β such that $|\alpha| + |\beta| \leqslant k$ with $k \in \mathbb{N}^*$. It follows that if $|\alpha| + |\beta| \leqslant k$ then

$$\sup_{\mathbb{R}^{2n}} \langle \xi \rangle^{|\beta|} \left|\partial_x^\alpha \partial_\xi^\beta S_t(x, \xi)\right| < +\infty. \tag{12.8}$$

We want to prove a similar estimate for A. We claim that if $|\alpha| + |\beta| \leqslant k$ then

$$\forall t \in [0, t_0], \quad \sup_{\mathbb{R}^{2n}} \langle \xi \rangle^{|\beta|} \left|\partial_x^\alpha \partial_\xi^\beta A(t, x, \xi)\right| < +\infty. \tag{12.9}$$

To see this, we first observe that, for any function $F \in C^\infty(\mathbb{R}^{2n})$, and for all multi-indices α, β, $\partial_x^\alpha \partial_\xi^\beta F(\Phi^t(x, \xi))$ is a linear combination of terms of the form

$$\Pi(x, \xi)(\partial_x^{\alpha'} \partial_\xi^{\beta'} F)(\Phi^t(x, \xi)),$$

where the factor Π is a product of the form:

$$\Pi = \left(\partial_x^{a_1} \partial_\xi^{b_1} X_{i_1}^t\right) \cdots \left(\partial_x^{a_{|\alpha'|}} \partial_\xi^{b_{|\alpha'|}} X_{i_{|\alpha'|}}^t\right) \left(\partial_x^{a'_1} \partial_\xi^{b'_1} X_{j_1}^t\right) \cdots \left(\partial_x^{a'_{|\alpha'|}} \partial_\xi^{b'_{|\alpha'|}} \Xi_{j_{|\beta'|}}^t\right),$$

with

$$a_1 + \cdots + a_{|\alpha'|} + a'_1 + \cdots + a'_{|\beta'|} = \alpha, \quad b_1 + \cdots + b_{|\alpha'|} + b'_1 + \cdots + b'_{|\beta'|} = \beta.$$

Moreover, for any symbol $r \in S^0$, we have

$$\left|(\partial_x^{\alpha'} \partial_\xi^{\beta'} r)(\Phi^t(x,\xi))\right| \leqslant C\langle\Xi^t(x,\xi)\rangle^{-|\beta'|} \leqslant C'\langle\xi\rangle^{-|\beta'|},$$

where we used (12.7). Using the previous inequality with

$$r = \partial_{\xi_j} \partial_{x_k} b, \quad r = \langle\xi\rangle^{-1} \partial_{x_j x_k}^2 b \quad \text{or} \quad r = \langle\xi\rangle \partial_{\xi_j \xi_k}^2 b,$$

(which are symbols of order 0) and combining this with the induction hypothesis, we obtain the desired result (12.9).

It follows that

$$\frac{\partial}{\partial t} \partial_x^\alpha \partial_\xi^\beta S_t(x,\xi) = R(t,x,\xi) + A(t,x,\xi)\partial_x^\alpha \partial_\xi^\beta S_t(x,\xi),$$

$$\partial_x^\alpha \partial_\xi^\beta S_0(x,\xi) = \delta_a^0 \delta_b^0 I_{\mathbb{R}^{2n}},$$

where $R(t,x,\xi)$ is a linear combination of terms of the form

$$(\partial_x^{\alpha-\alpha'} \partial_\xi^{\beta-\beta'} A)(\partial_x^{\alpha'} \partial_\xi^{\beta'} S_t)$$

with $|\alpha'| + |\beta'| < |\alpha'| + |\beta'| \leqslant k$. In particular, it follows from (12.8) and (12.9) that

$$\forall t \in [0,t_0], \quad \sup_{\mathbb{R}^{2n}} \langle\xi\rangle^{|\beta|} |R(t,x,\xi)| < +\infty.$$

Then Gronwall's lemma implies that

$$\sup \langle\xi\rangle^{|\beta|} \left|\partial_x^\alpha \partial_\xi^\beta S_t(x,\xi)\right| < +\infty.$$

Now, directly from the definition of S_t, we deduce that the induction hypothesis holds at rank $k + 1$.

Now consider a symbol p in S^0, a time t in $[0,t_0]$ and let $q(x,\xi) = p(\Phi^t(x,\xi))$. It follows from the estimates (12.4) and (12.5), and the arguments used to estimate A above, that q is a symbol of order 0. The argument above is valid for any sufficiently small time t, namely for $t \in [0,t_0]$. To conclude that the result is valid for all times, we will see that it suffices to iterate. To start, let us first prove the desired result over the time interval $[0,2t_0]$. For this, let us define $q = p \circ \Phi^{t_0}$ and consider $\tau \in [0,t_0]$. Since $\Phi^{t_0+\tau} = \Phi^{t_0} \circ \Phi^\tau$, we have $p \circ \Phi^{t_0+\tau} = q \circ \Phi^\tau$. We apply the previous result twice, to successively obtain that q is a symbol of order 0 then that $q \circ \Phi^\tau$ is also a symbol of order 0. More generally, by induction we successively prove that, for any integer N, $p \circ \Phi^t$ belongs to S^0 for any $t \in [0, Nt_0]$. This concludes the proof. $\square$

Now, consider a symbol $a \in S^1(\mathbb{R}^n)$. We assume that a can be written in the form $a^1 + a^0$ where

1. $a^0 \in S^0(\mathbb{R}^n)$;
2. $a^1 \in S^1(\mathbb{R}^n)$ is a symbol with purely imaginary values and homogeneous in ξ of order 1.

For example, $a(x, \xi) = iV(x)\xi$ with $V \in C_b^\infty(\mathbb{R}^n)$ a real-valued function.

We have shown in the previous chapter how to solve the Cauchy problem for the equation

$$\frac{\partial u}{\partial t} + \mathrm{Op}(a)u = 0.$$

We denote by $S(t, s) = e^{(s-t)\,\mathrm{Op}(a)} : L^2 \to L^2$ the solution operator that associates a given function $u_0 \in L^2(\mathbb{R}^n)$ with the value at time t of the unique solution of the Cauchy problem that equals u_0 at time s. That is to say: $u(t) = S(t, s)u_0$ is the unique function $u \in C^0(\mathbb{R}; L^2(\mathbb{R}^n))$ such that

$$\frac{\partial u}{\partial t} + \mathrm{Op}(a)u = 0, \quad u(s) = u_0.$$

Theorem 12.12 *Let $P_0 = \mathrm{Op}(p_0) \in \mathrm{Op}\, S^0(\mathbb{R}^n)$ be a pseudo-differential operator. Then, for all $t \in \mathbb{R}$, modulo a regularizing operator, $S(t, 0)P_0S(0, t)$ is a pseudo-differential operator: there exists a symbol $q_t \in S^0(\mathbb{R}^n)$ such that, for all $u_0 \in L^2(\mathbb{R}^n)$,*

$$S(t, 0)P_0S(0, t)u_0 - \mathrm{Op}(q_t)u_0 \in H^\infty(\mathbb{R}^n).$$

Furthermore,

$$q_t(x, \xi) - p_0(\Phi_H^{-t}(x, \xi)) \in S^{-1}(\mathbb{R}^n),$$

where Φ_H^t is the flow associated with the vector field

$$H = \frac{1}{i}\sum_{j=1}^n \left(\frac{\partial a^1}{\partial \xi_j}\frac{\partial}{\partial x_j} - \frac{\partial a^1}{\partial x_j}\frac{\partial}{\partial \xi_j} \right).$$

Proof By differentiating with respect to t, we find that $P(t) = S(t, 0)P_0S(0, t)$ satisfies

$$P'(t) + \left[\mathrm{Op}(a), P(t)\right] = 0, \quad P(0) = P_0.$$

We will construct an approximate solution $Q(t)$ of this equation and then show that $P(t) - Q(t)$ is a regularizing operator. We therefore seek $Q(t) = \mathrm{Op}(q_t)$ with $q \in S^0(\mathbb{R}^n)$ a solution of

$$Q'(t) + \left[\mathrm{Op}(a_t), Q(t)\right] = R(t), \quad Q(0) = P_0,$$

where $R(t)$ is a family of regularizing operators. We will construct q in the form

$$q(t, x, \xi) \sim q^{(0)}(t, x, \xi) + q^{(-1)}(t, x, \xi) + \cdots$$

where $q^{(-k)}$ is a symbol of order $-k$. Then the symbol of $[\mathrm{Op}(a_t), Q(t)]$ is of the form

$$Hq + \frac{1}{i}\{\mathrm{Op}(a^0), q\} + \sum_{|\alpha| \geqslant 2} \frac{1}{i^{|\alpha|}\alpha!} \left[(\partial_\xi^\alpha a)(\partial_x^\alpha q) - (\partial_\xi^\alpha q)(\partial_x^\alpha a)\right]. \qquad (12.10)$$

This suggests defining $q^{(0)}$ by

$$\left(\frac{\partial}{\partial t} + H\right)q^{(0)}(t, x, \xi) = 0, \quad q^{(0)}(0, x, \xi) = p_0(x, \xi).$$

Thus, $q^{(0)}(t, x, \xi) = p_0(\Phi_H^{-t}(x, \xi))$, the symbol given by the statement of the theorem. We have $q^{(0)}(t, x, \xi) \in S^0(\mathbb{R}^n)$. By induction, we solve

$$\left(\frac{\partial}{\partial t} + H\right)q^{(-j)}(t, x, \xi) = b^{(-j)}(t, x, \xi), \quad q^{(-j)}(0, x, \xi) = 0,$$

where b^{-j} is determined by induction, in order to obtain a solution of (12.10).

Finally, it remains to show that $P(t) - Q(t)$ is a regularizing operator. Equivalently, we will show that

$$v(t) - w(t) = S(t, 0)P_0 f - Q(t)S(t, 0)f \in H^\infty(\mathbb{R}^n).$$

Note that

$$\frac{\partial v}{\partial t} + \mathrm{Op}(a)v = 0, \quad v(0) = P_0 f,$$

while

$$\frac{\partial w}{\partial t} + \mathrm{Op}(a)w = g, \quad w(0) = P_0 f,$$

with $g = R(t)S(t, 0)w \in C^0(\mathbb{R}; H^\infty(\mathbb{R}^n))$. Taking the difference of the two equations, we find

$$\frac{\partial}{\partial t}(v - w) + \mathrm{Op}(a)(v - w) = -g, \quad v(0) - w(0) = 0.$$

Then the theorem on the resolution of hyperbolic equations implies that $v(t) - w(t) \in H^\infty(\mathbb{R}^n)$ for all t and all $f \in H^{-\infty}(\mathbb{R}^n)$. This completes the proof. $\qquad \square$

We can now calculate the action of the solution operator $\exp(t\,\mathrm{Op}(a))$ on the wave front set of the initial data.

Let us recall that we have shown the following proposition.

Proposition 12.13 *Let $m \in \mathbb{R}$ and let $a \in S^m$. Consider $u \in S'(\mathbb{R}^n)$ such that $\mathrm{Op}(a)u \in C^\infty(\mathbb{R}^n)$ and $|a(x, \xi)| \geqslant |\xi|^m$ for all $(x, \xi) \in \omega \times \Gamma$ where ω is a neighborhood of x_0 and Γ is a cone containing ξ_0. Then $(x_0, \xi_0) \notin \mathrm{WF}(u)$.*

Theorem 12.14 *If u satisfies*

$$\frac{\partial u}{\partial t} + \mathrm{Op}(a)u = 0, \quad u_{|t=0} = u_0$$

with $u_0 \in L^2(\mathbb{R}^n)$ then $u \in C^0([0,T]; L^2(\mathbb{R}^n))$ and, for all $t \in [0,T]$,

$$\mathrm{WF}(u(t,\cdot)) = \Phi_H^t(\mathrm{WF}(u_0)).$$

Proof Suppose that $(x_0, \xi_0) \notin \mathrm{WF}(u_0)$. Then there exists a neighborhood ω of x_0, a cone Γ containing ξ_0 and a symbol $p_0 \in S^0$ such that $|p_0(x,\xi)| \geqslant 1$ (for all $(x,\xi) \in \omega \times \Gamma$) and satisfying $\mathrm{Op}(p_0)u_0 \in S(\mathbb{R}^n)$.

Using the operator Q introduced in the proof of the previous theorem, we obtain that

$$(\partial_t + \mathrm{Op}(a))Qu = Ru \in C^0([0,T]; H^\infty), \quad Qu|_{t=0} \in S(\mathbb{R}^n).$$

We deduce that $Q(t)u(t) \in H^\infty(\mathbb{R}^n) \subset C^\infty(\mathbb{R}^n)$ and therefore $\Phi_H^t(x_0, \xi_0) \notin \mathrm{WF}\, u(t,\cdot)$. As we can reverse the direction of time, we find the desired result. $\qquad \square$

12.4 Nonlinear Problems

In this section, we present the paradifferential calculus introduced by Jean-Michel Bony (see [11, 134]) to study the singularities of nonlinear partial differential equations, of the form

$$F((\partial_x^\alpha f)_{|\alpha| \leqslant m}) = 0.$$

We do not know how to describe for such a general equation the set of points where the function is not microlocally C^∞. In particular, we do not have an analogue of Theorem 12.7. However, thanks to the paradifferential calculus, we can say things if we replace C^∞ with a space of functions of limited regularity (for example H^r with $r < +\infty$). For this, we will use the following definition. We say that u is microlocally H^r at (x_0, ξ_0) if there exists a symbol $\varphi = \varphi(x,\xi)$ homogeneous of order 0 in ξ, satisfying $\varphi(x_0, \xi_0) \neq 0$, such that $\mathrm{Op}(\varphi)u \in H^r$.

Theorem 12.15 *Let $n, m \geqslant 1$ and $s_0 = n/2 + m$. Consider a solution $f \in H^s(\mathbb{R}^n)$ with $s > s_0$ of*

$$F((\partial_x^\alpha f)_{|\alpha| \leqslant m}) = 0. \tag{12.11}$$

Let us introduce the symbol

$$p_m(x,\xi) = \sum_{|\alpha|=m} \frac{\partial F}{\partial f_\alpha}((\partial_x^\alpha f(x))_{|\alpha| \leqslant m})(i\xi)^\alpha.$$

At any point $(x_0, \xi_0) \in \mathbb{R}^n \times (\mathbb{R}^n \setminus \{0\})$ such that $p_m(x_0, \xi_0) \neq 0$, f is microlocally twice as regular: it is microlocally of class H^t for all $t < 2s - s_0$.

For the proof of this result, we refer to the original article by Bony [11], as well as the books by Hörmander [78], Métivier [134] and Taylor [187].

In the previous statement, the symbol p_m depends on the unknown f and this explains why the proof of this result requires working with symbols of limited regularity.

Definition 12.16 Let $m \in \mathbb{R}$ and $k \in \mathbb{N}$. The class of symbols of order m and regularity C^k in x, denoted $\Gamma_\rho^m(\mathbb{R}^n)$, is the set of functions $a = a(x, \xi)$ such that, for all multi-indices β in $\mathbb{N}^n$, there exists a constant C_β such that

$$\left\| \partial_\xi^\beta a(\cdot, \xi) \right\|_{C^k} \leqslant C_\beta (1 + |\xi|)^{m - |\beta|}.$$

Bony's paradifferential quantification associates with a symbol a the operator T_a defined by

$$\widehat{T_a u}(\xi) = (2\pi)^{-n} \int \chi(\xi - \eta, \eta) \widehat{a}(\xi - \eta, \eta) \psi(\eta) \widehat{u}(\eta) \, d\eta,$$

where $\widehat{a}(\theta, \xi) = \int e^{-ix \cdot \theta} a(x, \xi) \, dx$, and the truncation functions ψ and χ satisfy

$$\psi(\eta) = 0 \quad \text{for } |\eta| \leqslant 1, \quad \psi(\eta) = 1 \quad \text{for } |\eta| \geqslant 2,$$

and, for $\varepsilon_1, \varepsilon_2$ small enough,

$$\chi(\theta, \eta) = 1 \quad \text{if } |\theta| \leqslant \varepsilon_1 |\eta|, \quad \chi(\theta, \eta) = 0 \quad \text{if } |\theta| \geqslant \varepsilon_2 |\eta|.$$

The central point in the proof of the previous theorem is to show that, if f satisfies equation (12.11), then

$$T_{p_m} f \in H^t(\mathbb{R}^n),$$

which is a paradifferential equation. We say that we have paralinearized (12.11).

12.5 Exercises

Exercise 12.1 (Another characterization of the wave front) Let $s \in \mathbb{R}$. Consider a given function $f \in H^s(\mathbb{R}^n)$. We want to show the equivalence between the following two properties:

1. $(x_0, \xi_0) \notin WF(f)$;
2. there exists a conical neighborhood Γ of (x_0, ξ_0) such that, for any symbol $a \in S^\infty$ satisfying $\mathrm{supp}(a) \subset \Gamma$, we have $\mathrm{Op}(a)f \in H^\infty(\mathbb{R}^n)$.

1. Show $(2) \Rightarrow (1)$.
2. Conversely, let $(x_0, \xi_0) \notin WF(f)$. (a) Show that there exists a symbol $a \in S^0$ satisfying the following two properties:

- $\text{Op}(a)f \in H^{\infty}$.
- There exists a conical neighborhood Γ of (x_0, ξ_0) and $R > 0$ such that:

$$\forall (x, \xi) \in \Gamma, \quad |\xi| \geqslant R, \quad |a(x, \xi)| \geqslant 1.$$

(b) Let a, Γ be as before. Let Γ' be an open conical neighborhood of (x_0, ξ_0) such that $\overline{\Gamma}' \subset \mathring{\Gamma}$ and let $b \in S^{\infty}$ such that $\text{supp}(b) \subset \Gamma'$. Show that there exists a $c \in S^{\infty}$ such that:

$$\text{Op}(b) - \text{Op}(c)\,\text{Op}(a) \in \text{Op}(S^{-\infty}).$$

(c) Conclude.

Exercise 12.2 (Microlocal defect measures) The purpose of this difficult exercise is to demonstrate the construction of microlocal defect measures, which is inspired by the study of microlocal singularities. These measures were independently introduced by Patrick Gérard [57] and Luc Tartar [185]. They play a fundamental role in many fields. The aim of this problem is to study the construction of these measures. For this, we follow the approach of Patrick Gérard.

Notation

We consider complex-valued functions defined on any open set Ω of $\mathbb{R}^d$ with $d \geqslant 1$. Specifically, we consider a sequence $(u_n)_{n \in \mathbb{N}}$ of functions belonging to $L^2_{\text{loc}}(\Omega)$ (u_n belongs to $L^2(K)$ for all compact $K \subset \Omega$) which converges weakly to 0 in the sense that:

$$\forall \varphi \in L^2_{\text{comp}}(\Omega), \quad \int_{\Omega} u_n(x)\varphi(x)\,dx \xrightarrow[n \to +\infty]{} 0,$$

where $L^2_{\text{comp}}(\Omega)$ denotes the space of functions $\varphi \in L^2(\Omega)$ with compact support in Ω.

The aim of this problem is to characterize the lack of strong convergence using pseudo-differential operators.

Throughout the problem, χ denotes a real-valued function such that $\chi \in C_0^{\infty}(\Omega)$. If $u \in L^2_{\text{loc}}(\Omega)$, then $\chi u \in L^2(\Omega)$ and we can extend χu by 0 on $\mathbb{R}^d \setminus \Omega$ to obtain a function belonging to $L^2(\mathbb{R}^d)$. We will always call this function χu. This allows us to define the Fourier transform of χu, denoted $\widehat{\chi u}$, as well as $\text{Op}(a)(\chi u)$ for any symbol $a \in S^m(\mathbb{R}^d)$ (where m is any real number).

Given $s \in \mathbb{R}$, we denote by $H^s(\mathbb{R}^d)$ the Sobolev space of order s and $\langle D_x \rangle^s$ the pseudo-differential operator of symbol $\langle \xi \rangle^s = (1 + |\xi|^2)^{s/2}$. We denote by (u, v) the scalar product on $L^2(\mathbb{R}^d)$, defined by $(u, v) = \int_{\mathbb{R}^d} u(x)\overline{v(x)}\,dx$. Remember to use:

- the Cauchy–Schwarz inequality;
- the fact that $\|u\|_{H^s} = \|\langle D_x \rangle^s u\|_{L^2}$ by definition of the norm $\|\cdot\|_{H^s}$;
- the duality: for all $s \geqslant 0$ and for all $(u, v) \in H^s(\mathbb{R}^d) \times L^2(\mathbb{R}^d)$, we have $|(u, v)| \leqslant \|u\|_{H^s}\|v\|_{H^{-s}}$.

Problem

1. Let $(u_n)_{n \in \mathbb{N}}$ be a sequence that converges weakly to 0 in $L^2_{loc}(\Omega)$.

   (a) Let $\chi \in C_0^\infty(\Omega)$. Show that: (i) the sequence (χu_n) is bounded in $L^2(\mathbb{R}^d)$; (ii) the sequence $(\widehat{\chi u_n})$ is bounded in $L^\infty(\mathbb{R}^d)$ and (iii) for all $\xi \in \mathbb{R}^d$, $\lim_{n \to +\infty} \widehat{\chi u_n}(\xi) = 0$.

   (b) Let $\chi \in C_0^\infty(\Omega)$. Show that $\|\chi u_n\|_{H^{-1/2}(\mathbb{R}^d)}$ tends to 0 when n tends to $+\infty$.
   Hint: write

   $$\|\chi u_n\|^2_{H^{-1/2}(\mathbb{R}^d)} = \int_{|\xi| \leqslant R} (1 + |\xi|^2)^{-1/2} |\widehat{\chi u_n}(\xi)|^2 \, d\xi$$
   $$+ \int_{|\xi| \geqslant R} (1 + |\xi|^2)^{-1/2} |\widehat{\chi u_n}(\xi)|^2 \, d\xi,$$

   where R is a large parameter.

2. Let $\chi \in C_0^\infty(\Omega)$ and let $(u_n)_{n \in \mathbb{N}}$ be a sequence that converges weakly to 0 in $L^2_{loc}(\Omega)$.

   (a) Show that, for every symbol $a \in S^0(\mathbb{R}^d)$, the sequence

   $$(\mathrm{Op}(a)(\chi u_n), \chi u_n)$$

   is bounded in $\mathbb{C}$.

   (b) Show that, for every symbol $\tau \in S^{-1}(\mathbb{R}^d)$, we have

   $$(\mathrm{Op}(\tau)(\chi u_n), \chi u_n) \underset{n \to +\infty}{\longrightarrow} 0.$$

Reminder: Gårding's Inequality

Let $\varepsilon > 0$. Consider a symbol $a \in S^0(\mathbb{R}^d)$ such that $\mathrm{Re}\, a(x, \xi) \geqslant \varepsilon$ for all (x, ξ) in $\mathbb{R}^{2d}$. We have seen Gårding's inequality, which states that there exist two positive constants c_ε and C_ε such that for all $u \in L^2(\mathbb{R}^d)$,

$$\mathrm{Re}(\mathrm{Op}(a)u, u) \geqslant c_\varepsilon \|u\|^2_{L^2} - C_\varepsilon \|u\|^2_{H^{-1/2}}.$$

In particular, we have $\mathrm{Re}(\mathrm{Op}(a)u, u) \geqslant -C_\varepsilon \|u\|^2_{H^{-1/2}}$.

3. Let $\chi \in C_0^\infty(\Omega)$ and let $(u_n)_{n \in \mathbb{N}}$ be a sequence that converges weakly to 0 in $L^2_{loc}(\Omega)$. Consider a symbol $a \in S^0(\mathbb{R}^d)$ such that $\mathrm{Re}\, a(x, \xi) \geqslant 0$ for all (x, ξ) in $\mathbb{R}^{2n}$. Show that

   $$\liminf_{n \to +\infty} \mathrm{Re}(\mathrm{Op}(a)(\chi u_n), \chi u_n) \geqslant 0.$$

4. Let $a \in S^0(\mathbb{R}^d)$. We assume that a can be written in the form $a = a_0 + a_{-1}$, where a_{-1} belongs to $S^{-1}(\mathbb{R}^d)$ and $a_0 \in S^0(\mathbb{R}^d)$ satisfies $\mathrm{Im}\, a_0 = 0$. Show that

   $$\mathrm{Im}(\mathrm{Op}(a)(\chi u_n), \chi u_n) \underset{n \to +\infty}{\longrightarrow} 0.$$

5. Let $a \in S^0(\mathbb{R}^d)$. We set $M = \sup_{(x,\xi) \in \mathbb{R}^{2n}} |a(x,\xi)|$. Show that

$$\limsup_{n \to +\infty} \|\mathrm{Op}(a)(\chi u_n)\|_{L^2(\mathbb{R}^d)} \leqslant MC(\chi) \quad \text{where} \quad C(\chi) = \limsup_{n \to +\infty} \|\chi u_n\|_{L^2(\mathbb{R}^d)}.$$

Hint: you may introduce a function $\chi' \in C_0^\infty(\Omega)$ such that $\chi'\chi = \chi$ as well as the operator B defined by $Bv = M^2\chi^2 v - \chi \,\mathrm{Op}(a)^* \,\mathrm{Op}(a)(\chi v)$.

6. (difficult) Consider a compact $K \subset \Omega$. Let us introduce:
   – the set $C_K^0(\Omega)$ of continuous functions with support contained in K;
   – the set $C_K^\infty(\Omega \times S^{d-1})$ of functions of class C^∞ on $\Omega \times S^{d-1}$ with support in $K \times S^{d-1}$;
   – the space $S_K^0(\mathbb{R}^d)$ of symbols $a \in S^0(\mathbb{R}^d)$ such that $a(x,\xi)$ is homogeneous in ξ of order 0 for $|\xi| \geqslant 1/2$ and furthermore $\mathrm{supp}\, a \subset K \times \mathbb{R}^d$.
   We assume that $C_K^\infty(\Omega \times S^{d-1})$ is separable (i.e., there exists a countable dense subset).
   Deduce that there exists a continuous linear form $\Lambda_{K,\chi} : S_K^0(\mathbb{R}^d) \to \mathbb{C}$ such that, for all $a \in S_K^0(\mathbb{R}^d)$,

$$(\mathrm{Op}(a)(\chi u_n), \chi u_n) \xrightarrow[n \to +\infty]{} \Lambda_{K,\chi}(a).$$

7. Show that $\Lambda_{K,\chi}$ does not depend on χ then prove the following theorem.

Theorem 12.17 *Let $(u_n)_{n\in\mathbb{N}}$ be a sequence that converges weakly to 0 in $L^2_{\mathrm{loc}}(\Omega)$. There exists a subsequence $(u_{\theta(n)})$ and a positive Radon measure μ on $\Omega \times S^{d-1}$ such that the following result is true: if $a \in S^0(\mathbb{R}^d)$ and $\chi \in C_0^\infty(\Omega)$ satisfy:*

- *a is homogeneous in ξ of order 0 for $|\xi| \geqslant 1/2$;*
- *for some compact $K \subset \Omega$ we have $\mathrm{supp}\, a \subset K \times \mathbb{R}^d$ and $\chi = 1$ on K,*

then

$$(\mathrm{Op}(a)(\chi u_{\theta(n)}), \chi u_{\theta(n)}) \xrightarrow[n \to +\infty]{} \int_{\Omega \times S^{d-1}} a(x,\xi)\, \mathrm{d}\mu(x,\xi). \tag{12.12}$$

Definition 12.18 We say that μ is the microlocal defect measure of $(u_{\theta(n)})$.

8. Let $(u_n)_{n\in\mathbb{N}}$ be a sequence that converges weakly to 0 in $L^2_{\mathrm{loc}}(\Omega)$ and has a microlocal defect measure μ. Consider a differential operator of order m, $P = \sum_{|\alpha| \leqslant m} a_\alpha(x) \partial_x^\alpha$, with $a_\alpha \in C_b^\infty(\mathbb{R}^d)$. We denote by $p_m = \sum_{|\alpha|=m} a_\alpha(x)(i\xi)^\alpha$ the principal symbol of P. Suppose that, for all $\chi \in C_0^\infty(\Omega)$, the sequence $\chi P(u_n)$ converges strongly to 0 in $H^{-m}(\mathbb{R}^d)$. Deduce that

$$\mathrm{supp}\, \mu \subset \{(x,\xi) \in \Omega \times S^{d-1} : p_m(x,\xi) = 0\}.$$

Hint: write $\|\chi P u_n\|^2_{H^{-m}(\mathbb{R}^d)}$ in the form $(\mathrm{Op}(a)(\chi' u_n), \chi' u_n)$.

9. (difficult) Let $(u_n)_{n\in\mathbb{N}}$ be a sequence that converges weakly to 0 in $L^2_{\mathrm{loc}}(\Omega)$ and has a microlocal defect measure μ. Consider a differential operator of order m, $P = \sum_{|\alpha| \leqslant m} a_\alpha(x) \partial_x^\alpha$ with $a_\alpha \in C_b^\infty(\mathbb{R}^d)$. We further assume that $P^* = P$

and that, for all $\chi \in C_0^\infty(\Omega)$, the sequence $\chi P(u_n)$ converges to 0 strongly in $H^{1-m}(\mathbb{R}^d)$. Show that, for any function $a \in C^\infty(\Omega \times (\mathbb{R}^d \setminus \{0\}))$ homogeneous of order $1 - m$ in ξ for $|\xi| \geq 1/2$ and with compact support in x, we have

$$\int_{\Omega \times S^{d-1}} \{a, p_m\}(x, \xi) \, d\mu(x, \xi) = 0.$$

Hint: introduce χ such that $\chi a = a$.

Part IV
Analysis of Partial Differential Equations

Chapter 13
The Calderón Problem

13.1 Introduction

Given a bounded C^∞ open set $\Omega \subset \mathbb{R}^n$, a function $V \in L^\infty(\Omega)$ such that $V \geqslant 0$ and a function $f \in H^1(\Omega)$, we are interested in the following problem, with the unknown function $u \colon \overline{\Omega} \to \mathbb{R}$,

$$\begin{cases} -\Delta u + Vu = 0 & \text{in } \Omega, \\ u = f & \text{on } \partial\Omega. \end{cases}$$

This problem admits a unique solution u belonging to $H^1(\Omega)$. A very important operator in applications is the Dirichlet–Neumann operator, defined by

$$\Lambda_V(f) = \partial_\nu u \big|_{\partial\Omega},$$

where $\partial_\nu u$ denotes the normal derivative of u. Sylvester and Uhlmann proved (cf. [180]) that in dimension $n \geqslant 3$ the operator Λ_V determines the potential V. More precisely, their result states that the map $V \mapsto \Lambda_V$ is injective. This theorem answers a question posed and studied for the first time by Calderón (see [24]) and which is now at the heart of current concerns in the study of inverse problems.

Sylvester and Uhlmann's method of proof follows a surprising idea introduced by Calderón. It consists of working with harmonic functions (solutions of $\Delta u = 0$) of the form $x \mapsto \exp(\zeta \cdot x)$, where $\zeta = (\zeta_1, \ldots, \zeta_n) \in \mathbb{C}^n$ is such that $\zeta_1^2 + \cdots + \zeta_n^2 = 0$. Calderón had shown, thanks to these functions, that the vector space generated by the products of harmonic functions is dense in L^2. The proof of this result relies on the injectivity theorem of Fourier.

Sylvester and Uhlmann's construction of functions will play an analogous role in the case where the equation $\Delta u = 0$ is replaced by $-\Delta u + Vu = 0$. The fundamental difference is that it is an equation with variable coefficients. One of the objectives of this chapter is precisely to show how to approach a problem with variable coefficients using the notion of approximate solution. The proof given by Sylvester and Uhlmann contained a rather technical passage which was simplified by Hähner [66].

© The Author(s), under exclusive license to Springer Nature Switzerland AG 2024
T. Alazard, *Analysis and Partial Differential Equations*, Universitext,
https://doi.org/10.1007/978-3-031-70909-8_13

We will follow his very elegant proof, which will allow us to give an original illustration of the decomposition on a Hilbert basis. We also follow the presentation by Gunther Uhlmann [192] and Mikko Salo [167], to which we refer for many extensions.

Finally, in the last part, we will formally explain how these constructions allow us to prove the injectivity of $V \mapsto \Lambda_V$.

13.2 Density of the Products of Harmonic Functions

In this chapter, we mainly consider functions with real values. When we consider spaces of functions with complex values, we will indicate it explicitly (using the notation $L^2(\Omega, \mathbb{C})$ for example).

Theorem 13.1 (Calderón) *Let Ω be a bounded open set of $\mathbb{R}^n$ with $n \geqslant 2$. The vector space generated by the products of harmonic functions is dense in $L^2(\Omega)$. Similarly, the vector space generated by the scalar products of gradients of harmonic functions is dense in $L^2(\Omega)$.*

Proof Both results are proved in a similar way and we will only consider the case of scalar products of gradients of harmonic functions. Let us introduce

$$\mathcal{H} = \{u \in C^\infty(\overline{\Omega}) : \Delta u = 0\},$$
$$\Pi = \{\nabla u \cdot \nabla v : (u, v) \in \mathcal{H} \times \mathcal{H}\},$$
$$X = \text{Vect}\,\Pi.$$

We want to show that if $\varphi \in L^2(\Omega)$ and $\int_\Omega \varphi f \, dx = 0$ for all $f \in X$, then $\varphi = 0$. This will imply that $X^\perp = \{0\}$, hence the fact that X is dense in $L^2(\Omega)$ (see Corollary 3.8).

Let us consider the set

$$A := \{\rho \in \mathbb{C}^n : \rho \cdot \rho = \rho_1^2 + \cdots + \rho_n^2 = 0\}.$$

Note that

$$\rho \in A \quad \Longleftrightarrow \quad |\text{Re}\,\rho|^2 - |\text{Im}\,\rho|^2 + 2i\,\text{Re}\,\rho \cdot \text{Im}\,\rho = 0$$
$$\Longleftrightarrow \quad |\text{Re}\,\rho| = |\text{Im}\,\rho| \quad \text{and} \quad \text{Re}\,\rho \cdot \text{Im}\,\rho = 0.$$

The essential observation is that if $\rho \in A$, then $u_1(x) = e^{ix\cdot\rho}$ and $u_2 = e^{ix\cdot\overline{\rho}}$ are harmonic functions. Indeed, $\nabla u_1 = i\rho e^{ix\cdot\rho}$ hence $\Delta u_1 = -(\rho \cdot \rho)e^{ix\cdot\rho} = 0$ and similarly $\Delta u_2 = 0$. We will use these remarks with ρ chosen as follows: given $\eta \in \mathbb{R}^n$ and $k \in \mathbb{R}^n$ such that $\eta \cdot k = 0$ and $|\eta| = |k|$, we observe that $\rho = -i\eta + k$ satisfies $\rho \cdot \rho = 0$.

Let us consider $\varphi \in L^2(\Omega)$ such that $\int_\Omega \varphi f \, dx = 0$ for all $f \in X$. We must be careful here because X is a space of real-valued functions while the functions $u(x) = e^{ix\cdot\rho}$ and $v(x) = e^{ix\cdot\overline{\rho}}$ are complex-valued. But we directly verify that if

$\Delta u = 0$, then $\operatorname{Re} u$ and $\operatorname{Im} u$ are harmonic functions. Therefore, the assumption on φ implies that $\int_\Omega \varphi f \, dx = 0$ for all f in the set

$$\left\{ \nabla \operatorname{Re} u_1 \cdot \nabla \operatorname{Re} u_2, \ \nabla \operatorname{Re} u_1 \cdot \nabla \operatorname{Im} u_2, \ \nabla \operatorname{Im} u_1 \cdot \nabla \operatorname{Re} u_2, \ \nabla \operatorname{Im} u_1 \cdot \nabla \operatorname{Im} u_2 \right\}.$$

From this, we deduce that

$$\int_\Omega \varphi \nabla u_1 \cdot \nabla u_2 \, dx = 0.$$

Recalling that $\rho = -i\eta + k$ with $|\eta| = |k|$, we deduce that

$$
\begin{aligned}
0 &= \int_\Omega \varphi(x) \rho \cdot (-\overline{\rho}) e^{ix \cdot \rho} e^{ix \cdot \overline{\rho}} \, dx \\
&= \int_\Omega \varphi(x)(-2|k|^2) e^{ix \cdot (-i\eta + k) + ix \cdot (i\eta + k)} \, dx \\
&= \int_\Omega -2|k|^2 \varphi(x) e^{2ix \cdot k} \, dx = \int_{\mathbb{R}^n} -2|k|^2 (\chi_\Omega \varphi)(x) e^{2ix \cdot k} \, dx,
\end{aligned}
$$

where χ_Ω is the indicator function of Ω. The function $\chi_\Omega \varphi$ belongs to $L^1(\mathbb{R}^n)$ and therefore its Fourier transform $\widehat{\chi_\Omega \varphi}$ is a continuous function. The previous calculation implies that $\widehat{\chi_\Omega \varphi} = 0$. To conclude the proof, it remains only to observe that $\varphi = 0$ according to the injectivity theorem of the Fourier transform (see Corollary 5.14), the statement of which we recall here: if $f \in L^1(\mathbb{R}^n)$ is such that $\int e^{-i\xi \cdot x} f(x) \, dx = 0$ for all $\xi \in \mathbb{R}^n$, then $f = 0$. $\qquad\qquad\square$

13.3 Equations with Variable Coefficients

The proof of Theorem 13.1 relies on the use of exponentials $e^{ix \cdot \rho}$ with

$$\rho \in A := \{\zeta \in \mathbb{C}^n : \zeta \cdot \zeta = 0\}.$$

These exponentials are examples of harmonic functions and we propose to study the existence of similar functions for equations with variable coefficients. Consider a function $V \in L^\infty(\Omega, \mathbb{R})$ and an operator P defined by[1] $Pu = -\Delta u + Vu$. The function $e^{ix \cdot \rho}$ with $\rho \in A$ is not in the kernel of P if $V \neq 0$, but we will see that we can find an *approximate solution* to the equation $Pu = 0$ of the form

$$u(x) = e^{i\rho \cdot x}(1 + r(x)),$$

where r is a corrective term that will be small (of size $O(|\rho|^{-1})$) for a certain norm.

---

[1] The terminology, derived from Quantum Mechanics, is as follows: we say that V is a potential and we call P the Schrödinger operator associated with the potential V. We simply write $P = -\Delta + V$ by identifying the function V with the multiplication operator by the function V.

The idea of seeking an approximate solution rather than seeking an exact solution is essential when studying a problem with variable coefficients (and even more so a nonlinear problem).

Another essential idea is to relax the problem by seeking a *generalized solution*. We are not going to seek a solution to the equation $-\Delta u + V u = 0$ in the classical sense. Instead, we are going to seek a solution in the so-called weak sense (see Definition 7.14).

Let $u \in H^1(\Omega, \mathbb{C})$ and $V \in L^\infty(\Omega)$. Then the function $f = Vu$ satisfies $f \in L^2(\Omega)$ and therefore $f \in L^1_{loc}(\Omega)$. From Definition 7.14 we can deduce that u is a weak solution of the equation $-\Delta u + V u = 0$ if and only if

$$\forall \phi \in C_0^1(\Omega, \mathbb{C}), \quad \sum_{1 \leq i \leq n} \int_\Omega (\partial_i u)(\partial_i \phi)\, dx + \int_\Omega V u \phi\, dx = 0.$$

We are going to seek a weak solution to the equation $-\Delta u + V u = 0$ of the form

$$u = e^{i\rho \cdot x}(1 + r(x)),$$

with r in the Sobolev space $H^1(\Omega)$.

It is convenient to conjugate the equation by $e^{ix \cdot \rho}$. To do this, we formally write[2]

$$
\begin{aligned}
(-\Delta + V)\{e^{i\rho \cdot x}(1 + r)\} = 0 &\iff e^{-i\rho \cdot x}(-\Delta + V)\{e^{i\rho \cdot x}(1 + r)\} = 0 \\
&\iff (-\Delta - 2i\rho \cdot \nabla + V)(1 + r) = 0 \qquad (13.1) \\
&\iff (-\Delta - 2i\rho \cdot \nabla + V)r = -V.
\end{aligned}
$$

Note that if $r \in H^1(\Omega, \mathbb{C})$ and $V \in L^\infty(\Omega)$, then $-2i\rho \cdot \nabla r + Vr + V$ belongs to $L^1_{loc}(\Omega, \mathbb{C})$. Therefore, Definition 7.14 implies that r is a weak solution of

$$(-\Delta - 2i\rho \cdot \nabla + V)r = -V$$

if and only if, for all $\phi \in C_0^1(\Omega, \mathbb{C})$, we have

$$\sum_{1 \leq j \leq n} \int_\Omega (\partial_j r)(\partial_j \phi)\, dx + \int_\Omega (-2i\rho \cdot \nabla r + Vr)\phi\, dx + \int_\Omega V\phi\, dx = 0.$$

The following lemma states that the formal calculation (13.1) makes sense for weak solutions (the proof is left to the reader).

Lemma 13.2 *Let* $r \in H^1(\Omega, \mathbb{C})$. *Then* $u = e^{i\rho \cdot x}(1 + r)$ *belongs to* $H^1(\Omega, \mathbb{C})$ *and* u *is a weak solution of* $(-\Delta + V)u = 0$ *if and only if* r *is a weak solution of the equation* $(-\Delta - 2i\rho \cdot \nabla + V)r = -V$.

It remains now to solve the conjugate equation, which is the object of the following proposition.

---

[2] That is, we calculate as if the functions were regular, without worrying about the fact that we are actually manipulating weak solutions.

Proposition 13.3 *Let $V \in L^\infty(\Omega)$. There exists a constant C_0 that depends only on Ω and n such that, for all*

$$\rho \in A = \{\zeta \in \mathbb{C}^n : \zeta \cdot \zeta = 0\}$$

satisfying $|\rho| = \sqrt{\rho \cdot \overline{\rho}} \geqslant \max(C_0 \|V\|_{L^\infty}, 1)$, and for all $F \in L^2(\Omega, \mathbb{C})$, the equation

$$(-\Delta - 2i\rho \cdot \nabla + V)r = F$$

has a weak solution $r \in H^1(\Omega, \mathbb{C})$ that satisfies

$$\|r\|_{L^2} \leqslant \frac{C_0}{|\rho|} \|F\|_{L^2},$$

$$\|\nabla r\|_{L^2} \leqslant C_0 \|F\|_{L^2(\Omega)}.$$

(13.2)

Proof The proof is divided into two steps.

Step 1: solvability of the conjugate PDE in the case $V = 0$

First, we want to find a weak solution $r \in H^1(\Omega, \mathbb{C})$ of the equation

$$(-\Delta - 2i\rho \cdot \nabla)r = F,$$

satisfying the estimates (13.2).

Lemma 13.4 *For all $\rho \in A$ such that $|\rho| \geqslant 1$, there exists a bounded operator $Q_\rho : L^2(\Omega, \mathbb{C}) \to H^1(\Omega, \mathbb{C})$ such that*

$$(-\Delta - 2i\rho \cdot \nabla)Q_\rho = I$$

and satisfying, for all $F \in L^2(\Omega, \mathbb{C})$,

$$\|Q_\rho F\|_{L^2} \leqslant \frac{C_0}{|\rho|} \|F\|_{L^2},$$

$$\|\nabla Q_\rho F\|_{L^2} \leqslant C_0 \|F\|_{L^2(\Omega)},$$

where C_0 is a constant that depends only on Ω and n.

Proof We follow a very clever proof of this result which is due to Hähner [66]. Write $\rho = s(\omega_1 + i\omega_2)$ where $s = |\rho|/\sqrt{2}$ and ω_1, ω_2 are orthogonal vectors of $\mathbb{R}^n$. We can without loss of generality assume that $\omega_1 = (1, 0, \dots, 0)$ and $\omega_2 = (0, 1, 0, \dots, 0)$ are the first two vectors of the canonical basis of $\mathbb{R}^n$. We can also assume that Ω is included in the cube $\Pi = [-\pi, \pi]^n$. Extend F by 0 in Π. We are reduced to solving the equation

$$(-\Delta - 2is(\partial_1 + i\partial_2))r = F \quad \text{in } \Pi.$$

The clever idea is to introduce a well-chosen Hilbert basis (we refer to Section 3.2 of Chapter 3 for the study of Hilbert basis). Let us introduce

$$e_k(x) = \exp\left(i(k + \tfrac{1}{2}\omega_2) \cdot x\right) \quad (k \in \mathbb{Z}^n).$$

We directly verify that

$$(e_k, e_l) = \frac{1}{(2\pi)^n} \int_\Pi e_k(x)\overline{e_l(x)} \, dx = \delta_k^l,$$

so $(e_k)_{k\in\mathbb{Z}^n}$ is an orthonormal family of $L^2(\Pi, \mathbb{C})$ for the previous scalar product (the usual scalar product on $L^2(\Pi)$, normalized by the measure of Π). Moreover, if $v \in L^2(\Pi, \mathbb{C})$ satisfies $(v, e_k) = 0$ for all $k \in \mathbb{Z}$, then $(ve^{-\frac{1}{2}ix_2}, e^{ik \cdot x}) = 0$ for all $k \in \mathbb{Z}^n$, which implies $ve^{-\frac{1}{2}ix_2} = 0$, according to the uniqueness theorem of Fourier series. We deduce that $v = 0$. According to Theorem 3.17, this proves that $(e_k)_{k\in\mathbb{Z}^n}$ is a Hilbert basis. Consequently, we can then develop F on this basis: we have $F = \sum_{k\in\mathbb{Z}^n} F_k e_k$ with $F_k = (F, e_k)$ and

$$\|F\|_{L^2}^2 = \sum_{k\in\mathbb{Z}^n} |F_k|^2.$$

We also seek r in the form $r = \sum_{k\in\mathbb{Z}^n} r_k e_k$. Note that

$$\nabla e_k = i\left(k + \tfrac{1}{2}\omega_2\right)e_k,$$

from which

$$(-\Delta - 2is(\partial_1 + i\partial_2))e_k = p_k e_k,$$

where

$$p_k := \left(\left(k + \tfrac{1}{2}\omega_2\right)^2 - 2is\left(ik_1 - \left(k_2 + \tfrac{1}{2}\right)\right)\right).$$

Thus, the equation $(-\Delta - 2is(\partial_1 + i\partial_2))r = F$ suggests defining r_k by

$$p_k r_k = F_k.$$

Note that $|\mathrm{Im}\, p_k| = |2s(k_2 + \tfrac{1}{2})| \geqslant s > 0$ so p_k never cancels and we can solve the previous equation. Moreover,

$$|r_k| \leqslant \frac{1}{|p_k|}|F_k| \leqslant \frac{1}{s}|F_k|,$$

which implies that $r \in L^2(\Pi, \mathbb{C})$.

To show that r belongs to the Sobolev space $H^1(\Pi)$, we must show that r has a weak derivative in L^2. For this, we will use the fact that r is the sum of a series of terms which are C^∞ and therefore belong to $H^1(\Pi)$. It suffices to show that the series converges normally, that is, that $\sum |r_k| \|\nabla e_k\|_{L^2} < +\infty$. This will be a consequence of the following inequality

$$\left|\left(k + \frac{1}{2}e_2\right)r_k\right| \leqslant 4|F_k|, \quad k \in \mathbb{Z}^n,$$

which will imply $\|\nabla r\|_{L^2} \leqslant 4\|F\|_{L^2}$. To obtain this inequality, we consider two cases: if $|k + \frac{1}{2}e_2| \leqslant 4s$ then

$$\left|\left(k + \frac{1}{2}e_2\right)r_k\right| \leqslant \frac{4s}{s}|F_k| \leqslant 4|F_k|,$$

and if $|k + \frac{1}{2}e_2| \geqslant 4s$, then

$$|\mathrm{Re}\, p_k| = \left\|k + \frac{1}{2}e_2\right|^2 + 2sk_1\right| \geqslant \left|k + \frac{1}{2}e_2\right|^2 - 2s\,|k_1|$$

$$\geqslant \left|k + \frac{1}{2}e_2\right|^2 - 2s\left|k + \frac{1}{2}e_2\right|$$

$$\geqslant 2s\left|k + \frac{1}{2}e_2\right|,$$

from which

$$\left|\left(k + \frac{1}{2}e_2\right)r_k\right| \leqslant \frac{|k + \frac{1}{2}e_2|}{|\mathrm{Re}\, p_k|}|F_k| \leqslant \frac{|k + \frac{1}{2}e_2|}{2s|k + \frac{1}{2}e_2|}|F_k| \leqslant \frac{1}{2s}|F_k|.$$

This implies that the series $\sum r_k e_k$ converges normally in $H^1(\Pi)$. As $H^1(\Pi)$ is a Hilbert space, this series converges in $H^1(\Pi)$ and therefore its sum r belongs to $H^1(\Pi)$. Then r is a weak solution in $H^1(\Pi)$, which directly implies that r is a weak solution in $H^1(\Omega)$. Indeed, if the weak formulation is true for all test functions with supports in Π, then it is true *a fortiori* for any test function with support in Ω.

This concludes the proof of the lemma. □

Step 2: solvability of the conjugate PDE in the case $V \neq 0$

We now consider the equation

$$(-\Delta - 2i\rho \cdot \nabla + V)r = F,$$

in the general case $V \in L^\infty(\Omega)$. We are looking for a weak solution $r \in H^1(\Omega, \mathbb{C})$ satisfying the estimates (13.2).

For $V = 0$, the solution is given by $r = Q_\rho F$. For $V \neq 0$, inspired by the method of variation of the constant, we look for r in the form $Q_\rho \widetilde{F}$, where $\widetilde{F} \in L^2(\Omega, \mathbb{C})$ is a function to be determined. Note that

$$(-\Delta - 2i\rho \cdot \nabla + V)Q_\rho \widetilde{F} = \widetilde{F} + VQ_\rho \widetilde{F},$$

and we have therefore reduced to solving the equation

$$\widetilde{F} + VQ_\rho\widetilde{F} = F.$$

Recall that $\left\|Q_\rho F\right\|_{L^2} \leqslant (C_0/|\rho|)\|F\|_{L^2}$. Let T_ρ be the operator defined by $T_\rho f = -VQ_\rho f$. We deduce that, if $|\rho| \geqslant \max(2C_0\,\|V\|_{L^\infty}\,, 1)$,

$$\left\|T_\rho\widetilde{F}\right\|_{L^2} \leqslant \frac{1}{2}\left\|\widetilde{F}\right\|_{L^2},$$

which shows that $I - T_\rho$ is an invertible operator on $L^2(\Omega)$ thanks to a classical result on Neumann series (see Lemma 2.7).

Then

$$r = Q_\rho(I - T_\rho)^{-1}F$$

satisfies the equation; we deduce the estimates on r from those demonstrated previously. This concludes the proof of Proposition 13.3. □

We will also need a variant of Proposition 13.3 that gives solutions of the form $u(x) = e^{i\rho\cdot x}(a(x) + r(x))$ (we have so far only considered the case $a = 1$). We will see that if a is regular enough, then we can directly obtain such solutions from Proposition 13.3. The notion of regularity that suits us is that of belonging to the space $H^2(\Omega)$.

Corollary 13.5 *Let $V \in L^\infty(\Omega)$. There exists a constant C_0 that depends only on Ω and n such that, for all $\rho \in A = \{\zeta \in \mathbb{C}^n : \zeta \cdot \zeta = 0\}$ satisfying $|\rho| \geqslant \max(C_0\,\|V\|_{L^\infty}\,, 1)$, and for any function $a \in H^2(\Omega, \mathbb{C})$ satisfying $\rho \cdot \nabla a = 0$, the equation*

$$(-\Delta + V)u = 0$$

has a solution u of the form $u = e^{i\rho\cdot x}(a + r)$, where $r \in H^1(\Omega, \mathbb{C})$ satisfies

$$\|r\|_{L^2} \leqslant \frac{C_0}{|\rho|}\,\|(-\Delta + V)a\|_{L^2},$$

$$\|\nabla r\|_{L^2} \leqslant C_0\|(-\Delta + V)a\|_{L^2(\Omega)}.$$

Proof The function $u = e^{i\rho\cdot x}(a + r)$ is a weak solution of $(-\Delta + V)u = 0$ if and only if

$$(-\Delta - 2i\rho \cdot \nabla + V)(a + r) = 0.$$

Since $\rho \cdot \nabla a = 0$, we deduce the following equation on r:

$$(-\Delta - 2i\rho \cdot \nabla + V)r = -(-\Delta + V)a.$$

The existence of a solution to this problem is obtained by applying the previous proposition. □

We are now in a position to prove a result which, in dimensions greater than 3, generalizes the Calderón theorem stated in Section 13.2.

Theorem 13.6 *Let Ω be a bounded open set of $\mathbb{R}^n$. Suppose that the space dimension is $n \geqslant 3$. Consider two functions V_1 and V_2 belonging to $L^\infty(\Omega, \mathbb{R})$ and introduce the sets*

$$A_j = \{u \in H^1(\Omega, \mathbb{R}) : -\Delta u + V_j u = 0\}.$$

Suppose that $\varphi \in L^\infty(\Omega, \mathbb{R})$ is such that, for all u_j in A_j ($j = 1, 2$), we have

$$\int_\Omega u_1(x) u_2(x) \varphi(x) \, dx = 0,$$

then $\varphi = 0$.

Remark 13.7 The proof shows that we have a similar result for complex-valued functions.

Proof Note that if $\varphi \in L^\infty(\Omega, \mathbb{R})$, then the integral $\int_\Omega u_1 u_2 \varphi \, dx$ is well defined according to the Cauchy–Schwarz inequality.

We seek to show that $\varphi = 0$. For this, we will use an injectivity theorem. As in the proof of the Calderón theorem, we will use the injectivity theorem for the Fourier transform (which states that if the Fourier transform is equal to zero, then the function is equal to zero). The idea of the proof is then to try to approximate $e^{ix \cdot \xi}$ (ξ arbitrary in $\mathbb{R}^n$) by products $u_1 u_2$ with $u_j \in A_j$. This will be possible, thanks to the hypothesis $n \geqslant 3$ and a rather clever reasoning.

Fix $\xi \in \mathbb{R}^n$. Using the hypothesis $n \geqslant 3$, we introduce two vectors ζ_1, ζ_2 of $\mathbb{R}^n$ such that

$$\xi \perp \zeta_1, \quad \xi \perp \zeta_2, \quad \zeta_1 \perp \zeta_2, \quad |\zeta_1| = |\zeta_2| = 1.$$

Given $s > 0$, let

$$\rho_s = s(\zeta_1 + i\zeta_2),$$

so that $\rho_s \cdot \rho_s = 0$. Note that $\rho_s \cdot \xi = 0$ and therefore $\rho_s \cdot \nabla e^{ix \cdot \xi} = 0$. Moreover, since Ω is bounded, the functions $e^{ix \cdot \xi}$ and the constant function 1 belong to $H^2(\Omega)$. We can therefore apply the previous result with $a = e^{ix \cdot \xi}$ or $a = 1$. We deduce that if s is large enough, there exist two functions $u_{1,s}$ and $u_{2,s}$ in $H^1(\Omega, \mathbb{C})$ which are weak solutions of $(-\Delta + V_j) u_{j,s} = 0$ and which are of the form

$$u_{1,s} = e^{i\rho_s \cdot x} (e^{ix \cdot \xi} + r_{1,s}),$$
$$u_{2,s} = e^{-i\rho_s \cdot x} (1 + r_{2,s}),$$

where $\|r_{j,s}\|_{L^2} \leqslant C/s$ for $j = 1, 2$. Then the real and imaginary parts of $u_{1,s}$ (resp. $u_{2,s}$) belong to A_1 (resp. A_2). By hypothesis on φ, the integral $\int_\Omega \varphi f \, dx$ vanishes when f is the product of functions from A_1 and A_2 and therefore

$$\int_\Omega \varphi \operatorname{Re} u_{1,s} \operatorname{Re} u_{2,s} \, dx = \cdots = \int_\Omega \varphi \operatorname{Im} u_{1,s} \operatorname{Im} u_{2,s} \, dx = 0.$$

From this, we deduce that $\int_\Omega \varphi u_{1,s} u_{2,s} \, dx = 0$, which gives

$$\int_\Omega \varphi(e^{ix\cdot\xi} + r_{1,s})(1 + r_{2,s}) \, dx = 0.$$

By taking the limit as $s \to +\infty$, it follows that

$$\int_\Omega \varphi(x)e^{ix\cdot\xi} \, dx = 0.$$

Since this is true for all $\xi \in \mathbb{R}^n$, we have shown that $\widehat{\chi_\Omega \varphi} = 0$. This implies that $\varphi = 0$, which concludes the proof. $\square$

13.4 The Sylvester–Uhlmann Theorem

In this part, we want to explain how to use the previous result to solve the problem that motivated Calderón's work. We will limit ourselves to formal calculations and we will not seek to justify them rigorously.

Let us recall the context. We consider an open set Ω which is of class C^∞ and bounded. Given a function $V \in L^\infty(\Omega)$ such that $V \geqslant 0$ and a function $f : \Omega \to \mathbb{R}$, we are interested in the Dirichlet problem with unknown function $u : \overline{\Omega} \to \mathbb{R}$,

$$\begin{cases} -\Delta u + Vu = 0 & \text{in } \Omega, \\ u = f & \text{on } \partial\Omega. \end{cases}$$

This problem admits a unique solution u belonging to $H^1(\Omega)$. To see this, we look for u in the form $v + f$ with $v \in H^1_0(\Omega)$ a weak solution of $-\Delta v + Vv = \Delta f + Vf$. We then show that the right-hand side belongs to the topological dual of $H^1_0(\Omega)$, then we repeat the arguments used in Section 7.2.2.

We can then define the Dirichlet–Neumann operator by

$$\Lambda_V(f) = \partial_\nu u|_{\partial\Omega}.$$

If we have the right functional framework, we can show that it is a continuous linear self-adjoint operator (from the Sobolev space $H^{1/2}(\partial\Omega)$ to its dual, see for example the books by Brezis [15] and Evans [49]). An important theorem, due to Sylvester and Uhlmann, states that, in dimension $n \geqslant 3$, the operator Λ_V determines the potential V. More precisely, their result states that the map $V \mapsto \Lambda_V$ is injective.

We will explain, through formal calculations, how this result is related to what we have done previously. For this, consider two functions u_1, u_2 satisfying, in the weak sense,

$$-\Delta u_1 + V_1 u_1 = 0, \qquad -\Delta u_2 + V_2 u_2 = 0.$$

Let us introduce the traces of u_1 and u_2 on the boundary, denoted f_1, f_2, defined by

$$f_1 = u_1|_{\partial\Omega}, \quad f_2 = u_2|_{\partial\Omega}.$$

So $\Lambda_{V_1} f_1 = \partial_\nu u_1|_{\partial\Omega}$ and therefore, denoting the scalar product on $L^2(\partial\Omega)$ by $\langle \cdot, \cdot \rangle$ (to be rigorous, the following calculation should be understood in the sense of duality),

$$\langle \Lambda_{V_1} f_1, f_2 \rangle = \int_{\partial\Omega} \frac{\partial u_1}{\partial \nu} u_2 \, dS.$$

If u_1 and u_2 are regular functions, then we can write

$$\int_{\partial\Omega} \frac{\partial u_1}{\partial \nu} u_2 \, dS = \int_{\partial\Omega} X \cdot \nu \, dS \quad \text{with } X = u_2 \nabla u_1$$

$$= \int_\Omega \operatorname{div} X \, dx$$

$$= \int_\Omega \operatorname{div} (u_2 \nabla u_1) \, dx$$

$$= \int_\Omega u_2 \Delta u_1 + \nabla u_2 \cdot \nabla u_1 \, dx,$$

and as $\Delta u_1 = V_1 u_1$, we deduce that

$$\langle \Lambda_{V_1} f_1, f_2 \rangle = \int_\Omega \nabla u_1 \cdot \nabla u_2 + V_1 u_1 u_2 \, dx.$$

Similarly,

$$\langle \Lambda_{V_2} f_2, f_1 \rangle = \int_\Omega \nabla u_1 \cdot \nabla u_2 + V_2 u_1 u_2 \, dx.$$

As Λ_{V_2} is self-adjoint, we deduce

$$\langle (\Lambda_{V_1} - \Lambda_{V_2}) f_1, f_2 \rangle = \int_\Omega (V_1 - V_2) u_1 u_2 \, dx.$$

If $\Lambda_{V_1} = \Lambda_{V_2}$, then it follows that

$$0 = \int_\Omega (V_1 - V_2) u_1 u_2 \, dx,$$

for all functions u_1, u_2 in $H^1(\Omega)$ satisfying $-\Delta u_j + V_j u_j = 0$. Theorem 13.6 then leads to the desired result: $V_1 = V_2$.

For a complete proof we refer to the original article as well as the lectures notes by Gunther Uhlmann [192] and Mikko Salo [167].

13.5 An Exercise

Exercise (solved) 13.1 (Dirichlet Energy) Let Ω be a bounded open set with a regular boundary. Consider the map $Q : \gamma \in C^\infty(\overline{\Omega}) \mapsto Q_\gamma$ defined by

$$Q_\gamma : f \in H^{1/2}(\partial\Omega) \mapsto \int_\Omega \gamma |\nabla u|^2 \, dx$$

with u the solution in $H^1(\Omega)$ of the following Dirichlet problem:

$$\begin{cases} \mathrm{div}(\gamma \nabla u) = 0 \text{ in } \Omega, \\ \quad\quad u = f \text{ on } \partial\Omega. \end{cases}$$

Show that, for any function f, we have

$$dQ_{|\gamma=1}(h)(f) = \int_\Omega h|\nabla v|^2 \, dx$$

where v is the solution of:

$$\begin{cases} \Delta v = 0 \text{ in } \Omega, \\ v = f \text{ on } \partial\Omega. \end{cases}$$

Chapter 14
De Giorgi's Theorem

We will study the main theorem of the De Giorgi–Nash–Moser theory.[1] This is a spectacular result which states that weak solutions to an elliptic equation always belong to a Hölder space.

14.1 Introduction

Let $n \geqslant 2$ and Ω be a bounded open set of $\mathbb{R}^n$. Consider an operator L of order 2, of the following form

$$Lu = \sum_{1 \leqslant i,j \leqslant n} \partial_i(a_{ij}\partial_j u),$$

where the coefficients $a_{ij} \colon \Omega \to \mathbb{R}$ are measurable functions. We say that L is in divergence form because it can be written as

$$Lu = \operatorname{div}(A\nabla u), \quad A = (a_{ij})_{i,j}.$$

Assumption 14.1 We assume that L is an elliptic operator which means that there exist two positive constants λ, Λ such that

$$\forall (x,\xi) \in \Omega \times \mathbb{R}^n, \quad \lambda |\xi|^2 \leqslant \sum_{1 \leqslant i,j \leqslant n} a_{ij}(x)\xi_i\xi_j, \tag{14.1}$$

and

$$\sup_{i,j} \|a_{ij}\|_{L^\infty(\Omega)} \leqslant \Lambda. \tag{14.2}$$

---

[1] For a detailed exposition of this theory, we refer to the original articles by De Giorgi [45], Nash [148] and Moser [141] as well as the books by Han and Lin [68], Jost [88], and Taylor [187, 189] and the lecture notes by Ambrosio [4] and Smets [173].

© The Author(s), under exclusive license to Springer Nature Switzerland AG 2024
T. Alazard, *Analysis and Partial Differential Equations*, Universitext,
https://doi.org/10.1007/978-3-031-70909-8_14

We will be interested in the regularity of the solutions of the equation $Lu = 0$. Since we only assume that the coefficients belong to $L^\infty(\Omega)$, the equation $Lu = 0$ is to be understood in the weak sense.

Definition 14.2 A weak solution of the equation $Lu = 0$ is a function $u \in H^1(\Omega)$ satisfying

$$\forall \phi \in C_0^\infty(\Omega), \quad \sum_{1 \leqslant i,j \leqslant n} \int_\Omega a_{ij}(\partial_j u)(\partial_i \phi) \, dx = 0. \tag{14.3}$$

The following proposition states that if L is elliptic, then such weak solutions always exist.

Proposition 14.3 *For any operator L as above and for any bounded open set $\Omega \subset \mathbb{R}^n$, there exist non-zero weak solutions.*

Proof Let $B(0, R)$ be an open ball that contains $\overline{\Omega}$. We will prove a stronger result: namely, we will prove that there are as many weak solutions as there exist functions $f \in L^2(B(0, R))$ such that $f(x) = 0$ for $x \in \Omega$.

Let $\widetilde{A}$ be the extension of the coefficient matrix $A = (a_{ij})_{i,j}$ to $\mathbb{R}^n$, obtained by setting $A(x) = I$ outside of Ω. The hypotheses (14.1) and (14.2) imply that, if necessary by changing the values of the constants λ and Λ, the extension satisfies

$$\inf_{(x,\xi) \in \mathbb{R}^n \times (\mathbb{R}^n \setminus \{0\})} \frac{(\widetilde{A}(x)\xi) \cdot \xi}{|\xi|^2} \geqslant \lambda, \quad \left\| \widetilde{A} \right\|_{L^\infty(\mathbb{R}^n)} \leqslant \Lambda.$$

Let us introduce the bilinear form B defined by

$$B(u, \phi) = \int_{B(0,R)} (\widetilde{A}\nabla u) \cdot \nabla \phi \, dx.$$

Then the previous properties of $\widetilde{A}$ imply that B is a continuous and coercive bilinear form on $H_0^1(B(0, R))$.

Now consider any function $f \in L^2(B(0, R))$. Then the map

$$\phi \longmapsto -\int_{B(0,R)} f\phi \, dx$$

is a continuous linear form on $L^2(B(0, R))$. The Poincaré inequality (see Theorem 7.19) then implies that it is and a continuous linear form on the Hilbert space $H_0^1(B(0, R))$. The Riesz–Fréchet theorem (see Theorem 3.10) then implies that there exists a unique $U \in H_0^1(B(0, R))$ such that

$$\forall v \in H_0^1(B(0, R)), \quad B(U, v) = -\int_{B(0,R)} fv \, dx.$$

Set $u = U|_\Omega$. Since $U \in H_0^1(B(0, R)) \subset H^1(B(0, R))$, we trivially have $u \in H^1(\Omega)$. If in addition we choose f to vanish on Ω, then we directly verify that u satisfies (14.3). □

Remark 14.4

1. If the boundary is regular enough, we can also construct weak solutions by solving a Dirichlet problem (that is, construct u satisfying (14.3) by seeking a function u such that $Lu = 0$ in Ω and $u|_{\partial\Omega} = f$; see the books by Brezis [15] and Evans [49]).
2. We note that Assumption 14.1 is very general in the sense that the hypotheses on the coefficients are necessary to guarantee that B is a continuous and coercive bilinear form on $H_0^1(\Omega)$.

In this chapter and the following one, we will study the regularity of weak solutions. Here, our goal is to prove the main theorem of the De Giorgi–Nash–Moser theory, which states that weak solutions always belong to a Hölder space.[2]

Definition 14.5 (Hölder Spaces) Let O be an open set in $\mathbb{R}^n$, $u\colon O \to \mathbb{R}$ and $\alpha \in (0, 1]$. We define

$$\|u\|_{\alpha, O} := \sup_{x \neq y \in O} \frac{|u(x) - u(y)|}{|x - y|^\alpha}.$$

By definition, the Hölder space $C^{0,\alpha}(O)$ is the set of continuous and bounded functions u on O such that $\|u\|_{\alpha, O} < \infty$. It is a Banach space (exercise) for the norm

$$\|u\|_{C^{0,\alpha}(O)} := \sup_{x \in O} |u(x)| + \sup_{x \neq y \in O} \frac{|u(x) - u(y)|}{|x - y|^\alpha}.$$

Theorem 14.6 (De Giorgi) *For every ball $B \subset \Omega$, there exist $\alpha \in (0, 1]$ and $C > 0$ such that*

$$\|u\|_{C^{0,\alpha}(B)} \leqslant C\|u\|_{L^2(\Omega)},$$

for every weak solution $u \in H^1(\Omega)$ of $Lu = 0$.

It is interesting to compare this theorem and the results of the previous chapters on microlocal analysis, which also dealt with the regularity of solutions to a linear elliptic problem. There are two fundamental differences:

1. *Difference in context.* The De Giorgi theorem applies under very general assumptions about the coefficients. No regularity assumption is made on the coefficients a_{ij}, whereas in the previous chapters on microlocal analysis we considered equations with regular coefficients.
2. *Difference in the method of proof.* The fact that no regularity assumption is made on the coefficients means that we cannot use an approach that consists of modifying the equation (by conjugation or by symbolic calculation). Instead of modifying the equation, the proof of the De Giorgi theorem will consist of transforming the unknown. The key element to remember is that this result, which is a result on the solutions of a *linear* equation, relies on *nonlinear* changes of unknowns.

---

[2] We have already introduced and studied Hölder spaces in Section 5.4.2, thanks to the Fourier transform, in the case of functions defined on the entire space. We will study them again by other methods in Section 15.2.

14.2 Subsolutions and Nonlinear Transformations

The proof of the De Giorgi theorem consists of introducing nonlinear changes of
unknowns, of the form $v = \Phi(u)$. This will require several ideas. Of course, these
functions Φ will have to be well chosen. The main obstacle is that $v = \Phi(u)$ is not a
weak solution of $Lv = 0$. To be able to work with nonlinear expressions we will rely
on two principles: (i) relax the notion of solution (we will consider *subsolutions*,
which means that $Lv \geq 0$) and (ii) consider specific nonlinear change of variables
compatible with the notion of subsolution (this will be the case if Φ is an *increasing
convex* function). As an introduction to this approach, let us start by considering
regular solutions.

Definition 14.7 Let $u \in C^2(\Omega)$ and suppose that $a_{ij} \in C^1(\Omega)$. We say that u is a
subsolution if $Lu \geq 0$.

   Similarly, we say that $u \in C^2(\Omega)$ is a supersolution if $Lu \leq 0$.

Proposition 14.8 *Let $\Phi \in C^2(\mathbb{R})$ be an increasing convex function. If $u \in C^2(\Omega)$ is
a subsolution then $\Phi(u)$ is a subsolution.*

Proof We have $L(\Phi(u)) = \Phi''(u) \sum a_{ij}(\partial_i u)(\partial_j u) + \Phi'(u) Lu$. Now, by hypothesis
(see Assumption 14.1), we have $\sum a_{ij}(\partial_i u)(\partial_j u) \geq \lambda |\nabla u|^2$. Since $\Phi''(u) \geq 0$ and
$\Phi'(u) \geq 0$ by hypothesis on Φ and since $Lu \geq 0$ by hypothesis on u, we verify that
$L(\Phi(u)) \geq 0$, which is the desired result.   □

   The previous proposition concerns the regular solutions of $Lu = 0$ while a crucial
aspect of De Giorgi's theorem is that it concerns weak solutions. We will therefore
give a definition of weak subsolution, then show that we can extend the previous
proposition to this context.

Definition 14.9 A weak subsolution is a function $u \in H^1(\Omega)$ such that, for all
$\phi \in C_0^1(\Omega)$ with $\phi \geq 0$,

$$\sum_{1 \leq i,j \leq n} \int a_{ij}(\partial_j u)(\partial_i \phi)\, dx \leq 0.$$

A non-negative weak subsolution is a weak subsolution satisfying $u \geq 0$.

   A weak supersolution is a function $u \in H^1(\Omega)$ such that $-u$ is a weak subsolution:
for all $\phi \in C_0^1(\Omega)$ with $\phi \geq 0$, we have

$$\sum_{1 \leq i,j \leq n} \int a_{ij}(\partial_j u)(\partial_i \phi)\, dx \geq 0.$$

Remark 14.10 Assume that $u \in H^1(\Omega)$ is a weak subsolution. Since $C_0^\infty(\Omega)$ is dense, by definition, in the space $H_0^1(\Omega)$, we also have

$$\sum_{1 \leq i,j \leq n} \int a_{ij}(\partial_j u)(\partial_i \phi) \, dx \leq 0,$$

for any non-negative function ϕ belonging to $H_0^1(\Omega)$.

Proposition 14.11 *Let* $\Phi: \mathbb{R} \to [0, +\infty)$ *be an increasing, convex* C^2 *function with* Φ' *bounded and such that* $\Phi''(y) = 0$ *if* $|y| \geq R$ *for a certain* $R > 0$. *Suppose that* $u \in H^1(\Omega)$ *is a weak subsolution. Then* $\Phi(u) \in H^1(\Omega)$ *and* $\Phi(u)$ *is also a subsolution.*

Proof Recall that Ω is a bounded open set by hypothesis.

Lemma 14.12 *Let* $G \in C^1(\mathbb{R})$ *be a function whose derivative* G' *is a bounded function on* $\mathbb{R}$. *If* $u \in H^1(\Omega)$, *then the composite function* $G(u)$ *belongs to* $H^1(\Omega)$ *and moreover the weak derivatives of* $G(u)$ *satisfy*

$$\partial_j G(u) = G'(u)\partial_j u, \quad j = 1, \ldots, n.$$

Proof By hypothesis on G, by setting $M = \sup_{\mathbb{R}} |G'|$, we have

$$\forall (s,t) \in \mathbb{R}^2, \quad |G(t) - G(s)| \leq M|t - s|. \tag{14.4}$$

We deduce that $|G(u(x))| \leq |G(0)| + M |u(x)|$, which implies that $G(u)$ belongs to $L^2(\Omega)$ (recall that Ω is bounded). Furthermore, as G' is a bounded function, the function $G'(u)\partial_j u$ belongs to $L^2(\Omega)$ for all j. It remains to show that, for all indices j and for any test function $\phi \in C_0^1(\Omega)$, we have

$$\int_\Omega G(u)\partial_j \phi \, dx = -\int_\Omega \phi G'(u)\partial_j u \, dx. \tag{14.5}$$

Note that this formula is true if $u \in C^1(\overline{\Omega})$. Indeed, in this case it is a corollary of the integration by parts formula. We therefore consider a sequence (u_n) of C^1 functions with compact support that converges to u in $L^2(\Omega)$, which converges to u almost everywhere and such that ∇u_n converges to ∇u in $L^2(\omega)$ whenever $\overline{\omega} \subset \Omega$ (we do not have convergence in $L^2(\Omega)$ in general). Then, using the inequality (14.4), we deduce that $G(u_n)$ tends to $G(u)$ in $L^2(\Omega)$ so that $\int_\Omega G(u_n)\partial_j \phi \, dx$ converges to $\int_\Omega G(u)\partial_j \phi \, dx$. To study the integral $\int_\Omega \phi G'(u_n)\partial_j u_n \, dx$, we use the fact that ϕ has compact support and we note that, extracting a subsequence if necessary, we can assume that u_n converges almost everywhere to u and that $\partial_j u_n$ converges almost everywhere to $\partial_j u$, so that the dominated convergence theorem leads to the convergence of the integrals to $\int_\Omega \phi G'(u)\partial_j u \, dx$. We obtain (14.5) by passing to the limit. $\qquad\square$

The previous lemma and the assumptions on Φ guarantee that $\Phi(u) \in H^1(\Omega)$. Consider $\phi \in C_0^1(\Omega)$ with non-negative values. We want to show that

$$\int_\Omega \sum_{i,j} a_{ij}(\partial_j \Phi(u))(\partial_i \phi)\, dx \leqslant 0. \tag{14.6}$$

Lemma 14.12 implies that, for all j such that $1 \leqslant j \leqslant n$,

$$\partial_j \Phi(u) = \Phi'(u)\partial_j u, \quad \partial_j \Phi'(u) = \Phi''(u)\partial_j u.$$

Using Proposition 7.23, we can then write that

$$\int_\Omega \sum_{i,j} a_{ij}(\partial_j \Phi(u))(\partial_i \phi)\, dx = \int_\Omega \sum_{i,j} a_{ij}\Phi'(u)(\partial_j u)(\partial_i \phi)\, dx = (\mathrm{I}) + (\mathrm{II}),$$

$$\text{where} \quad (\mathrm{I}) = \int_\Omega \sum_{i,j} a_{ij}(\partial_j u)\partial_i(\Phi'(u)\phi)\, dx,$$

$$(\mathrm{II}) = -\int_\Omega \Phi''(u)\phi \sum_{i,j} a_{ij}(\partial_i u)(\partial_j u)\, dx.$$

Since $\Phi'(u)$ belongs to $H^1(\Omega)$ (according to Lemma 14.12) and since $\phi \in C_0^1(\Omega)$, we have $\Phi'(u)\phi \in H^1(\Omega)$ according to Proposition 7.23. Moreover, using that ϕ is a function with compact support, we show that $\Phi'(u)\phi$ belongs to the space $H_0^1(\Omega)$ (because this function is the limit for the norm $\|\cdot\|_{H^1(\Omega)}$ of a sequence of functions belonging to $C_0^1(\Omega)$). In addition, $\Phi'(u)\phi \geqslant 0$ because $\phi \geqslant 0$ and because Φ is increasing. Then, as mentioned in Remark 14.10, we have $(\mathrm{I}) \leqslant 0$. We still have to show that $(\mathrm{II}) \leqslant 0$. For this, we use on the one hand the convexity assumption, to obtain that $\Phi''(u) \geqslant 0$, and on the other hand the assumption of ellipticity (14.1), which implies that $\sum_{i,j} a_{ij}(\partial_i u)(\partial_j u) \geqslant 0$. We therefore obtain $(\mathrm{II}) \leqslant 0$, which completes the proof of (14.6). The proposition is proven. $\square$

The essential argument for studying elliptic regularity is the Caccioppoli inequality. We have already seen this inequality in Section 8.3 dedicated to the regularity of harmonic functions. We will now see that this inequality remains true for subsolutions of a general problem with variable coefficients. Consider two open sets ω, Ω with $\omega \subset \Omega$ and a weak subsolution $u \in H^1(\Omega)$ of the equation $Lu = 0$. The Caccioppoli inequality allows us to estimate $\|\nabla u\|_{L^2(\omega)}$ as a function of $\|u\|_{L^2(\Omega)}$.

Lemma 14.13 (Caccioppoli's Lemma for subsolutions) *Consider a non-negative weak subsolution $v \in H^1(\Omega)$ and an open set $\omega \subset \Omega$ such that $\overline{\omega} \subset \Omega$. There exists a constant C, depending only on $\Omega, \omega, n, \lambda, \Lambda$, such that*

$$\int_\omega |\nabla v|^2\, dx \leqslant C \int_\Omega v^2\, dx. \tag{14.7}$$

In the case where $\Omega = B(x_0, \rho)$ and $\omega = B(x_0, r)$ with $0 < r < \rho$, we have

$$C \leqslant \frac{K}{(\rho - r)^2}, \tag{14.8}$$

where K is a constant that depends only on n, λ, Λ.

Proof As we mentioned in Remark 14.10, we have

$$\sum_{1 \leqslant i,j \leqslant n} \int a_{ij}(\partial_j v)(\partial_i \phi) \, dx \leqslant 0, \tag{14.9}$$

for any non-negative function ϕ belonging to $H_0^1(\Omega)$.

Now consider a function $\psi \in C_0^1(\Omega)$, with values in $[0, +\infty)$, and which equals 1 over ω. As we recalled in the proof of the previous proposition, the function $\phi = \psi^2 v$ belongs to $H_0^1(\Omega)$. Moreover, this function is non-negative because v is non-negative by assumption. We can use Proposition 7.23 and the relation (14.9) to write that

$$\int_\Omega \psi^2 \sum_{i,j} a_{ij}(\partial_i v)(\partial_j v) \, dx \leqslant -2 \int_\Omega \psi v \sum_{i,j} a_{ij}(\partial_j v)(\partial_i \psi) \, dx.$$

Introduce the matrix $A = (a_{ij})_{1 \leqslant i,j \leqslant n}$. According to the Cauchy–Schwarz inequality, we have

$$\left| \int_\Omega \psi v \sum_{i,j} a_{ij}(\partial_j v)(\partial_i \psi) \, dx \right| \leqslant \left(\int_\Omega \psi^2 |A \nabla v|^2 \, dx \right)^{1/2} \left(\int_\Omega v^2 |\nabla \psi|^2 \, dx \right)^{1/2}.$$

Furthermore, the assumptions (14.1) and (14.2) on the coefficients a_{ij} imply that $|A \nabla v| \leqslant n\Lambda |\nabla v|$ and

$$\lambda \int_\Omega \psi^2 |\nabla v|^2 \, dx \leqslant \int_\Omega \psi^2 \sum_{i,j} a_{ij}(\partial_i v)(\partial_j v) \, dx.$$

By combining the previous inequalities, we verify that there exists a constant C, depending only on $\Omega, \omega, n, \lambda, \Lambda$, such that

$$\int_\Omega \psi^2 |\nabla v|^2 \, dx \leqslant C \left(\int_\Omega \psi^2 |\nabla v|^2 \, dx \right)^{1/2} \left(\int_\Omega v^2 |\nabla \psi|^2 \, dx \right)^{1/2},$$

from which we deduce that

$$\int_\Omega \psi^2 |\nabla v|^2 \, dx \leqslant C^2 \int_\Omega v^2 |\nabla \psi|^2 \, dx.$$

From this, we deduce the desired result (14.7). To obtain the estimate (14.8), it suffices to consider a function $\psi \in C_0^1(B(x_0, \rho))$ which equals 1 on $B(x_0, r)$ and which tends to 0 almost linearly between $\partial B(x_0, r)$ andx $\partial B(x_0, \rho)$. $\qquad\square$

We will now combine the Caccioppoli lemma and Proposition 14.11. The goal is to be able to consider more general changes of unknowns, of the form $v = \Phi(u)$ where Φ is any increasing convex function such that $\Phi(u)$ belongs to $L^2(\Omega)$.

Proposition 14.14 *Let $\Phi \colon \mathbb{R} \to [0, +\infty)$ be a convex and increasing function. Consider a weak subsolution $u \in H^1(\Omega)$ and consider an open set $\omega \subset \Omega$ such that $\bar{\omega} \subset \Omega$. If $v = \Phi(u)$ belongs to $L^2(\Omega)$, then $v \in H^1(\omega)$ and v is a non-negative subsolution in ω, which means that, for all $\phi \in C_0^1(\omega)$ with $\phi \geqslant 0$,*

$$\sum_{1 \leqslant i, j \leqslant n} \int_\omega a_{ij}(\partial_j v)(\partial_i \phi)\, dx \leqslant 0.$$

Proof Let us start by showing that we can approximate the function Φ by a sequence of functions Φ_p that satisfy the assumptions of Proposition 14.11.

Lemma 14.15 *There exists a sequence of functions $\Phi_p \colon \mathbb{R} \to [0, +\infty)$, convex, increasing, of class C^2, such that $\Phi_p''(y) = 0$ for $|y| \geqslant R_p$ for a certain constant R_p depending on p and satisfying*

$$\forall y \in \mathbb{R}, \quad 0 \leqslant \Phi_p(y) \leqslant \Phi(y) \quad \text{and} \quad \lim_{p \to +\infty} \Phi_p(y) = \Phi(y).$$

Proof Recall that if $f \colon \mathbb{R} \to \mathbb{R}$ is convex then the rate of increase

$$\tau(x) = (f(x) - f(a))/(x - a)$$

is an increasing function for all $a \in \mathbb{R}$. This property allows us to define a convex and increasing function, which coincides with Φ on $[-p, p]$ and which is affine outside of this interval, in the following way:

$$\widetilde{\Phi}_p(y) = \begin{cases} \alpha(y + p) + \Phi(-p) & \text{if } y \leqslant -p, \quad \text{where } \alpha = \lim_{\substack{y \to -p \\ y < -p}} \dfrac{\Phi(y) - \Phi(-p)}{y + p}, \\[2em] \Phi(y) & \text{if } -p \leqslant y \leqslant p, \\[1em] \beta(y - p) + \Phi(p) & \text{if } y \geqslant p, \quad \text{where } \beta = \lim_{\substack{y \to p \\ y > p}} \dfrac{\Phi(y) - \Phi(p)}{y - p}. \end{cases}$$

Note that $\widetilde{\Phi}_p(y) \leqslant \Phi(y)$. We then set

$$\widetilde{\Phi}_p^+(y) = \max\{0, \widetilde{\Phi}_p(y)\}.$$

Then $y \mapsto \widetilde{\Phi}_p^+(y)$ is convex, affine for $|y|$ large enough and $0 \leqslant \widetilde{\Phi}_p^+ \leqslant \Phi$.

To conclude, it remains to regularize $\widetilde{\Phi}_p^+$. Consider a C^∞ function $\rho \colon \mathbb{R} \to [0, +\infty)$ satisfying $\int_\mathbb{R} \rho(t)\, dt = 1$ and such that $\operatorname{supp} \rho \subset [0, 1]$. We set

$$\Phi_p(y) = \rho_p * \widetilde{\Phi}_p^+(y) \quad \text{with} \quad \rho_p(y) = p\rho\,(py).$$

It directly follows from the definition of the convolution product that Φ_p is regular, convex and satisfies $\Phi_p \geqslant 0$. Moreover, as ρ_p is supported in $[0,1]$, recalling that $\widetilde{\Phi}_p^+$ is increasing, we verify that

$$\Phi_p(x) = \int_{\mathbb{R}} \rho_p(y) \widetilde{\Phi}_p^+(x-y) \, dy \leqslant \int_{\mathbb{R}} \rho_p(y) \widetilde{\Phi}_p^+(x) \, dy = \widetilde{\Phi}_p^+(x)$$

so $\Phi_p \leqslant \widetilde{\Phi}_p^+ \leqslant \Phi$. Finally, as $\widetilde{\Phi}_p^+$ coincides with Φ on $[-p,p]$, the pointwise convergence of Φ_p to Φ comes from the results on the convergence of identity approximations seen in Chapter 6. □

As $\Phi(u) \in L^2(\Omega)$ by hypothesis and $0 \leqslant \Phi_p(u) \leqslant \Phi(u)$, the dominated convergence theorem implies that $\Phi_p(u)$ converges to $\Phi(u)$ in $L^2(\Omega)$. Moreover, the Caccioppoli lemma and Proposition 14.11 imply that $\Phi_p(u)$ is bounded in $H^1(\omega)$ whenever $\overline{\omega} \subset \Omega$. As $H^1(\omega)$ is a Hilbert space, we can extract a subsequence $(\Phi_{p'}(u))$ that converges weakly in $H^1(\omega)$. By uniqueness of the limit, we deduce that $v = \Phi(u)$ belongs to $H^1(\omega)$. Moreover, v is a subsolution because the weak limit of a sequence of subsolutions is always a subsolution. □

When Φ is a convex function, which is not necessarily increasing, we have a similar result but it only applies to weak solutions (it is generally false for weak subsolutions).

Proposition 14.16 *Let* $\Phi \colon \mathbb{R} \to [0, +\infty)$ *be a monotone convex function. Consider a weak solution* $u \in H^1(\Omega)$ *and an open set* $\omega \subset \Omega$ *such that* $\overline{\omega} \subset \Omega$. *If* $v = \Phi(u)$ *belongs to* $L^2(\Omega)$, *then* $v \in H^1(\omega)$ *and* v *is a non-negative subsolution in* ω.

Proof If Φ is increasing then the result is a direct corollary of the previous proposition. Let us then consider the case where Φ is decreasing. In this case, the function $\widetilde{\Phi} \colon \mathbb{R} \to [0, +\infty)$ defined by $\widetilde{\Phi}(t) = \Phi(-t)$ is an increasing convex function and we have $v = \widetilde{\Phi}(-u)$. As u is a weak solution, $-u$ is also a weak solution, therefore *a fortiori* a weak subsolution. The desired result is therefore a consequence of the previous proposition. □

14.3 Moser Iterations

We denote by $B(x_0, \rho)$ the open ball of center x_0 and radius $\rho > 0$.

Theorem 14.17 *Fix* $x_0 \in \Omega$ *and consider* r, ρ *with* $0 < r < \rho$ *such that* $B(x_0, \rho) \subset \Omega$. *There exists a constant* $c > 0$ *such that for all non-negative subsolutions* $v \in H^1(\Omega)$,

$$\|v\|_{L^\infty(B(x_0,r))} \leqslant c \|v\|_{L^2(B(x_0,\rho))}.$$

Proof The proof we follow is due to Moser. Fix $x_0 \in \Omega$ and consider r, ρ with $0 < r < \rho$ such that $B(x_0, \rho) \subset \Omega$. We will allow all constants to depend on x_0, r, ρ.

Consider a sequence of balls $B_j = B(x_0, R_j)$ with $R_j = r + (\rho - r)2^{-j}$ so that

$$B_{j+1} \subset B_j \subset \cdots \subset B_0 = B(x_0, \rho) \quad \text{and} \quad B_\infty := \bigcap_{j \in \mathbb{N}} B_j = \overline{B(x_0, r)}.$$

The principle of the proof is that there exists $\kappa > 1$ such that we can estimate $\|v\|_{L^{2\kappa^{j+1}}(B_{j+1})}$ in terms of $\|v\|_{L^{2\kappa^j}(B_j)}$. The existence of the constant κ comes from the following corollary of the Sobolev embedding theorem.

Lemma 14.18 *Let $\kappa \in [1, n/(n-2))$ for $n \geqslant 3$ and $\kappa \in (1, +\infty)$ if $n = 2$. There exists a constant γ such that, for all $j \in \mathbb{N}$ and all $v \in H^1(B_{j+1})$, we have*

$$\|v^\kappa\|_{L^2(B_{j+1})}^2 \leqslant \gamma \|\nabla v\|_{L^2(B_{j+1})}^{2\kappa} + \gamma \|v\|_{L^2(B_{j+1})}^{2\kappa}. \tag{14.10}$$

Proof The Sobolev inequality (see Theorem 7.51) implies that

$$\|v^\kappa\|_{L^2(B_{j+1})}^2 \leqslant C \|v\|_{H^1}^{2\kappa} = C \left(\|\nabla v\|_{L^2(B_{j+1})} + \|v\|_{L^2(B_{j+1})} \right)^{2\kappa}.$$

We then use that the function $t \mapsto t^{2\kappa}$ is increasing to obtain that

$$(s + t)^{2\kappa} \leqslant (2 \max\{s, t\})^{2\kappa} \leqslant 2^{2\kappa}(s^{2\kappa} + t^{2\kappa}),$$

and we deduce the desired inequality.   □

Lemma 14.19 *Let us fix $\kappa \in (1, n/(n-2)]$ for $n \geqslant 3$ and $\kappa \in [1, +\infty)$ if $n = 2$. Suppose that $v \in H^1(B_j)$ is a non-negative weak subsolution. Then v^κ belongs to $H^1(B_{j+1})$ and is a non-negative weak subsolution in B_{j+1}. Moreover,*

$$\|v^\kappa\|_{L^2(B_{j+1})}^2 \leqslant C(2^{2\kappa j} + 1) \|v\|_{L^2(B_j)}^{2\kappa},$$

where $C > 0$ is a constant that only depends on $n, \lambda, \Lambda, r, \rho$.   □

Proof Since $B_{j+1} \subset B_j$, we have $v \in H^1(B_{j+1})$ and we may use the Sobolev embedding theorem to get that

$$\|v^\kappa\|_{L^2(B_{j+1})}^2 \leqslant \gamma \|\nabla v\|_{L^2(B_{j+1})}^{2\kappa} + \gamma \|v\|_{L^2(B_{j+1})}^{2\kappa}. \tag{14.11}$$

Now, consider the convex function $\Phi \colon \mathbb{R} \to [0, +\infty)$ defined by

$$\Phi(t) = \begin{cases} 0 & \text{if } t \leqslant 0, \\ t^\kappa & \text{if } t \geqslant 0. \end{cases}$$

As v is non-negative by hypothesis, note that $\Phi(v) = v^\kappa$. Again, since, by hypothesis, we have $v \in H^1(B_j)$, the Sobolev embedding theorem implies that $\Phi(v)$ belongs to $L^2(B_j)$. We can then apply Proposition 14.14 with $\omega = B_{j+1}$. We deduce that

$v^\kappa = \Phi(v)$ belongs to $H^1(B_{j+1})$ and that v^κ is a non-negative weak subsolution in B_{j+1}. Moreover, we have the Caccioppoli inequality

$$\|\nabla v\|_{L^2(B_{j+1})} \leqslant C_j \|v\|_{L^2(B_j)}.$$

To control the constant C_j, we will use property (14.8) (being careful that in (14.8) we have an inequality for the square of the norms).

As the difference between the radius of B_j and that of B_{j+1} is proportional to 2^{-j}, we obtain that we can choose C_j such that

$$C_j \leqslant K 2^j,$$

where $K > 0$ is a constant that depends only on $n, \lambda, \Lambda, r, \rho$.

By combining the previous inequalities, we get

$$\|v^\kappa\|^2_{L^2(B_{j+1})} \leqslant \gamma \left[C_j^{2\kappa} \|v\|^{2\kappa}_{L^2(B_j)} + \|v\|^{2\kappa}_{L^2(B_{j+1})} \right] \leqslant \gamma (C_j^{2\kappa} + 1) \|v\|^{2\kappa}_{L^2(B_j)},$$

equivalent to the desired result.   □

We now introduce a sequence of functions defined by

$$v_j = v^{\kappa^j}.$$

Then, if $v \in H^1(B_0)$ is a non-negative weak subsolution, we obtain from the previous lemma, by induction, that v_j belongs to $H^1(B_j)$ and is also a weak subsolution in B_j. Note that $v_{j+1} = (v_j)^\kappa$ and let

$$N_j = \|v\|_{L^{2\kappa^j}(B_j)} = \|v_j\|^{1/\kappa^j}_{L^2(B_j)}.$$

We have

$$N_{j+1}^{2\kappa^{j+1}} = \|v_{j+1}\|^2_{L^2(B_{j+1})} \leqslant C(2^{2\kappa j} + 1) \|v_j\|^{2\kappa}_{L^2(B_j)} = C(2^{2\kappa j} + 1) N_j^{2\kappa^{j+1}}.$$

So

$$N_{j+1}^2 \leqslant \left(C(2^{2\kappa j} + 1) \right)^{1/\kappa^{j+1}} N_j^2.$$

Since

$$c = \prod_{j=0}^{\infty} \left(C(2^{2\kappa j} + 1) \right)^{1/\kappa^{j+1}} < +\infty,$$

as it is easily verified by taking the logarithm of the product and noticing that $\sum_{j=0}^{+\infty} \frac{j}{\kappa^j} < +\infty$, we find

$$\limsup_{j \to +\infty} N_j^2 \leqslant c N_0^2.$$

This shows that the sequence $(N_j)_{j \in \mathbb{N}}$ is bounded. We will conclude by deducing that v belongs to $L^\infty(B(x_0, r))$. Indeed, let $M = \sup_{j \in \mathbb{N}} N_j$. Then, by definition of N_j, we have

$$\int_{B(x_0, r)} |v|^{2\kappa^j} \, dx \leqslant \int_{B_j} |v|^{2\kappa^j} \, dx \leqslant M^{2\kappa^j}.$$

Now, set $A = \{x \in B(x_0, r) : |v(x)| > 2M\}$. Then

$$|A| (2M)^{2\kappa^j} \leq \int_{B(x_0,r)} |v|^{2\kappa^j} \, dx.$$

So, by combining the two previous inequalities, it follows that $|A| \leq 2^{-2\kappa^j}$ for all $j \in \mathbb{N}$, which implies $|A| = 0$. This shows that v is bounded by $2M$ outside a set of measure zero; which proves that v belongs to $L^\infty(B(x_0, r))$. Moreover, we have proved the desired estimate of $\|v\|_{L^\infty(B(x_0,r))}$ by a multiple of N_0, which is the L^2 norm of v on the ball $B(x_0, \rho)$. This concludes the proof.   □

14.4 The Harnack Inequality

We have already seen the Harnack inequality for harmonic functions, that is, functions u such that $\Delta u = 0$ (see Theorem 8.5). This inequality states that if Ω is connected and bounded, and u is a positive harmonic function on Ω, then

$$\sup_{x \in \Omega} u(x) \leq C \inf_{x \in \Omega} u(x),$$

where C is a universal constant. In fact this holds for any non-negative function: indeed, the maximum principle implies that, if $u(x_0) = 0$ for some $x_0 \in \Omega$, then $u(x) = 0$ for all $x \in \Omega$ and the inequality is then obvious.

The proof of the Harnack inequality for harmonic functions relied on the mean value formula. Here we will consider the much more difficult case of an equation with variable coefficients. In this case, we no longer have the mean value formula, nor any other representation formula. However, we will see that we can prove, for an equation with L^∞ coefficients, a weakened version of the Harnack inequality.

Theorem 14.20 *Suppose that $u \in H^1(B(x_0, 4R))$ is a non-negative weak solution satisfying*

$$|\{x \in B(x_0, 2R) : u(x) \geq 1\}| \geq \varepsilon |B(x_0, 2R)|, \tag{14.12}$$

with $\varepsilon > 0$. Then there exists a constant $c > 0$ that depends only on $\varepsilon, n, \lambda, \Lambda, R$ such that

$$\inf_{B(x_0, R)} u \geq c.$$

To prove this theorem, we use the following version of the Poincaré inequality.

Lemma 14.21 *Consider a ball $B \subset \mathbb{R}^n$ of radius R. For all $\varepsilon > 0$, there exists a constant $C = C(\varepsilon, n)$ such that, for all $u \in H^1(B)$ satisfying*

$$|\{x \in B : u = 0\}| \geq \varepsilon |B|,$$

we have

$$\int_B u^2 \, dx \leq CR \int_B |\nabla u|^2 \, dx.$$

Proof By contradiction, if this is false we can construct a sequence (u_m) of elements of $H^1(B)$ such that

$$|\{x \in B : u_m = 0\}| \geqslant \varepsilon |B|, \quad \int_B u_m^2 \, dx = 1, \quad \int_B |\nabla u_m|^2 \, dx \longrightarrow 0.$$

Then we can suppose that (u_m) converges to $u_0 \in H^1(B)$, strongly in $L^2(B)$ and weakly in $H^1(B)$. Then u_0 is a non-zero constant. Moreover,

$$0 = \lim_{m \to +\infty} \int_B |u_m - u_0|^2 \, dx$$
$$\geqslant \lim_{m \to +\infty} \int_{\{u_m = 0\}} |u_m - u_0|^2 \, dx \geqslant |u_0|^2 \inf_m |\{u_m = 0\}| > 0,$$

from which the sought contradiction follows.   □

Proof (of Theorem 14.20) We can suppose that $u \geqslant \delta > 0$ (if necessary, apply the result to $u + \delta$, then let δ tend to 0). The result will be obtained by examining $v = \max(-\log u, 0)$.

Let us introduce $\Phi \colon \mathbb{R} \to [0, +\infty)$, of the form

$$\Phi(t) = \alpha t + \beta \quad \text{if } t \leqslant \delta,$$
$$\Phi(t) = -\log(t) \quad \text{if } \delta \leqslant t \leqslant 1,$$
$$\Phi(t) = 0 \quad \text{if } 1 \leqslant t,$$

where α and β are chosen to obtain a convex function. As $u \geqslant \delta$ we have $v = \Phi(u)$. We want to show that v is a subsolution. Note that Φ is decreasing. As u is a weak solution, and not just a weak subsolution, we can use Proposition 14.16. Thus, to show that v belongs to $H^1(B(x_0, 2R))$ and that v is a subsolution on $B(x_0, 2R)$, it is sufficient to know that $v \in L^2(B(x_0, 3R))$. However, this is an immediate corollary of the fact that u belongs to $L^\infty(B(x_0, 3R))$ according to Theorem 14.17. We can also apply the L^∞ estimate given by Theorem 14.17 to v, which implies

$$\|v\|_{L^\infty(B(x_0, R))} \leqslant c \|v\|_{L^2(B(x_0, 2R))}. \tag{14.13}$$

Note that the assumption (14.12) implies that

$$|\{x \in B(x_0, 2R) : v = 0\}| = |\{x \in B(x_0, 2R) : u \geqslant 1\}| \geqslant \varepsilon |B(x_0, 2R)|.$$

The previous Poincaré inequality implies that

$$\|v\|_{L^2(B(x_0, 2R))} \leqslant C \|\nabla v\|_{L^2(B(x_0, 2R))}. \tag{14.14}$$

We want to show that the right-hand side is bounded. For this, we will use a variant of the Caccioppoli inequality obtained by considering the test function $\phi = \zeta^2 / u$ where $\zeta \in C_0^1(B(x_0, 4R))$. Since $u \geqslant \delta$, the function ζ is well defined and belongs to H_0^1. Since u is a weak solution, we obtain that

$$0 = \int_{B(x_0,4R)} \sum a_{ij}(\partial_j u)(\partial_i \phi)\, dx$$

$$= -\int_{B(x_0,4R)} \frac{\zeta^2}{u^2} \sum_{ij} a_{ij}(\partial_i u)(\partial_j u)\, dx + 2\int_{B(x_0,4R)} \frac{\zeta}{u} \sum_{ij} a_{ij}(\partial_j u)(\partial_i \zeta)\, dx,$$

which implies (using the ellipticity condition and the Hölder inequality)

$$\int_{B(x_0,4R)} \zeta^2 |\nabla \log u|^2\, dx \leqslant C \int_{B(x_0,4R)} |\nabla \zeta|^2\, dx.$$

So, by choosing ζ equal to 1 on $B(x_0, 2R)$ and then linearly going to 0 on $\partial B(x_0, 4R)$,

$$\int_{B(x_0,2R)} |\nabla \log u|^2\, dx \leqslant C.$$

By combining this inequality with (14.13) and (14.14) (and the fact that $f \mapsto \max\{f, 0\}$ is bounded from H^1 to H^1), we deduce

$$\sup_{B(x_0,R)} v \leqslant C,$$

which implies that $\inf_{B(x_0,R)} u \geqslant e^{-C} > 0$. The theorem is proven. $\qquad\square$

14.5 Hölder Regularity

We are now in a position to prove the De Giorgi theorem 14.6, whose statement is equivalent to the following result.

Theorem 14.22 (De Giorgi) *Consider an open set Ω, a point x_1 in Ω and a radius $R > 0$ such that $B(x_1, 5R) \subset \Omega$. Then there exist $\alpha \in (0, 1]$ and $C > 0$ such that every weak solution $u \in H^1(\Omega)$ of $Lu = 0$ belongs to $C^{0,\alpha}(B(x_1, R/2))$ and moreover*

$$\|u\|_{C^{0,\alpha}(B(x_1,R/2))} \leqslant C\|u\|_{L^2(B(x_1,5R))}.$$

Remark 14.23 The same result is true if we replace in the above statement $5R$ (resp. $R/2$) by R' (resp. R'') where R' (resp. R'') is any real number strictly greater (resp. smaller) than R. To see this, it suffices to use a scaling argument, as will be done in the proof of Lemma 14.25.

Proof We begin by showing that u belongs to L^∞ on a smaller ball.

Lemma 14.24 *We have $u \in L^\infty(B(x_1, 2R))$ and moreover*

$$\|u\|_{L^\infty(B(x_1,2R))} \leqslant C\|u\|_{L^2(B(x_1,4R))},$$

where the constant C depends only on n, λ, Λ, R.

Proof Consider the convex functions

$$y \longmapsto \max\{y, 0\}, \quad y \longmapsto \max\{-y, 0\}.$$

Since u is a weak solution (and not just a subsolution), we can apply Proposition 14.16 to deduce that $\max\{u, 0\}$ and $\max\{-u, 0\}$ are positive weak subsolutions. We then use the L^∞ estimate for non-negative subsolutions given by Theorem 14.17. We deduce that $\max\{u, 0\}$ and $\max\{-u, 0\}$ belong to $L^\infty(B(x_1, 2R))$, which implies the desired result.   □

Now that we know that u belongs to $L^\infty(B(x_1, 2R))$, it remains to show that

$$\sup_{x \neq y \in B(x_1, R/2)} \frac{|u(x) - u(y)|}{|x - y|^\alpha} \leqslant C\|u\|_{L^2(B(x_1, 5R))}. \tag{14.15}$$

For this, it will be convenient to fix $x_0 \in B(x_1, R/2)$ and to show that there exists a constant C, independent of x_0, such that

$$\frac{|u(x) - u(x_0)|}{|x - x_0|^\alpha} \leqslant C\|u\|_{L^2(B(x_0, 4R))}, \tag{14.16}$$

for all $x \in B(x_0, R) \setminus \{x_0\}$. We will explain later how to easily deduce inequality (14.15) from this.

To prove (14.16), we will study the oscillation of u on the balls $B(x_0, r)$ with $r \leqslant 2R$. By definition, the oscillation of u on the ball $B(x_0, r)$ is the quantity

$$\omega(r) = \sup_{B(x_0, r)} u(x) - \inf_{B(x_0, r)} u(x).$$

Lemma 14.25 *There exists a* $\gamma \in [0, 1)$ *such that, for all* $r \in (0, R]$,

$$\omega(r) \leqslant \gamma\omega(2r).$$

Proof The proof is in two steps. We start by proving the result in the particular case where $r = R$. Then, we will deduce the result for all $r \in (0, R]$ by a scaling argument.

Step 1: case $r = R$

Without loss of generality, we can add a constant to u, so that

$$\sup_{B(x_0, 2R)} u(x) = - \inf_{B(x_0, 2R)} u(x) = \frac{1}{2}\omega(2R).$$

Let $M = \omega(2R)/2$ and

$$u_+ = 1 + \frac{u}{M}, \quad u_- = 1 - \frac{u}{M}.$$

Then u_- and u_+ are weak solutions of $Lu_+ = Lu_- = 0$ and moreover these functions are non-negative on $B(x_0, 2R)$. Note that

$$|\{x \in B(x_0, 2R) : u_+ \geqslant 1\}| + |\{x \in B(x_0, 2R) : u_- \geqslant 1\}|$$
$$= |\{x \in B(x_0, 2R) : u \geqslant 0\}| + |\{x \in B(x_0, 2R) : u \leqslant 0\}| \geqslant |B(x_0, 2R)|.$$

Therefore, either

$$|\{x \in B(x_0, 2R) : u_+ \geqslant 1\}| \geqslant \frac{1}{2}|B(x_0, 2R)|,$$

or

$$|\{x \in B(x_0, 2R) : u_- \geqslant 1\}| \geqslant \frac{1}{2}|B(x_0, 2R)|.$$

Assume that u_+ satisfies this condition (otherwise we can reduce to this case by changing u to $-u$). Then the Harnack inequality given by Theorem 14.20 implies that

$$u_+(x) > c \quad \text{in } B(x_0, R),$$

for some constant $c > 0$. By definition of $u_+ = 1 + u/M$, it follows that

$$-M(1 - c) \leqslant u(x) \quad \text{in} \quad B(x_0, R).$$

On the other hand, by definition of M, we also have $u(x) \leqslant M$ in $B(x_0, 2R)$. Since, by construction, we have $\omega(2R) = 2M$, we conclude that

$$\omega(R) = \sup_{B(x_0,R)} u - \inf_{B(x_0,R)} u$$

$$\leqslant M - (-M(1 - c)) = 2M - cM = 2M\left(1 - \frac{c}{2}\right)$$

$$\leqslant \gamma\omega(2R) \quad \text{with} \quad \gamma := \left(1 - \frac{c}{2}\right).$$

Note that γ depends on n, λ, Λ and possibly R. We will see in the next step that γ is actually independent of R.

Step 2: case $r \in (0, R)$

Introduce

$$\widetilde{u}(x) = u\left(x_0 + \frac{r}{R}(x - x_0)\right), \quad \widetilde{a}_{ij}(x) = a_{ij}\left(x_0 + \frac{r}{R}(x - x_0)\right).$$

Since u is a weak solution of $\sum \partial_i(a_{ij}\partial_j u) = 0$, it can be directly verified that $\widetilde{u}$ is a weak solution of the equation $\sum \partial_i(\widetilde{a}_{ij}\partial_j\widetilde{u}) = 0$. Moreover, the coefficients $\widetilde{a}_{ij}$ satisfy the assumptions (14.1) and (14.2) with the same positive constants λ and Λ.

Therefore, we can apply the result of the previous step to obtain that

$$\widetilde{\omega}(R) = \sup_{B(x_0,R)} \widetilde{u}(x) - \inf_{B(x_0,R)} \widetilde{u}(x) \leqslant \gamma\widetilde{\omega}(2R).$$

However, $\widetilde{\omega}(R) = \omega(r)$ and $\widetilde{\omega}(2R) = \omega(2r)$. Therefore, the desired result follows. □

Consider x_0 in the ball $B(x_1, R/2)$ and now consider $x \in B(x_0, R)$ with $x \neq x_0$. Let $r = |x - x_0|$. Then there exists $n \in \mathbb{N}^*$ such that $2^{-n+1}R > r \geqslant 2^{-n}R$. We can then write that

$$|u(x) - u(x_0)| \leqslant \sup_{B(x_0,r)} u - \inf_{B(x_0,r)} u = w(r) \leqslant \gamma^{n-1}\omega(2^{n-1}r).$$

Moreover, since $2^{n-1}r \leqslant R$ we have $\omega(2^{n-1}r) \leqslant 2\|u\|_{L^\infty(B(x_0,R))}$ (directly by definition of ω) and therefore, according to the L^∞ estimate given by Lemma 14.24, we have

$$\omega(2^{n-1}r) \leqslant K\|u\|_{L^2(B(x_0,4R))}.$$

As $r \geqslant 2^{-n}R$ we have $r^\alpha \geqslant 2^{-n\alpha}R^\alpha$ for all $\alpha \in (0,1)$. Then

$$\frac{|u(x) - u(x_0)|}{|x - x_0|^\alpha} \leqslant \gamma^{n-1}2^{n\alpha}R^{-\alpha}K\|u\|_{L^2(B(x_0,4R))}.$$

We now choose α such that $2^\alpha\gamma \leqslant 1$ (then $\alpha \in (0, 1/2)$). Then

$$\frac{|u(x) - u(x_0)|}{|x - x_0|^\alpha} \leqslant R^{-\alpha}K\|u\|_{L^2(B(x_0,4R))},$$

for all $x \in B(x_0, R) \setminus \{x_0\}$.

We now take the upper bound as x_0 traverses $B(x_1, R/2)$, to obtain that

$$\sup_{\substack{x_0,x\in B(x_1,R/2)\\x_0\neq x}} \frac{|u(x) - u(x_0)|}{|x - x_0|^\alpha} \leqslant \sup_{x_0\in B(x_1,R/2)} \sup_{\substack{x\in B(x_0,R)\\x\neq x_0}} \frac{|u(x) - u(x_0)|}{|x - x_0|^\alpha}$$

$$\leqslant \sup_{x_0\in B(x_1,R/2)} R^{-\alpha}K\|u\|_{L^2(B(x_0,4R))}$$

$$\leqslant R^{-\alpha}K\|u\|_{L^2(B(x_1,5R))}.$$

As Lemma 14.24 already gave us an L^∞ estimate, this concludes the proof of the De Giorgi theorem.   □

14.6 Exercises

Exercise 14.1 (Nash's Theorem) In this exercise, we study a general parabolic equation and show a result analogous to Theorem 14.17. This proof is taken from Taylor's book [187] to which we refer for the solution.

Let us fix $T > 0$ and $R > 0$. We denote by B_R the ball of center 0 and radius R, and we set $Q = [0, T] \times \Omega$. Consider the operator

$$L = \sum_{1 \leqslant j, k \leqslant n} \partial_{x_j}(a_{jk}\partial_{x_k} \cdot),$$

where the coefficients a_{jk} satisfy $a_{jk} = a_{kj}$ and $a_{jk} \in L^\infty(B_R)$ for all j, k such that $1 \leqslant j, k \leqslant n$. We also assume that there exist two constants $\lambda, \Lambda > 0$ such that

$$\lambda \sum_{1 \leqslant j \leqslant n} \xi_j^2 \leqslant \sum_{1 \leqslant j, k \leqslant n} a_{jk}(x)\xi_j\xi_k \leqslant \Lambda \sum_{1 \leqslant j \leqslant n} \xi_j^2$$

for all ξ in $\mathbb{R}^n$ and almost every $x \in B_R$.

We have already studied the regularity of solutions of the elliptic equation $Lu = 0$ and proved in particular the De Giorgi theorem. The goal of this problem is to prove Nash's theorem, which concerns the solutions $u : Q \to \mathbb{R}$ of the parabolic equation

$$\partial_t u - Lu = 0.$$

In this problem, we will only show *a priori* estimates: we prove estimates for regular functions; obtaining these estimates for less regular functions will not be asked.

1. Let $v = v(t, x)$ be a function belonging to $C^2(Q)$. We say that v is a subsolution if $(\partial_t - L)v \leqslant 0$.
   Let $\kappa \in [1, +\infty)$ and $v = v(t, x)$ be a positive function belonging to $C^2(Q)$. Show that if v is a subsolution then v^κ is also a subsolution.
2. Let $V = (V_1, \ldots, V_n)(t, x)$ and $W = (W_1, \ldots, W_n)(t, x)$. We denote by $\langle V, W \rangle$ the function defined by

$$\langle V, W \rangle(t, x) = \sum_{1 \leqslant j, k \leqslant n} a_{jk}(x)V_j(t, x)W_k(t, x).$$

We denote by $\|V\|$ the function $\sqrt{\langle V, V \rangle}$. Show that $\langle V, W \rangle \leqslant \|V\| \|W\|$.
Let $v \in C^2(Q)$. Suppose that $w \in C^2(Q)$ vanishes in the neighborhood of $[0, T] \times \partial\Omega$. Show that

$$\iint_Q w(\partial_t - L)v \, dx \, dt = \iint_Q \langle \nabla_x v, \nabla_x w \rangle \, dx \, dt + \iint_Q w\partial_t v \, dx \, dt.$$

Let us set $g = (\partial_t v - Lv)$ and consider a function $\psi \in C^\infty(Q)$ that vanishes in the neighborhood of $[0, T] \times \partial\Omega$. Using the previous identity (applied with w chosen appropriately), show that for all $(T_1, T_2) \in [0, T]^2$,

$$\int_{T_1}^{T_2} \int_{\Omega} \psi^2 \, \|\nabla_x v\|^2 \, dx \, dt$$

$$= -2 \int_{T_1}^{T_2} \int_{\Omega} \langle \psi \nabla_x v, v \nabla_x \psi \rangle \, dx \, dt + \int_{T_1}^{T_2} \int_{\Omega} \psi^2 gv \, dx \, dt$$

$$+ \frac{1}{2} \int_{T_1}^{T_2} \int_{\Omega} (\partial_t \psi^2) v^2 \, dx \, dt - \frac{1}{2} \int_{\Omega} \psi^2 v^2 (T_2, x) \, dx + \frac{1}{2} \int_{\Omega} \psi^2 v^2 (T_1, x) \, dx.$$

Deduce

$$\int_{T_1}^{T_2} \int_{\Omega} \psi^2 \, \|\nabla_x v\|^2 \, dx \, dt + \int_{\Omega} \psi^2 v^2 (T_2, x) \, dx$$

$$\leqslant \int_{T_1}^{T_2} \int_{\Omega} v^2 \left(4 \, \|\nabla_x \psi\|^2 + \partial_t \psi^2 \right) dx \, dt$$

$$+ 2 \int_{T_1}^{T_2} \int_{\Omega} \psi^2 gv \, dx \, dt + \int_{\Omega} \psi^2 v^2 (T_1, x) \, dx.$$

(Hint: $2xy \leqslant \frac{1}{2} x^2 + 2y^2$.)

3. Let $(T_j)_{j \in \mathbb{N}}$ and $(R_j)_{j \in \mathbb{N}}$ be two sequences such that

$$T_0 = 0, \qquad T_{j+1} \geqslant T_j, \qquad T_{j+1} - T_j = A j^{-2} T, \qquad \lim T_j = T/2,$$
$$R_0 = R, \qquad R_{j+1} \leqslant R_j, \qquad R_j - R_{j+1} = B j^{-2} R, \qquad \lim R_j = R/2,$$

with $A, B > 0$ two given constants. We set $Q_j = [T_j, T] \times B_{R_j}$. Note that $Q_{j+1} \subset Q_j$ (make a drawing in dimension 1).

Let $v \in C^2(Q)$ be a positive subsolution of $\partial_t - L$. Show that

$$\|\nabla_x v\|_{L^2(Q_{j+1})} + \sup_{t \in [T_{j+1}, T]} \|v(t, \cdot)\|_{L^2(B_{R_{j+1}})} \leqslant C_j \|v\|_{L^2(Q_j)},$$

where the constant C_j satisfies $C_j \leqslant C j^2 + C$ with C independent of j.

4. Let us fix $\kappa = 1 + 2/n$.

• Let $R/2 \leqslant r \leqslant R$ and B_r be the ball of center 0 and radius r. Let $w = w(x)$ belong to $H^1(B_r)$. We admit that there exists a constant $\gamma = \gamma(R)$ such that

$$\|w\|_{L^{2\kappa}(B_r)}^{2\kappa} \leqslant \gamma \, \|\nabla_x w\|_{L^2(B_r)}^2 \, \|w\|_{L^2(B_r)}^{4/n} + \gamma \|w\|_{L^2(B_r)}^{2\kappa}.$$

(Explain why, if you know how to prove this inequality.)

• Show that there exists a constant γ such that,

$$\forall j \in \mathbb{N}, \quad \|v^\kappa\|_{L^2(Q_j)}^2 \leqslant \gamma \sigma_j(v)^{4/n} \, \|\nabla_x v\|_{L^2(Q_j)}^2 + \gamma \|v\|_{L^2(Q_j)}^{2\kappa},$$

where

$$\sigma_j(v) = \sup_{t \in [T_j, T]} \|v(t, \cdot)\|_{L^2(B_{R_j})}^2.$$

5. Let $v \in C^2(Q)$ be a positive subsolution of $\partial_t - L$. Show that

$$\|v^\kappa\|_{L^2(Q_{j+1})}^2 \leqslant \gamma(C_j^{2\kappa} + 1)\|v\|_{L^2(Q_j)}^{2\kappa}.$$

6. Deduce that there exists $K > 0$ such that, for all positive subsolutions $v \in C^2(Q)$ of $\partial_t - L$, we have

$$\|v\|_{L^\infty(Q')} \leqslant K\|v\|_{L^2(Q)},$$

where $Q' = [T/2, T] \times B_{R/2}$.

Chapter 15
Schauder's Theorem

I have read your paper on the relationship between existence and uniqueness of the solution of a nonlinear equation. I know now that existence is independent of uniqueness. I admire your topological methods. In my opinion they ought to be useful for establishing an existence theorem independent of any uniqueness and assuming only some a priori estimates.

Letter from Jean Leray to Juliusz Schauder [116, 131].

Das wäre ein Satz. [That would be a theorem!]

Reply from Juliusz Schauder.

Juliusz Schauder devised several methods for solving partial differential equations (see [168, 169]), with a notable contribution being a compelling continuity argument (whose simplest version is given by Exercise 1.7). This argument demonstrates the ability to simplify the proof of the existence of solutions for equations with variable coefficients to that of equations with constant coefficients. The approach relies on *a priori* estimates, which involve proving certain estimates under the assumption that solutions exist. In many situations, these estimates affirm that, given appropriately smooth coefficients and solutions, the Hölder norm of the solution can be controlled in relation to the Hölder norms of the coefficients and the data. Subsequently, utilizing these estimates for supposed solutions allows one to establish the actual existence of solutions. This line of research culminated in the proof of a famous result by Leray and Schauder [117]. We will not study this method from functional analysis. Instead, our main goal in this chapter is to study the Hölder regularity for variational solutions to elliptic equations, following the work of De Giorgi and Campanato [28], as well as the lecture notes by Ambrosio [4] and Smets [173]. The primary outcome generalizes a theorem by Schauder applicable to classical solutions. Through a combination of this result with De Giorgi's theorem, we will then derive a C^∞ regularity result for minimal surfaces.

T. Alazard, *Analysis and Partial Differential Equations*, Universitext,
https://doi.org/10.1007/978-3-031-70909-8_15

15.1 Local Averages and Elliptic Equations

Definition 15.1 Let A be a bounded measurable subset of Ω and $u \in L^1_{loc}(\Omega)$. In the following, we will use the notations $\fint_A$ and u_A defined by

$$u_A = \fint_A u \, dx = \frac{1}{|A|} \int_A u \, dx.$$

We say that u_A is the average of u over A.

One of the objectives of this chapter is to explain how to use the notion of local average to study the regularity of solutions to elliptic equations.

We saw in Chapter 13 that harmonic functions, i.e., functions $u \in C^2(\Omega)$ such that $\Delta u = 0$, satisfy the mean value property:

$$u(x) = u_{B(x,r)} \quad \text{for all } r \text{ such that } B(x,r) \subset \Omega.$$

In this section, we will start by presenting two inequalities due to De Giorgi (see also Campanato [28, section 7]), that reflect this averaging effect for solutions to an elliptic problem with constant coefficients.

Proposition 15.2 *Consider a matrix A with constant coefficients satisfying*

$$\exists \lambda > 0 / \ \forall \xi \in \mathbb{R}^n, \quad \lambda |\xi|^2 \leq A\xi \cdot \xi.$$

Suppose that $u \in H^2(B(x_0, R))$ is a weak solution of $\operatorname{div}(A\nabla u) = 0$. Then, for all r such that $0 < r < R$,

$$\int_{B(x_0,r)} |u|^2 \, dx \leq c \left(\frac{r}{R}\right)^n \int_{B(x_0,R)} |u|^2 \, dx, \tag{15.1}$$

$$\int_{B(x_0,r)} |u - u_{B(x_0,r)}|^2 \, dx \leq c \left(\frac{r}{R}\right)^{n+2} \int_{B(x_0,R)} |u - u_{B(x_0,R)}|^2 \, dx. \tag{15.2}$$

Proof Let us prove (15.1). By a simple scaling argument, it suffices to assume that $R = 1$ and $r < 1$. Note that the result is trivial if $1/2 \leq r \leq 1$ (indeed, the inequality is then trivially true with $c = 2^n$ for any function $u \in L^2$). Consider the case $r \leq 1/2$. By writing

$$\int_{B(x_0,r)} |u|^2 \, dx \leq 2^n r^n \sup_{B(x_0,r)} |u|^2 \leq 2^n r^n \sup_{B(x_0,1/2)} |u|^2,$$

we see that it suffices to show that

$$\sup_{B(x_0,1/2)} |u|^2 \leq C \int_{B(x_0,1)} |u|^2 \, dx,$$

which has already been seen in the previous chapter in a much more general setting (*cf.* Theorem 14.17).

Let us prove (15.2). Suppose first that $r < R/2$. As A is a constant matrix, we verify that $\partial_i u \in H^1(B(x_0, R/2))$ is a weak solution of $\mathrm{div}(A\nabla\partial_i u) = 0$. We can then use (15.1) with u replaced by ∇u, which gives

$$\int_{B(x_0,r)} |\nabla u|^2 \, dx \leqslant c_1 \left(\frac{r}{R}\right)^n \int_{B(x_0,R/2)} |\nabla u|^2 \, dx.$$

Furthermore, we have the Poincaré inequality (cf. Exercise 7.3 and its solution 12)

$$\int_{B(x_0,r)} |u - u_{B(x_0,r)}|^2 \, dx \leqslant c_2 r^2 \int_{B(x_0,r)} |\nabla u|^2 \, dx$$

and the Caccioppoli inequality (cf. Lemma 14.13), which states that for[1] any $c \in \mathbb{R}$, we have

$$\int_{B(x_0,R/2)} |\nabla u|^2 \, dx \leqslant \frac{c_3}{R^2} \int_{B(x_0,R)} |u - c|^2 \, dx.$$

Using this inequality with $c = u_{B(x_0,R)}$ and combining the result with the Poincaré inequality, we find that

$$\int_{B(x_0,r)} |u - u_{B(x_0,r)}|^2 \, dx \leqslant c_1 c_2 c_3 \left(\frac{r}{R}\right)^{n+2} \int_{B(x_0,R)} |u - u_{B(x_0,R)}|^2 \, dx,$$

which is the desired result. It remains to consider the case $R/2 \leqslant r \leqslant R$. For this, let us first verify that $u_{B(x_0,r)}$ is a minimizer of the function

$$m \longmapsto \int_{B(x_0,r)} |u(x) - m|^2 \, dx.$$

To see this, we observe that

$$\int_{B(x_0,r)} |u - m|^2 \, dx = \int_{B(x_0,r)} |u|^2 \, dx - 2m \int_{B(x_0,r)} u \, dx + m^2 |B(x_0,r)|,$$

so

$$\frac{d}{dm} \int_{B(x_0,r)} |u - m|^2 \, dx = 0 \quad \text{for } m = \fint_{B(x_0,r)} u \, dx$$

and we deduce the announced result. This implies that (recalling that $1/2 \leqslant r/R \leqslant 1$)

$$\int_{B(x_0,r)} |u(x) - u_{B(x_0,r)}|^2 \, dx \leqslant \int_{B(x_0,r)} |u(x) - u_{B(x_0,R)}|^2 \, dx$$

$$\leqslant 2^{n+2} \left(\frac{r}{R}\right)^{n+2} \int_{B(x_0,R)} |u - u_{B(x_0,R)}|^2 \, dx,$$

since the factor $2^{n+2}(r/R)^{n+2}$ is larger than 1. This concludes the proof. $\qquad\square$

---

[1] The fact that the inequality is true for any $c \in \mathbb{R}$ comes simply from the fact that if u is a solution of $\mathrm{div}(A\nabla u)$ then $v = u - c$ is a solution of the same equation. Moreover, we have $\nabla v = \nabla u$.

15.2 Local Averages and Hölder Spaces

In the following, we consider an open Ω of $\mathbb{R}^n$ and we denote by $d_\Omega = \sup_{x,y\in\Omega} |x-y|$ the diameter of Ω.

Let us recall Jensen's inequality.

Lemma 15.3 (Jensen) *Let $(X, \mathcal{A}, \mu)$ be a measure space with $\mu(X) = 1$, g a μ-integrable function with values in an interval I, and ϕ a convex function from I to $\mathbb{R}$. Then*

$$\phi\left(\int_X g\,d\mu\right) \leqslant \int_X \phi(g)\,d\mu.$$

In particular, for all $p \in [1, +\infty)$,

$$|u_\Omega|^p \leqslant \fint_\Omega |u|^p\,dx.$$

The definition of Campanato spaces involves an L^p analogue of variance. To motivate this definition, let us start by studying

$$J_p(u) = \inf_{m\in\mathbb{R}} \fint_\Omega |u - m|^p\,dx \qquad (u \in L^p(\Omega)).$$

In the proof of Proposition 15.2, we saw that for $p = 2$, we have

$$J_2(u) = \fint_\Omega |u - u_\Omega|^2\,dx.$$

We will see a similar result for the general case $1 \leqslant p < \infty$.

Lemma 15.4 *For all $p \in [1, +\infty)$, we have*

$$J_p(u) \leqslant \fint_\Omega |u - u_\Omega|^p\,dx \leqslant 2^p J_p(u).$$

Proof The inequality $J_p(u) \leqslant \fint_\Omega |u - u_\Omega|^p\,dx$ is trivial.

Consider $v \in L^p(\Omega)$. We use the inequality $|a + b|^p \leqslant 2^{p-1}(|a|^p + |b|^p)$ to deduce

$$\fint_\Omega |v - v_\Omega|^p\,dx \leqslant 2^{p-1} \fint_\Omega |v|^p\,dx + 2^{p-1} \fint_\Omega |v_\Omega|^p\,dx.$$

Noting that

$$\fint_\Omega |v_\Omega|^p\,dx \leqslant \fint_\Omega |v|^p\,dx$$

(according to Jensen's inequality), we deduce that

$$\fint_\Omega |v - v_\Omega|^p\,dx \leqslant 2^p \fint_\Omega |v|^p\,dx.$$

By applying this inequality with $v = u - m$ (then $v - v_\Omega = u - u_\Omega$) and taking the infimum for $m \in \mathbb{R}$, we obtain the desired result. □

We will now see the link between Hölder regularity and variance. For this, we will focus on the local averages of a function rather than its point values. Let us start by introducing a notation.

Definition 15.5 Let Ω be an open set of $\mathbb{R}^n$, $x_0 \in \Omega$ and $r > 0$. We define

$$\Omega(x_0, r) = \Omega \cap B(x_0, r).$$

If $f : \Omega \to \mathbb{R}$, we set

$$f_{x_0,r} = f_{\Omega(x_0,r)} = \fint_{\Omega(x_0,r)} f \, dx.$$

Lemma 15.6 Suppose that $f \in C^{0,\alpha}(\Omega)$. Then, for all x_0 and all $r > 0$,

$$\int_{\Omega(x_0,r)} |f(x) - f_{x_0,r}|^p \, dx \leqslant 2^{\alpha p} \|f\|_{C^{0,\alpha}}^p \omega_n r^{n+\alpha p},$$

where ω_n is the volume of the unit ball (note that the integral on the left-hand side is a usual integral and not a mean value $\fint_{\Omega(x_0,r)}$).

Proof As

$$f(x) - f_{x_0,r} = \frac{1}{|\Omega(x_0, r)|} \int_{\Omega(x_0,r)} (f(x) - f(y)) \, dy,$$

we have

$$|f(x) - f_{x_0,r}| \leqslant \frac{1}{|\Omega(x_0, r)|} \int_{\Omega(x_0,r)} \|f\|_{C^{0,\alpha}} |x - y|^\alpha \, dy \leqslant \|f\|_{C^{0,\alpha}} (2r)^\alpha.$$

We conclude by raising the left-hand side to the power p and integrating over $\Omega(x_0, r)$. □

Definition 15.7 (Campanato Spaces) Let Ω be an open set in $\mathbb{R}^n$, $\lambda > 0$, $p \in [1, \infty)$. A function $f \in L^p(\Omega)$ belongs to the Campanato space $\mathcal{L}^{p,\lambda}(\Omega)$ if

$$\|f\|_{\mathcal{L}^{p,\lambda}} := \sup_{0 < r} \sup_{x_0 \in \Omega} \left(r^{-\lambda} \int_{\Omega(x_0,r)} |f(x) - f_{x_0,r}|^p \, dx \right)^{1/p} < \infty.$$

Remark 15.8 For $0 \leqslant \lambda < n$, these spaces coincide with the spaces previously introduced by Morrey [140]. When $\lambda = n$, we obtain the space of functions with bounded mean oscillations, which was studied by John and Nirenberg [85]. In what follows, we will only be interested in the case $\lambda > n$.

15.3 Campanato's Theorem

Lemma 15.6 implies that

$$C^{0,\alpha}(\Omega) \subset \mathcal{L}^{p,n+\alpha p}(\Omega).$$

The study of the reverse inclusion is the subject of the Campanato theorem, which will be proved in this section, following Campanato's original proof (see [27]) and also [4]. We begin by introducing a generalized notion of Lipschitz domain (called a domain satisfying the (A)-property by Campanato).

Definition 15.9 (Lipschitz Domain) A bounded open set Ω in $\mathbb{R}^n$ is Lipschitz (in the generalized sense) if it satisfies the following property: there exists $c_* > 0$ such that, for every $x_0 \in \overline{\Omega}$ and every $r \in (0, d_\Omega)$,

$$|\Omega \cap B(x_0, r)| \geqslant c_* r^n. \tag{15.3}$$

Theorem 15.10 (Campanato) *Consider a bounded Lipschitz open set* $\Omega \subset \mathbb{R}^n$, *that is, satisfying condition* (15.3). *Let* $p \in [1, \infty)$ *and let* $\lambda \in (n, n + p)$. *Then*

$$\mathcal{L}^{p,\lambda}(\Omega) \subset C^{0,\alpha}(\Omega) \quad \text{with } \alpha = \frac{\lambda - n}{p}.$$

Proof We denote by c various constants that depend only on p, λ, n and c_*.

Lemma 15.11 *Consider a Lipschitz open set* Ω. *Let* $x_0 \in \Omega$ *and* r, ρ *with*

$$0 < r < \rho < d_\Omega.$$

Then

$$|f_{x_0,r} - f_{x_0,\rho}| \leqslant \frac{2}{c_*^{1/p}} \|f\|_{\mathcal{L}^{p,\lambda}} r^{-n/p} \rho^{\lambda/p}.$$

Proof We have

$$c_* r^n |f_{x_0,r} - f_{x_0,\rho}|^p \leqslant |\Omega(x_0, r)| |f_{x_0,r} - f_{x_0,\rho}|^p = \int_{\Omega(x_0,r)} |f_{x_0,r} - f_{x_0,\rho}|^p \, dx.$$

Using the inequality $|a + b|^p \leqslant 2^{p-1}(|a|^p + |b|^p)$ we deduce

$$c_* r^n |f_{x_0,r} - f_{x_0,\rho}|^p$$
$$\leqslant 2^{p-1} \left(\int_{\Omega(x_0,r)} |f_{x_0,r} - f(x)|^p \, dx + \int_{\Omega(x_0,r)} |f(x) - f_{x_0,\rho}|^p \, dx \right)$$
$$\leqslant 2^{p-1} \left(\int_{\Omega(x_0,r)} |f_{x_0,r} - f(x)|^p \, dx + \int_{\Omega(x_0,\rho)} |f(x) - f_{x_0,\rho}|^p \, dx \right)$$
$$\leqslant 2^{p-1} \|f\|_{\mathcal{L}^{p,\lambda}}^p (r^\lambda + \rho^\lambda) \leqslant 2^p \|f\|_{\mathcal{L}^{p,\lambda}}^p \rho^\lambda,$$

from which the result follows.   □

Let $R < d_\Omega/2$. Let us set $R_i = R/2^i$ for $i \in \mathbb{N}$ and apply the inequality of Lemma 15.11 with $r = R_{i+1}$ and $\rho = R_i$. We find that

$$\left|f_{x_0,R_{i+1}} - f_{x_0,R_i}\right| \leqslant \frac{2}{c_*^{1/p}}\|f\|_{\mathcal{L}^{p,\lambda}}R_{i+1}^{-n/p}R_i^{\lambda/p} = c'2^{i(n-\lambda)/p}R^{(\lambda-n)/p}\|f\|_{\mathcal{L}^{p,\lambda}}.$$

Since $\lambda > n$ the series $\sum 2^{i(n-\lambda)/p}$ converges and we deduce that, for all $j > i \geqslant 0$,

$$\left|f_{x_0,R_j} - f_{x_0,R_i}\right| \leqslant cR_i^{(\lambda-n)/p}\|f\|_{\mathcal{L}^{p,\lambda}}. \tag{15.4}$$

Consequently, $(f_{x_0,R_i})_{i \in \mathbb{N}}$ is a Cauchy sequence. The Lebesgue differentiation theorem (*cf.* Corollary 6.17) implies that this Cauchy sequence converges to $f(x_0)$ for almost every x_0 in Ω. Then, we deduce from (15.4) (applied with $i = 0$ and letting j tend to $+\infty$) that

$$\left|f(x_0) - f_{x_0,R}\right| \leqslant c\|f\|_{\mathcal{L}^{p,\lambda}}R^\alpha \quad \text{with } \alpha = \frac{\lambda - n}{p}. \tag{15.5}$$

Let us fix $R > 0$. Since $x_0 \mapsto f_{x_0,R}$ is bounded (because f belongs to $L^p(\Omega)$), we deduce that f is also bounded. It remains to prove that f is α-Hölderian. Given x, y in Ω and $R := 2|x - y|$, we seek to estimate $|f(x) - f(y)|$ in terms of R^α. From (15.5), we have

$$|f(x) - f(y)| \leqslant \left|f(x) - f_{x,R}\right| + \left|f_{x,R} - f_{y,R}\right| + \left|f_{y,R} - f(y)\right|$$
$$\leqslant 2c\|f\|_{\mathcal{L}^{p,\lambda}}R^\alpha + |f_{x,R} - f_{y,R}|,$$

and it only remains to estimate $\left|f_{x,R} - f_{y,R}\right|$. For this, we write that

$$c_*2^{-n}R^n\left|f_{x,R} - f_{y,R}\right|^p$$
$$\leqslant \int_{\Omega(y,R/2)}\left|f_{x,R} - f_{y,R}\right|^p \, ds$$
$$\leqslant 2^{p-1}\left(\int_{\Omega(x,R)}\left|f(s) - f_{x,R}\right|^p \, ds + \int_{\Omega(y,R)}\left|f(s) - f_{y,R}\right|^p \, ds\right)$$
$$\leqslant 2^p\|f\|_{\mathcal{L}^{p,\lambda}}^p R^\lambda,$$

from which we get

$$\left|f_{x,R} - f_{y,R}\right| \leqslant c\|f\|_{\mathcal{L}^{p,\lambda}}R^{(\lambda-n)/p} = c\|f\|_{\mathcal{L}^{p,\lambda}}R^\alpha.$$

Recalling that $R = 2|x - y|$ and combining the previous identities, we have shown that

$$|f(x) - f(y)| \leqslant c\|f\|_{\mathcal{L}^{p,\lambda}}|x - y|^\alpha,$$

which concludes the proof. $\qquad\square$

15.4 The Schauder–Campanato Theorem

Theorem 15.12 (Schauder–Campanato) *Let $\alpha \in (0, 1)$ and $u \in H^1(\Omega)$ be a weak solution of the equation $\mathrm{div}(A(x)\nabla u) = 0$, where $A \in C^{0,\alpha}$ is a symmetric matrix satisfying*

$$\exists c_0 > 0 / \; \forall x \in \Omega, \; \forall \xi \in \mathbb{R}^n, \quad A(x)\xi \cdot \xi \geqslant c_0|\xi|^2.$$

Then $u \in C^{1,\alpha}_{\mathrm{loc}}(\Omega)$ and for every compact $F \subset \Omega$, there exists a constant $C > 0$ such that

$$\|u\|_{C^{1,\alpha}(F)} \leqslant C\|u\|_{H^1(\Omega)}.$$

Proof Let us fix a bounded set $K \subset \Omega$ such that $\mathrm{dist}(K, \partial\Omega) > 0$. We make this assumption to guarantee that there exists $r_0 > 0$ such that $B(x, r_0) \subset \Omega$ for all $x \in K$.

Lemma 15.13 *Let $\lambda \in (0, n)$. There exists $R_0 > 0$ such that, for all $x_0 \in K$,*

$$\sup_{0 < r < R_0} r^{-\lambda} \int_{B(x_0,r)} |\nabla u|^2 \, dx < +\infty. \tag{15.6}$$

Proof Let $x_0 \in K$. We use the Korn method, the idea of which is to reduce to a problem with constant coefficients in the neighborhood of x_0. For this, we write

$$\mathrm{div}(A(x)\nabla u) = \mathrm{div}(A(x_0)\nabla u) + \mathrm{div}\big((A(x) - A(x_0))\nabla u\big).$$

We deduce that

$$\mathrm{div}(A(x_0)\nabla u) = \mathrm{div}\big((A(x_0) - A(x))\nabla u\big).$$

We then decompose u as the solution of two distinct problems: $u = v + w$ where

1. v is the unique solution of

$$\mathrm{div}(A(x_0)\nabla v) = 0 \text{ in } B(x_0, R), \quad v|_{\partial B(x_0,R)} = u|_{\partial B(x_0,R)}.$$

The key point is that we can apply Proposition 15.2, because v is the solution of an equation with constant coefficients. Then

$$\int_{B(x_0,r)} |\nabla v|^2 \, dx \leqslant C\left(\frac{r}{R}\right)^n \int_{B(x_0,R)} |\nabla v|^2 \, dx. \tag{15.7}$$

The constant C depends only on the norm of $A(x_0)$, which is bounded on K.

2. Moreover, $w = u - v$ is the solution of

$$\mathrm{div}(A(x_0)\nabla w) = \mathrm{div}\big((A(x_0) - A(x))\nabla u\big), \quad w|_{\partial B(x_0,R)} = 0.$$

The key point is that the source term is small because $|A(x_0) - A(x)| = O(|x_0 - x|^\alpha)$ by assumption on A. We will deduce later that

$$\|\nabla w\|_{L^2(B(x_0,R))} \leqslant CR^\alpha \|\nabla u\|_{L^2(B(x_0,R))}. \tag{15.8}$$

Let us fix $R \in (0, \min\{1, d_\Omega\}]$. We want to compare ∇u and ∇v. To do this, the first observation is the following: as $u - v$ belongs to $H_0^1(B(x_0, R))$, the weak formulation of $\mathrm{div}(A(x_0)\nabla v) = 0$ gives

$$\int_{B(x_0,R)} A(x_0)\nabla v \cdot \nabla(v - u)\, dx = 0,$$

therefore

$$\int_{B(x_0,R)} A(x_0)\nabla v \cdot \nabla v\, dx = \int_{B(x_0,R)} A(x_0)\nabla u \cdot \nabla v\, dx.$$

Using the Cauchy–Schwarz inequality and the ellipticity of $A(x_0)$, we deduce that

$$c_0 \int_{B(x_0,R)} |\nabla v|^2\, dx \leq \|A(x_0)\|^2 \int_{B(x_0,R)} |\nabla u|^2\, dx.$$

The inequality (15.7) then implies that

$$\int_{B(x_0,r)} |\nabla v|^2\, dx \leq C'\left(\frac{r}{R}\right)^n \int_{B(x_0,R)} |\nabla u|^2\, dx.$$

We now use the inequality $|x + y|^2 \leq 2|x|^2 + 2|y|^2$ for x, y in $\mathbb{R}^n$:

$$\int_{B(x_0,r)} |\nabla u|^2\, dx \leq 2 \int_{B(x_0,r)} |\nabla v|^2\, dx + 2\int_{B(x_0,r)} |\nabla u - \nabla v|^2\, dx.$$

By combining the two previous inequalities, we deduce that, for all r such that $0 < r < R$,

$$\int_{B(x_0,r)} |\nabla u|^2\, dx \leq C'\left(\frac{r}{R}\right)^n \int_{B(x_0,R)} |\nabla u|^2\, dx + 2\int_{B(x_0,R)} |\nabla u - \nabla v|^2\, dx. \quad (15.9)$$

We will now estimate the contribution of $\int_{B(x_0,R)} |\nabla u - \nabla v|^2\, dx$. For this, let us recall that, by construction, $u - v \in H_0^1(B(x_0, R))$. As already mentioned, the weak formulations of the equations

$$\mathrm{div}(A(x_0)\nabla v) = 0,$$

and

$$\mathrm{div}(A\nabla u) = 0,$$

imply that

$$\int_{B(x_0,R)} A(x_0)\nabla v \cdot (\nabla u - \nabla v)\, dx = 0,$$

$$\int_{B(x_0,R)} A(x)\nabla u \cdot (\nabla u - \nabla v)\, dx = 0.$$

Then we have

$$c_0 \int_{B(x_0,R)} |\nabla u - \nabla v|^2 \, dx$$

$$\leqslant \int_{B(x_0,R)} A(x_0) \nabla(u-v) \cdot \nabla(u-v) \, dx$$

$$= \underbrace{\int_{B(x_0,R)} A(x_0) \nabla v \cdot (\nabla v - \nabla u) \, dx}_{=0} + \int_{B(x_0,R)} A(x_0) \nabla u \cdot (\nabla u - \nabla v) \, dx$$

$$= \underbrace{\int_{B(x_0,R)} A(x) \nabla u \cdot (\nabla u - \nabla v) \, dx}_{=0}$$

$$+ \int_{B(x_0,R)} (A(x_0) - A(x)) \nabla u \cdot (\nabla u - \nabla v) \, dx,$$

thus $\int_{B(x_0,R)} |\nabla u - \nabla v|^2 \, dx$ is bounded by

$$\frac{\|A - A(x_0)\|_{L^\infty(B(x_0,R))}}{c_0} \|\nabla u\|_{L^2(B(x_0,R))} \|\nabla u - \nabla v\|_{L^2(B(x_0,R))},$$

from which

$$\|\nabla u - \nabla v\|_{L^2(B(x_0,R))} \leqslant C R^\alpha \|\nabla u\|_{L^2(B(x_0,R))}.$$

Obviously, this implies that

$$\|\nabla u - \nabla v\|_{L^2(B(x_0,R))}^2 \leqslant C^2 R^{2\alpha} \|\nabla u\|_{L^2(B(x_0,R))}^2. \tag{15.10}$$

Let us introduce the function $f \colon [0, R] \to \mathbb{R}_+$ defined by

$$f(r) = r^{-\lambda} \int_{B(x_0,r)} |\nabla u|^2 \, dx.$$

From (15.9) and (15.10), we deduce that

$$f(r) \leqslant C \left(\left(\frac{r}{R}\right)^{n-\lambda} + \left(\frac{r}{R}\right)^{-\lambda} R^{2\alpha} \right) f(R).$$

As $\alpha > 0$ and $\lambda < n$ by assumption, there exist two real numbers $\theta > 0$ and $R_1 > 0$ such that, if $r/R \leqslant \theta$ and $R \leqslant R_1$, we have

$$C \left(\left(\frac{r}{R}\right)^{n-\lambda} + \left(\frac{r}{R}\right)^{-\lambda} R^{2\alpha} \right) \leqslant 1.$$

Then $f(r) \leqslant f(R)$ for all $(r, R) \in [0, \theta R_1] \times [0, R_1]$ and hence $f(r) \leqslant f(R_1)$ for all $r \leqslant R_0 = \theta R_1$. Since $f(R_1) < +\infty$, this concludes the proof. $\qquad \square$

Lemma 15.14 *Let $\lambda \in (0, n]$. If there exists $R_0 > 0$ such that*

$$\sup_{x_0 \in K} \sup_{0 < r < R_0} r^{-\lambda} \int_{B(x_0, r)} |\nabla u|^2 \, dx < +\infty, \tag{15.11}$$

then there exists $R_1 > 0$ such that

$$\sup_{x_0 \in K} \sup_{0 < r < R_1} r^{-\lambda - 2\alpha} \int_{B(x_0, r)} |\nabla u - (\nabla u)_r|^2 \, dx < +\infty,$$

where $(\nabla u)_r$ denotes the average of ∇u over the ball of center x_0 and radius r.

Proof Let $x_0 \in K$. Without loss of generality, by diagonalizing $A(x_0)$ and making a change of variables, we can assume that $A(x_0)$ is the identity matrix. As in the proof of the previous lemma, we will start by showing that, for all r such that $0 < r < R$,

$$\int_{B(x_0, r)} |\nabla u - (\nabla u)_r|^2 \, dx$$

$$\leqslant C \left(\frac{r}{R}\right)^{n+2} \int_{B(x_0, R)} |\nabla u - (\nabla u)_R|^2 \, dx + C \int_{B(x_0, R)} |\nabla u - \nabla v|^2 \, dx, \tag{15.12}$$

where v is as before. For this, we write (*cf.* Proposition 15.2)

$$\int_{B(x_0, r)} |\nabla v - (\nabla v)_r|^2 \, dx \leqslant C \left(\frac{r}{R}\right)^{n+2} \int_{B(x_0, R)} |\nabla v - (\nabla v)_R|^2 \, dx.$$

Using the already seen result

$$\int_{B(x_0, r)} |f - (f)_r|^2 \, dx = \inf_{m \in \mathbb{R}} \int_{B(x_0, r)} |f - m|^2 \, dx,$$

we deduce that

$$\int_{B(x_0, r)} |\nabla v - (\nabla v)_r|^2 \, dx \leqslant C \left(\frac{r}{R}\right)^{n+2} \int_{B(x_0, R)} |\nabla v - (\nabla u)_R|^2 \, dx. \tag{15.13}$$

We will now compare

$$\int_{B(x_0, R)} |\nabla v - (\nabla u)_R|^2 \, dx \quad \text{and} \quad \int_{B(x_0, R)} |\nabla u - (\nabla u)_R|^2 \, dx$$

using the fact that $\Delta v = 0$ in $B(x_0, R)$ and the fact that $u - v \in H_0^1(B(x_0, R))$ can be used as a test function. We write

$$\int_{B(x_0, R)} |\nabla v - (\nabla u)_R|^2 \, dx = \int_{B(x_0, R)} (\nabla v - (\nabla u)_R) \cdot (\nabla u - (\nabla u)_R) \, dx$$

$$+ \int_{B(x_0, R)} (\nabla v - (\nabla u)_R) \cdot \nabla (v - u) \, dx.$$

We will see that the second term on the right-hand side is equal to zero. Indeed, in the weak formulation of $\Delta v = 0$, we see that

$$\int_{B(x_0,R)} \nabla v \cdot \nabla(v - u) \, dx = 0.$$

Furthermore, $\int_{B(x_0,R)} (\nabla u)_R \cdot \nabla(v - u) \, dx = 0$ by integrating by parts (using again that $u - v \in H_0^1(B(x_0, R))$ and that $\nabla(\nabla u)_R = 0$).

The cancellation of the second term and the Cauchy–Schwarz inequality gives us directly

$$\int_{B(x_0,R)} |\nabla v - (\nabla u)_R|^2 \, dx \leqslant \int_{B(x_0,R)} |\nabla u - (\nabla u)_R|^2 \, dx.$$

By combining this inequality with (15.13), we find

$$\int_{B(x_0,r)} |\nabla v - (\nabla v)_r|^2 \, dx \leqslant C \left(\frac{r}{R}\right)^{n+2} \int_{B(x_0,R)} |\nabla u - (\nabla u)_R|^2 \, dx. \tag{15.14}$$

Finally, we have

$$\int_{B(x_0,r)} |\nabla u - (\nabla u)_r|^2 \, dx \leqslant 3 \int_{B(x_0,r)} |\nabla v - (\nabla v)_r|^2 \, dx$$

$$+ 3 \int_{B(x_0,r)} |\nabla u - \nabla v|^2 \, dx$$

$$+ 3 \int_{B(x_0,r)} |(\nabla u)_r - (\nabla v)_r|^2 \, dx.$$

The last expression is controlled by Jensen's inequality:

$$\int_{B(x_0,r)} |(\nabla u)_r - (\nabla v)_r|^2 \, dx \leqslant \int_{B(x_0,r)} |\nabla u - \nabla v|^2 \, dx.$$

By combining the above, we obtain the desired result (15.12).

We then use the inequality (15.8) to estimate $\nabla(u - v)$ then the hypothesis (15.11) to obtain, for $r \leqslant R_0$,

$$\int_{B(x_0,r)} |\nabla u - (\nabla u)_r|^2 \, dx$$

$$\leqslant C \left(\frac{r}{R}\right)^{n+2} \int_{B(x_0,R)} |\nabla u - (\nabla u)_R|^2 \, dx + C_\lambda R^{\lambda+2\alpha}. \tag{15.15}$$

Let us introduce the function

$$F(r) = r^{-\lambda-2\alpha} \int_{B(x_0,r)} |\nabla u - (\nabla u)_r|^2 \, dx.$$

Then,

$$F(\theta r) \leqslant C\theta^{n+2-\lambda-2\alpha} F(R) + C_{\lambda,\theta}$$

for all θ such that $0 < \theta < 1$ and all $R \leqslant R_0$. We then choose θ so that

$$C\theta^{n+2-\lambda-2\alpha} = 1/2$$

and we deduce that

$$r^{-\lambda-2\alpha} \int_{B(x_0,r)} |\nabla u - (\nabla u)_r|^2 \, dx \leqslant C_\lambda$$

for $0 < r \leqslant R_1$. This concludes the proof of the lemma. $\square$

We are now ready to prove the theorem. Consider $z \in \Omega$ and $\theta > 0$ such that $B(z,\theta) \subset \Omega$. We will prove that ∇u belongs to $C^{0,\alpha}(B(z,\theta/2))$, which will conclude the proof of Theorem 15.12.

To shorten notation, set $B = B(z,\theta)$. Lemma 15.13 implies that the hypothesis (15.11) of Lemma 15.14 is satisfied for all $\lambda \in (0,n)$. From Lemma 15.14, we deduce that there exists $R_1 > 0$ such that

$$\sup_{x_0 \in B} \sup_{0<r<R_1} r^{-\lambda-2\alpha} \int_{B(x_0,r)} |\nabla u - (\nabla u)_r|^2 \, dx < +\infty.$$

Furthermore, for any function u such that ∇u belongs to $L^2(\Omega)$, it is easy to show that

$$\sup_{x_0 \in B} \sup_{R_1 \leqslant r} r^{-\lambda-2\alpha} \int_{\Omega(x_0,r)} |\nabla u - (\nabla u)_r|^2 \, dx < +\infty.$$

It follows that ∇u belongs to the Campanato space $\mathcal{L}^{2,\lambda+2\alpha}(B)$. We then choose λ such that $\lambda > n - 2\alpha$. We then deduce from Campanato's theorem that $\nabla u \in C^{0,\beta}(B)$ with

$$\beta = \frac{\lambda + 2\alpha - n}{2}.$$

Then $\nabla u \in C^{0,\beta}(B)$ for all $\beta < \alpha$ and hence $\nabla u \in L^\infty(B)$. As $B = B(z,\theta)$, the fact that $\nabla u \in L^\infty(B)$ implies immediately that

$$\sup_{x_0 \in B(z,\theta/2)} \sup_{0<r<\theta/2} r^{-n} \int_{B(x_0,r)} |\nabla u|^2 \, dx \leqslant 2^n \|\nabla u\|^2_{L^\infty(B)}.$$

The hypothesis (15.11) of Lemma 15.14 is therefore satisfied with $\lambda = n$. Lemma 15.14 applies and gives that there exists $\varepsilon_1 > 0$ such that

$$\sup_{x_0 \in B(z,\theta/2)} \sup_{0<r<\varepsilon_1} r^{-n-2\alpha} \int_{B(x_0,r)} |\nabla u - (\nabla u)_r|^2 \, dx < +\infty.$$

As before, we deduce that $\nabla u \in \mathcal{L}^{2,n+2\alpha}(B(z,\theta/2))$. Therefore, thanks to Campanato's theorem, we have $\nabla u \in C^{0,\alpha}(B(z,\theta/2))$. This is the desired result. $\square$

The previous proof allows us to consider the case of an equation with a source term and we have the following result.

Theorem 15.15 *Let* $\alpha \in (0, 1)$ *and* $u \in H^1(\Omega)$ *be a weak solution of the equation*

$$\mathrm{div}(A(x)\nabla u) = \mathrm{div}\, F$$

where $A \in C^{0,\alpha}$ *is a symmetric elliptic matrix and* $F \in C^{0,\alpha}_{\mathrm{loc}}(\Omega)$. *Then* $\nabla u \in C^{0,\alpha}_{\mathrm{loc}}(\Omega)$.

15.5 H^2 Regularity

The lemmas that we have proved to study elliptic regularity all concern inequalities for functions $u \in H^1(\Omega)$. We will need to know that we can apply these inequalities for ∇u instead of u. For this, we will state in this section an H^2 regularity result inside Ω (far from the boundary).

Let us start with an elementary observation.

Lemma 15.16 *If* $v \in C^3_0(\Omega)$ *then the* L^2 *norm of the Laplacian of* v *controls the* L^2 *norm of all the second derivatives of* v:

$$\|D^2 v\|^2_{L^2(\Omega)} = \|\Delta v\|^2_{L^2(\Omega)} .$$

Proof By integrating by parts, we find

$$\|D^2 v\|^2_{L^2(\Omega)} = \int_\Omega \sum_{i,j=1}^n (\partial_i \partial_j v)^2 \, dx = -\int_\Omega \sum_{i,j=1}^n (\partial_j v)(\partial_i^2 \partial_j v) \, dx$$

$$= \int_\Omega \sum_{i,j=1}^n (\partial_j^2 v)(\partial_i^2 v) \, dx = \|\Delta v\|^2_{L^2(\Omega)} ,$$

which is the desired result.   □

The following result is much stronger. It corresponds to the previous lemma but:

- it applies to weak solutions (H^1 and not C^3);
- there is no need to make a hypothesis about the behavior at the boundary;
- it is true for operators with variable coefficients, like

$$\mathrm{div}(A\nabla \cdot) = \sum_{i,j=1}^n \partial_i(a_{ij}(x)\partial_j \cdot),$$

where $A = (a_{ij})_{1\leqslant i,j\leqslant n}$ is an elliptic matrix (such that $A \in L^\infty(\Omega)$ and $\langle A\xi, \xi\rangle \geqslant \lambda |\xi|^2$ for a certain constant $\lambda > 0$).

Theorem 15.17 *Let Ω be a regular domain of $\mathbb{R}^n$ and let $f \in L^2_{loc}(\Omega)$. Consider an elliptic matrix $A \in C^{0,1}_{loc}(\Omega)$ and a weak solution $u \in H^1_{loc}(\Omega)$ of $\mathrm{div}(A\nabla u) = f$. Then, for every $\Omega'' \Subset \Omega' \Subset \Omega$, there exists a constant c such that*

$$\int_{\Omega''} |\nabla^2 u|^2 \, dx \leqslant c \int_{\Omega'} \left(|u|^2 + |f|^2 \right) dx.$$

The proof is the subject of the solved Exercise 15.1. The proof is based on the Caccioppoli inequality and the Nirenberg method: the idea is to consider discrete partial derivatives. Given $y \in \mathbb{R}^n$, we introduce the operator τ_y and Δ_y defined on $L^2(\Omega)$ by

$$\tau_y u(x) = u(x + y), \quad \Delta_y u(x) = \frac{u(x + y) - u(x)}{|y|}.$$

The elementary properties of differentiation are still true: the Leibniz rule

$$\Delta_y(ab) = (\tau_y a)\Delta_y b + (\Delta_y a)b,$$

and integration by parts

$$\int_\Omega \varphi(x)\Delta_y u(x) \, dx = - \int_\Omega u(x)(\Delta_{-y}\varphi)(x) \, dx,$$

(for all $\varphi \in C^1_0(\Omega)$ and all $|y| < \mathrm{dist}(\mathrm{supp}\,\varphi, \partial\Omega)$). We are then able to conclude the proof thanks to the following result.

Lemma 15.18 *Let $u \in L^p_{loc}(\Omega)$ with $1 < p \leqslant \infty$. Then $\nabla u \in L^p_{loc}(\Omega)$ if and only if, for every $\Omega' \Subset \Omega$, there exists a constant $C > 0$ such that,*

$$\forall y \in B(0, 1) \smallsetminus \{0\}, \ \forall \varphi \in C^1_0(\Omega'), \quad \left| \int_{\Omega'} (\Delta_y u)\varphi \, dx \right| \leqslant C\|\varphi\|_{L^{p'}(\Omega')},$$

where $1/p + 1/p' = 1$.

15.6 Regularity of Minimal Surfaces

Let Ω be a bounded regular connected open set and $g \colon \partial\Omega \to \mathbb{R}$ a given regular function. We introduce the affine space

$$H^1_g(\Omega) = \left\{ u \in H^1(\Omega) : u|_{\partial\Omega} = g \right\}.$$

Consider a function $u \in H^1_g(\Omega)$. We denote by $S(u)$ the surface

$$S(u) = \{(x, u(x)) : x \in \Omega\},$$

and by $A(u)$ the area of $S(u)$, defined by $A(u) = \int_\Omega \sqrt{1 + |\nabla u|^2} \, dx$.

Definition 15.19 Let $u \in H_g^1(\Omega)$. We say that $S(u)$ is a minimal surface if

$$A(u) = \inf_{v \in H_g^1(\Omega)} A(v).$$

The main result of this chapter states that a minimal surface is of class C^∞ inside its domain of definition. To obtain this result, we start with the following proposition.

Proposition 15.20 1. *Let* $u \in H_g^1(\Omega)$. *If $S(u)$ is a minimal surface then*

$$H(u) = \operatorname{div}\left(\frac{\nabla u}{\sqrt{1 + |\nabla u|^2}}\right) = 0$$

in the sense where

$$\forall \varphi \in H_0^1(\Omega), \quad \int_\Omega \frac{\nabla u \cdot \nabla \varphi}{\sqrt{1 + |\nabla u|^2}} \, dx = 0. \tag{15.16}$$

2. *If $u \in W^{1,\infty}(\Omega)$ is a solution of (15.16) then $u \in H_{loc}^2(\Omega)$ and, for all k such that $1 \leqslant k \leqslant n$,*

$$\operatorname{div}(A \nabla \partial_k u) = 0 \tag{15.17}$$

where $A = (\partial_i \partial_j F(\nabla u))_{1 \leqslant i, j \leqslant n}$ with $F(\zeta) = \sqrt{1 + |\zeta|^2}$.

Proof 1. Let us consider the function $\theta : t \mapsto A(u + t\varphi)$, defined for $t \in [-1, 1]$. Then $\theta(t) = \int_\Omega f(t, x) \, dx$ with $f = \sqrt{1 + |\nabla u + t\nabla \varphi|^2}$. We have

$$|\partial_t f| = \left|\frac{\nabla u \cdot \nabla \varphi + t |\nabla \varphi|^2}{\sqrt{1 + |\nabla u + t\nabla \varphi|^2}}\right| \leqslant |\nabla u| |\nabla \varphi| + |\nabla \varphi|^2 \in L^1(\Omega),$$

where we used the Cauchy–Schwarz inequality. We can then apply the Lebesgue differentiation theorem, which implies that θ is differentiable. We obtain equation (15.16) by writing that, if u is a minimizer, we have $\theta'(0) = 0$.

2. Let us set $G(\zeta) = \nabla_\zeta \sqrt{1 + |\zeta|^2} = \zeta/\sqrt{1 + |\zeta|^2}$ so that equation (15.16) can be written as

$$\int_\Omega G(\nabla u) \cdot \nabla \varphi \, dx = 0.$$

Let $k \in \{1, \ldots, n\}$ and h such that $|h| < \operatorname{dist}(\operatorname{supp} \varphi, \partial\Omega)$. Then

$$\int_\Omega (F(\nabla u(x + he_k)) - F(\nabla u(x))) \cdot \nabla \varphi(x) \, dx = 0.$$

Let us denote by G^i the coordinates of G (thus $G = (G^1, \ldots, G^n)$) and write

$$G^i(\nabla u(x + he_k)) - G^i(\nabla u(x)) = \sum_{j=1}^n c_{ij}^h(x)\big(\partial_j u(x + he_k) - \partial_j u(x)\big),$$

where
$$c_{ij}^h(x) = \int_0^1 (\partial_{\zeta_j} G^i)\Big(t\nabla u(x + he_k) + (1 - t)\nabla u(x) \Big) \, dt.$$

We find that
$$\Delta_{he_k} u = \frac{1}{h}\Big(u(x + he_k) - u(x) \Big)$$

satisfies
$$\int_\Omega \sum_{i,j} c_{ij}^h(x) \partial_j (\Delta_{he_k} u) \partial_i \varphi \, dx = 0.$$

We will now see that the matrix $(c_{ij}^h(x))_{1 \leqslant i,j \leqslant n}$ is elliptic, uniformly in h. We need to show that there exists $\lambda > 0$ such that, for all h, all ξ and all x, we have

$$\sum_{1 \leqslant i,j \leqslant n} c_{ij}^h(x)\xi_i \xi_j \geqslant \lambda |\xi|^2.$$

By differentiating G, we calculate that

$$\partial_{\zeta_j} G^i(\zeta) = \frac{\delta_i^j}{\sqrt{1 + |\zeta|^2}} - \frac{\zeta_i \zeta_j}{(1 + |\zeta|^2)^{3/2}}.$$

Let us set
$$Y(t,x) = t\nabla u(x + he_k) + (1 - t)\nabla u(x).$$

Then

$$\sum_{1 \leqslant i,j \leqslant n} c_{ij}^h(x)\xi_i \xi_j = \int_0^1 \frac{(1 + |Y(t,x)|^2)|\xi|^2 - (Y(t,x) \cdot \xi)^2}{(1 + |Y(t,x)|^2)^{3/2}} \, dt.$$

Now, observe that, according to the Cauchy–Schwarz inequality, we have

$$(1 + |Y(t,x)|^2)|\xi|^2 - (Y(t,x) \cdot \xi)^2 = |\xi|^2 + \Big[|Y(t,x)|^2|\xi|^2 - (Y(t,x) \cdot \xi)^2 \Big] \geqslant |\xi|^2,$$

so

$$\sum_{1 \leqslant i,j \leqslant n} c_{ij}^h(x)\xi_i \xi_j \geqslant |\xi|^2 \int_0^1 \frac{1}{(1 + |Y(t,x)|^2)^{3/2}} \, dt.$$

Recall that, by hypothesis, $\nabla u \in L^\infty(\Omega)$. This implies that $(1 + |Y(t,x)|^2)^{-3/2}$ is bounded from below by a positive constant and therefore c is an elliptic matrix, uniformly in h. We can then apply (the proof of) the Caccioppoli inequality. Consider $\Omega'' \Subset \Omega' \Subset \Omega$ with

$$|h| < \operatorname{dist}(\Omega', \partial\Omega).$$

We can estimate the $H^1(\Omega'')$-norm of $\Delta_{he_k} u$ in terms of the $L^2(\Omega')$-norm of $\Delta_{he_k} u$ which is itself controlled by the $L^2(\Omega)$-norm of ∇u. We then deduce that u belongs to $H^2(\Omega'')$ and that we have the desired equation (*cf.* lemma 15.18). $\square$

We can now prove the following result.

Theorem 15.21 *If $u \in W^{1,\infty}(\Omega)$ satisfies $H(u) = 0$, then $u \in C^{\infty}(\Omega)$.*

Proof We give only a sketch of the proof. We have $u \in H^2_{\text{loc}}(\Omega)$ and moreover $\text{div}(A\nabla\partial_k u) = 0$. As $\nabla u \in L^{\infty}(\Omega)$, the matrix A satisfies the hypotheses of the De Giorgi theorem, which implies that $\partial_k u$ belongs to $C^{0,\alpha}_{\text{loc}}(\Omega)$ for all k and consequently $u \in C^{1,\alpha}_{\text{loc}}(\Omega)$. Then we deduce that A has coefficients in $C^{0,\alpha}_{\text{loc}}(\Omega)$. The Schauder theorem then implies that $\partial_k u \in C^{1,\alpha}_{\text{loc}}$ for all k. We then deduce that $u \in C^{2,\alpha}_{\text{loc}}$. We can differentiate the equation and verify that $\partial_j \partial_k u$ satisfies an equation of the same type (with a source term), and we use Theorem 15.15. The Schauder theorem leads to $u \in C^{3,\alpha}_{\text{loc}}$. Reasoning by induction, we obtain the desired result. $\qquad\square$

15.7 Exercises

Exercise (solved) 15.1 Let $n \geqslant 1$ be any integer. In this exercise, $\langle \cdot, \cdot \rangle$ denotes the scalar product in $\mathbb{R}^n$. Let Ω be an open set of $\mathbb{R}^n$ and let $A \in C^{0,1}(\Omega, \mathcal{M}_n(\mathbb{R}))$ be a matrix-valued Lipschitz function on Ω such that, for a certain $\lambda > 0$:

$$\forall x \in \Omega, \ \forall \xi \in \mathbb{R}^n, \quad \langle A(x)\xi, \xi \rangle \geqslant \lambda |\xi|^2.$$

For all $h \in \mathbb{R}^n$, and for any function $u = u(x)$, we define

$$\tau_h u = u(x + h) \quad \text{and} \quad \Delta_h u = \frac{u(x + h) - u(x)}{|h|}.$$

Consider two open sets Ω' and Ω'' such that $\overline{\Omega'} \subset \Omega''$ and $\overline{\Omega''} \subset \Omega$.

1. Show that, for $u \in H^1(\Omega)$ and h small enough, we have

$$\|\tau_h u - u\|_{L^2(\Omega'')} \leqslant |h| \, \|\nabla u\|_{L^2(\Omega)}.$$

2. Let $f \in L^2(\Omega)$. Let $u \in H^1(\Omega)$ be a weak solution of $-\text{div}(A\nabla u) = f$. Show that for all h small enough and $\phi \in H^1(\Omega)$, with support in Ω'', we have

$$\int_{\Omega} (\tau_h A)\nabla(\Delta_h u) \cdot \nabla\phi \, dx = \int_{\Omega} (\Delta_h f)\phi \, dx - \int_{\Omega} (\Delta_h A)\nabla u \cdot \nabla\phi \, dx.$$

3. Drawing inspiration from the proof of the Caccioppoli inequality, show that

$$\int_{\Omega'} |\nabla\Delta_h u|^2 \, dx \leqslant C \int_{\Omega} (u^2 + f^2) \, dx.$$

[Hint: one can use the fact that $\int_{\Omega''} |\nabla u|^2 \, dx \leqslant c \int_{\Omega} (u^2 + f^2)$.]

4. Deduce that $u \in H^2(\Omega')$ and that

$$\int_{\Omega'} |\nabla^2 u|^2 \, dx \leqslant C' \int_{\Omega} (u^2 + f^2) \, dx.$$

5. Consider the rectangle $R = [0,1] \times [-1,1]^{n-1}$ and the following Dirichlet problem:

$$\begin{cases} -\operatorname{div}(A\nabla u) = f & \text{on } R, \\ u = 0 & \text{on } \partial R. \end{cases}$$

Let $R' = [0, 1/2] \times [-1/2, 1/2]^{n-1}$. Show that

$$\int_{R'} |\nabla^2 u|^2 \, dx \leqslant C'' \int_R (u^2 + f^2) \, dx.$$

Chapter 16
Dispersive Estimates

This chapter is composed of two distinct parts. We begin with an introduction to the study of the Cauchy problem for the nonlinear Schrödinger equation. We will state many results without proof[1] to try to give an overview of recent results in this field. This will allow us to introduce the Strichartz–Bourgain estimates. In the second part, we will prove one of these estimates.

16.1 The Schrödinger Equation

The linear Schrödinger equation is written as:

$$i\partial_t u + \Delta u = 0,$$

where the unknown is $u \colon \mathbb{R} \times \mathbb{R}^n \to \mathbb{C}$, depending on the time variable $t \in \mathbb{R}$ and the space variable $x \in \mathbb{R}^n$. This equation can be solved by Fourier transform.

Proposition 16.1 *Let $t \in \mathbb{R}$. By definition, $S(t) = e^{it\Delta}$ is the Fourier multiplier of symbol $e^{-it|\xi|^2}$, such that*

$$\widehat{S(t)u_0}(\xi) = e^{-it|\xi|^2}\,\widehat{u_0}(\xi).$$

For all $t \in \mathbb{R}$, the operator $S(t)$ is an isometry of $H^s(\mathbb{R}^n)$ into itself for all $s \in \mathbb{R}$. Moreover, we have $S(t) \circ S(t') = S(t+t')$ for all t, t' in $\mathbb{R}$.

If u_0 belongs to $H^s(\mathbb{R}^n)$ with $s \in \mathbb{R}$, then the map $u \colon t \mapsto S(t)u_0$ satisfies

$$u \in C^0(\mathbb{R}; H^s(\mathbb{R}^n)) \cap C^1(\mathbb{R}; H^{s-2}(\mathbb{R}^n)),$$

and it is a solution of the Schrödinger equation.

---

[1] For proofs and a detailed exposition of the study of dispersive equations, we refer the reader to the books of Bahouri, Chemin and Danchin [9], Cazenave and Haraux [32], Klainerman [100], Linares and Ponce [125], Muscalu and Schlag [145], and Tao [181].

© The Author(s), under exclusive license to Springer Nature Switzerland AG 2024
T. Alazard, *Analysis and Partial Differential Equations*, Universitext,
https://doi.org/10.1007/978-3-031-70909-8_16

Therefore, it is very easy to solve the Cauchy problem for the linear Schrödinger equation in Sobolev spaces. One might wonder if a similar result is true in other spaces. The following proposition shows that the operator $S(t)$ is unbounded on Hölder spaces (and even on Zygmund spaces).

Proposition 16.2 Let $s, \sigma \in \mathbb{R}$. Suppose there exists a $t_0 \neq 0$ such that $S(t_0)$ is continuous from $C_*^{\sigma}(\mathbb{R}^n)$ to $C_*^s(\mathbb{R}^n)$. Then $s \leqslant \sigma - n$.

In fact, the operator $S(t)$ is unbounded on all Sobolev spaces built on $L^p(\mathbb{R}^n)$ with $p \neq 2$. This explains why it is natural to seek a solution, continuous in time, with values in a Sobolev space $H^s(\mathbb{R}^n)$ built on $L^2(\mathbb{R}^n)$.

We now consider the Cauchy problem for the nonlinear Schrödinger equation: given an initial data u_0, we seek a solution u of the problem

$$i\partial_t u + \Delta u = |u|^2 u, \quad u_{|t=0} = u_0.$$

A first difficulty is to give meaning to the equation. We say that u is a solution of the nonlinear Schrödinger equation with initial data u_0 if

$$u(t) = S(t)u_0 - i \int_0^t S(t - t')\big(|u(t')|^2 u(t')\big)\, dt'. \tag{16.1}$$

This identity is called Duhamel's formula. For this formula to make sense, we need to find a framework in which the nonlinearity $|u|^2 u$ has values in a Sobolev space. This is why the study of the Cauchy problem depends heavily on the regularity of the initial data.

The simplest situation is where u is continuous in time with values in a Sobolev space $H^s(\mathbb{R}^n)$ with $s > n/2$. Indeed, in this case, we know that $H^s(\mathbb{R}^n)$ is a Banach algebra: the product of two elements of $H^s(\mathbb{R}^n)$ belongs to $H^s(\mathbb{R}^n)$ if $s > n/2$. The previous formula therefore has a clear meaning and we can state the following result.

Proposition 16.3 Let $n \geqslant 1$ and $s > n/2$. For any initial data $u_0 \in H^s(\mathbb{R}^n)$, there exists $T > 0$ such that there is a unique function $u \in C^0([0, T]; H^s(\mathbb{R}^n))$ satisfying

$$u(t) = S(t)u_0 - i \int_0^t S(t - t')\big(|u(t')|^2 u(t')\big)\, dt'.$$

Let T^* be the lifespan of the maximal solution. Then, either the maximal solution is defined for all time, that is $T^* = +\infty$, or

$$\limsup_{t \to T^*} \|u(t)\|_{L^\infty} = +\infty. \tag{16.2}$$

The proof of this proposition relies on the fixed point theorem, and the fact that the product of two elements of $H^s(\mathbb{R}^n)$ belongs to $H^s(\mathbb{R}^n)$ if $s > n/2$. The blow-up criterion (16.2) is a consequence of the Sobolev embedding: $H^s(\mathbb{R}^n) \subset L^\infty(\mathbb{R}^n)$ for $s > n/2$.

It is desirable to go further and extend this result by considering less regular initial data, belonging to Sobolev spaces of index $s \leqslant n/2$. There are several motivations. First, such a result provides information about the nature of possible singularities. Also, in order to obtain global existence results in time, it is paradoxically easier to work in spaces of low regularity associated with a scale invariance or the conservation of natural quantities (mass and energy). Let us introduce the mass M and energy E defined by

$$M(t) = \int_{\mathbb{R}^n} |u(t,x)|^2 \, dx,$$

$$E(t) = \frac{1}{2} \int_{\mathbb{R}^n} |\nabla u(t,x)|^2 \, dx + \frac{1}{2} \int_{\mathbb{R}^d} |u(t,x)|^4 \, dx.$$

Then

$$\frac{dM}{dt}(t) = 0, \qquad \frac{dE}{dt}(t) = 0.$$

For example, for $n = 1$, to solve the Cauchy problem globally in time in the class of functions of class C^∞, we use the Sobolev embedding $H^1(\mathbb{R}) \subset L^\infty(\mathbb{R})$, which allows us to show in a first step that the Cauchy problem is globally well posed on $H^1(\mathbb{R})$.

Proposition 16.4 *Assume that $n = 1$. For any initial data $u_0 \in H^1(\mathbb{R})$, there exists a unique solution $u \in C^0(\mathbb{R}_+; H^1(\mathbb{R}))$ of (16.1).*

We have the following result on propagation of regularity.

Proposition 16.5 *Let $u \in C^0(\mathbb{R}_+; H^1(\mathbb{R}))$ be a solution of (16.1). If $u_0 \in H^s(\mathbb{R})$ with $s > 1$, then $u \in C^0(\mathbb{R}_+; H^s(\mathbb{R}))$.*

By combining the two previous propositions, we obtain that

Corollary 16.6 *Assume that $n = 1$ and let $s \geqslant 1$. For any initial data $u_0 \in H^s(\mathbb{R})$, there exists a unique solution $u \in C^0(\mathbb{R}_+; H^s(\mathbb{R}))$ of (16.1).*

In dimension 2, $H^1(\mathbb{R}^2) \not\subset L^\infty(\mathbb{R}^2)$ and the problem is more difficult. However, the lack of inclusion is weak and we have the logarithmic estimate

$$\|f\|_{L^\infty(\mathbb{R}^2)} \leqslant C\|f\|_{H^1(\mathbb{R}^2)} \left(\log\left(1 + \frac{\|f\|_{H^2(\mathbb{R}^2)}}{\|f\|_{H^1(\mathbb{R}^2)}} \right) \right)^{1/2}.$$

This estimate allowed Brézis and Gallouët ([16]) to show that the Cauchy problem is globally well posed in $H^2(\mathbb{R}^2)$. This functional analysis method does not however allow us to deal with the case of higher order nonlinearities, nor to consider data in the energy space, that is to say to solve the Cauchy problem for an initial data belonging to $H^1(\mathbb{R}^2)$.

From dimension 2, to solve the Cauchy problem in spaces of low regularity, we must use a fundamental property of the equation. Specifically, the fact that it is a dispersive equation, which means that the characteristic variety of the free equation is curved. This leads to dispersion and Strichartz inequalities.

Theorem 16.7 *For any triplet* (n, p, q) *satisfying*

$$\frac{2}{p} + \frac{n}{q} = \frac{n}{2}, \qquad (p, q) \neq (2, +\infty), \qquad p \geqslant 2,$$

there exists a constant $C(p, q, n)$ *such that*

$$\left\| e^{it\Delta} u_0 \right\|_{L^p(\mathbb{R}_+; L^q(\mathbb{R}^n))} \leqslant C(p, q, n) \| u_0 \|_{L^2(\mathbb{R}^n)},$$

for all $u_0 \in L^2(\mathbb{R}^n)$.

Strichartz [179] had proven the estimate for $(n, p, q) = (2, 4, 4)$. Keel and Tao [98] obtained the limit case $(p, q) = (2, 2n/(n-2))$ for $n \geqslant 3$. The other estimates are due to Ginibre and Velo [60] as well as Yajima [198].

The Strichartz estimates provide an improvement over the Sobolev embedding which can be seen as a gain of derivative, which explains why these estimates are at the origin of a very clear improvement of the understanding of the Cauchy problem.

To clarify the meaning of the solution for initial data of low regularity, we use the following definition, borrowed from [58].

Definition 16.8 Let $s \geqslant 0$ and $n \geqslant 1$. We say that the Cauchy problem for the nonlinear Schrödinger equation is well-posed on $H^s(\mathbb{R}^n)$ if, for any open ball B of $H^s(\mathbb{R}^n)$, there exists a time $T > 0$ and a Banach space $X_T^s \hookrightarrow C^0([0, T]; H^s(\mathbb{R}^n))$ such that

1. for all $v \in X_T^s$ the function $|v|^2 v$ is well-defined and belongs to $L^1((0, T); H^s(\mathbb{R}^n))$,
2. for all $u_0 \in B$, there exists a unique solution $u \in X_T^s$ of the equation

$$u(t) = S(t)u_0 - i \int_0^t S(t - t')\left(|u(t')|^2 u(t')\right) dt'.$$

3. If $u_0 \in H^\sigma(\mathbb{R}^n)$ for a $\sigma > s$, then $u \in C^0([0, T]; H^\sigma(\mathbb{R}^n))$.

To clarify the statement, we sometimes say that the Cauchy problem is well-posed locally in time. We say that the Cauchy problem is globally well-posed if the previous result is true for all $T > 0$.

Theorem 16.9 *Assume that* $n = 2$. *The Cauchy problem for the nonlinear Schrödinger equation is globally well-posed on* $H^1(\mathbb{R}^2)$.

The proof of Strichartz's inequalities relies on an interpolation argument applied on one hand with the inequality $\left\| e^{it\Delta} u_0 \right\|_{L^2(\mathbb{R}^n)} \leqslant \| u_0 \|_{L^2(\mathbb{R}^n)}$, and on the other hand with

$$\left\| e^{it\Delta} u_0 \right\|_{L^\infty(\mathbb{R}^n)} \leqslant \frac{K}{t^{n/2}} \| u_0 \|_{L^1(\mathbb{R}^n)}. \tag{16.3}$$

We have already seen the first inequality (which is in fact an equality).

The second equality is obtained by a direct calculation of the fundamental solution, which is written:

$$(e^{it\Delta}u_0)(x) = \frac{1}{(i\pi t)^{n/2}} \int_{\mathbb{R}^n} e^{i|x-y|^2/(2t)} u_0(y)\, dy.$$

Let us now consider the same problem but in the case where the initial data u_0 and the solution u are periodic in the spatial variable x.

Definition 16.10 (Periodic Sobolev spaces) We denote by $L^2_{\mathrm{per}}(\mathbb{R}^n)$ the space of functions $u \in L^2_{\mathrm{loc}}(\mathbb{R}^n)$ which are 2π-periodic with respect to each variable, which is equivalent to $u(\cdot + \gamma) = u$ for all $\gamma \in (2\pi\mathbb{Z})^n$. Let us recall that the functions $u \in L^2_{\mathrm{per}}(\mathbb{R}^n)$ are represented by Fourier series by (see Theorem 4.3):

$$u = \sum_{k \in \mathbb{Z}^n} c_k e^{ik \cdot x}.$$

Let $s \in [0, +\infty)$. By definition,

$$H^s_{\mathrm{per}}(\mathbb{R}^n) = \left\{ u \in L^2_{\mathrm{per}}(\mathbb{R}^n) : \textstyle\sum_{k \in \mathbb{Z}^n}(1 + |k|^2)^s\, |c_k|^2 < +\infty \right\}.$$

Remark 16.11 If $s \in \mathbb{N}$, then we have

$$H^s_{\mathrm{per}}(\mathbb{R}^n) = \left\{ u \in L^2_{\mathrm{per}}(\mathbb{R}^n) : \forall \alpha \in \mathbb{N}^n,\ |\alpha| \leqslant s,\ \partial_x^\alpha u \in L^2_{\mathrm{per}}(\mathbb{R}^n) \right\},$$

where ∂_x^α denotes the weak derivative of u (cf. Definitions 7.6 and 7.12).

The results that are based solely on functional analysis or Fourier analysis arguments are still true. For example, if $s > n/2$, the Cauchy problem is well posed on $H^s_{\mathrm{per}}(\mathbb{R}^n)$. But we no longer have the dispersion estimate (16.3). Indeed, we have

$$1 \lesssim \|u_0\|_{L^2_{\mathrm{per}}(\mathbb{R}^n)} = \left\|e^{it\Delta}u_0\right\|_{L^2_{\mathrm{per}}(\mathbb{R}^n)} \lesssim \left\|e^{it\Delta}u_0\right\|_{L^\infty_{\mathrm{per}}(\mathbb{R}^n)}.$$

This shows that, if $u_0 \in L^1_{\mathrm{per}}(\mathbb{R}^n)$ is non-zero, then $\|e^{it\Delta}u_0\|_{L^\infty_{\mathrm{per}}(\mathbb{R}^n)}$ cannot tend to 0 when t tends to $+\infty$. Therefore, the inequality (16.3) cannot be true for all large t. In fact, the obstruction is much deeper.

Proposition 16.12 For all $t \geqslant 0$, $S(t)$ does not map $L^1_{\mathrm{per}}(\mathbb{R}^n)$ into $L^\infty_{\mathrm{per}}(\mathbb{R}^n)$.

However, we have an analogue of the Strichartz inequalities, which allowed Bourgain to show that the Cauchy problem is well posed even for initial data that are periodic.

Theorem 16.13 If $n = 2$ or $n = 3$, then the Cauchy problem for the nonlinear Schrödinger equation is well posed on $H^1_{\mathrm{per}}(\mathbb{R}^n)$ locally in time.

16.2 A Strichartz–Bourgain Estimate for KdV

Let us now consider the Korteveg–De Vries equation

$$u_t + u_{xxx} + uu_x = 0. \tag{16.4}$$

A major result of Bourgain [13, 14] is that the Cauchy problem for this equation is globally well posed on $L^2_{\mathrm{per}}(\mathbb{R})$. We will present in this section an element of the proof. The idea is to introduce spaces whose definition is based on the study of the group $(e^{-t\partial_{xxx}})_{t\in\mathbb{R}}$ associated with the linear equation $u_t + u_{xxx} = 0$.

Consider a solution $u = u(t,x)$, 2π-periodic in t and in x, of the linear equation $u_t + u_{xxx} = 0$. Consider its Fourier transform, defined by

$$\widetilde{u}(\tau, k) = \iint_{\mathbb{T}^2} e^{-it\tau - ixk} u(t,x)\, dx\, dt \qquad (\tau, k \in \mathbb{Z}).$$

So $\widetilde{u}$ is supported in the characteristic variety:

$$\operatorname{supp}\widetilde{u} \subset \left\{ (\tau, k) \in \mathbb{Z} \times \mathbb{Z} : \tau = k^3 \right\}.$$

This property of exact spectral localization is not true for the solutions of the non-linear equation (16.4). It is therefore important to introduce function spaces that are approximately spectrally localized near the characteristic variety.

Definition 16.14 Let $s \in \mathbb{R}_+$, $b \in \mathbb{R}_+$ and $u = u(t,x) \in L^2(\mathbb{T}^2)$ be an L^2 function, 2π-periodic in t and x. We introduce

$$\|u\|^2_{X^{s,b}(\mathbb{T}^2)} := \sum_{\tau \in \mathbb{Z}} \sum_{k \in \mathbb{Z}} \left(1 + |\tau - k^3|^2\right)^b \left(1 + |k|^2\right)^s |\widetilde{u}(\tau, k)|^2,$$

as well as the Bourgain space

$$X^{s,b}(\mathbb{T}^2) = \left\{ u \in L^2(\mathbb{T}^2) : \|u\|_{X^{s,b}(\mathbb{T}^2)} < +\infty \right\}.$$

Remark 16.15 In particular, if u satisfies $u_t + u_{xxx} = 0$ with $u(0) \in H^s_{\mathrm{per}}(\mathbb{R})$, then $u \in X^{s,b}(\mathbb{T}^2)$ for all $b \geqslant 0$.

Proposition 16.16 *There exists a constant C such that*

$$\|u\|_{L^4(\mathbb{T}^2)} \leqslant C \|u\|_{X^{0,1/3}(\mathbb{T}^2)}.$$

Remark 16.17 This result plays a fundamental role in the study of the Cauchy problem for the nonlinear equation (16.4). To understand why this result is surprising, consider a solution u of the linear equation $\partial_t u + \partial_{xxx} u = 0$. If the initial data u_0 belongs to $L^2_{\mathrm{per}}(\mathbb{R})$, then $u \in X^{0,b}(\mathbb{T}^2)$ for all $b \geqslant 0$ as we have already said. We can therefore apply this proposition to deduce that $\|u\|_{L^4(\mathbb{T}^2)} \leqslant C\|u\|_{L^2(\mathbb{T}^2)}$. This is a clear improvement of the Sobolev embedding theorem, which only implies in dimension 2 that $\|u\|_{L^4(\mathbb{T}^2)} \leqslant C\|u\|_{H^{1/4}(\mathbb{T}^2)}$.

Proof This estimate was proved by Bourgain in [14]. The following proof is borrowed from Tao [181] where it is mentioned that it relies on an argument due to Tzvetkov.

Decompose

$$u = \sum_{p \in \mathbb{N}} u_p, \quad \text{where} \quad u_p = (2\pi)^{-2} \sum_{\substack{\tau, k \in \mathbb{Z} \\ 2^p \leqslant \langle \tau - k^3 \rangle < 2^{p+1}}} e^{i\tau t + ikx} \widetilde{u}(\tau, k).$$

Then, for all $b \geqslant 0$,

$$\sum_{p \in \mathbb{N}} 2^{2bp} \|u_p\|_{L^2(\mathbb{T}^2)}^2 \leqslant \|u\|_{X^{0,b}(\mathbb{T}^2)}^2.$$

In particular, for $b = 1/3$, we have

$$\sum_{p \in \mathbb{N}} 2^{2p/3} \|u_p\|_{L^2(\mathbb{T}^2)}^2 \leqslant \|u\|_{X^{0,1/3}(\mathbb{T}^2)}^2.$$

Furthermore,

$$\|u\|_{L^4(\mathbb{T}^2)}^2 = \|uu\|_{L^2(\mathbb{T}^2)} \leqslant \sum_{p \in \mathbb{N}} \sum_{p' \in \mathbb{N}} \|u_p u_{p'}\|_{L^2(\mathbb{T}^2)}$$

$$\leqslant 2 \sum_{m \in \mathbb{N}} \sum_{p \in \mathbb{N}} \|u_p u_{p+m}\|_{L^2(\mathbb{T}^2)},$$

so it suffices to show that

$$\sum_{m \in \mathbb{N}} \sum_{p \in \mathbb{N}} \|u_p u_{p+m}\|_{L^2(\mathbb{T}^2)} \leqslant K \sum_{p \in \mathbb{N}} 2^{2p/3} \|u_p\|_{L^2(\mathbb{T}^2)}^2.$$

Of course, it is sufficient to show that there exists $\varepsilon > 0$ such that

$$\sum_{p \in \mathbb{N}} \|u_p u_{p+m}\|_{L^2(\mathbb{T}^2)} \leqslant K 2^{-\varepsilon m} \sum_{p \in \mathbb{N}} 2^{2p/3} \|u_p\|_{L^2(\mathbb{T}^2)}^2,$$

for a constant K independent of m. Using the Cauchy–Schwarz inequality, we see that this inequality will be a consequence of

$$\|u_p u_{p+m}\|_{L^2(\mathbb{T}^2)} \leqslant K 2^{-\varepsilon m} 2^{2p/3} \|u_p\|_{L^2(\mathbb{T}^2)} (2^m 2^p)^{1/3} \|u_{p+m}\|_{L^2(\mathbb{T}^2)}, \qquad (16.5)$$

because

$$\left(\sum_p 2^{2p/3} \|u_p\|_{L^2(\mathbb{T}^2)}^2 \right)^{\frac{1}{2}} \left(\sum_p (2^m 2^p)^{2/3} \|u_{p+m}\|_{L^2(\mathbb{T}^2)}^2 \right)^{\frac{1}{2}} \leqslant \sum_p 2^{2p/3} \|u_p\|_{L^2(\mathbb{T}^2)}^2.$$

Our goal is therefore to prove the inequality (16.5). Note that we can assume without loss of generality that $\|u_p\|_{L^2(\mathbb{T}^2)} = 1 = \|u_{p+m}\|_{L^2(\mathbb{T}^2)}$.

Let us write u_p and u_{p+m} in the form

$$u_p = \sum_{\tau_1, k_1 \in \mathbb{Z}} e^{i\tau_1 t + ik_1 x} \widetilde{u}_p(\tau_1, k_1), \quad u_{p+m} = \sum_{\tau_2, k_2 \in \mathbb{Z}} e^{i\tau_2 t + ik_2 x} \widetilde{u}_{p+m}(\tau_2, k_2),$$

with $\widetilde{u}_p(\tau, k) = \varphi_p(\tau - k^3)\widetilde{u}(\tau, k)$, where $\varphi_p(y)$ is the function that equals 1 if $2^p \leqslant \langle y \rangle < 2^{p+1}$ and 0 otherwise. Then

$$u_p u_{p+m} = \sum_{\tau, k} c(\tau, k) e^{i\tau t + ikx}$$

with

$$c(\tau, k) = \sum_{\tau_1 + \tau_2 = \tau} \sum_{k_1 + k_2 = k} \widetilde{u}_p(\tau_1, k_1)\widetilde{u}_{p+m}(\tau_2, k_2).$$

We decompose this sum into two parts. We introduce

$$c_1(\tau, k) = \begin{cases} c(\tau, k) & \text{if } |k| \leqslant 2^{\frac{1}{3}(p+m)}, \\ 0 & \text{if } |k| > 2^{\frac{1}{3}(p+m)}, \end{cases}$$

and

$$c_2(\tau, k) = \begin{cases} 0 & \text{if } |k| \leqslant 2^{\frac{1}{3}(p+m)}, \\ c(\tau, k) & \text{if } |k| > 2^{\frac{1}{3}(p+m)}. \end{cases}$$

Let us set for $j = 1, 2$,

$$U_j = \sum_{\tau, k} c_j(\tau, k) e^{i\tau t + ikx}.$$

To prove (16.5), we will show that

$$\|U_j\|_{L^2(\mathbb{T}^2)} \leqslant K 2^{-\varepsilon m} 2^{p/3} \|u_p\|_{L^2(\mathbb{T}^2)} (2^m 2^p)^{1/3} \|u_{p+m}\|_{L^2(\mathbb{T}^2)}. \tag{16.6}$$

Let us start by studying U_2. According to the Cauchy–Schwarz inequality

$$\left(\sum_{\alpha \in A} X_\alpha \right)^2 \leqslant (\#A) \sum_{\alpha \in A} X_\alpha^2$$

and the localization property, we obtain

$$|c_2(\tau, k)|^2 \leqslant a(\tau, k) \sum_{k_1 + k_2 = k} \sum_{\tau_1 + \tau_2 = \tau} |\widetilde{u}_p(\tau_1, k_1)|^2 |\widetilde{u}_{p+m}(\tau_2, k_2)|^2,$$

where $a(\tau, k)$ is the cardinality of the set $A(\tau, k)$ of quadruplets $(k_1, k_2, \tau_1, \tau_2)$ in $\mathbb{Z}^4$ that satisfy

$$k_1 + k_2 = k, \quad \tau_1 + \tau_2 = \tau, \quad \tau_1 = k_1^3 + O(2^p), \quad \tau_2 = k_2^3 + O(2^{m+p}). \tag{16.7}$$

To prove (16.6) for $j = 1$, using the Plancherel inequality, we reduce to showing that

$$\sum_{\tau,k} |c_2(\tau,k)|^2 \leqslant K2^{((2/3)-2\varepsilon)m}2^{4/p3}.$$

Furthermore, as $\|u_p\|_{L^2(\mathbb{T}^2)} = 1 = \|u_{p+m}\|_{L^2(\mathbb{T}^2)}$, using Fubini we obtain

$$\sum_{\tau,k} \sum_{k_1+k_2=k} \sum_{\tau_1+\tau_2=\tau} |\widetilde{u}_p(\tau_1,k_1)|^2 |\widetilde{u}_{p+m}(\tau_2,k_2)|^2 = 1.$$

Thus, it suffices to show that

$$a(\tau,k) \leqslant C2^{((2/3)-2\varepsilon)m}2^{4/p3}, \qquad (16.8)$$

for all $\tau \in \mathbb{Z}$ and all $k \in \mathbb{Z}$ satisfying $|k| \geqslant 2^{(p+m)/3}$.

Let us fix (k,τ) and estimate $a(\tau,k)$. Under the previous constraints (16.7), we have

$$\tau = k_1^3 + k_2^3 + O(2^{m+p}).$$

As $k = k_1 + k_2$, we deduce $\tau = k^3 - 3kk_1k_2 + O(2^{m+p})$, which implies

$$3k\left(k_1 - \frac{k}{2}\right)^2 = -3kk_1k_2 + \frac{3}{4}k^3 = \tau - \frac{1}{4}k^3 + O(2^{m+p}).$$

By hypothesis, if c_1 is non-zero we have $|k| \geqslant (2^{m+p})^{1/3}$. We then deduce that

$$\left(k_1 - \frac{k}{2}\right)^2 = e(\tau,k) + O\left(2^{2(m+p)/3}\right),$$

where $e(\tau,k)$ is a fixed number that only depends on τ and k. In particular, there are at most $O(2^{m/3}2^{p/3})$ possible solutions for k_1 (and therefore for (k_1,k_2)). Once k_1 is chosen, there are at most $O(2^p)$ integers τ_1 that satisfy $\tau_1 = k_1^3 + O(2^p)$. So there are at most $O(2^p)$ pairs (τ_1,τ_2) that satisfy the constraints

$$\tau_1 + \tau_2 = \tau, \quad \tau_1 = k_1^3 + O(2^p), \quad \tau_2 = k_2^3 + O(2^{m+p}).$$

By combining these two remarks, we find that $|a(\tau,k)| = O(2^{m/3}2^{4/p3})$. Therefore, we obtain the desired inequality (16.8), in which we can then take $\varepsilon = 1/6$.

It remains to prove (16.6) for $j = 1$. We want to show that

$$\|U_1\|_{L^2(\mathbb{T}^2)} \leqslant 2^{((1/3)-\varepsilon)m}2^{2p/3}\|u_p\|_{L^2(\mathbb{T}^2)} \|u_{p+m}\|_{L^2(\mathbb{T}^2)}. \qquad (16.9)$$

Let $J = [-2^{(m+p)/3}, 2^{(m+p)/3}]$ and denote by $\|f\|_{\ell_k^2(J;\ell_\tau^2(\mathbb{Z}))}$ the semi-norm defined by

$$\|f\|_{\ell_k^2(J;\ell_\tau^2(\mathbb{Z}))}^2 = \sum_{k\in J} \sum_{\tau\in\mathbb{Z}} |f(\tau,k)|^2.$$

Using Plancherel and Minkowski's inequality, we find

$$
\|U_1\|_{L^2(\mathbb{T}^2)} = \left\| \sum_{k_1+k_2=k} \sum_{\tau_1} \tilde{u}_p(\tau_1, k_1)\tilde{u}_{p+m}(\tau - \tau_1, k_2) \right\|_{\ell^2_k(J;\ell^2_\tau(\mathbb{Z}))}
$$

$$
\leqslant \left\| \sum_{k_1+k_2=k} \sum_{\tau_1} |\tilde{u}_p(\tau_1, k_1)| \, \|\tilde{u}_{p+m}(\cdot, k_2)\|_{\ell^2_\tau(\mathbb{Z})} \right\|_{\ell^2_k(J)}.
$$

Using the inequality $(\sum_{\alpha \in A} X_\alpha)^2 \leqslant (\#A) \sum_{\alpha \in A} X_\alpha^2$ and observing as before that, for all k_1, there are $O(2^p)$ elements τ_1 satisfying $\tau_1 = k_1^3 + O(2^p)$, we obtain

$$
\sum_{\tau_1} |\tilde{u}_p(\tau_1, k_1)| \leqslant 2^{p/2} \left(\sum_{\tau_1} |\tilde{u}_p(\tau_1, k_1)|^2 \right)^{1/2}.
$$

Thus,

$$
\|U_1\|_{L^2(\mathbb{T}^2)} \leqslant 2^{p/2} \left\| \sum_{k_1+k_2=k} \|\tilde{u}_p(\cdot, k_1)\|_{\ell^2_\tau} \|\tilde{u}_{p+m}(\cdot, k_2)\|_{\ell^2_\tau} \right\|_{\ell^2_k}
$$

$$
\leqslant 2^{m/6} 2^{2p/3} \|u_p\|_{L^2(\mathbb{T}^2)} \|u_{p+m}\|_{L^2(\mathbb{T}^2)},
$$

where we used again the inequality $(\sum_{\alpha \in A} X_\alpha)^2 \leqslant (\#A) \sum_{\alpha \in A} X_\alpha^2$ and the fact that if $k \in J$ then $|k| \leqslant C 2^{(m+p)/3}$. Therefore, we obtain (16.9) with $\varepsilon = 1/6$. $\square$

16.3 Exercises

Exercise (solved) 16.1 We are interested in the behavior, as the parameter λ tends to $+\infty$, of the oscillatory integrals

$$
I_{a,b,\phi}(\lambda) = \int_a^b e^{i\lambda\phi(x)} \, dx,
$$

where $(a, b) \in \mathbb{R}^2$ and $\phi \in C^2(\mathbb{R})$ is a real-valued function.

1. Suppose there exist two constants $c, C > 0$ such that,

$$
\forall x \in [a, b], \quad |\phi'(x)| \geqslant c \quad \text{and} \quad |\phi''(x)| \leqslant C.
$$

Using the relation

$$
e^{i\lambda\phi} = \frac{1}{i\lambda\phi'} \frac{d}{dx} (e^{i\lambda\phi}),
$$

show that, for all $\lambda > 0$, we have

$$
|I_{a,b,\phi}(\lambda)| \leqslant \frac{1}{\lambda} \left(\frac{2}{c} + \frac{C(b-a)}{c^2} \right).
$$

2. Show that for all $(a, b) \in \mathbb{R}^2$, for all $\lambda > 0$ and for any function $\phi \in C^2(\mathbb{R})$ such that ϕ' is monotone and does not vanish on $[a, b]$,

$$\left| I_{a,b,\phi}(\lambda) \right| \leqslant \frac{4}{\inf_{a \leqslant x \leqslant b} |\phi'(x)|} \frac{1}{\lambda}.$$

(We can use that $\int |(f(g(x))'| \, dx = |\int (f(g(x)))' \, dx|$ if f and g are two monotone functions.)

3. Show that, for all $(a, b) \in \mathbb{R}^2$, all $\lambda > 0$ and any function $\phi \in C^2(\mathbb{R})$ satisfying $\phi'' \geqslant 1$ on $[a, b]$, we have

$$\left| I_{a,b,\phi}(\lambda) \right| \leqslant \frac{10}{\lambda^{1/2}}.$$

(We can use that $\{x \in [a, b] : |\phi'(x)| \leqslant \lambda^{-1/2}\}$ is an interval of length at most equal to $2\lambda^{-1/2}$.)

Exercise (solved) 16.2

1. Let $\psi \in C^1(\mathbb{R})$ be a real or complex-valued function. Show that

$$\int_a^b e^{i\lambda\phi(x)} \psi(x) \, dx = \psi(b) I_{a,b,\phi}(\lambda) - \int_a^b \psi'(x) I_{a,x,\phi}(\lambda) \, dx.$$

2. From this, deduce that, for all $(a, b) \in \mathbb{R}^2$, all $\lambda > 0$, any phase such that $\phi'' \geqslant 1$ on $[a, b]$, and any $\psi \in C^1(\mathbb{R})$,

$$\left| \int_a^b e^{i\lambda\phi(x)} \psi(x) \, dx \right| \leqslant \frac{10}{\lambda^{1/2}} \left(|\psi(b)| + \int_a^b |\psi'(x)| \, dx \right).$$

3. (Application) Show that there exists a constant C such that, for all $t \in \mathbb{R}$ and for all $R > 0$,

$$\left| \int_{-R}^R e^{i(\xi+t\xi^2)} |\xi|^{-1/2} \, d\xi \right| \leqslant C.$$

Exercise 16.3 The goal of this exercise is to prove the inequality

$$\left\| \int_{\mathbb{R}} |D_x|^{-1/4} e^{-it\partial_x^2} g(t, x) \, dt \right\|_{L_x^2} \leqslant C \|g\|_{L_x^{4/3} L_t^1}.$$

To simplify, we can assume that $g \in S(\mathbb{R} \times \mathbb{R})$ and that its Fourier transform $\widehat{g}(s, \xi)$ with respect to x is supported in a compact $A \subset \mathbb{R}$ independent of s.

1. Show that it is sufficient to prove that

$$\left\| \int_{\mathbb{R}} |D_x|^{-1/4} e^{i(t-s)\partial_x^2} g(s, x) \, ds \right\|_{L_x^4 L_t^\infty} \leqslant C \|g\|_{L_x^{4/3} L_t^1}.$$

2. Show that

$$\int_{\mathbb{R}} |D_x|^{-1/4} e^{i(t-s)\partial_x^2} g(s,x)\, ds = \iint K(s-t, x-y) g(s,y)\, dy\, ds,$$

with

$$K(t,x) = \int_A e^{i(x\xi + t\xi^2)} |\xi|^{-1/2}\, d\xi.$$

3. Conclude by using exercise 16.2 and the Hardy–Littlewood–Sobolev theorem 6.20.
4. Show that the inequality we have proved implies that the solution $u = u(t,x)$ of

$$i\partial_t u + \partial_x^2 u = 0, \quad u_{|t=0} = u_0 \in \mathcal{S}(\mathbb{R}),$$

satisfies

$$\|u\|_{L_x^4 L_t^\infty} \leqslant C \||D_x|^{1/4} u_0\|_{L^2}. \tag{$*$}$$

5. Using the TT^* argument, deduce the estimate $(*)$ from the estimate given in question (1).
6. Compare this estimate with the one obtained by energy estimation.

Exercise 16.4 (Regularizing effect for Schrödinger and Airy) Let $n \geqslant 1$. We denote by $L^2(\mathbb{R}^n)$ the space of square integrable complex-valued functions, equipped with the scalar product

$$(f,g) = \int_{\mathbb{R}^n} f(x)\overline{g(x)}\, dx.$$

Given two operators A and B, we denote by $AB = A \circ B$ their product (composition) and by $[A, B] = AB - BA$ the commutator of A and B.

Let $m \in \mathbb{N}$. Consider a symbol $a \in S^m(\mathbb{R}^n)$ and let $A = \mathrm{Op}(a)$. We denote by A^* the adjoint of A and we assume that $A - A^*$ is an operator of order 0, so that

$$\forall f \in H^m(\mathbb{R}^n), \quad \|Af - A^*f\|_{L^2(\mathbb{R}^n)} \leqslant K_0 \|f\|_{L^2(\mathbb{R}^n)}. \tag{16.10}$$

We fix a time $T > 0$ and we consider a solution $u \in C^1([0,T]; H^m(\mathbb{R}^n))$ of

$$\partial_t u = iAu. \tag{16.11}$$

We admit the existence of such a solution.

1. Consider the following operators:

   • $A_1 = \Delta$;
   • $A_2 = \mathrm{div}(\gamma(x)\nabla\cdot)$ where $\gamma \in C_b^\infty(\mathbb{R}^n; \mathbb{R})$ (which means that γ is a real-valued function, C^∞ and bounded on $\mathbb{R}^n$, as well as all its derivatives);
   • $n = 1$ and $A_3 = i\partial_x^3 + iV(x)\partial_x$ where $V \in C_b^\infty(\mathbb{R}; \mathbb{R})$.

   Write these operators in the form $A_j = \mathrm{Op}(a_j)$, where a_j is a symbol of order m_j (for an m_j to be specified). Then verify that these operators satisfy the hypothesis (16.10).

2. Let $f, g \in C^1([0,T]; H)$ where H is a Hilbert space equipped with the scalar product $(\cdot, \cdot)_H$. Show that the function $(f, g)_H : t \mapsto (f(t), g(t))_H$ is C^1 and that

$$\frac{d}{dt}(f(t), g(t))_H = \left(\frac{df}{dt}(t), g(t)\right)_H + \left(f(t), \frac{dg}{dt}(t)\right)_H.$$

Consider a symbol $b = b(x, \xi)$ belonging to $S^0(\mathbb{R}^n)$. We let $B = \mathrm{Op}(b)$. Show that

$$\frac{d}{dt}(Bu(t), u(t)) = (i[B, A]u(t), u(t)) + (Bu(t), i(A - A^*)u(t)).$$

3. By applying this with $b = 1$, show that

$$\frac{d}{dt}\|u(t)\|^2_{L^2(\mathbb{R}^n)} \leqslant K_0\|u(t)\|^2_{L^2(\mathbb{R}^n)},$$

where K_0 is defined by (16.10), and then that there exists a constant K_1 depending only on T and K_0 such that

$$\sup_{t \in [0,T]} \|u(t)\|^2_{L^2(\mathbb{R}^n)} \leqslant K_1\|u(0)\|^2_{L^2(\mathbb{R}^n)}.$$

4. Let $b \in S^0(\mathbb{R}^n)$ be arbitrary. We set

$$C = i[B, A].$$

Deduce from the previous questions that there exists a constant K_2 (depending only on T, A, B) such that

$$\int_0^T (Cu(t), u(t)) \, dt \leqslant K_2\|u(0)\|^2_{L^2(\mathbb{R}^n)}.$$

5. Suppose $n = 1$ and $A = \partial_x^2$.
   (a) Write C in the form $\mathrm{Op}(p) + R$, where $p \in S^1(\mathbb{R})$ is a symbol depending on b to be calculated and R is an operator of order 0. Deduce from the above that there exists a constant K_3 depending only on T, A, B such that

$$\int_0^T (\mathrm{Op}(p)u(t), u(t)) \, dt \leqslant K_3\|u(0)\|^2_{L^2(\mathbb{R})}.$$

   (b) Let us choose

$$b(x, \xi) = -\frac{1}{2}\frac{\xi}{\langle\xi\rangle}\int_0^x \frac{dy}{\langle y\rangle^2} \quad \text{where} \quad \langle\zeta\rangle = (1 + |\zeta|^2)^{1/2}.$$

Verify that $b \in S^0(\mathbb{R})$ and then that $\mathrm{Op}(p) = -\langle x \rangle^{-2} \Lambda^{-1} \partial_x^2$, where $\Lambda^{-1} = \mathrm{Op}(\langle \xi \rangle^{-1})$.

(c) Deduce from this that there exists a constant K_4 (depending only on T, A) such that

$$\int_0^T \left\| \partial_x \Lambda^{-1/2} (\langle x \rangle^{-1} u(t)) \right\|_{L^2(\mathbb{R})}^2 dt \leqslant K_4 \| u(0) \|_{L^2(\mathbb{R})}^2 .$$

Then show that

$$\int_0^T \left\| \langle x \rangle^{-1} u(t) \right\|_{H^{1/2}(\mathbb{R})}^2 dt \leqslant K_4 \| u(0) \|_{L^2(\mathbb{R})}^2 . \tag{16.12}$$

6. Suppose $n = 1$ and $A = i \partial_x^3 + iV(x)\partial_x$. Let $M > 0$ and consider an increasing function $\varphi \in C^\infty(\mathbb{R}; \mathbb{R})$ such that $\varphi(x) = x$ if $|x| \leqslant M$ and $\varphi'(x) = 0$ if $|x| \geqslant 2M$. We set $b(x, \xi) = \varphi(x)$ (independent of ξ). Write C in the form $\mathrm{Op}(p) + R$ where R is of order 0 (be careful to use symbolic calculation with the correct order) and verify that $C = 3\partial_x(\varphi'(x)\partial_x \cdot) + R$ where R is of order 0. Deduce from this that

$$\int_0^T \int_{-M}^M |\partial_x u(t, x)|^2 \, dx \, dt \leqslant K \int_{\mathbb{R}} |u(0, x)|^2 \, dx,$$

for a constant K depending only on T and M.

7. (∗) Show that the estimate (16.12) holds for $A = \Delta$ in any dimension n.

Part V
Recap and Solutions to the Exercises

Chapter 17
Recap on General Topology

In this chapter, we will recall some notions of General Topology[1] and prove Baire's theorem as well as the existence of smooth functions with compact support.

17.1 Topological Spaces

Consider a set X and denote by $\mathcal{P}(X)$ the set of all subsets of X.

Definition 17.1 A *topology* on X is a collection $\mathcal{T} \subset \mathcal{P}(X)$ of subsets of X such that:

1. $\mathcal{T}$ contains the empty set and the set X;
2. any union of elements of $\mathcal{T}$ belongs to $\mathcal{T}$;
3. any finite intersection of elements of $\mathcal{T}$ belongs to $\mathcal{T}$.

We simply say that $\mathcal{T}$ is stable under arbitrary union and finite intersection. The pair $(X, \mathcal{T})$ is called a *topological space*. By abuse of notation, we generally denote the pair $(X, \mathcal{T})$ by X. In this chapter, and only in this chapter, we will use the notation $X = (X, \mathcal{T})$ to distinguish the set X from the topological space $(X, \mathcal{T})$.

By definition, the elements of $\mathcal{T}$ are the *open* sets of the topological space X. Any complement of an open set is said to be *closed*. It follows from the above properties that any intersection and any finite union of closed sets is a closed set. Similarly, the empty set and the set X are closed.

Each non-empty set X has at least two trivial topologies:

- the topology $\mathcal{T}_1 = \{\emptyset, X\}$ (called the trivial or indiscrete topology);
- the topology $\mathcal{T}_2 = \mathcal{P}(X)$ (called the discrete topology).

The set of all possible topologies on a given set forms a partially ordered set. Given two topologies $\mathcal{T}_1$ and $\mathcal{T}_2$ on the same set X, we will say that $\mathcal{T}_2$ is finer (one also says

---

[1] See Dixmier [46], Kelley [99] and Rudin [164].

© The Author(s), under exclusive license to Springer Nature Switzerland AG 2024
T. Alazard, *Analysis and Partial Differential Equations*, Universitext,
https://doi.org/10.1007/978-3-031-70909-8_17

stronger or larger) than $\mathcal{T}_1$ if $\mathcal{T}_1 \subset \mathcal{T}_2$. Then the topology $\mathcal{T}_1$ is said to be a coarser (weaker or smaller) topology than $\mathcal{T}_2$. A finer topology therefore contains more open sets.

Let $X = (X, \mathcal{T})$ be a topological space and E a subset of X. The topology induced by $\mathcal{T}$ on E is

$$\mathcal{T}_E := \{U \cap E : U \in \mathcal{T}\}.$$

Definition 17.2 Let A be a subset of X.

- The *interior* of A is by definition the union of all open sets of $\mathcal{T}$ contained in A; this set is denoted A°.
- The *closure* of A, also called the *adherence* of A, is the intersection of all closed sets containing A, denoted $\overline{A}$.
- The *boundary* of A is the set $\partial A = \overline{A} \setminus A^\circ$.
- A *neighborhood* of A is any set that contains an open set that contains A. A neighborhood of an element x of X is by definition a neighborhood of $\{x\}$.

It follows directly from these definitions that the interior of A is the largest open set contained in A while the closure of A is the smallest closed set containing A. Note that ∂A is a closed set because it is the intersection of two closed sets: $\overline{A}$ and the complement of A°.

A subset A of X is said to be *dense* as soon as its closure is the set X itself, while it is said to be nowhere dense if $(\overline{A})^\circ$ is the empty set. If there exists a subset of X that is countable and dense, we say that the topological space is *separable*. Let A be a subset of X and x a point of X, then x is said to be *accumulation point* of A if $A \cap (U \setminus \{x\})$ is non-empty for every neighborhood U of x. It follows that the closure of a set A is the union of A and its accumulation points. Also, A is closed if and only if it contains all of its accumulation points.

Consider a topology $\mathcal{T}$ on X and a point $x \in X$. A *basis of open neighborhoods of* x for $\mathcal{T}$ is a collection $\mathcal{B} \subset \mathcal{T}$ of open sets containing x, such that every neighborhood V of x contains an element U of $\mathcal{B}$ (that is, $U \subset V$), or in other words: a set $\mathcal{B}$ of subsets of X such that the neighborhoods of x are exactly the subsets of X that contain an element of $\mathcal{B}$. A *basis* (or base) of $\mathcal{T}$ is a collection of open sets that contains a basis of open neighborhoods of every point of X. A family $\mathcal{B} \subset \mathcal{P}(X)$ is a basis of $\mathcal{T}$ if and only if every open set of $\mathcal{T}$ is a union of elements of $\mathcal{B}$. Also, the specification of a basis of a topology completely defines the topology.

The concept of topology allows us to define the continuity of functions.

Definition 17.3 Let $(X, \mathcal{T}_X)$ and $(Y, \mathcal{T}_Y)$ be two topological spaces, $f : X \to Y$ a function and x a point of X. The function f is *continuous at the point* x if and only if $f^{-1}(V)$ is a neighborhood of x for every neighborhood V of $f(x)$. We say that f is *continuous* on X if and only if $f^{-1}(V)$ belongs to $\mathcal{T}_X$ for every $V \in \mathcal{T}_Y$.

We denote by $C(\mathcal{X}; \mathcal{Y})$ the set of functions $f : X \to Y$ that are continuous. In fact, by abuse, we will always denote this set by $C(X; Y)$ because the underlying topologies are generally clearly identified by the context.

The following statements are easily proven:

- a function is continuous if and only if the preimage of a closed set is closed;
- a function is continuous on X if and only if it is continuous at every point;
- the composition of two continuous functions is continuous.

Consider a bijection $f : X \to Y$. We say that it is a *homeomorphism* if f and f^{-1} are continuous. We say that two topological spaces are *homeomorphic* if there exists a homeomorphism between them.

17.1.1 Metric spaces

Definition 17.4 A *distance* on a set X is a function

$$d : X \times X \longrightarrow [0, +\infty)$$

that satisfies the following three properties: for all x, y, z in X,

1. (separation) $d(x, y) = 0$ if and only if $x = y$;
2. (symmetry) $d(x, y) = d(y, x)$;
3. (triangle inequality) $d(x, z) \leqslant d(x, y) + d(y, z)$.

The pair (X, d) is a *metric space*. Given a point x in X and a positive real number r, we call the *open ball* of center x and radius r the set

$$B(x, r) := \{y \in X : d(x, y) < r\}.$$

By definition, the *closed ball* with center x and radius r is

$$B_f(x, r) := \{y \in X : d(x, y) \leqslant r\}.$$

Be careful: this is not necessarily the closure of the open ball with the same center and radius (we can very well have $B_f(x, r) \neq \overline{B(x, r)}$).

The collection $\mathcal{T}_d$ of subsets U of X satisfying

$$\forall x \in U, \ \exists r > 0 \ / \quad B(x, r) \subset U$$

is a topology on X, called the topology induced by the distance d. It is shown that the set of open balls is a basis of open sets. In general, we will say that a topological space $X = (X, \mathcal{T})$ is *metrizable* when there exists a distance d on X such that $\mathcal{T} = \mathcal{T}_d$ (that is, when there exists a distance on X such that every open set of $\mathcal{T}$ can be written as a union of open balls for this distance).

Consider two metric spaces (X, d) and (Y, ρ) and an element x of X. A function $f : X \to Y$ is continuous at the point x if and only if,

$$\forall \varepsilon > 0, \ \exists \delta > 0 \ / \ \forall y \in X, \quad d(x, y) \leqslant \delta \implies \rho(f(x), f(y)) \leqslant \varepsilon.$$

Definition 17.5 Consider a sequence $(x_n)_{n \in \mathbb{N}}$ in a metric space (X, d).

1. We say that this sequence converges to $x \in X$ if, for every $\varepsilon > 0$, there exists $N \in \mathbb{N}$ such that $d(x_n, x) < \varepsilon$ for all $n \geqslant N$.
2. We say that it is a *Cauchy sequence* if, for every $\varepsilon > 0$, there exists $N \in \mathbb{N}$ such that $d(x_p, x_n) < \varepsilon$ when n and p are greater than N.
3. We say that ℓ is a subsequential limit of this sequence if, for every $\varepsilon > 0$, and for every integer $N \in \mathbb{N}$, there exists an $n \geqslant N$ such that $d(x_n, \ell) < \varepsilon$.

For example, every convergent sequence is a Cauchy sequence.

Definition 17.6 We say that a metric space (X, d) is *complete* if and only if every Cauchy sequence is convergent.

We note that every subsequence of a Cauchy sequence is a Cauchy sequence. A Cauchy sequence that has a subsequential limit converges to this value (in other words: if a subsequence of a Cauchy sequence converges then the entire sequence converges).

The property of being complete often allows objects to be constructed by continuous extension, using the following elementary lemma.

Lemma 17.7 *Let (X, d) and (Y, δ) be two metric spaces, D a dense subset of X, and g a uniformly continuous function from (D, d) to (Y, δ). If Y is complete, then there exists a unique continuous function $\widetilde{g}$ from (X, d) to (Y, δ) such that $\widetilde{g}|_D = g$ and such that $\widetilde{g}$ is also uniformly continuous. And moreover, if g is Λ-Lipschitz, then $\widetilde{g}$ is also Λ-Lipschitz.*

Proof By density, such an extension is necessarily unique if it exists. To show existence, consider $x \in X$ and a sequence $(x_n)_{n \in \mathbb{N}}$ of elements of D that converges to x. Let $\varepsilon > 0$. By uniform continuity of g, there exists $\eta > 0$ such that

$$\forall y, y' \in D, \quad d(y, y') < \eta \implies \delta(g(y), g(y')) < \varepsilon.$$

Furthermore, since the sequence $(x_n)_{n \in \mathbb{N}}$ converges, it is a Cauchy sequence and hence there exists $N \in \mathbb{N}$ such that $d(x_n, x_p) < \eta$ if $n, p \geqslant N$. Then we get that $\delta(g(x_n), g(x_p)) < \varepsilon$ for all $n, p \geqslant N$. The sequence $(g(x_n))_{n \in \mathbb{N}}$ is therefore a Cauchy sequence in the complete metric space Y, and therefore it converges to an element, denoted $\widetilde{g}(x)$.

We verify that this limit does not depend on the chosen sequence $(x_n)_{n \in \mathbb{N}}$. Indeed, if $(x_n)_{n \in \mathbb{N}}$ and $(x'_n)_{n \in \mathbb{N}}$ are two sequences that converge to x, then we can introduce a third sequence $(x''_n)_{n \in \mathbb{N}}$ defined by $x''_{2n} = x_n$ and $x''_{2n+1} = x'_n$, which also converges to x. The sequence $(g(x''_n))_{n \in \mathbb{N}}$ is a Cauchy sequence, therefore it converges, and as it admits two convergent subsequences, we deduce the uniqueness of their limits. In particular, if $x \in D$, we can consider the constant sequence $(x_n)_{n \in \mathbb{N}}$ defined by $x_n = x$, which shows that $\widetilde{g}(x) = g(x)$ and therefore that $\widetilde{g}$ is an extension of g.

Finally, we verify that $\widetilde{g}$ is uniformly continuous (or Λ-Lipschitzian) if g is uniformly continuous (or Λ-Lipschitzian), directly by taking the limit in the inequalities that characterize these properties.   □

17.1.2 Induced Topology, Weak Topology

We now want to introduce the fundamental concept of a induced topology, the most important examples being the product and weak topologies. For this, we begin with two remarks. The first is that if $\{\mathcal{T}_\beta\}_{\beta \in B}$ is a collection of topologies defined on the same set X then the intersection of the $\mathcal{T}_\beta$ is still a topology on X. The second remark is that, since $\mathcal{P}(X)$ is a topology on X, for any collection $\mathcal{A} \subset \mathcal{P}(X)$, there always exists a topology $\mathcal{T}$ on X such that $\mathcal{A} \subset \mathcal{T}$. This allows us to define the topology induced by $\mathcal{A}$ as the intersection of all topologies containing $\mathcal{A}$.

Proposition 17.8 *Let $\mathcal{A} \subset \mathcal{P}(X)$. The topology induced by $\mathcal{A}$ consists of the empty set, the set X and all unions of finite intersections of elements of $\mathcal{A}$.*

Proof Let $\mathcal{B}$ be the set formed by X and all possible finite intersections of elements of $\mathcal{A}$. Consider the family $\mathcal{T}$ of all unions of sets taken from $\mathcal{B}$, to which we add the empty set, which can also be defined by

$$\mathcal{T} := \{U \in \mathcal{P}(X) : \forall x \in U, \exists V \in \mathcal{B} \,/\, x \in V \subset U\}.$$

Then we can verify (exercise) that $\mathcal{T}$ is a topology. This topology is finer than the topology induced by $\mathcal{A}$ (because this one must, by the axioms defining a topology, contain the unions of finite intersections of elements of $\mathcal{A}$). We deduce that $\mathcal{T}$ is the topology induced by $\mathcal{A}$. □

Product topology

Let $\{(X_\alpha, \mathcal{T}_\alpha)\}_{\alpha \in A}$ be any family of topological spaces. We want to equip the product set $X = \prod_{\alpha \in A} X_\alpha$ with a topology. For this, given $a \in A$, we consider the projection π_a defined by

$$\pi_a : X \longrightarrow X_a, \quad x = (x_\alpha)_{\alpha \in A} \longmapsto x_a.$$

By definition, the product topology on X is the topology induced by

$$\mathcal{A} = \{\pi_a^{-1}(U_a) : a \in A, U_a \in \mathcal{T}_a\}.$$

Proposition 17.9 *Let A be any set, $\{X_\alpha\}_{\alpha \in A}$ a family of topological spaces and X the product topological space. Let $\mathcal{Y}$ be a topological space and $\mathcal{F} = \{f_\alpha : \mathcal{Y} \to X_\alpha\}_{\alpha \in A}$ a family of mappings. The mapping*

$$\varphi : Y \longrightarrow X, \quad y \longmapsto (f_\alpha(y))_{\alpha \in A}$$

is continuous if and only if all the mappings f_α are continuous.

Proof Let V be an open set of X. By definition of the product topology, V contains an open set U of the form $U = \prod_{\alpha \in A} U_\alpha$ where $U_\alpha \in \mathcal{T}_\alpha$ is equal to X_α for all indices α except a finite number of them, denoted $\alpha_1, \ldots, \alpha_n$.

Then

$$\varphi^{-1}(U) = \bigcap_{\alpha \in A} f_\alpha^{-1}(U_\alpha) = \bigcap_{1 \leqslant \ell \leqslant n} f_{\alpha_\ell}^{-1}(U_{\alpha_\ell}).$$

All the elements of these intersections are open because the f_α are continuous. Therefore, $\varphi^{-1}(U)$ is an open subset of $\mathcal{Y}$ as a *finite* intersection of open sets. It follows that the preimage of an open set by φ is an open set, which proves that φ is continuous.   □

In particular, the sum, the usual product, the difference, and the quotient (when it makes sense) of continuous mappings with numerical (or vectorial) values are continuous.

Topology induced by a family of mappings

Let X be a set and $\mathcal{F} = \{f_\alpha : X \to Y_\alpha\}_{\alpha \in A}$ a family of mappings from X to topological spaces $\mathcal{Y}_\alpha = (Y_\alpha, \mathcal{T}_\alpha)$. The topology induced by $\mathcal{F}$ on X is the coarsest topology that makes all the mappings f_α continuous. More precisely, it is the topology induced by

$$\mathcal{A} := \{f_\alpha^{-1}(U_\alpha) : \alpha \in A, U_\alpha \in \mathcal{T}_\alpha\}.$$

Thus, the product topology is the topology induced by the family of projections π_α.

Weak topology

Consider a normed vector space E over the field $\mathbb{K} = \mathbb{R}$ or $\mathbb{C}$. Recall that we denote by E' its topological dual, which is the set of continuous linear forms on E. By definition, the weak topology on E, denoted $\sigma(E, E')$, is the topology induced by E' (that is the topology induced by the family of mappings $\mathcal{F} = \{f : E \to \mathbb{K} : f \in E'\}$).

17.2 Separability, Compactness and Completeness

17.2.1 Separability

By definition:

- a topological space is *separable* if it has a countable dense subset;
- a topology has a *countable basis of open sets* when it has a countable basis of open sets.

17.2.2 Separated Spaces and Normal Spaces

A topological space is said to be a separated space (also called a *Hausdorff space*) if it satisfies the following property: if $x \neq y$, there exist two disjoint open sets U, V such that $x \in U$ and $y \in V$. Note that a product of separated spaces is still a separated space. A topological space is said to be *normal* if the following two properties are satisfied:

1. if $x \neq y$, there exists an open set that contains x and does not contain y;
2. for every pair (F_1, F_2) of disjoint closed sets, there exist two disjoint open sets U_1, U_2 such that $F_1 \subset U_1$ and $F_2 \subset U_2$.

For example, it is verified (exercise) that metric spaces are normal spaces. The following proposition gives an equivalent formulation of the first property. In particular, it is deduced that a normal space is separated.

Proposition 17.10 *Consider a normal space X. Then the following two statements are equivalent:*

1. *if $x \neq y$, there exists an open set that contains x and does not contain y;*
2. *the singletons $\{x\}$ are closed for every $x \in X$.*

Proof If X satisfies property (1), for every $x \in X$ and for every $y \neq x$, there exists an open set U_y such that $x \notin U_y$. We can then write $\{x\}^c = \bigcup_{y \neq x} U_y$, which shows that $\{x\}^c$ is open as a union of open sets, therefore $\{x\}$ is closed. Conversely, suppose the singletons are closed. Let x and y be two distinct elements of X, then y belongs to the open set $\{x\}^c$, which does not contain x.   □

Proposition 17.11 *A topological space X is separated if and only if the diagonal $\Delta := \{(x, x) : x \in X\}$ is closed in $X \times X$.*

Proof First, we suppose that X is separated and show that Δ is closed. If $(x, y) \in \Delta^c$, then $x \neq y$ and there exist two disjoint open sets U and V such that $(x, y) \in U \times V$. Since U and V are disjoint, $U \times V \subset \Delta^c$. Therefore $(x, y) \in (\Delta^c)^\circ$ and $\Delta^c = (\Delta^c)^\circ$ is an open set.

Now we suppose that Δ is closed and show that X is separated. If $x \neq y$, then $(x, y) \in \Delta^c$, which is an open set. By definition of the product topology, Δ^c contains a neighborhood of x of the form $U \times V$, where U and V are two open sets of $\mathcal{T}$, necessarily disjoint.   □

The following proposition is easily proven.

Proposition 17.12 *Any topological space homeomorphic to a subspace of a separated (resp. normal) topological space is also separated (resp. normal).*

Proposition 17.13 *Let X be a topological space and $\mathcal{Y}$ a separated topological space. Consider two continuous functions f and g from X to Y. Then the set $\mathcal{E} := \{x \in X : f(x) = g(x)\}$ is closed.*

Proof According to Proposition 17.11, the diagonal

$$\Delta_Y = \{(y, y) : y \in Y\}$$

is closed since $\mathcal{Y}$ is separated. Let $\varphi : X \to Y \times Y$ be defined by $\varphi(x) = (f(x), g(x))$. This function is continuous because f and g are. Then $\mathcal{E} = \varphi^{-1}(\Delta_Y)$ and therefore $\mathcal{E}$ is closed as the preimage of a closed set by a continuous function.   □

From this proposition, under these same assumptions, if $f = g$ on a dense subset of X, then $f = g$ on X.

17.2.3 Compactness

For this part, we are given a topological space X. We call a *cover* of X any collection of subsets of X, $\{U_\alpha\}_{\alpha \in A}$ (A any set), such that

$$X = \bigcup_{\alpha \in A} U_\alpha.$$

We say that it is an open cover if the sets U_α are open. A sub-cover of $\{U_\alpha\}_{\alpha \in A}$ is a sub-family $\{U_\beta\}_{\beta \in B}$ with $B \subset A$, such that $\{U_\beta\}_{\beta \in B}$ is a cover of X. A cover is said to be *finite* if the set of indices A is finite, and *countable* if the set A is countable.

Definition 17.14 A topological space X is compact if it is separated and if from any open cover of X we can extract a finite sub-cover.

We now want to define the notion of a compact subset of a topological space. Let $X = (X, \mathcal{T})$ be a topological space and E a subset of X. Recall that the topology induced by $\mathcal{T}$ on E, denoted $\mathcal{T}_E$, is defined by $\mathcal{T}_E := \{U \cap E : U \in \mathcal{T}\}$. A subset E of X is said to be compact if $(E, \mathcal{T}_E)$ is a compact topological space.

Proposition 17.15 *Consider a compact topological space X and a separated topological space $\mathcal{Y}$. For any continuous function $f : X \to Y$, the image $f(X)$ is compact. In particular, the property of being compact is invariant under homeomorphism.*

Proof Let us give ourselves an open cover of $f(X)$. It follows that the set of pre-images by f of the elements of this cover is an open cover of X (indeed, f being continuous, the pre-image of an open set is open). Therefore, by compactness of X, we can extract a finite sub-cover of X. The images by f of the elements of this sub-cover of X then form a finite sub-cover of $f(X)$.   □

However, the pre-image of a compact set by a continuous function f is not necessarily compact. When this is the case, we say that f is a *proper* function.

We now give a characterization of compactness that involves a criterion based on intersections of closed sets, rather than on unions of open sets.

Proposition 17.16 *A topological space X is compact if and only if every family of closed sets with empty intersection has a finite sub-family with empty intersection.*

Proof This proposition follows from Definition 17.14 in terms of open covers by passing to the complement.   □

Corollary 17.17 *Consider a compact topological space X and a sequence $\{F_p\}_{p\in\mathbb{N}}$ of non-empty closed sets, which is decreasing in the sense that $F_{p+1} \subset F_p$ for every integer p. Then the intersection $\bigcap_{p\in\mathbb{N}} F_p$ is non-empty.*

Proof Suppose, by contradiction, that the intersection $\bigcap_{p\in\mathbb{N}} F_p$ is empty. Then the preceding proposition implies that there would exist a finite subset $J \subset \mathbb{N}$ such that $\bigcap_{j\in J} F_j = \emptyset$. As the sequence $\{F_p\}_{p\in\mathbb{N}}$ is decreasing, this leads to $F_{\sup J} = \emptyset$. This contradicts the assumption that the closed sets F_p are non-empty.   □

Consider a sequence $(x_n)_{n\in\mathbb{N}}$ of points in a topological space. Recall that we say that ℓ is a subsequential limit of this sequence if, for every neighborhood V of ℓ, and for every integer $N \in \mathbb{N}$, there exists an $n \geq N$ such that $x_n \in V$.

Corollary 17.18 *In a compact topological space, every sequence has a subsequential limit.*

Proof Consider a sequence $(x_n)_{n\in\mathbb{N}}$ and introduce the closed sets F_p defined by

$$F_p = \overline{\{x_n : n \geq p\}}.$$

This is a decreasing sequence of closed sets whose intersection is non-empty according to Corollary 17.17. Then we verify that every element $\ell \in \bigcap_{p\in\mathbb{N}} F_p$ is a subsequential limit.   □

Proposition 17.19 *Consider a separated topological space X.*

1. *If K is a compact subset and x belongs to $X \setminus K$. Then there exist two disjoint open sets V and W such that $x \in V$ and $K \subset W$.*
2. *If K is a compact subset of X, then K is closed in X.*
3. *If X is compact and F is a closed subset of X, then F is compact.*

Proof 1. Let y be a point in K. Since $x \neq y$ and X is separated, there exist two disjoint open sets V_y and W_y such that $x \in V_y$ and $y \in W_y$. The collection $\{W_y\}_{y\in K}$ is a cover of K so by compactness of K we can extract a finite subcover indexed by a finite number of points in K: $y_1, \ldots, y_n$. We then set

$$V := \bigcap_{1\leq\ell\leq n} V_{y_\ell}, \qquad W := \bigcup_{1\leq\ell\leq n} W_{y_\ell},$$

and we obtain the desired result.

2. Let us show that the complement of K is open. This is equivalent to showing that $X \setminus K$ is a neighborhood of each of its points. Therefore, it suffices to show that for every $x \in X \setminus K$, there exists an open set $V \subset X \setminus K$ such that $x \in V$. This is a direct consequence of point (1).

3. Let us first show that F is separated. Let $\mathcal{T}$ be the topology on X and $\mathcal{T}_F$ the topology induced by $\mathcal{T}$ on F. Consider two distinct points x and y in F. As X is separated (by definition of compactness), there exist two disjoint open sets U, V in $\mathcal{T}$ such that $x \in U$ and $y \in V$. Then $U \cap F$ and $V \cap F$ are two disjoint open sets in the induced topology $\mathcal{T}_F$. This shows that $\mathcal{T}_F$ is separated.
To show that F is compact, it is now sufficient to show that every family $\{F_\alpha\}_{\alpha \in A}$ of closed sets (for the induced topology $\mathcal{T}_F$) with empty intersection has a finite subfamily with empty intersection. By compactness of X, it is therefore sufficient to show that the sets F_α are also closed for the topology $\mathcal{T}$. For this, note that $F \setminus F_\alpha$ being an open set of $\mathcal{T}_F$ implies that there exists an open set U_α in $\mathcal{T}$ such that $F \setminus F_\alpha = U_\alpha \cap F$. We then verify (exercise) that $X \setminus F_\alpha = (X \setminus F) \cup U_\alpha$. Since $X \setminus F$ and U_α are open, the set $X \setminus F_\alpha$ is open, therefore F_α is closed. $\square$

A subset E of X is said to be *relatively compact* if its closure is a compact subset of X. Since every closed set of a compact topological space X is compact according to the previous proposition, it follows that every subset of a compact topological space is relatively compact.

Proposition 17.20 *Let X be a separated topological space, and K_1 and K_2 two disjoint compact subsets. Then there exist two disjoint open sets V_1 and V_2 such that $K_i \subset V_i$ ($i = 1, 2$).*

Proof According to Proposition 17.19, for every x in K_1, there exist two disjoint open sets V_x and W_x such that $x \in V_x$ and $K_2 \subset W_x$. The open sets $\{V_x\}_{x \in K_1}$ form a cover of K_1, from which we can extract a finite subcover: $\{V_{x_\ell}\}_{1 \leqslant \ell \leqslant n}$. We set

$$V_1 := \bigcup_{1 \leqslant \ell \leqslant n} V_{x_\ell}, \qquad V_2 := \bigcap_{1 \leqslant \ell \leqslant n} W_{x_\ell},$$

and we easily verify that $V_1 \cap V_2 = \varnothing$ and $K_i \subset V_i$ ($i = 1, 2$). $\square$

Corollary 17.21 *Every compact topological space is normal.*

Proof This is a consequence of Proposition 17.20 and the fact that every closed set of a compact space is compact. $\square$

Definition 17.22 A topological space X is sequentially compact if from any sequence of elements of X, we can extract a convergent subsequence. A subset E of X is said to be sequentially compact if $(E, \mathcal{T}_E)$ is sequentially compact.

Definition 17.23 A metric space (X, d) is pre-compact if, for every $\varepsilon > 0$, X can be covered by a finite number of balls of radius ε.

We recall without proof the following equivalences.

Theorem 17.24 *Consider a metric space (X, d). Then the following properties are equivalent:*

1. *X is compact;*
2. *X is pre-compact and complete;*
3. *X is sequentially compact.*

Proposition 17.25 *Every compact metric space is separable.*

Proof Let (X, d) be a metric space. For every $n \in \mathbb{N}^*$, the family $\{B(x, 1/n)\}_{x \in X}$ covers X so we can extract a finite subcover $\{B(x_n^p, 1/n)\}_{p \in I_n}$, where I_n is finite. The set $\mathcal{D} := \{x_n^p : n \in \mathbb{N}^*, \ p \in I_n\}$ is then a countable dense set. □

17.3 Baire's Theorem

Definition 17.26 A topological space is said to be a Baire space if every countable intersection of open dense sets is a dense subset. It is equivalent to say that every countable union of closed sets with empty interior is a set with empty interior.

Theorem 17.27 *Every complete metric space is a Baire space.*

Remark 17.28 Consider the metric space $(\mathbb{Q}, d)$ with $d(x, y) = |x - y|$. As $\mathbb{Q}$ is countable, we see that $\mathbb{Q}$ is a countable union of singletons. These singletons are closed because a metric space is always separated, and singletons are closed in a separated space. Moreover, these singletons have an empty interior. However, the union, equal to $\mathbb{Q}$, has a non-empty interior. This shows that $(\mathbb{Q}, d)$ is not a Baire space. The result can therefore be false in a metric space that is not complete.

Proof Consider a metric space X and a sequence $(\Omega_n)_{n \in \mathbb{N}}$ of open dense sets. We want to show that the intersection $\Omega = \bigcap_{n \in \mathbb{N}} \Omega_n$ is a dense subset of X. To do this, we need to show that, for every open set V, the intersection $V \cap \Omega$ is non-empty. We will verify this property by constructing a Cauchy sequence, denoted $(x_n)_{n \in \mathbb{N}}$, which will converge by completeness to an element of $V \cap \Omega$.

To define the first term, we use the fact that Ω_0 is dense to deduce that $V \cap \Omega_0$ is non-empty. This allows us to choose a point, denoted x_0, in $V \cap \Omega_0$. Since $V \cap \Omega_0$ is open (intersection of two open sets), there exists $\rho_0 \in (0, 1]$ such that $B(x_0, \rho_0) \subset V \cap \Omega_0$. By setting $r_0 = \rho_0/2$, we deduce that $\overline{B(x_0, r_0)} \subset V \cap \Omega_0$. Then we use the fact that Ω_1 is dense to deduce that $B(x_0, r_0) \cap \Omega_1$ is non-empty. From this, we deduce that there exist $x_1 \in B(x_0, r_0) \cap \Omega_1$ and a radius $\rho_1 \in (0, 1/2]$ such that $B(x_1, \rho_1) \subset B(x_0, r_0) \cap \Omega_1$. By setting $r_1 = \rho_1/2$, we get $\overline{B(x_1, r_1)} \subset B(x_0, r_0) \cap \Omega_1$. By induction, we construct a sequence $(x_n)_{n \in \mathbb{N}}$ of points in X and a sequence of radii $(r_n)_{n \in \mathbb{N}}$ such that

$$\overline{B(x_{n+1}, r_{n+1})} \subset \overline{B(x_n, r_n)} \cap \Omega_{n+1}, \quad 0 < r_n \leqslant \frac{1}{n+1}.$$

Consider an integer N and two integers n, p greater than N. Then x_n and x_p belong to the closed ball $\overline{B(x_{N-1}, r_{N-1})}$, and therefore the distance between x_n and x_p is bounded by $2r_{N-1}$, which shows that this sequence is Cauchy. It converges to an element x, which belongs to $\overline{B(x_{N-1}, r_{N-1})}$ for all N. We deduce that x belongs to $V \cap \Omega$, which is the desired result.   □

We conclude this section by giving two definitions associated with Baire's theorem, which are frequently used when trying to characterize the size of a set.

Definition 17.29

1. A property is said to be generic if it holds at least for all elements of a countable intersection of dense open sets.
2. A set is said to be meager (or meagre or of first category) in the sense of Baire if it is a countable union of nowhere dense subsets. Recall that a subset of a topological space is called nowhere dense if its closure has empty interior.

To conclude, we will show that a set can be meager and yet be of full measure (in the sense that the complement is of measure zero). The example we give is from the study of Diophantine approximation. Informally, this field can be presented as a branch of number theory that studies quantitatively the density of $\mathbb{Q}$ in $\mathbb{R}$, that is to say the property that for every real x and every $\varepsilon > 0$, there exists a pair of integers (p, q) in $\mathbb{Z} \times \mathbb{N}^*$ such that

$$|x - p/q| < \varepsilon.$$

We would like to have information about the minimum size that we can impose on ε. We easily verify that we can take $\varepsilon = 1/q$. The following theorem is a classical result of Dirichlet that improves this bound.

Theorem 17.30 (Dirichlet) *For every real number x and every non-zero integer N, there exists a rational number p/q such that*

$$\left| x - \frac{p}{q} \right| \leq \frac{1}{Nq} \quad \text{and} \quad 1 \leq q \leq N.$$

Proof We can assume without loss of generality that x is a positive irrational number. Let $\{y\}$ be the fractional part of a positive real number y (the difference between y and its integer part; it is an element of $[0, 1)$). Now consider the $N + 1$ numbers (pairwise distinct because $x \notin \mathbb{Q}$)

$$0, \{x\}, \{2x\}, \ldots, \{Nx\},$$

and the N intervals

$$\left[0, \frac{1}{N} \right), \left[\frac{1}{N}, \frac{2}{N} \right), \ldots, \left[\frac{N-1}{N}, 1 \right).$$

Since x is irrational, the $N + 1$ numbers necessarily belong to these open intervals. Moreover, two such numbers necessarily belong to the same interval according to the pigeonhole principle.

From this we deduce that there exist two distinct integers $0 \leqslant i, j \leqslant N$ such that $|\{ix\} - \{jx\}| \leqslant 1/N$. This implies the desired result with $q = |i - j|$.   □

We can now introduce the set of Diophantine numbers, which are irrational real numbers that are difficult to approximate by rationals.

Definition 17.31 Consider two positive real numbers K, ν. An irrational number ρ is of type (K, ν) if for every rational p/q with $q > 0$ we have

$$\left| \rho - \frac{p}{q} \right| > \frac{K}{q^{\nu}}.$$

An irrational number ρ is Diophantine if it is of type (K, ν) for some $\nu > 2$.

Proposition 17.32 *The set of Diophantine irrational numbers is both meager in the sense of Baire and of full Lebesgue measure.*

Proof Let D be the set of Diophantine numbers. Then

$$D = \left\{ \rho \in \mathbb{R} : \exists r \in \mathbb{N}, \ \exists n \in \mathbb{N}, \ \forall (p, q) \in \mathbb{Z} \times \mathbb{N}^*, \ |\rho - p/q| \geqslant \frac{1}{nq^r} \right\}.$$

We see that D is a countable union, indexed by r and n, of closed sets that clearly have empty interior (because these closed sets cannot contain rationals). This shows that D is meager in the sense of Baire, by definition of this property.

To show that the complement of D is of measure zero, it suffices to show that this is the case for $D \cap [0, 1]$. Fix $\nu > 2$ and introduce

$$D_\nu = \left\{ \rho \in [0, 1] : \exists K \in \mathbb{R}_+^*, \ \forall (p, q) \in \mathbb{Z} \times \mathbb{N}^* / |\rho - p/q| \geqslant K/q^\nu \right\}.$$

The complement is

$$\bigcap_{K>0} \bigcup_{q=1}^{+\infty} \bigcup_{p=0}^{q} \left(\frac{p}{q} - \frac{K}{q^\nu}, \frac{p}{q} + \frac{K}{q^\nu} \right).$$

We have

$$\left| \bigcup_{q=1}^{+\infty} \bigcup_{p=0}^{q} \left(\frac{p}{q} - \frac{K}{q^\nu}, \frac{p}{q} + \frac{K}{q^\nu} \right) \right| \leqslant 2K \sum_q \frac{q+1}{q^\nu} = O(K)$$

for $\nu > 2$. Therefore, the complement of D_ν is an intersection indexed by $K > 0$ of sets of measure $O(K)$. It is therefore negligible.   □

17.4 Regular Functions with Compact Support

We will focus on smooth functions $f : \mathbb{R}^n \to \mathbb{R}$, of class C^∞, which have compact support. We denote the set of these functions by $C_0^\infty(\mathbb{R}^n)$.

We begin by studying in this section the question of the existence of functions with compact support. The first result states that, for any closed set F of $\mathbb{R}^n$, there exists a function of class C^∞ that vanishes exactly on F, which is equivalent to saying that, for any open set Ω of $\mathbb{R}^n$, there exists a function of class C^∞ that does not vanish on Ω, and which is zero on its complement $F = \mathbb{R}^n \setminus \Omega$.

Proposition 17.33 *For any open set $\Omega \subset \mathbb{R}^n$, there exists a function $\varphi_\Omega \in C^\infty(\mathbb{R}^n)$, satisfying $0 \leqslant \varphi_\Omega \leqslant 1$ and such that*

$$\Omega = \{x \in \mathbb{R}^n : \varphi_\Omega(x) > 0\}.$$

Proof Let us start by recalling that, for any open set Ω of $\mathbb{R}^n$, there exists a countable collection of open balls such that

$$\Omega = \bigcup_{m \in \mathbb{N}} B(a_m, r_m) \qquad (a_m \in \mathbb{R}^n, \; r_m \in \mathbb{R}_*^+).$$

Indeed, for every $x \in \Omega$, by density of $\mathbb{Q}$ in $\mathbb{R}$, there exist $q(x) \in \mathbb{Q}^n$ and $r(x) \in \mathbb{Q}_+$ such that $x \in B(q, r) \subset \Omega$. Therefore $\Omega = \bigcup_{x \in \Omega} B(q(x), r(x))$. Let us now introduce the set $A = \{(q(x), r(x)) : x \in \Omega\}$. This set is countable, as it is a subset of $\mathbb{Q}^n \times \mathbb{Q}_+$, so we can write it as $A = \{(a_m, r_m) : m \in \mathbb{N}\}$.

We then define

$$\varphi_\Omega(x) = \sum_{m \in \mathbb{N}} \frac{e^{-1/r_m}}{2^{m+1}} \phi\left(1 - \frac{|x - a_m|^2}{r_m^2}\right), \tag{17.1}$$

where $\phi : \mathbb{R} \to \mathbb{R}_+$ is defined by:

$$\text{for } y > 0, \quad \phi(y) = \exp(-1/y); \qquad \text{for } y \leqslant 0, \quad \phi(y) = 0.$$

We verify that $\varphi_\Omega(x)$ is well defined because $0 \leqslant \phi(y) \leqslant 1$, so the series converges absolutely at each point. Moreover, $\varphi_\Omega(x) = 0$ if x does not belong to Ω (because all terms of the series are zero) and $\varphi_\Omega(x) > 0$ if x belongs to Ω (because at least one of the terms of the series is strictly positive and all the others are non-negative). To conclude, we need to show that φ_Ω is of class C^∞ on $\mathbb{R}^n$.

We start by studying the function ϕ. It is clear that ϕ is continuous at 0, therefore continuous on $\mathbb{R}$. Then we verify that the right derivative of ϕ at 0 is zero, as is the left derivative, which implies that ϕ is of class C^1. More generally, we show by induction that for any integer $k \geqslant 1$ and any $y > 0$, we have

$$\partial_y^k \phi(y) = P_k(1/y) \exp(-1/y),$$

where P_k is a polynomial. This implies that $\partial_y^k \phi(y)$ tends to 0 when y tends to 0 from the right. As the derivative $\partial_y^k \phi(y)$ on the left is trivially zero (because ϕ is zero on the left), we deduce that ϕ is of class C^∞ on $\mathbb{R}$.

To see that φ_Ω is of class C^∞ on $\mathbb{R}^n$, it suffices to show that, for all $\alpha \in \mathbb{N}^n$, the series

$$\sum_{m \in \mathbb{N}} \frac{e^{-1/r_m}}{2^{m+1}} \sup_{x \in \mathbb{R}^n} \left| \partial_x^\alpha \left(\phi \left(1 - \frac{|x - a_m|^2}{r_m^2} \right) \right) \right|$$

is convergent. However, if $|\alpha| \leq p$ this series is bounded by $C(p) \sum\limits_{m \in \mathbb{N}} \dfrac{e^{-1/r_m}}{2^{m+1}} r_m^{-p}$,

where $C(p)$ is a constant that depends only on p. To deduce the desired result, we start by noting that

$$e^{-1/r} r^{-p} \leq e^{-p} p^p$$

(which is shown by verifying that the function $r \mapsto e^{-1/r} r^{-p}$ reaches its maximum at the point $r = 1/p$). It follows that

$$\sum_{m \in \mathbb{N}} \frac{e^{-1/r_m}}{2^{m+1}} \sup_{\mathbb{R}^n} \left| \partial_x^\alpha \left(\phi \left(1 - \frac{|x - a_m|^2}{r_m^2} \right) \right) \right| \leq C(p) \sum_m 2^{-m+1} e^{-p} p^p < +\infty.$$

This concludes the proof. $\qquad\qquad\qquad\qquad\qquad\qquad\qquad\qquad\qquad\qquad\qquad$ $\square$

The aim of the following two propositions is to deduce from the above the existence of particular functions of class C^∞, which are very useful when we need to use localization arguments.

Proposition 17.34 *Consider two open sets Ω and Ω' of $\mathbb{R}^n$, such that $\overline{\Omega'} \subset \Omega$. Then there exists a function $\chi \in C^\infty(\mathbb{R}^n)$ such that $0 \leq \chi \leq 1$ and*

$$\chi(x) = 1 \text{ if } x \in \Omega', \qquad \chi(x) = 0 \text{ if } x \notin \Omega.$$

Proof It suffices to apply Proposition 17.33 with the open sets Ω and $\mathbb{R}^n \setminus \overline{\Omega'}$. We deduce that there exist two functions φ, ψ in $C^\infty(\mathbb{R}^n)$ such that $0 \leq \varphi, \psi \leq 1$ and

$$\Omega = \{x \in \mathbb{R}^n : \varphi(x) > 0\}, \quad \mathbb{R}^n \setminus \overline{\Omega'} = \{x \in \mathbb{R}^n : \psi(x) > 0\}.$$

Then the function $\varphi + \psi$ is C^∞ on $\mathbb{R}^n$ and strictly positive at every point (because every point x of $\mathbb{R}^n$ belongs either to Ω or to $\mathbb{R}^n \setminus \overline{\Omega'}$). It follows that the function

$$\chi(x) = \frac{\varphi(x)}{\varphi(x) + \psi(x)}$$

is well defined and C^∞ on $\mathbb{R}^n$. Moreover, if $x \in \Omega'$, we have $\psi(x) = 0$, so $\chi(x) = 1$. $\square$

Proposition 17.35 *Let K be a compact set of $\mathbb{R}^n$ and let $\{U_i\}_{1 \leq i \leq N}$ be an open cover of K. Then, there exists a finite family $\{\zeta_i\}_{1 \leq i \leq N}$ of functions of class C^∞ on $\mathbb{R}^n$ such that $0 \leq \zeta_i \leq 1$ and $\operatorname{supp} \zeta_i \subset U_i$ for every index i such that $1 \leq i \leq N$, and satisfying*

$$\sum_{1 \leq i \leq N} \zeta_i(x) = 1, \qquad \forall x \in K.$$

Remark 17.36

1. We say that $\{\zeta_i\}_{1 \leqslant i \leqslant N}$ is a C^∞ partition of unity that is subordinate to the cover $\{U_i\}_{1 \leqslant i \leqslant N}$.
2. We recall that $\operatorname{supp} \zeta_i = \overline{\{x \in \mathbb{R}^n : \zeta_i(x) \neq 0\}}$.

Proof It is done in two steps.

First step: construction of a better cover.

We start by constructing a collection of N open sets $\{U_i'\}_{1 \leqslant i \leqslant N}$ such that

$$K \subset \bigcup_{i=1}^N U_i' \quad \text{and} \quad \overline{U_i'} \subset U_i \quad (\forall i \in \{1, \ldots, N\}). \tag{17.2}$$

For this, let us introduce the following notation: if $i \in \{1, \ldots, N\}$ and $\varepsilon \in (0, +\infty)$, we define

$$U_i^\varepsilon = \{x \in U_i : \operatorname{dist}(x, \partial U_i) > \varepsilon\}.$$

Note that, for all $\varepsilon > 0$ and any index i, we have $\overline{U_i^\varepsilon} \subset U_i$. It is therefore sufficient to show that there exists $\varepsilon > 0$ such that $K \subset \bigcup_{i=1}^N U_i^\varepsilon$. For this, note that for every element $x \in K$, as $\{U_i\}_{1 \leqslant i \leqslant N}$ is a cover of K, there exists $i(x) \in \{1, \ldots, N\}$ such that $x \in U_{i(x)}$. Moreover, since $U_{i(x)}$ is open, there exists $\varepsilon(x) > 0$ such that $x \in U_{i(x)}^{\varepsilon(x)}$. We can then write that

$$K \subset \bigcup_{x \in K} U_{i(x)}^{\varepsilon(x)}.$$

This gives an open cover of K, from which we can extract by compactness a finite subcover $K \subset \bigcup_{1 \leqslant \ell \leqslant p} U_{i(x_\ell)}^{\varepsilon(x_\ell)}$. This verifies the desired property by taking for ε the minimum value of the $\varepsilon(x_\ell)$.

Second step: construction of the partition of unity.

Consider the subcover constructed in the previous step, satisfying (17.2).

According to Proposition 17.33, for all $i \in \{1, \ldots, N\}$, there exists a function $\varphi_i : \mathbb{R}^n \to [0, 1]$, of class C^∞ and such that

$$\{x \in \mathbb{R}^n : \varphi_i(x) > 0\} = U_i'.$$

Then we have $\varphi_i(x) = 0$ if $x \notin U_i'$, which directly implies that

$$\operatorname{supp} \varphi_i \subset \overline{U_i'} \subset U_i.$$

Furthermore, by applying Proposition 17.33 to the open set $\mathbb{R}^n \setminus K$, we deduce that there exists a function Φ_{comp} such that

$$\{x \in \mathbb{R}^n : \Phi_{\mathrm{comp}}(x) > 0\} = \mathbb{R}^n \setminus K.$$

Now let us define

$$\Phi = \Phi_{\mathrm{comp}} + \sum_{i=1}^N \varphi_i.$$

Then Φ is C^∞ on $\mathbb{R}^n$ (because it is a finite sum of functions of class C^∞) and $\Phi(x) > 0$ for all x in $\mathbb{R}^n$ (because, either $x \in \mathbb{R}^n \setminus K$ and then $\Phi_{\text{comp}}(x) > 0$, or there exists an index $1 \leqslant i \leqslant N$ such that $x \in U'_i$ and therefore such that $\varphi_i(x) > 0$). We deduce that the function $1/\Phi$ is well defined and C^∞ on $\mathbb{R}^n$. We then define

$$\zeta_i(x) = \frac{\varphi_i(x)}{\Phi(x)} = \frac{\varphi_i(x)}{\Phi_{\text{comp}}(x) + \sum_{i=1}^{N} \varphi_i(x)}.$$

These functions are C^∞ and we directly verify that, by construction, $\text{supp}\,\zeta_i \subset \text{supp}\,\varphi_i \subset \overline{U'_i} \subset U_i$. Moreover, if $x \in K$ then $\Phi_{\text{comp}}(x) = 0$ so

$$\sum_{1 \leqslant i \leqslant N} \zeta_i(x) = \sum_{i=1}^{N} \frac{\varphi_i(x)}{\sum_{i=1}^{N} \varphi_i(x)} = 1.$$

This concludes the proof.

17.5 Exercises

Exercise 17.1 We use the definitions from Section 17.2.1.

1. Show that a topological space with a countable basis of open sets is separable.
2. Show that a separable metric space has a countable basis of open sets.

Exercise 17.2 Let (X, d) and (Y, δ) be two metric spaces. Assume that (X, d) is complete and (Y, δ) is separable. Consider a sequence of continuous functions $f_n : X \to Y$ that converges pointwise to a function $f : X \to Y$. We want to show that f is continuous on a dense set.

1. Show that there exists a countable family $\mathcal{B} = \{B_p\}_{p \in \mathbb{N}}$ of closed balls in (Y, δ) such that every open set in Y can be written as a union of elements of $\mathcal{B}$.
2. Show that the set D of discontinuity points of f satisfies $D \subset \bigcup_{p=1}^{\infty} W_p \setminus \overset{\circ}{W}_p$ with $W_p := f^{-1}(B_p)$.
3. Show that each set W_p is a countable union of closed sets.
4. Deduce that f is continuous at every point of a dense set.

Exercise (solved) 17.3 (Divergence theorem) Let $n \geqslant 2$. Let $\Omega \subset \mathbb{R}^n$ be a bounded open set. We assume that, for every $x_0 \in \partial\Omega$, there exists a neighborhood $\mathcal{V}$ of x_0 and a neighborhood $\mathcal{U}$ of 0, as well as a C^∞-diffeomorphism $\phi : \mathcal{U} \to \mathcal{V}$ satisfying:

$$\phi(\mathcal{U} \cap (\mathbb{R}^*_- \times \mathbb{R}^{n-1})) = \mathcal{V} \cap \Omega. \tag{17.3}$$

We then denote[2] by $n(x_0)$ the *normal* to $\partial\Omega$ at x_0, that is, the unique unit vector orthogonal to $d\phi_0(\{0\} \times \mathbb{R}^{n-1})$ whose dot product with $d\phi_0(1, 0, \dots)$ is positive.

---

[2] For consistency with the notations used in this book, we will use the same notation n for the normal and for the space dimension.

If $f \in C^0(\overline{\Omega}, \mathbb{R}^n)$ has support included in an open set $\mathcal{V}$ as above, we define the integral of f over $\partial\Omega$ by:

$$\int_{\partial\Omega} f \, dS = \int_{\widetilde{\mathcal{U}}} f \circ \widetilde{\phi}(y) \, |\det(d\widetilde{\phi}(y))| \, dy,$$

where $\widetilde{\mathcal{U}} = \{(\widetilde{x}_1, \ldots, \widetilde{x}_{n-1}) \in \mathbb{R}^{n-1} : (0, \widetilde{x}_1, \ldots, \widetilde{x}_{n-1}) \in \mathcal{U}\}$ and $\widetilde{\phi}(\widetilde{x}) = \phi(0, \widetilde{x})$ for every $\widetilde{x} \in \widetilde{\mathcal{U}}$.

If $f \in C^0(\overline{\Omega}, \mathbb{R}^n)$ vanishes on $\partial\Omega$, we define $\int_{\partial\Omega} f \, dS = 0$.

Finally, if $f \in C^0(\overline{\Omega}, \mathbb{R}^n)$ is in neither of these two situations, we use Proposition 17.35 to define the integral by:

$$\int_{\partial\Omega} f \, dS = \sum_{i=1}^N \int_{\partial\Omega} (\phi_i f) \, dS,$$

where the C^∞ functions ϕ_i are chosen such that $\sum_{i=1}^N \phi_i = 1$ on $\overline{\Omega}$ and, for all i, the support of ϕ_i either has empty intersection with $\partial\Omega$ or is included in an open set $\mathcal{V}$ of the previous form.

1. Let $f : \overline{\Omega} \to \mathbb{R}^n$ be a function of class C^1. Assume that the support of f is included in a compact set of the form $[a_1, b_1] \times \cdots \times [a_n, b_n]$, which is itself included in Ω. Show that:
$$\int_\Omega \text{div}(f) \, dx = 0.$$

2. Assume that the support of f is included in an open set $\mathcal{V}$, which satisfies the property (17.3) for a $\mathcal{U}$ of the form $[a_1, b_1] \times \cdots \times [a_n, b_n]$ (with $a_1 < 0 < b_1$). Show that:
$$\int_\Omega \text{div}(f) \, dx = \int_{\partial\Omega} f \cdot n \, dS.$$

3. Show that, in the general case, we have the divergence formula:
$$\int_\Omega \text{div}(f) \, dx = \int_{\partial\Omega} f \cdot n \, dS.$$

4. Assume that $u, v : \overline{\Omega} \to \mathbb{R}$ are respectively of class C^2 and C^1. Prove the Green's formula:
$$\int_{\partial\Omega} v \frac{\partial u}{\partial n} \, dS = \int_\Omega (\Delta u)v \, dx + \int_\Omega \nabla u \cdot \nabla v \, dx \quad \text{with} \quad \frac{\partial u}{\partial n} = n \cdot \nabla u.$$

Chapter 18
Inequalities in Lebesgue Spaces

The aim of this chapter is to recall several fundamental inequalities concerning Lebesgue spaces and to prove the Marcinkiewicz interpolation theorem. We assume known the basic properties of these spaces[1] of which we will only recall the statements in Section 18.1.

18.1 The L^p Spaces

Let $n \geqslant 1$ and Ω be an open set of $\mathbb{R}^n$. For $p \in [1, +\infty)$, we denote by $\mathcal{L}^p(\Omega)$ the set of functions $f \colon \Omega \to \mathbb{C}$ that are Lebesgue measurable and such that

$$\|f\|_{L^p} := \left(\int_{\Omega} |f(x)|^p \, \mathrm{d}x \right)^{1/p} < +\infty.$$

The map $f \mapsto \|f\|_{L^p}$ is a semi-norm (the triangle inequality is proved in the following section) but not a norm on $\mathcal{L}^p(\Omega)$. Indeed, the relation $\|f\|_{L^p} = 0$ only implies that f is equal to 0 almost everywhere.

We denote by $L^p(\Omega) = \mathcal{L}^p(\Omega)/\mathcal{N}$ the quotient of the space $\mathcal{L}^p(\Omega)$ by the linear subspace of functions equal to 0 almost everywhere

$$\mathcal{N} = \{ f \in \mathcal{L}^p(\Omega) : f = 0 \text{ almost everywhere} \}.$$

In other words, $L^p(\Omega)$ is the set of equivalence classes of functions of $\mathcal{L}^p(\Omega)$ for the equivalence relation $f \sim g$ if and only if f and g are equal almost everywhere. In practice, we confuse the elements of L^p (which are equivalence classes of functions) with their representatives (which are functions).

---

[1] We refer to the books of Lieb and Loss [123], Rudin [163], Stein [176] and Tao [182] as well as the lecture notes of Danchin [42].

© The Author(s), under exclusive license to Springer Nature Switzerland AG 2024
T. Alazard, *Analysis and Partial Differential Equations*, Universitext,
https://doi.org/10.1007/978-3-031-70909-8_18

The space $\mathcal{L}^\infty(\Omega)$ is by definition the vector space of functions that are essentially bounded, that is, the functions bounded on the complement of a set of measure zero, equipped with the semi-norm called the essential upper bound, defined by

$$\|f\|_{L^\infty} = \inf\{C \geqslant 0 : |f(x)| \leqslant C \text{ for almost all } x\}.$$

The normed vector space $L^\infty(\Omega)$ is the quotient of this space by the subspace of functions that are almost everywhere zero. Let us recall the following result (*cf.* [42, 123, 163, 182]).

Theorem 18.1 (Riesz–Fischer) *For all $p \in [1, +\infty]$, the space $L^p(\Omega)$ is a Banach space for the norm $\|\cdot\|_{L^p}$.*

We refer to Section 3.4 for results concerning the topological dual of the space $L^p(\Omega)$. We recall that:

1. if $1 < p < \infty$, then $L^p(\Omega)$ is a separable reflexive Banach space whose dual can be identified with $L^{p'}(\Omega)$ with $p' = p/(p-1)$;
2. the space $L^1(\Omega)$ is separable, non-reflexive and its dual can be identified with the space $L^\infty(\Omega)$;
3. the space $L^\infty(\Omega)$ is neither reflexive nor separable and its dual strictly contains $L^1(\Omega)$.

18.2 Hölder, Minkowski, and Hardy Inequalities

In this section, we recall some fundamental inequalities in the study of Lebesgue spaces L^p. In the following, p denotes a Lebesgue exponent belonging to $[1, +\infty]$, and Ω an open set of $\mathbb{R}^n$, with $n \geqslant 1$. It is to be noted that we allow the value $p = +\infty$ and in this case $1/p = 0$.

Proposition 18.2 (Hölder) *Consider three real numbers p, q, r in $[1, +\infty]$, satisfying*

$$\frac{1}{p} + \frac{1}{q} = \frac{1}{r}.$$

For every pair of functions $(f, g) \in L^p(\Omega) \times L^q(\Omega)$, the product fg belongs to $L^r(\Omega)$ and moreover

$$\|fg\|_{L^r} \leqslant \|f\|_{L^p} \|g\|_{L^q}. \tag{18.1}$$

Proof The result is trivial if $f = 0$ or if $g = 0$. It is therefore sufficient to assume that f and g are non-zero and, in this case, we can assume without loss of generality that $\|f\|_{L^p} = 1$ and $\|g\|_{L^q} = 1$. Our goal is then to show that $fg \in L^r(\Omega)$ and that $\|fg\|_{L^r} \leqslant 1$.

The proof is based on a convexity argument. Specifically, as

$$\frac{r}{p} + \frac{r}{q} = 1,$$

by concavity of the logarithm, for every pair $(a, b) \in (0, +\infty)^2$, we have

$$\frac{r}{p} \log(a) + \frac{r}{q} \log(b) \leqslant \log\left(\frac{r}{p}a + \frac{r}{q}b\right).$$

The exponential function being increasing, it follows that

$$a^{r/p} b^{r/q} \leqslant \frac{r}{p}a + \frac{r}{q}b.$$

By applying this inequality with $A = a^{1/p}$ and $B = b^{1/q}$, it follows that, for every $(A, B) \in (0, +\infty)^2$, we have

$$A^r B^r \leqslant \frac{r}{p}A^p + \frac{r}{q}B^q.$$

In particular, for every x in $\mathbb{R}^n$, we have

$$|f(x)g(x)|^r \leqslant \frac{r}{p}|f(x)|^p + \frac{r}{q}|g(x)|^q.$$

By integrating this inequality, we then verify that

$$\|fg\|_{L^r}^r = \int |f(x)g(x)|^r \, dx \leqslant \frac{r}{p}\int |f(x)|^p \, dx + \frac{r}{q}\int |g(x)|^q \, dx = \frac{r}{p} + \frac{r}{q} = 1.$$

This concludes the proof. □

In particular, if p and q are such that

$$\frac{1}{p} + \frac{1}{q} = 1,$$

we have

$$\|fg\|_{L^1} \leqslant \|f\|_{L^p}\|g\|_{L^q}. \tag{18.2}$$

We say that q is the conjugate exponent of p, which we generally denote as p'. If $p = +\infty$, then $p' = 1$ and if $p \in [1, +\infty)$, we have

$$p' = \frac{p}{p-1}.$$

Proposition 18.3 (Minkowski Inequality) *Let $p \in [1+\infty]$. For any pair of functions* $(f, g) \in L^p(\Omega) \times L^p(\Omega)$, *we have*

$$\|f + g\|_{L^p} \leqslant \|f\|_{L^p} + \|g\|_{L^p}. \tag{18.3}$$

Proof The result is a direct consequence of the definition of the norm $\|\cdot\|_{L^\infty}$ if $p = +\infty$. We therefore assume that $p \in [1, +\infty)$. We can then use the triangle inequality in $\mathbb{R}$ to write that

$$|f(x) + g(x)|^p \leqslant |f(x)|\,|f(x) + g(x)|^{p-1} + |g(x)|\,|f(x) + g(x)|^{p-1}. \qquad (18.4)$$

By applying the Hölder inequality (18.2), we have

$$\int |f(x)|\,|f(x) + g(x)|^{p-1}\,dx$$
$$\leqslant \left(\int |f(x)|^p\,dx\right)^{1/p}\left(\int \left(|f(x) + g(x)|^{p-1}\right)^{p/(p-1)}dx\right)^{1/p'}$$
$$\leqslant \|f\|_{L^p}\|f+g\|_{L^p}^{p-1},$$

as well as a similar estimate for the second term on the right-hand side of (18.4). By combining these two estimates, we obtain

$$\|f+g\|_{L^p}^p = \int |f(x) + g(x)|^p\,dx \leqslant \left(\|f\|_{L^p} + \|g\|_{L^p}\right)\|f+g\|_{L^p}^{p-1},$$

which directly implies (18.3). □

Then, following Stein [176], we prove the Minkowski integral inequality and the Hardy inequality.

Proposition 18.4 (Minkowski Integral Inequality) *Consider two σ-finite spaces (S_1, dx) and (S_2, dy) and a measurable function $F: S_1 \times S_2 \to \mathbb{R}$. Then, for any $p \in [1, +\infty)$,*

$$\left(\int_{S_2}\left|\int_{S_1} F(x, y)\,dy\right|^p dx\right)^{1/p} \leqslant \int_{S_1}\left(\int_{S_2} |F(x, y)|^p\,dx\right)^{1/p} dy. \qquad (18.5)$$

Proof Thanks to the triangle inequality, we can assume without loss of generality that F is a positive function (if necessary, prove the inequality for $|F|$), which will allow us to apply Fubini's theorem for positive functions. We then introduce the positive functions

$$f(x) = \int_{S_1} F(x, y)\,dy, \qquad \varphi(x) = \frac{1}{\|f\|_{L^p}^{p-1}}f(x)^{p-1},$$

so that, on the one hand,

$$\left(\int_{S_2}\left|\int_{S_1} F(x, y)\,dy\right|^p dx\right)^{1/p} = \left(\int_{S_2}\left(\int_{S_1} F(x, y)\,dy\right)^p dx\right)^{1/p} = \|f\|_{L^p}.$$

On the other hand,

$$\|\varphi\|_{L^{p'}} = 1 \quad \text{and} \quad \|f\|_{L^p} = \int_{S_2} f(x)\varphi(x)\,dx.$$

Consequently, we have

$$
\begin{aligned}
\|f\|_{L^p} &= \int_{S_2} f(x)\varphi(x)\,dx \\
&= \int_{S_2}\int_{S_1} F(x,y)\varphi(x)\,dy\,dx \qquad \text{(by definition of } f) \\
&= \int_{S_1}\int_{S_2} F(x,y)\varphi(x)\,dx\,dy \qquad \text{(by Fubini)} \\
&\leqslant \int_{S_2}\left(\int_{S_1} F(x,y)^p\,dx\right)^{1/p}\left(\int \varphi(x)^{p'}\,dx\right)^{1/p'}dy \quad \text{(Hölder)} \\
&\leqslant \int_{S_2}\left(\int_{S_1} F(x,y)^p\,dx\right)^{1/p}dy \qquad \text{(because } \|\varphi\|_{L^{p'}} = 1).
\end{aligned}
$$

This proves (18.5). $\qquad\qquad\qquad\qquad\qquad\qquad\qquad\qquad\qquad\qquad\qquad\qquad$ □

Proposition 18.5 (Hardy) *For any real number $p \in (1, +\infty)$ and for any function f in $L^p((0, +\infty))$, we have*

$$\int_0^{+\infty}\left|\frac{1}{x}\int_0^x f(y)\,dy\right|^p dx \leqslant \left(\frac{p}{p-1}\right)^p \int_0^{+\infty} |f(x)|^p\,dx.$$

Proof See Exercise 18.3 and its solution on page 424. $\qquad\qquad\qquad\qquad\qquad$ □

18.3 Distribution Functions and Marcinkiewicz's Theorem

It is often useful to study a function by measuring its level sets (see [20, 123]). More precisely, to study the L^p norms of a measurable function $f: \mathbb{R}^n \to \mathbb{C}$, a fruitful point of view is to consider the Lebesgue measure of the sets

$$\{x \in \mathbb{R}^n : |f(x)| > \lambda\},$$

where λ is a positive real number.

Notation 18.6 We will denote by $\{|f| > \lambda\}$ the set $\{x \in \mathbb{R}^n : |f(x)| > \lambda\}$. Also, let us recall that we denote by $|A|$ the Lebesgue measure of a measurable set A.

Definition 18.7 The distribution function of f is the function $F: \mathbb{R} \to [0, +\infty)$ defined by

$$F(\lambda) = |\{|f| > \lambda\}|.$$

We start with two lemmas that relate the distribution function F to the L^p norms .

Lemma 18.8 *For all $p \in [1, +\infty)$, there holds*

$$\|f\|_{L^p}^p = p \int_0^\infty \lambda^{p-1} F(\lambda) \, d\lambda.$$

Proof This formula is obtained by writing

$$\|f\|_{L^p}^p = \int_{\mathbb{R}^n} \int_0^{|f(x)|} p\lambda^{p-1} \, d\lambda \, dx,$$

then using Fubini's theorem.   □

Lemma 18.9 (Chebyshev's Inequality) *For all $p \in [1, +\infty)$ and all $\lambda > 0$, we have*

$$F(\lambda) \leqslant \lambda^{-p} \|f\|_{L^p}^p. \tag{18.6}$$

Proof Indeed, we have

$$\|f\|_{L^p}^p = \int |f(x)|^p \, dx \geqslant \int_{\{|f| > \lambda\}} \lambda^p \, dx = \lambda^p F(\lambda),$$

which directly implies the desired result.   □

This inequality suggests introducing the following spaces.

Definition 18.10 Let $p \in [1, +\infty)$. We define the weak Lebesgue space $L_w^p(\mathbb{R}^n)$ as the set of measurable functions $f \colon \mathbb{R}^n \to \mathbb{C}$ such that

$$\|f\|_{L_w^p} := \sup_{\lambda > 0} \left(\lambda \, |\{|f| > \lambda\}|^{1/p} \right) < +\infty,$$

quotiented by the equivalence relation of almost everywhere equality.

Remark 18.11 The Chebyshev inequality (18.6) implies that $L^p(\mathbb{R}^n) \subset L_w^p(\mathbb{R}^n)$ but the converse is not true because, for example, the function $1/|x|$ belongs to the space $L_w^1(\mathbb{R})$.

The most important result concerning these spaces is the following interpolation theorem from Marcinkiewicz [129] (see Maligranda [127] for many historical comments).

Theorem 18.12 (Marcinkiewicz) *Let $q \in (1, +\infty]$. Consider a linear map T defined on $L^1(\mathbb{R}^n) + L^q(\mathbb{R}^n)$ with values in $L_w^1(\mathbb{R}^n) + L^q(\mathbb{R}^n)$ and suppose that there exist two constants C_1 and C_q such that*

$$\forall f \in L^1(\mathbb{R}^n), \quad \|T(f)\|_{L_w^1} \leqslant C_1 \|f\|_{L^1},$$
$$\forall f \in L^q(\mathbb{R}^n), \quad \|T(f)\|_{L^q} \leqslant C_q \|f\|_{L^q}. \tag{18.7}$$

Then, for all $p \in (1, q]$, there is a constant C_p such that $\|T(f)\|_{L^p} \leqslant C_p \|f\|_{L^p}$.

Proof We follow the proof by Stein [176]. We will prove this result under weaker assumptions: specifically, we will not assume that T is linear but only that T satisfies a sub-additivity property (this version is used in the proof of Theorem 6.12).

Assume there exists a constant $A > 0$ such that

$$|T(f_1 + f_2)(x)| \leqslant A\,|T(f_1)(x)| + A\,|T(f_2)(x)|\,.$$

Let us start by considering the case $q < +\infty$. The idea of the proof is that $L^p(\mathbb{R}^n) \subset L^1(\mathbb{R}^n) + L^q(\mathbb{R}^n)$. Indeed, for all $\gamma > 0$, we can decompose f as $f = f^\gamma + f_\gamma$ where

$$f^\gamma(x) = \begin{cases} f(x) & \text{if } |f(x)| > \gamma, \\ 0 & \text{if } |f(x)| \leqslant \gamma, \end{cases} \qquad f_\gamma(x) = \begin{cases} 0 & \text{if } |f(x)| > \gamma, \\ f(x) & \text{if } |f(x)| \leqslant \gamma. \end{cases}$$

We directly verify that

$$\|f^\gamma\|_{L^1} \leqslant \gamma^{1-p}\|f\|_{L^p}^p, \qquad \|f_\gamma\|_{L^q}^q \leqslant \gamma^{q-p}\|f\|_{L^p}^p\,.$$

We will use this decomposition by varying the parameter γ.

Let $f \in L^p(\mathbb{R}^n)$. For all $\lambda > 0$, by the sub-additivity assumption, we have

$$|T(f)(x)| \leqslant A\left|T(f^\lambda)(x)\right| + A\,|T(f_\lambda)(x)|\,,$$

therefore, according to the triangle inequality,

$$\{|T(f)| > \lambda\} \subset \left\{|T(f^\lambda)| > \lambda/(2A)\right\} \cup \{|T(f_\lambda)| > \lambda/(2A)\}\,.$$

Consequently, using Chebyshev's inequality (see Lemma 18.9), the assumptions (18.7) imply that

$$|\{|T(f)| > \lambda\}| \leqslant \frac{2AC_1}{\lambda}\,\|f^\lambda\|_{L^1} + \frac{(2A)^q C_q^q}{\lambda^q}\,\|f_\lambda\|_{L^q}^q\,.$$

Using Lemma 18.8, it follows that

$$\|T(f)\|_{L^p}^p = p\int_0^\infty \lambda^{p-1}\,|\{|T(f)| > \lambda\}|\,\mathrm{d}\lambda$$

$$\leqslant 2AC_1 p\int_0^\infty \lambda^{p-2}\,\|f^\lambda\|_{L^1}\,\mathrm{d}\lambda + (2A)^q C_q^q p\int_0^\infty \lambda^{p-1-q}\,\|f_\lambda\|_{L^q}^q\,\mathrm{d}\lambda.$$

However, by definition of f^λ,

$$\int_0^\infty \lambda^{p-2}\,\|f^\lambda\|_{L^1}\,\mathrm{d}\lambda = \int_0^\infty \lambda^{p-2}\left(\int_{\{|f|>\lambda\}} |f(x)|\,\mathrm{d}x\right)\mathrm{d}\lambda$$

$$= \int_{\mathbb{R}^n} |f(x)|\left(\int_0^{|f(x)|} \lambda^{p-2}\mathrm{d}\lambda\right)\mathrm{d}x = \frac{1}{p-1}\int_{\mathbb{R}^n} |f(x)|^p\,\mathrm{d}x,$$

where we used Fubini's theorem.

By reasoning similarly, we obtain that

$$\int_0^\infty \lambda^{p-1-q} \|f_\lambda\|_{L^q}^q \, d\lambda = \int_0^\infty \lambda^{p-1-q} \left(\int_{\{|f|\leqslant\lambda\}} |f(x)|^q \, dx \right) d\lambda$$

$$= \int_{\mathbb{R}^n} |f(x)|^q \left(\int_{|f(x)|}^\infty \lambda^{p-1-q} d\lambda \right) dx$$

$$= \frac{1}{q-p} \int_{\mathbb{R}^n} |f(x)|^p \, dx.$$

We have shown that there exists a constant C_p such that,

$$\|T(f)\|_{L^p} \leqslant C_p \|f\|_{L^p}.$$

This concludes the proof in the case $q < +\infty$.

Now suppose that $q = +\infty$. Then

$$\|Tf_\lambda\|_{L^\infty} \leqslant C_\infty \|f_\lambda\|_{L^\infty} \leqslant C_\infty \lambda.$$

Therefore, we have

$$|\{|Tf| > 2AC_\infty\lambda\}| \leqslant |\{|Tf^\lambda| > C_\infty\lambda\}|,$$

and we conclude as before.   □

18.4 Exercises

Exercise (solved) 18.1 Construct a sequence $(f_n)_{n\in\mathbb{N}}$ that converges to 0 in $L^p([0,1])$ for all p such that $1 \leqslant p < +\infty$ and such that, for all $x \in [0,1]$, the sequence $(f_n(x))_{n\in\mathbb{N}}$ does not have a limit.

Exercise (solved) 18.2 This exercise is about an alternative proof of the Hölder inequality, obtained by using the tensor power trick, as popularized by Terence Tao in his research blog.

1. Let p and q be such that $1/p + 1/q = 1$. Suppose there exists a constant $C \geqslant 1$ such that, for any dimension $n \geqslant 1$, for any $f \in L^p(\mathbb{R}^n)$ and for any $g \in L^q(\mathbb{R}^n)$, we have

$$\|fg\|_{L^1(\mathbb{R}^n)} \leqslant C\|f\|_{L^p(\mathbb{R}^n)} \|g\|_{L^q(\mathbb{R}^n)}.$$

Let $k \geqslant 1$ be an integer. By applying this inequality to the functions $F: (\mathbb{R}^n)^k \to \mathbb{C}$ and $G: (\mathbb{R}^n)^k \to \mathbb{C}$ defined by

$$F(x_1,\ldots,x_k) = f(x_1)\cdots f(x_k), \quad G(x_1,\ldots,x_k) = g(x_1)\cdots g(x_k)$$

(that is, F and G are the tensor products of k copies of f and g), deduce that the Hölder inequality is true with the constant $C = 1$.

2. Let $p \in [1, +\infty)$. Prove the Hölder inequality $L^p \cdot L^q \hookrightarrow L^1$ from the representation of the L^1 norm from the level sets (see Lemma 18.8):

$$\|fg\|_{L^1} = \int_0^\infty |\{x \in \mathbb{R}^n : |f(x)g(x)| > \lambda\}| \, d\lambda.$$

Hint: start by verifying that

$$|\{|f(x)g(x)| > \lambda\}| \leq \left|\left\{|f(x)| > \lambda^{1/p}\right\}\right| + \left|\left\{|g(x)| > \lambda^{1/q}\right\}\right|.$$

Exercise (solved) 18.3 The goal of this exercise is to prove a lemma due to Stein, then deduce the Hardy inequality. Consider $p \in (1, +\infty)$ and a mesurable function $K: (0, +\infty)^2 \to \mathbb{R}$, homogeneous of degree -1, in the sense that

$$\forall \lambda > 0, \ \forall (x, y) \in (0, +\infty)^2, \quad K(\lambda x, \lambda y) = \frac{1}{\lambda} K(x, y),$$

and satisfying the integrability condition $\int_0^{+\infty} |K(1, y)| \, y^{-1/p} \, dy < +\infty$.

1. Show that for any continuous function $f \in C_0^0((0, +\infty))$ with compact support, and for any $x \in (0, +\infty)$, the following integral is well defined

$$T(f)(x) = \int_0^{+\infty} K(x, y) f(y) \, dy.$$

2. Verify that

$$T(f)(x) = \int_0^{+\infty} K(1, r) \, f(rx) \, dr,$$

then, using the Minkowski inequality, show that there exists a constant C such that, for any continuous function $f \in C_0^0(\mathbb{R})$ with compact support, we have

$$\left(\int_0^{+\infty} |T(f)|^p \, dx\right)^{1/p} \leq C \left(\int_0^{+\infty} |f(x)|^p \, dx\right)^{1/p}.$$

3. Prove the Hardy inequality: for any $p \in (1, +\infty)$ and for any function f in $L^p((0, +\infty))$, we have

$$\left(\int_0^{+\infty} \left|\frac{1}{x} \int_0^x f(y) \, dy\right|^p dx\right)^{1/p} \leq \frac{p}{p-1} \left(\int_0^{+\infty} |f(x)|^p \, dx\right)^{1/p}.$$

Exercise 18.4 Let $p, q \in (1, \infty)$ such that

$$\frac{1}{p} + \frac{1}{q} = 1.$$

Consider a continuous function $k : \mathbb{R}^n \times \mathbb{R}^n \to [0, +\infty)$ such that

$$\sup_{y \in \mathbb{R}^n} \int_{\mathbb{R}^n} k(x, y) \, dx \leqslant C_1 < +\infty, \quad \sup_{x \in \mathbb{R}^n} \int_{\mathbb{R}^n} k(x, y) \, dy \leqslant C_2 < +\infty.$$

Consider two continuous functions $f, g : \mathbb{R}^n \to [0, +\infty)$ with compact support.

1. Recall the proof of the following inequality: for all $(a, b) \in (0, +\infty)^2$ we have

$$ab \leqslant \frac{a^p}{p} + \frac{b^q}{q}.$$

2. Show that

$$\iint_{\mathbb{R}^n \times \mathbb{R}^n} k(x, y) f(y) g(x) \, dx \, dy \leqslant \frac{C_1}{p} \|f\|_{L^p}^p + \frac{C_2}{q} \|g\|_{L^q}^q.$$

3. By applying the above to λf and $\lambda^{-1} g$, deduce that

$$\iint_{\mathbb{R}^n \times \mathbb{R}^n} k(x, y) f(y) g(x) \, dx \, dy \leqslant C_1^{1/p} C_2^{1/q} \|f\|_{L^p} \|g\|_{L^q}.$$

4. Deduce that the operator K, defined on the space of continuous functions u with compact support by

$$Ku(x) = \int_{\mathbb{R}^n} k(x, y) u(y) \, dy,$$

extends by continuity to a continuous operator from $L^p(\mathbb{R}^n)$ to $L^p(\mathbb{R}^n)$.

Exercise 18.5 (Riesz–Thorin Interpolation Inequality) The goal is to prove a result first proved by Riesz [160] in the real case, and then extended to the complex case with a beautiful proof by Thorin [190] (see also [10] for many historical comments about this fundamental result).

Theorem 18.13 (Riesz–Thorin) *Let* (Ω, μ) *and* (Λ, ν) *be two* σ-*finite measure spaces, and consider four exponents* $1 \leqslant p_0, p_1, q_0, q_1 \leqslant +\infty$. *Let* T *be a continuous linear map from* $L^{p_0}(\Omega) + L^{p_1}(\Omega)$ *to* $L^{q_0}(\Lambda) + L^{q_1}(\Lambda)$. *Then, for all* $\theta \in (0, 1)$,

$$\|T\|_{L^{p_\theta} \to L^{q_\theta}} \leqslant \|T\|_{L^{p_0} \to L^{q_0}}^{1-\theta} \|T\|_{L^{p_1} \to L^{q_1}}^{\theta},$$

where

$$\frac{1}{p_\theta} = \frac{1-\theta}{p_0} + \frac{\theta}{p_1}, \quad \frac{1}{q_\theta} = \frac{1-\theta}{q_0} + \frac{\theta}{q_1}.$$

We will use the two elementary results recalled here:

1. *Density of simple functions.* Let (Ω, μ) be a σ-finite measure space. Then the set of finite linear combinations $\sum \alpha_j 1_{A_j}$ where the α_j are complex numbers and the A_j are measurable parts of Ω with finite measure, is dense in the space $L^p(\Omega, \mu)$ for all $p \in [1, +\infty)$.
2. *Representation of the norm by duality.* Let (Ω, μ) be a σ-finite measure space. Then, for all $q \in [1, +\infty]$ and all $f \in L^q(\Omega, \mu)$, we have

$$\|f\|_{L^q} = \sup_{g \in L^{q'}} \left(\frac{1}{\|g\|_{L^{q'}}} \int fg \right), \qquad \frac{1}{q} + \frac{1}{q'} = 1.$$

In addition, we can limit the supremum above to simple functions.

1. Treat the case where $p_\theta = +\infty$.
2. We now assume that $p_\theta \neq +\infty$. Let us set

$$\alpha(z) = p_\theta \left(\frac{1-z}{p_0} + \frac{z}{p_1} \right), \quad \beta(z) = q'_\theta \left(\frac{1-z}{q'_0} + \frac{z}{q'_1} \right),$$

and, for f and g simple functions (see the preamble), we introduce

$$f_z(x) = |f(x)|^{\alpha(z)} \frac{f(x)}{|f(x)|}, \quad g_z(x) = |g(x)|^{\beta(z)} \frac{g(x)}{|g(x)|}.$$

Show that the function

$$\varphi : z \longmapsto \int (Tf_z) g_z \, dx$$

is well defined, holomorphic on the strip S and continuous on $\overline{S}$.
3. By applying Hadamard's three lines lemma (see Exercise 8.1), deduce that

$$\left| \int (Tf)g \, dx \right| \leqslant \|T\|_{L^{p_0} \to L^{q_0}}^{1-\theta} \|T\|_{L^{p_1} \to L^{q_1}}^{\theta} \|f\|_{L^p} \|g\|_{L^{q'}}.$$

4. Conclude.

Chapter 19
Solutions

Solution 1 (Solution to Exercise 1.1) Let us use the sequence (u_n) defined in the proof of Theorem 1.15. Recall that it is a sequence of vectors of norm 1 that satisfies the following property:

$$\forall (n, m) \in \mathbb{N} \times \mathbb{N}, \quad n \neq m \implies \|u_n - u_m\| \geq \frac{1}{2}.$$

We introduce the function

$$F(x) = \sum_{n \in \mathbb{N}} n \max\left\{0, \frac{1}{5} - \|x - u_n\|\right\}.$$

This function is well defined because, for every x in B, at most one of the terms of the sum is non-zero. Indeed, if $\|x - u_n\| \leq 1/5$ and $\|x - u_m\| \leq 1/5$, then $\|u_n - u_m\| \leq 2/5$ therefore $n = m$. We directly deduce that F is continuous. Moreover, F is not bounded because it equals n on the ball centered at u_n with radius $1/5$.

Solution 2 (Solution to Exercise 1.2) Let $(P_n)_{n \in \mathbb{N}}$ be a sequence of polynomials that converges uniformly to f. Then there exists $N \in \mathbb{N}$ such that, for all $n \geq N$, we have

$$\sup_A |P_n(x) - P_N(x)| \leq 1.$$

Since $P_n - P_N$ is a polynomial and the only bounded polynomials on A are constant polynomials, we deduce that there exists $\alpha_n \in \mathbb{R}$ such that $P_n(x) = P_N(x) + \alpha_n$. As the sequence $(P_n - P_N)_{n \geq N}$ is a convergent sequence, it is a Cauchy sequence and we deduce that $(\alpha_n)_{n \geq N}$ is a Cauchy sequence in $\mathbb{R}$. It therefore converges to a real number denoted α. Therefore, for all $x \in \mathbb{R}$, we have

$$f(x) = \lim_{n \to \infty} (P_n(x)) = \lim_{n \to \infty} (P_N(x) + \alpha_n) = P_N(x) + \alpha.$$

This proves that f is a polynomial function.

© The Author(s), under exclusive license to Springer Nature Switzerland AG 2024
T. Alazard, *Analysis and Partial Differential Equations*, Universitext,
https://doi.org/10.1007/978-3-031-70909-8_19

Solution 3 (Solution to Exercise 1.3)

1. For all $x \in E$ and all $\xi \in E$, we have

$$T\xi = \frac{1}{2}T(x+\xi) - \frac{1}{2}T(x-\xi).$$

From the triangle inequality, we get

$$\|T\xi\|_F \leqslant \frac{1}{2}\|T(x+\xi)\|_F + \frac{1}{2}\|T(x-\xi)\|_F,$$

from which the desired result follows by using the elementary inequality $\frac{1}{2}a + \frac{1}{2}b \leqslant \max\{a,b\}$.

2. By definition of the operator norm, we have

$$r\,\|T\|_{\mathcal{L}(E,F)} = r \sup_{\|y\|_E=1} \|Ty\|_F = \sup_{\|y\|_E=1} \|T(ry)\|_F = \sup_{\|\xi\|_E=r} \|T\xi\|_F.$$

However, according to the first question, if $\|\xi\|_E = r$,

$$\|T\xi\|_F \leqslant \max\{\|T(x+\xi)\|_F, \|T(x-\xi)\|_F\} \leqslant \sup_{\{x' \in E : \|x'-x\|_E = r\}} \|Tx'\|_F,$$

which leads to the desired result.

3. We construct the sequence $(x_n)_{n \in \mathbb{N}}$ by induction. It follows from the result of the previous question applied with $r = 3^{-n}$ that

$$\sup_{\{x' \in E : \|x'-x_{n-1}\|_E = 3^{-n}\}} \|T_{\alpha_n}x'\|_F \geqslant 3^{-n}\|T_{\alpha_n}\|_{\mathcal{L}(E,F)},$$

which implies (by definition of the upper bound) that there exists x_n such that

$$\|x_n - x_{n-1}\|_E = 3^{-n}, \quad \|T_{\alpha_n}x_n\|_F \geqslant \frac{2}{3}3^{-n}\|T_{\alpha_n}\|_{\mathcal{L}(E,F)}. \tag{19.1}$$

By construction, the sequence $(x_n)_{n \in \mathbb{N}}$ is Cauchy, so it converges because E is a Banach space. Let us denote its limit by x. We verify that $\|x - x_n\|_E \leqslant \frac{1}{2}3^{-n}$, which implies that $\|T_{\alpha_n}(x - x_n)\|_F \leqslant \frac{1}{2}3^{-n}\|T_{\alpha_n}\|_{\mathcal{L}(E,F)}$. Then, we have

$$\|T_{\alpha_n}x\|_F$$
$$\geqslant \|T_{\alpha_n}x_n\|_F - \|T_{\alpha_n}(x - x_n)\|_F \qquad \text{(triangle inequality)}$$
$$\geqslant \frac{2}{3}3^{-n}\|T_{\alpha_n}\|_{\mathcal{L}(E,F)} - \frac{1}{2}3^{-n}\|T_{\alpha_n}\|_{\mathcal{L}(E,F)} \qquad \text{(from (19.1))}$$
$$\geqslant \frac{1}{6}3^{-n}4^n \qquad \text{(because } \|T_{\alpha_n}\|_{\mathcal{L}(E,F)} \geqslant 4^n\text{).}$$

This is absurd because $\|T_{\alpha_n}x\|_F$ is bounded by hypothesis.

Solution 4 (Solution to Exercise 2.1)

1. It is easy to verify that the space X_T is a Banach space for the norm

$$\|u\|_\infty = \sup_{t \in [0,T]} \|u(t)\|_E .$$

Moreover,

$$\|\Phi(u) - \Phi(v)\|_\infty \leqslant \sup_{t \in [0,T]} \int_0^t \|F(s,u(s)) - F(s,v(s))\| \, ds$$

$$\leqslant \sup_{t \in [0,T]} \int_0^t C \|u(s) - v(s)\| \, ds$$

$$\leqslant \int_0^T C \|u(s) - v(s)\| \, ds \leqslant TC \|u - v\|_\infty.$$

This shows that Φ is a contraction if $TC < 1$.

2. Let us set $T = 1/(2C)$ so that $TC < 1$. The fixed point theorem implies that there exists a unique solution $u \in X_T$ of

$$\forall t \in [0,T], \quad u(t) = u_0 + \int_0^t F(s,u(s)) \, ds.$$

It remains to define a solution over the time interval $[0, +\infty)$. For this, the main observation is that the time interval given by the previous argument does not depend on the initial data. Also, we will be able to construct a solution defined for all time by piecing together solutions defined over time intervals of size T. Let us start by setting $u_T = u(T)$. Then the previous argument implies that there exists a unique function, denoted U, belonging to $C^0([0,T];E)$ which is a solution of

$$\forall t \in [0,T], \quad U(t) = u_T + \int_0^t F(s+T, U(s)) \, ds.$$

Let us define $\tilde{u} \in C^0([0,2T];E)$ by $\tilde{u}(t) = u(t)$ if $t \in [0,T]$ and $\tilde{u}(t) = U(t-T)$ if $t \in [0,2T]$. Then $\tilde{u} \in C^0([0,2T];E)$ and moreover we verify that

$$\forall t \in [0,2T], \quad \tilde{u}(t) = u_0 + \int_0^t F(s, \tilde{u}(s)) \, ds.$$

By iterating this argument we construct the sought solution on $[0, +\infty)$. Finally, we need to verify that u is C^1 in time with values in E, which is a direct consequence of the fact that u is given by the time integral of a continuous function.

3. The space

$$X = \left\{ u \in C^0([0,+\infty); E) : \sup_{t \in [0,+\infty)} e^{-\lambda t} \|u(t)\|_E < +\infty \right\}$$

is a Banach space for the norm

$$\|u\|_X = \sup_{t\in[0,+\infty)} e^{-\lambda t}\|u(t)\|_E.$$

Let u belong to X. By writing $F(u) = F(u) - F(0) + F(0)$ and using the triangle inequality, we obtain that

$$e^{-\lambda t}\|\Phi(u)(t)\|_E \leqslant e^{-\lambda t}\left(\|u_0\|_E + t\,\|F(0)\|_E\right) + \int_0^t e^{-\lambda t}C\,\|u(s)\|_E\,ds.$$

Then we write that

$$\int_0^t e^{-\lambda t}C\,\|u(s)\|_E\,ds \leqslant C\|u\|_X \int_0^t e^{-\lambda(t-s)}\,ds \leqslant \frac{C}{\lambda}\|u\|_X,$$

to obtain that $\Phi(u)$ also belongs to X. Moreover,

$$\|\Phi(u) - \Phi(v)\|_X \leqslant \sup e^{-\lambda t}\int_0^t \|F(s,u(s)) - F(s,v(s))\|_E\,ds$$

$$\leqslant \sup e^{-\lambda t}\int_0^t C\,\|u(s) - v(s)\|_E\,ds,$$

so as above, we deduce that

$$\|\Phi(u) - \Phi(v)\|_X \leqslant \frac{C}{\lambda}\|u - v\|_X\,.$$

For $\lambda > C$, we verify that Φ is a contraction and we conclude with the help of the fixed point that there exists a solution $u \in X$ of $\Phi(u) = u$.

Solution 5 (Solution of Exercise 2.2)

1. Consider a compact set K homeomorphic to $\overline{B}$ and let $f : K \to K$ be a continuous function. Consider a homeomorphism $\phi : K \to \overline{B}$, that is, a bijective, continuous function, whose inverse is also continuous. Then the function $g = \phi \circ f \circ \phi^{-1}$ is continuous from the ball $\overline{B}$ into itself. According to Theorem 2.12, it has a fixed point, that is, an element $x \in \overline{B}$ such that $g(x) = x$. Then $f(\phi^{-1}(x)) = \phi^{-1}(x)$, which proves that $\phi^{-1}(x)$ is a fixed point of f.
2. The first point comes from the fact that there exist by hypothesis two positive numbers r, R such that
$$B(0,r) \subset C \subset B(0,R).$$

The following three points were proved in the first chapter, where we showed that the function μ is a norm. Let us show that μ is continuous. For this, let us use the triangle inequality to write

$$\mu(x + y) - \mu(x) \leqslant \mu(y), \quad \mu(x) \leqslant \mu(x + y) + \mu(-y).$$

Then the inequalities of the first point imply that

$$-\frac{|y|}{r} \leqslant -\mu(-y) \leqslant \mu(x+y) - \mu(x) \leqslant \mu(y) \leqslant \frac{|y|}{r}.$$

This directly implies that μ is continuous.

It remains only to show that $x \in C$ if and only if $\mu(x) \leqslant 1$. Suppose that $x \in C$. Then $1 \in \{t > 0 : x/t \in C\}$ so $\mu(x) \leqslant 1$ trivially. Conversely, suppose that $\mu(x) \leqslant 1$. Then there exists a decreasing sequence (t_n) that tends towards $\mu(x)$ and a sequence (c_n) of elements of C such that $x = t_n c_n$. If $\mu(x) < 1$, then $t_n < 1$ for n large enough, and as $0 \in C$, we deduce that $x \in C$ by convexity. If $\mu(x) = 1$, then the sequence t_n tends towards 1 so the identity $c_n = x/t_n$ implies that c_n tends towards x. As C is closed, this implies that $x \in C$.

3. Consider the function

$$\phi : C \longrightarrow \overline{B}, \quad x \longmapsto \begin{cases} \dfrac{\mu(x)}{|x|} x & \text{if } x \neq 0, \\ 0 & \text{if } x = 0. \end{cases}$$

Note that

$$\forall x \in C, \quad |\phi(x)| \leqslant \mu(x) \leqslant \frac{|x|}{r}. \tag{19.2}$$

(Note that this inequality holds for $x = 0$ and $x \neq 0$.) This inequality justifies the fact that ϕ has values in $\overline{B}$ because, if $x \in C$, we have seen that $\mu(x) \leqslant 1$. Furthermore, as $|\cdot|$ and μ are continuous, ϕ is continuous at every point different from 0. And the inequality (19.2) guarantees that ϕ is also continuous at 0. So ϕ is continuous at every point. Let us show that ϕ is bijective and that its inverse is continuous. To find the inverse of ϕ, we seek to solve the equation $\phi(y) = x$. As $\mu(\lambda x) = \lambda \mu(x)$, we are led to consider the function

$$\psi : \overline{B} \longrightarrow C, \quad x \longmapsto \begin{cases} \dfrac{|x|}{\mu(x)} x & \text{if } x \neq 0, \\ 0 & \text{if } x = 0. \end{cases}$$

It is easily verified that ϕ is bijective, with inverse ψ. By reasoning as previously, using the inequality $\mu(x) \geqslant |x|/R$ we verify that ψ is indeed a continuous function. We have shown that C is homeomorphic to $\overline{B}$, and as we saw in point 1, this proves that any continuous function from C to C has a fixed point.

4. The set $\overline{B} = \{x \in \mathbb{R}^n : N(x) \leqslant 1\}$ is closed and bounded in $(\mathbb{R}^n, N(\cdot))$. By equivalence of norms in finite-dimensional spaces, it is also closed and bounded in $(\mathbb{R}^n, |\cdot|)$, and therefore compact. Furthermore, the triangle inequality for N implies that

$$N(\lambda x + (1 - \lambda)y) \leqslant \lambda N(x) + (1 - \lambda)N(y),$$

which shows that $\overline{B}$ is convex.

Solution 6 (Solution to Exercise 3.1)

1. This is a direct consequence of the Riesz–Fréchet theorem.
2. The map A is linear. Since a is continuous, there exists a constant $C > 0$ such that, for all u:

$$\forall v \in H, \quad |\langle Au, v \rangle| = a(u, v) \leqslant C \|u\| \|v\|,$$

which implies $\|Au\| \leqslant C \|u\|$. Therefore, the map A is continuous.
3. By hypothesis, for all $u \in H$, we have

$$\|Au\| \, \|u\| \geqslant |\langle Au, u \rangle| = |a(u, u)| \geqslant c \|u\|^2,$$

therefore $\|Au\| \geqslant c \|u\|$. This implies that A is injective and therefore an isomorphism onto its image. It follows that $\mathrm{Im}(A)$ is a complete subspace, and therefore a closed subspace of H. To show $\mathrm{Im}(A) = H$, it is now sufficient to show that $(\mathrm{Im}(A))^{\perp} = \{0\}$. For this, let us assume $h \in (\mathrm{Im}(A))^{\perp}$ is fixed. Then we must have

$$0 = \langle Ah, h \rangle = a(h, h) \geqslant c \, |h|^2 .$$

Therefore $h = 0$.
4. Let ϕ be a continuous linear form. Using the Riesz–Fréchet theorem again, we see that there exists $U \in H$ such that:

$$\forall v \in H, \quad \phi(v) = \langle U, v \rangle.$$

It is therefore sufficient to show that there exists u such that $Au = U$. This is a consequence of the previous question, which shows that A is surjective.

Solution 7 (Solution to Exercise 3.2)

1. (a) The sequence E_1 is a frame because, if (e_n) is a Hilbert basis, then

$$\sum_{n \in \mathbb{N}} |(x, x_n)|^2 = \|x\|^2.$$

This also implies that the sequence E_3 is a frame. The sequence E_2 is not a frame because $(e_0, e_{n+1}) = 0$ for all integers n. The sequence E_4 is not a frame.
   (b) Suppose that $(x_n)_{n \in \mathbb{N}}$ is a frame and let

$$V = \mathrm{Vect}\{x_n : n \in \mathbb{N}\}.$$

We want to show that V is dense. For this, it is sufficient, by a general result on Hilbert spaces, to show that the orthogonal of V is reduced to $\{0\}$. Consider a vector x orthogonal to V. Then a fortiori $(x, e_n) = 0$ for all integers n, and therefore $\|x\| = 0$. This shows that $x = 0$.
2. (a) The function

$$U : H \longrightarrow \ell^2(\mathbb{N}), \quad x \longmapsto U(x) = \big((x, x_n)\big)_{n \in \mathbb{N}},$$

is a well-defined linear map. To show that it is continuous, we can use the closed graph theorem. According to this theorem, it is sufficient to show that if $(y_k)_{k\in\mathbb{N}}$ is a sequence in H that converges to $y \in H$ and also $(U(y_k))_{k\in\mathbb{N}}$ converges to a sequence $c = (c_n)_{n\in\mathbb{N}} \in \ell^2(\mathbb{N})$, then $c = U(y)$, that is, $c_n = (y, x_n)$ for all n. For this, note that $U(y_k) = ((y_k, x_n))_{n\in\mathbb{N}}$ and write

$$|c_n - (y_k, x_n)|^2 \leqslant \sum_{p\in\mathbb{N}} |c_p - (y_k, x_p)|^2 = \|c - U(y_k)\|_{\ell^2}^2 .$$

As $U(y_k)$ tends to c in $\ell^2(\mathbb{N})$, we deduce by taking the limit that $c_n = \lim_{k\to+\infty}(y_k, x_n)$ for all $n \in \mathbb{N}$. But $\lim_{k\to+\infty}(y_k, x_n) = (y, x_n)$ because (y_k) converges to y in H. This proves that $c = U(y)$ and therefore that the graph of U is closed.

The continuity of U implies that there exists a constant C such that $\|U(x)\|_{\ell^2} \leqslant C\|x\|$. From this, we deduce that

$$\forall x \in H, \quad \sum_{n\in\mathbb{N}} |(x, x_n)|^2 \leqslant B\|x\|^2, \tag{19.3}$$

with $B^2 = C$.

(b) Consider a sequence $(x_n)_{n\in\mathbb{N}}$ in H that satisfies property (19.3). Consider a finite set $F \subset \mathbb{N}$ and a sequence $(c_n)_{n\in\mathbb{N}} \in \ell^2(\mathbb{N})$. Recall that, for any vector $u \in H$, we have

$$\|u\| = \sup_{y\in H,\ \|y\|=1} |\langle u, y\rangle| .$$

In particular,

$$\left\| \sum_{n\in F} c_n x_n \right\|^2 = \sup_{\substack{y\in H \\ \|y\|=1}} \left| \left\langle \sum_{n\in F} c_n x_n, y \right\rangle \right|^2 .$$

Since F is finite, we can apply the Cauchy–Schwarz inequality in $\mathbb{R}^{\#F}$, to obtain

$$\left| \left\langle \sum_{n\in F} c_n x_n, y \right\rangle \right|^2 = \left| \sum_{n\in F} c_n \langle x_n, y\rangle \right|^2 \leqslant \left(\sum_{n\in F} |c_n|^2 \right)\left(\sum_{n\in F} |\langle x_n, y\rangle|^2 \right).$$

(c) It follows from the inequality of the previous question and the hypothesis (19.3) that

$$\left\| \sum_{n\in F} c_n x_n \right\|^2 \leqslant B \sum_{n\in F} |c_n|^2.$$

In particular, for all integers m, M such that $1 \leqslant m \leqslant M$, we have

$$\left\| \sum_{m\leqslant n\leqslant M} c_n x_n \right\|^2 \leqslant B \sum_{m\leqslant n\leqslant M} |c_n|^2.$$

By letting m tend to $+\infty$, we see that the series of partial sums is a Cauchy sequence, and therefore converges.

Solution 8 (Solution to Exercise 3.3)

1. Consider k linear forms $\varphi_1, \ldots, \varphi_k$. The map

$$\Phi: \begin{cases} E \longrightarrow \mathbb{R}^k \\ x \longmapsto (\varphi_1(x), \ldots, \varphi_k(x)), \end{cases}$$

is not injective, otherwise E would be finite-dimensional. From this, we deduce that there exists $x \in E \setminus \{0\}$ such that $\Phi(x) = 0$. By linearity, the line $\mathbb{R}x$ is included in the intersection $\bigcap_{1 \leqslant \ell \leqslant k} \ker \varphi_\ell$. It trivially follows that $\mathbb{R}x$ is included in the set $\bigcap_{k=1}^N \{y \in E : |\varphi_k(y)| < \varepsilon\}$ for all $\varepsilon > 0$.

2. (a) Let $n \in \mathbb{N}^*$. By hypothesis, the ball $B_d(0, 1/n)$ is open for the weak topology. According to the previous question, there exists a non-zero $y_n \in B_d(0, 1/n)$ such that $\mathbb{R}y_n \subset B_d(0, 1/n^2)$. This implies that $x_n := ny_n/\|y_n\|_E$ belongs to $B_d(0, 1/n)$. We have indeed $\|x_n\|_E = n$. Moreover, the fact that $x_n \in B_d(0, 1/n)$ implies that the sequence $(x_n)_{n \in \mathbb{N}}$ tends towards 0 sequentially, hence $x_n \rightharpoonup 0$ if we assume that the weak topology $\sigma(E, E')$ is induced by the distance d.

(b) We deduce a contradiction because a sequence that converges weakly is bounded according to Proposition 3.41.

Solution 9 (Solution to Exercise 4.1)

1. This is a consequence of the decomposition of the solution into Fourier series and the Plancherel identity.

2. (a) First, note that, due to periodicity in x,

$$\frac{d}{dt} \int_{-\pi}^{\pi} u(t, x) \, dx = \int_{-\pi}^{\pi} \partial_x(\gamma(x)\partial_x u) \, dx = 0.$$

We deduce that $\int_{-\pi}^{\pi} u(t, x) \, dx = 0$ for all time $t \geqslant 0$ since this property is initially true by hypothesis.

(b) We obtain the requested identity by multiplying the equation by u and integrating over $[-\pi, \pi]$.

(c) As γ is bounded below by a positive constant, the Poincaré–Wirtinger inequality (see Lemma 4.17) ensures that there exists a constant C such that

$$\frac{C}{2} \int_{-\pi}^{\pi} u(t, x)^2 \, dx \leqslant \int_{-\pi}^{\pi} \gamma(x)(\partial_x u(t, x))^2 \, dx.$$

We deduce that

$$\frac{1}{2}\frac{d}{dt} \int_{-\pi}^{\pi} u(t, x)^2 \, dx + \frac{C}{2} \int_{-\pi}^{\pi} u(t, x)^2 \, dx \leqslant 0.$$

(d) The desired inequality comes from Gronwall's lemma.

Solution 10 (Solution to Exercise 7.1) Assume that $I = (-1, 1)$ to fix the notation. For all ϕ in $C_0^\infty(I)$, we can write

$$\int_I |x| \, \phi'(x) \, dx = \int_0^1 x\phi'(x) \, dx + \int_{-1}^0 (-x)\phi'(x) \, dx.$$

As x vanishes at 0 and $\phi(x)$ vanishes in the vicinity of -1 and 1, by integrating by parts on each of the intervals, we find that

$$\int_I |x| \, \phi'(x) \, dx = -\int_0^1 \phi(x) \, dx + \int_{-1}^0 \phi(x) \, dx = -\int_I H(x)\phi(x) \, dx.$$

As $H|_I$ belongs to $L^1(I)$, this means that $|x|$ is weakly differentiable and its derivative is given by $H|_I$.

Solution 11 (Solution to Exercise 7.2) Let us prove this result by contradiction. Suppose there exists v in $L^p(\mathbb{R})$ such that

$$\int_{\mathbb{R}} v\phi \, dx = -\int_{\mathbb{R}} \mathbf{1}_{(0,1)} \frac{\partial \phi}{\partial x} \, dx, \qquad \forall \phi \in C_0^\infty(\mathbb{R}).$$

Then,

$$|\phi(1) - \phi(0)| = \left| \int_{\mathbb{R}} \mathbf{1}_{(0,1)} \frac{\partial \phi}{\partial x} \, dx \right| = \left| \int_{\mathbb{R}} v\phi \, dx \right| \le C\|\phi\|_{L^{p'}(\mathbb{R})},$$

with $C = \|v\|_{L^p(\mathbb{R})}$. We then show that we cannot have such an inequality by considering a sequence of functions ϕ_n belonging to $C_0^\infty(\mathbb{R})$ such that $\phi_n(1) = 1$, $\phi_n(0) = 0$ and $\|\phi_n\|_{L^{p'}(\mathbb{R})} = 1/n$.

Solution 12 (Solution to Exercise 7.3) We want to show that there exists a constant $C = C(n)$ such that, for all $r > 0$ and for every ball B of radius r in $\mathbb{R}^n$, for all $u \in H^1(B)$, we have

$$\left(\frac{1}{|B|} \int_B |u - u_B|^2 \, dx \right)^{1/2} \le Cr \left(\frac{1}{|B|} \int_B |\nabla u|^2 \, dx \right)^{1/2},$$

where u_B is the average of u over B.

Note that we can assume that u is real-valued (by considering separately $\mathrm{Re}(u)$ and $\mathrm{Im}(u)$). The proof is based on a duplication argument:

$$\frac{1}{|B|} \int_B u^2 \, dx = \frac{1}{2|B|} \int_B u^2(x) \, dx + \frac{1}{2|B|} \int_B u^2(y) \, dy.$$

Note that we can assume without loss of generality that $u_B = 0$, so that

$$\frac{1}{|B|} \int_B u^2 \, dx = \frac{1}{2|B|^2} \iint_{B^2} (u(x) - u(y))^2 \, dx \, dy.$$

Furthermore

$$(u(x) - u(y))^2 \leqslant |x - y|^2 \int_0^1 |\nabla u(tx + (1-t)y)|^2 \, dt$$

$$\leqslant (2r)^2 \int_0^1 |\nabla u(tx + (1-t)y)|^2 \, dt.$$

However, for all $t \in [0, 1]$, we have

$$\iint_{B^2} |\nabla u(tx + (1-t)y)|^2 \, dx \, dy = \frac{1}{t^n} \int_B \int_{tB+(1-t)y} |\nabla u(\sigma)|^2 \, d\sigma \, dy$$

$$= \frac{1}{t^n} \int_B \int_B 1_{tB+(1-t)y}(\sigma) \, |\nabla u(\sigma)|^2 \, d\sigma \, dy,$$

and

$$\int_B 1_{tB+(1-t)y}(\sigma) \, dy = \left| B \cap \frac{1}{1-t}(\sigma - tB) \right| \leqslant \min\left(1, \frac{t^n}{(1-t)^n}\right) |B|.$$

We deduce that

$$\frac{1}{|B|} \int_B u^2 \, dx \leqslant \frac{(2r)^2}{2|B|} \left\{ \int_0^1 \min\left(1, \frac{t^n}{(1-t)^n}\right) \frac{dt}{t^n} \right\} \int_B |\nabla u|^2 \, dx,$$

which concludes the proof.

Solution 13 (Solution to Exercise 7.4) Assume that $n = 1$, $\Omega_1 = (a, b)$, $\Omega_2 = \theta((a, b)) = (c, d)$ and let $\varphi = \phi \circ \theta^{-1}$. By writing

$$\phi'(\theta^{-1}(y))(\theta^{-1})'(y) = \varphi'$$

we verify that

$$\left| \int_a^b u(\theta(x))\phi'(x) \, dx \right| = \left| \int_c^d u(y)\varphi'(y) \, dy \right|.$$

However, as $u \in H^1(c, d)$, by definition of weak derivation and applying the Cauchy–Schwarz inequality, we have

$$\left| \int_c^d u(y)\varphi'(y) \, dy \right| = \left| \int_c^d u'(y)\varphi(y) \, dy \right| \leqslant \|u'\|_{L^2(c,d)} \|\varphi\|_{L^2(c,d)}.$$

This concludes the proof because $\|\varphi\|_{L^2(c,d)} \leqslant C\|\phi\|_{L^2(a,b)}$, as can be seen by using the change of variables formula for an integral and the assumption that the derivative of θ is bounded.

Solution 14 (Solution to Exercise 7.5) By definition of u_*, we have

$$\|u_*\|_{L^p(\mathbb{R}^n)}^p = \int_{x_1>0} |u(x_1,\ldots,x_n)|^p \, dx + \int_{x_1<0} |u(-x_1,\ldots,x_n)|^p \, dx$$

$$= 2\int_{x_1>0} |u(x)|^p \, dx = 2\|u\|_{L^p(\mathbb{R}_+^n)}^p.$$

So $\|u_*\|_{L^p} = 2^{1/p}\|u\|_{L^p}$. The preceding reasoning applies to $p \neq +\infty$ but the result is also valid for $p = +\infty$.

We will now show that u has partial derivatives in L^p, which are, if $i = 1$:

$$\partial_i u_*(x_1,\ldots,x_n) = \begin{cases} \partial_i u(x_1,x_2,\ldots,x_n) & \text{if } x_1 > 0, \\ -\partial_i u(-x_1,x_2,\ldots,x_n) & \text{if } x_1 < 0 \end{cases}$$

and, if $i \neq 1$:

$$\partial_i u_*(x_1,\ldots,x_n) = \begin{cases} \partial_i u(x_1,x_2,\ldots,x_n) & \text{if } x_1 > 0, \\ \partial_i u(-x_1,x_2,\ldots,x_n) & \text{if } x_1 < 0. \end{cases}$$

This will imply that, for all i, $\|\partial_i u_*\|_{L^p(\mathbb{R}^n)} = 2^{1/p}\|\partial_i u\|_{L^p(\mathbb{R}_+^n)}$.

Let $\phi \in C_0^\infty(\mathbb{R}^n)$. We will calculate $\int_{\mathbb{R}^n} u_* \partial_i \phi \, dx$. Let $\chi: \mathbb{R} \to \mathbb{R}$ be an even function, of class C^∞, which is 1 on $[-1,1]$ and 0 outside of $[-2,2]$. For all $k \in \mathbb{N}$, we set, for all $x \in \mathbb{R}^n$, $\chi_k(x) = \chi(2^k x_1)$. The function χ_k thus defined is of class C^∞ with support in $[-2^{1-k}, 2^{1-k}] \times \mathbb{R}^{n-1}$. We have

$$\int_{\mathbb{R}^n} u_* \partial_i \phi \, dx = \int_{\mathbb{R}^n} u_* \partial_i (\phi(1-\chi_k) + \phi\chi_k) \, dx$$

$$= \int_{\mathbb{R}^n} u_* \partial_i (\phi(1-\chi_k)) + \int_{\mathbb{R}^n} u_* (\partial_i \phi)\chi_k + \int_{\mathbb{R}^n} u_* \phi(\partial_i \chi_k) \, dx.$$

Since $|\int_{\mathbb{R}^n} u_*(\partial_i\phi)\chi_k| \leq \int_{|x_1|\leq 2^{1-k}} |u_*(\partial_i\phi)|$ and since $u_*(\partial_i\phi) \in L^1$, this term tends to 0 when k tends to $+\infty$.

For $i \neq 1$, $\partial_i \chi_k = 0$ so $\int_{\mathbb{R}^n} u_*\phi(\partial_i\chi_k) = 0$. For $i = 1$:

$$\int_{\mathbb{R}^n} u_*\phi(\partial_i\chi_k) = \int_{x_1>0} u(x_1,\ldots,x_n)\phi(x_1,\ldots,x_n)(\partial_1\chi_k)(x_1,\ldots,x_n) \, dx$$

$$+ \int_{x_1<0} u(-x_1,\ldots,x_n)\phi(x_1,\ldots,x_n)(-\partial_1\chi_k)(-x_1,\ldots,x_n) \, dx$$

$$= \int_{x_1>0} u(x_1,\ldots,x_n)\partial_1(\chi_k)(x_1,\ldots,x_n)(\phi(x_1,\ldots,x_n) - \phi(-x_1,\ldots,x_n)) \, dx.$$

So

$$\left| \int_{\mathbb{R}^n} u_* \phi (\partial_i \chi_k) \right|$$

$$\leqslant 2\|\partial_1 \phi\|_{L^\infty} \int_{x_1 > 0} |u(x_1, \ldots, x_n)| \, |\partial_1 (\chi_k)(x_1, \ldots, x_n)| \, |x_1| \, \mathbf{1}_{\mathrm{supp}(\phi)}(x) \, dx$$

$$\leqslant 2 \cdot 2^k \|\partial_1 \phi\|_{L^\infty} \|\chi'\|_{L^\infty} \int_{0 < x_1 < 2^{1-k}} |u(x_1, \ldots, x_n)| \, |x_1| \, \mathbf{1}_{\mathrm{supp}(\phi)}(x) \, dx$$

$$\leqslant 2 \cdot 2^k \|\partial_1 \phi\|_{L^\infty} \|\chi'\|_{L^\infty} \int_{0 < x_1 < 2^{1-k}} |u(x_1, \ldots, x_n)| \, 2^{1-k} \, \mathbf{1}_{\mathrm{supp}(\phi)}(x) \, dx$$

$$= 4\|\partial_1 \phi\|_{L^\infty} \|\chi'\|_{L^\infty} \int_{0 < x_1 < 2^{1-k}} |u(x_1, \ldots, x_n)| \, \mathbf{1}_{\mathrm{supp}(\phi)}(x) \, dx \xrightarrow[k \to +\infty]{} 0.$$

We have shown that, for all i, $\int_{\mathbb{R}^n} u_* \partial_i \phi \, dx = \int_{\mathbb{R}^n} u_* \partial_i (\phi(1 - \chi_k)) \, dx + o(1)$. Furthermore:

$$\int_{\mathbb{R}^n} u_* \partial_i (\phi(1 - \chi_k)) = \int_{x_1 > 0} u(x_1, \ldots, x_n) \partial_i (\phi(1 - \chi_k))(x_1, \ldots, x_n) \, dx$$

$$+ \int_{x_1 < 0} u(-x_1, \ldots, x_n) \partial_i (\phi(1 - \chi_k))(x_1, \ldots, x_n) \, dx$$

$$= \int_{x_1 > 0} u(x_1, \ldots, x_n) \partial_i (\phi(1 - \chi_k))(x_1, \ldots, x_n) \, dx$$

$$+ \int_{x_1 > 0} u(x_1, \ldots, x_n) \partial_i (g_k)(x_1, \ldots, x_n) \, dx,$$

where

$$g_k(x) = \begin{cases} (\phi(1 - \chi_k))(-x_1, \ldots, x_n) & \text{if } i \neq 1, \\ -(\phi(1 - \chi_k))(-x_1, \ldots, x_n) & \text{if } i = 1. \end{cases}$$

Now, since $\phi(1 - \chi_k)$ and g_k are C^∞ and vanishes for $|x_1| \leqslant 2^{-k}$:

$$-\int_{\mathbb{R}^n} u_* \partial_i (\phi(1 - \chi_k)) = \int_{x_1 > 0} \partial_i u(x_1, \ldots, x_n) \phi(1 - \chi_k)(x_1, \ldots, x_n) \, dx$$

$$+ \int_{x_1 > 0} \partial_i u(x_1, \ldots, x_n) g_k(-x_1, \ldots, x_n) \, dx$$

$$= \int_{x_1 > 0} \partial_i u(x_1, \ldots, x_n) \phi(1 - \chi_k)(x_1, \ldots, x_n) \, dx$$

$$+ \int_{x_1 < 0} \varepsilon_i \partial_i u(-x_1, \ldots, x_n) \phi(1 - \chi_k)(x_1, \ldots, x_n) \, dx$$

with $\varepsilon_i = 1$ if $i \neq 1$ and $\varepsilon_i = -1$ if $i = 1$.

By defining $\partial_i u_*$ as announced above, we thus have:

$$-\int_{\mathbb{R}^n} u_* \partial_i (\phi(1 - \chi_k)) = \int_{\mathbb{R}^n} \partial_i u_* \phi - \int_{\mathbb{R}^n} \partial_i u_* \phi \chi_k = \int_{\mathbb{R}^n} \partial_i u_* \phi + o(1).$$

By combining the above, we obtain, when k tends to $+\infty$, $\int_{\mathbb{R}^n} u_* \partial_i \phi = \int_{\mathbb{R}^n} \partial_i u_* \phi + o(1)$. So $\int_{\mathbb{R}^n} u_* \partial_i \phi = -\int_{\mathbb{R}^n} \partial_i u_* \phi$. This completes the proof.

Solution 15 (Solution to Exercise 7.6)

1. If $f \in W^{s,2}(\mathbb{R}^n)$ (with Definition 7.12), then $f \in L^2(\mathbb{R}^n)$, so $\widehat{f} \in L^2(\mathbb{R}^n)$ according to Proposition 5.23. Moreover, for any index j such that $1 \leqslant j \leqslant n$, $\partial_j^s f \in L^2(\mathbb{R}^n)$ so $\widehat{\partial_j^s f} = (i\xi_j)^s \widehat{f}$ belongs to $L^2(\mathbb{R}^n)$. By summing over j, we verify that $|\xi|^s \widehat{f}$ belongs to $L^2(\mathbb{R}^n)$ (where $|\xi|$ denotes the usual Euclidean norm, by equivalence of norms in finite dimension).
   Therefore, $(1 + |\xi|^2)^{s/2} \widehat{f}$ belongs to $L^2(\mathbb{R}^n)$ and $f \in H^s(\mathbb{R}^n)$.

   Conversely, if $f \in H^s(\mathbb{R}^n)$, then $f \in L^2(\mathbb{R}^n)$ and, for every multi-index α such that $|\alpha| \leqslant s$, $(i\xi)^\alpha \widehat{f}$ belongs to $L^2(\mathbb{R}^n)$. This function is the Fourier transform of $\partial_x^\alpha f$ (in the sense of tempered distributions), so this means that $\partial_x^\alpha f \in L^2(\mathbb{R}^n)$. Therefore, $f \in W^{s,2}(\mathbb{R}^n)$.

   Let us now show the equivalence of the two norms. For all $f \in W^{s,2}(\mathbb{R}^n)$ (with Definition 7.12), we have:

$$\|f\|_{W^{s,2}}^2 = \sum_{|\alpha| \leqslant k} \|\partial_x^\alpha f\|_{L^2}^2 = \frac{1}{(2\pi)^n} \sum_{|\alpha| \leqslant s} \|\widehat{\partial_x^\alpha f}\|_{L^2}^2 = \frac{1}{(2\pi)^n} \sum_{|\alpha| \leqslant s} \||\xi|^\alpha \widehat{f}(\xi)\|_{L^2}^2$$

$$\leqslant \frac{1}{(2\pi)^n} \sum_{|\alpha| \leqslant s} \|(1 + |\xi|^2)^{s/2} \widehat{f}(\xi)\|_{L^2}^2 = C\|f\|_{H^s}^2,$$

for $C = (1/(2\pi)^n) \operatorname{Card}\{\alpha : |\alpha| \leqslant s\}$.
We also have, for $s \geqslant 1$,

$$\|f\|_{W^{s,2}}^2 \geqslant \|f\|_{L^2}^2 + \sum_{j \leqslant n} \|\partial_j^s f\|_{L^2}^2 = \frac{1}{(2\pi)^n} \left(\|\widehat{f}\|_{L^2}^2 + \sum_{j \leqslant n} \||\xi_j|^s \widehat{f}\|_{L^2}^2 \right)$$

$$\geqslant \frac{C}{(2\pi)^n} \|(1 + |\xi|^2)^{s/2} \widehat{f}\|_{L^2}^2 = \frac{C}{(2\pi)^n} \|f\|_{H^s}^2,$$

for a certain constant $C > 0$.

2. We first notice:

$$
\iint_{\mathbb{R}^n \times \mathbb{R}^n} \frac{|f(x) - f(y)|^2}{|x - y|^{n+2s}} \, dx \, dy \iint_{\mathbb{R}^n \times \mathbb{R}^n} \frac{|f(x) - f(x+z)|^2}{|z|^{n+2s}} \, dx \, dz
$$

$$
= \frac{1}{(2\pi)^n} \int_{\mathbb{R}^n} \frac{\| f - f(\,.\,+z) \|_{L^2}^2}{|z|^{n+2s}} \, dz
$$

$$
= \frac{1}{(2\pi)^n} \iint_{\mathbb{R}^n \times \mathbb{R}^n} |\widehat{f}(\omega)|^2 \frac{|1 - e^{i\omega \cdot z}|^2}{|z|^{n+2s}} \, d\omega \, dz.
$$

If we set

$$
G(\omega) = \int_{\mathbb{R}^n} \frac{|1 - e^{i\omega \cdot z}|^2}{|z|^{n+2s}} \, dz,
$$

we verify that this function is well defined and (by change of variables) that it satisfies the following properties:

$$
\forall \omega \in \mathbb{R}^n,\ R \in O_n(\mathbb{R}), \quad G(R\omega) = G(\omega),
$$

$$
\forall \omega \in \mathbb{R}^n,\ a > 0, \quad G(a\omega) = a^{2s} G(\omega).
$$

Therefore, there exists a constant $C_s > 0$ such that, for all $\omega \in \mathbb{R}^n$, $G(\omega) = C_s |\omega|^{2s}$. It follows that

$$
\iint_{\mathbb{R}^n \times \mathbb{R}^n} \frac{|f(x) - f(y)|^2}{|x - y|^{n+2s}} \, dx \, dy = \frac{C_s}{(2\pi)^n} \int_{\mathbb{R}^n} |\omega|^{2s} |\widehat{f}(\omega)|^2 \, d\omega,
$$

which proves that the two norms are equivalent.

Solution 16 (Solution to Exercise 9.1)

1. We use elementary changes of variables to write

$$
\mathrm{Op}_h(a)u(x) = \frac{1}{(2\pi)^n} \iint e^{i(x-y) \cdot \xi} a(x, h\xi) u(y) \, dy \, d\xi
$$

$$
= \frac{1}{(2\pi h)^n} \iint e^{\frac{i}{h}(x-y) \cdot \xi} a(x, \xi) u(y) \, dy \, d\xi
$$

$$
= \frac{1}{(2\pi)^n} \iint e^{i(x'-y') \cdot \xi'} a(h^{\frac{1}{2}}x', h^{\frac{1}{2}}\xi') u(h^{\frac{1}{2}}y') \, dy' \, d\xi'.
$$

From this, we deduce that

$$
\mathrm{Op}_h(a)u(x) = (\mathrm{Op}(a_h)u_h)(h^{-\frac{1}{2}}x),
$$

where

$$
a_h(x, \xi) = a(h^{\frac{1}{2}}x, h^{\frac{1}{2}}\xi),\ u_h(y) = u(h^{\frac{1}{2}}y).
$$

2. Theorem 9.8 implies that

$$
\| \mathrm{Op}_h(a)u \|_{L^2} = h^{\frac{n}{4}} \| \mathrm{Op}(a_h)u_h \|_{L^2} \leqslant C h^{\frac{n}{4}} N(a_h) \| u_h \|_{L^2},
$$

where

$$N(a_h) = \sup_{|\alpha|+|\beta|\leqslant M} \sup_{(x,\xi)\in\mathbb{R}^{2n}} \left|\partial_x^\alpha \partial_\xi a_h\right|$$

for a certain sufficiently large M. We conclude the proof by noting that $h^{\frac{n}{4}} \|u_h\|_{L^2}$ is equal to $\|u\|_{L^2}$.

Solution 17 (Solution to Exercise 10.1)

1. Let $\chi \in C_0^\infty(\mathbb{R}^n \times \mathbb{R}^n)$ be a C^∞ function with compact support, such that $\chi(0,0) = 1$.

   From the definition of oscillatory integrals:

$$\int e^{-iy\cdot x} x_j a(x,y)\, dx\, dy = \lim_{\varepsilon\to 0} \int e^{-iy\cdot x} x_j a(x,y)\chi(\varepsilon x, \varepsilon y)\, dx\, dy$$

$$= \lim_{\varepsilon\to 0} i \int \partial_{y_j}[e^{-iy\cdot x}] a(x,y)\chi(\varepsilon x, \varepsilon y)\, dx\, dy$$

$$= \lim_{\varepsilon\to 0} -i \int e^{-iy\cdot x} \partial_{y_j}[a(x,y)\chi(\varepsilon x, \varepsilon y)]\, dx\, dy$$

$$= -i \lim_{\varepsilon\to 0} \int e^{-iy\cdot x} \partial_{y_j} a(x,y)\chi(\varepsilon x, \varepsilon y)\, dx\, dy$$

$$\quad - i \lim_{\varepsilon\to 0} \varepsilon \int e^{-iy\cdot x} a(x,y)\partial_{y_j}\chi(\varepsilon x, \varepsilon y)\, dx\, dy.$$

The first limit is $-i\int e^{-iy\cdot x}\partial_{y_j} a(x,y)\, dx\, dy$ (by definition of this oscillatory integral). For the second term, again by the definition of oscillatory integrals:

$$- i \lim_{\varepsilon\to 0} \varepsilon \int e^{-iy\cdot x} a(x,y)\partial_{y_j}\chi(\varepsilon x, \varepsilon y)\, dx\, dy$$

$$= -i \lim_{\varepsilon\to 0} \varepsilon\, \partial_{y_j}\chi(0,0) \int e^{-iy\cdot x} a(x,y)\, dx\, dy = 0,$$

which gives the desired result.

2. We only show

$$\frac{1}{(2\pi)^n} \int e^{-iy\cdot x} a(x)\, dy\, dx = a(0).$$

By symmetry between the variables y and x, this is sufficient.

When m is sufficiently negative, a is in L^1 and $\widehat{a}$ is also in L^1. We then have:

$$\frac{1}{(2\pi)^n} \int e^{-iy\cdot x} a(x)\, dy\, dx = \frac{1}{(2\pi)^n} \int \widehat{a}(y)\, dy = a(0).$$

To conclude, it is therefore sufficient to show that if the equality is true for $a \in A^m$, then it is true for $a \in A^{m+1}$. Suppose then that it is true on $a \in A^m$ and suppose fixed $a \in A^{m+1}$. Let us set

$$b(x) = a(x)(1 + |x|^2)^{-1}.$$

We have $b \in A^{m-1} \subset A^m$. Using question (1):

$$\int e^{-iy \cdot x} a(x) \, dy \, dx = \int e^{-iy \cdot x} b(x) \, dy \, dx + \sum_j \int e^{-iy \cdot x} x_j^2 b(x) \, dy \, dx$$

$$= b(0) - i \sum_j \int e^{-iy \cdot x} \partial_{y_j} \left[x_j b(x) \right] \, dy \, dx$$

$$= b(0) = a(0).$$

3. When $\beta = 0$, it is a consequence of question (2), applied to $a(y) = y^\alpha/(\alpha!)$. Let us now proceed by induction on $|\beta|$, using question (1). Let

$$\delta_j = (0, \ldots, 0, 1, 0, \ldots, 0)$$

(with the 1 in position j).

$$\frac{1}{(2\pi)^n} \int e^{-iy \cdot x} \frac{y^\alpha x^{\beta+\delta_j}}{\alpha!(\beta+\delta_j)!} \, dy \, dx = \frac{1}{(2\pi)^n} \int e^{-iy \cdot x} x_j \frac{y^\alpha x^\beta}{\alpha!(\beta+\delta_j)!} \, dy \, dx$$

$$= \frac{-i}{(2\pi)^n} \int e^{-iy \cdot x} \partial_{y_j} \left[\frac{y^\alpha x^\beta}{\alpha!(\beta+\delta_j)!} \right] \, dy \, dx$$

$$= \frac{-i}{(2\pi)^n} \int e^{-iy \cdot x} \alpha_j \frac{y^{\alpha-\delta_j} x^\beta}{\alpha!(\beta+\delta_j)!} \, dy \, dx$$

$$= 1_{\alpha_j>0} \frac{-i}{(2\pi)^n} \int e^{-iy \cdot x} \frac{y^{\alpha-\delta_j} x^\beta}{(\alpha-\delta_j)!(\beta+\delta_j)!} \, dy \, dx$$

$$= 1_{\alpha_j>0} \frac{(-i)^{|\alpha-\delta_j|+1}}{(\beta+\delta_j)!} 1_{\beta=\alpha-\delta_j} = \frac{(-i)^{|\alpha|}}{\alpha!} 1_{\beta+\delta_j=\alpha},$$

which is the desired result.

Solution 18 (Solution to Exercise 10.2)

1. Let $M > 0$ such that χ vanishes outside the ball $B(0, M)$ and let $M' > 0$ such that $\chi = 1$ on $B(0, M')$.
Let us fix $j \in \mathbb{N}$. For all multi-indices α, β in $\mathbb{N}^n$, we have:

$$\partial_x^\alpha \partial_\xi^\beta \tilde{a}_j = \partial_\xi^\beta \left[(1 - \chi(\varepsilon_j \xi)) \partial_x^\alpha a_j \right]$$

$$= \sum_{\gamma \leqslant \beta} n_\gamma \partial_\xi^\gamma [1 - \chi(\varepsilon_j \xi)] (\partial_\xi^{\beta-\gamma} \partial_x^\alpha a_j),$$

where the n_γ are integers.
For every multi-index $\gamma \neq 0$, we have $\partial_\xi^\gamma [1 - \chi(\varepsilon_j \xi)] = -\varepsilon_j^{|\gamma|} (\partial^\gamma \chi)(\varepsilon_j \xi)$, which implies, for $\varepsilon_j < M$:

$$\left|\partial_\xi^\gamma[1 - \chi(\varepsilon_j\xi)]\right| \leqslant \varepsilon_j^{|\gamma|} 1_{B(0,M/\varepsilon_j)}(\xi) \|\partial^\gamma \chi\|_{L^\infty}$$

$$\leqslant \varepsilon_j^{|\gamma|} \|\partial^\gamma \chi\|_{L^\infty} \left(\frac{1 + |\xi|}{1 + M/\varepsilon_j}\right)^{-|\gamma|+1}$$

$$\leqslant \varepsilon_j^{|\gamma|}(2M)^{|\gamma|-1} \|\partial^\gamma \chi\|_{L^\infty} (1 + |\xi|)^{-|\gamma|+1} .$$

On the other hand, we have

$$|1 - \chi(\varepsilon_j\xi)| \leqslant (1 + \|\chi\|_{L^\infty}) 1_{\mathbb{R}^n - B(0,M'/\varepsilon_j)}(\xi)$$

$$\leqslant (1 + \|\chi\|_{L^\infty})(1 + |\xi|)\left(\frac{\varepsilon_j}{M'}\right).$$

Therefore, using the fact that, for all γ and β, there exists a constant D such that $|\partial_\xi^{\beta-\gamma}\partial_x^\alpha a_j| \leqslant D(1 + |\xi|)^{m_j-|\beta|+|\gamma|}$, we obtain:

$$|\partial_x^\alpha \partial_\xi^\beta \tilde{a}_j|$$

$$\leqslant n_0 |1 - \chi(\varepsilon_j\xi)| (\partial_\xi^\beta \partial_x^\alpha a_j) + \varepsilon_j \sum_{0 \neq \gamma \leqslant \beta} n_\gamma C_\gamma (1 + |\xi|)^{-|\gamma|+1} |\partial_\xi^{\beta-\gamma}\partial_x^\alpha a_j|$$

$$\leqslant \varepsilon_j C (1 + |\xi|)^{m_j-|\beta|+1},$$

where C is a constant independent of ε_j (but dependent on α and β).
By taking ε_j small enough, we verify that if $|\alpha| \leqslant j$ and if $|\beta| \leqslant j$, then:

$$|\partial_x^\alpha \partial_\xi^\beta \tilde{a}_j| \leqslant \frac{1}{2^j}(1 + |\xi|)^{1+m_j-|\beta|}.$$

If, for all j, we choose ε_j in this way, then, for all α, β, the desired property is indeed satisfied as soon as $j \geqslant \max(|\alpha|, |\beta|)$.

2. According to the first question, for all α, β, the series $\sum_{j \in \mathbb{N}} \partial_x^\alpha \partial_\xi^\beta \tilde{a}_j$ converges uniformly on any compact set.

3. Let us first show that, for all k, $\sum_{j \geqslant k} \tilde{a}_j \in S^{m_k}$.
For all $j \geqslant k$, $\tilde{a}_j$ and a_j coincide outside a certain compact set therefore, since a_j is in S^{m_j}, so is $\tilde{a}_j$ (verification left to the reader). In particular, as $m_j \leqslant m_k$, this implies $a_j \in S^{m_k}$.
Consider two arbitrary multi-indices α, β. Let J be large enough so that the property of question (1) is satisfied if $j \geqslant J$ (for the fixed α and β). The function $\sum_{k \leqslant j < J} \tilde{a}_j$ is a symbol of S^{m_k} therefore $\sum_{k \leqslant j < J} \partial_x^\alpha \partial_\xi^\beta \tilde{a}_j$ decreases at worst as $(1 + |\xi|)^{m_k-|\beta|}$. Let us show that we obtain the same decrease for $\sum_{j \geqslant J} \partial_x^\alpha \partial_\xi^\beta \tilde{a}_j$. To see this, note that

$$\left| \sum_{j \geqslant J} \partial_x^\alpha \partial_\xi^\beta \tilde{a}_j \right| \leqslant \sum_{j \geqslant J} \frac{1}{2^j} (1 + |\xi|)^{1+m_j - |\beta|}$$

$$\leqslant \sum_{j \geqslant J} \frac{1}{2^j} (1 + |\xi|)^{1+m_{k+1} - |\beta|}$$

$$= \frac{1}{2^{J-1}} (1 + |\xi|)^{1+m_{k+1} - |\beta|}$$

$$\leqslant \frac{1}{2^{J-1}} (1 + |\xi|)^{m_k - |\beta|}.$$

The two previous remarks imply that, for a large enough constant C:

$$\left| \sum_{j \geqslant k} \partial_x^\alpha \partial_\xi^\beta \tilde{a}_j \right| \leqslant C(1 + |\xi|)^{m_k - |\beta|}.$$

As this is true for any α and β, we deduce that $\sum_{j \geqslant k} \tilde{a}_j$ is an element of S^{m_k}.

We conclude by noting that $\sum_{j < k} (\tilde{a}_j - a_j)$ is a symbol of S^{m_0} whose support is compact in ξ. Therefore, it is a symbol of $S^{-\infty}$. Thus, $a - \sum_{j < k} a_j$ is the sum of a symbol of $S^{-\infty}$ and a symbol of S^{m_k}. It is then an element of S^{m_k}.

Solution 19 (Solution to Exercise 10.3)

1. We find
$$K(x, y) = \frac{1}{(2\pi)^n} \int_{\mathbb{R}^n} a(x, \xi) e^{i(x-y)\cdot\xi} \, d\xi.$$

By integrating by parts, we can make sense of this integral for all (x, y) such that $x \neq y$.

2. For all multi-indices α and β, the function $(x, y, \xi) \to a(x, \xi) e^{i(x-y)\cdot\xi}$ is (α, β)-times differentiable with respect to (x, y). Moreover, the (α, β)-th derivative is a linear combination of terms of the form:

$$(\partial_x^\gamma a(x, \xi)) \xi^{(\alpha-\gamma)+\beta} e^{i(x-y)\cdot\xi}$$

with $\gamma \leqslant \alpha$. Since $a \in S^{-\infty}$, such a linear combination is bounded by $C(1 + |\xi|)^{-(n+1)}$ for some constant C, which is an integrable function in ξ. We can therefore differentiate under the sum sign and K is of class C^∞.

3. By symbolic calculation, we have $\mathrm{Op}(a) M_\psi = \mathrm{Op}(a) \, \mathrm{Op}(\psi) = \mathrm{Op}(c)$ with:

$$c \sim \sum_\alpha (\partial_\xi^\alpha a)(\partial_x^\alpha \psi).$$

Each term of this asymptotic expansion is supported in $\mathrm{supp}(\psi) \times \mathbb{R}^n$. We deduce that there exists $b \in S^m$ and $R \in \mathrm{Op}(S^{-\infty})$ such that $b(x, \xi) = 0$ for all (x, ξ) such that $x \notin \mathrm{supp}(\psi)$ and:
$$\mathrm{Op}(a) M_\psi = \mathrm{Op}(b) + R.$$

We also have $M_\phi \, \mathrm{Op}(b) = \mathrm{Op}(c)$ with:

$$c \sim \sum_\alpha \frac{1}{i^{|\alpha|}\alpha!} (\partial_\xi^\alpha \phi)(\partial_x^\alpha b).$$

For all α, $(\partial_\xi^\alpha \phi)(\partial_x^\alpha b) = 0$, since, for all ξ, supp $(\partial_x^\alpha b(.,\xi)) \subset$ supp(ψ) and supp $(\partial_\xi^\alpha \phi(.,\xi)) \subset$ supp(ϕ) and the two supports are disjoint.

This implies $c \in S^{-\infty}$. Therefore $M_\phi \operatorname{Op}(b) \in \operatorname{Op}(S^{-\infty})$ and $M_\phi \operatorname{Op}(a) M_\psi = M_\phi \operatorname{Op}(b) + M_\phi R \in \operatorname{Op}(S^{-\infty})$.

4. We denote by

$$K(x,y) = \frac{1}{(2\pi)^n} \int_{\mathbb{R}^n} a(x,\xi) e^{i(x-y)\cdot\xi} \, d\xi$$

the kernel of $\operatorname{Op}(a)$ and K' the kernel of $M_\phi \operatorname{Op}(a) M_\psi$. Intuitively, for a function $u \in S(\mathbb{R}^n)$:

$$\int_{\mathbb{R}^n} K'(x,y) u(y) \, dy = M_\phi \operatorname{Op}(a) M_\psi u(x) = \phi(x) \operatorname{Op}(a)(\psi u)(x)$$

$$= \phi(x) \int_{\mathbb{R}^n} K(x,y) \psi(y) u(y) \, dy$$

$$= \int_{\mathbb{R}^n} (\phi(x) K(x,y) \psi(y)) \, u(y) \, dy,$$

so we expect to have $K'(x,y) = \phi(x) K(x,y) \psi(y)$. Let us now rigorously justify this result.

Denoting by $b \in S^m$ the symbol such that $\operatorname{Op}(a) M_\psi = \operatorname{Op}(b)$ and using the fact that, for all functions f_1, f_2, $\widehat{f_1 f_2} = (2\pi)^{-n} \widehat{f_1} * \widehat{f_2}$, we have, if u is of Schwartz:

$$\frac{1}{(2\pi)^n} \int b(x,\xi) e^{ix\cdot\xi} \widehat{u}(\xi) \, d\xi = \operatorname{Op}(a) M_\psi u(x)$$

$$= \frac{1}{(2\pi)^n} \int a(x,\xi') e^{ix\cdot\xi'} \widehat{\psi u}(\xi') \, d\xi'$$

$$= \frac{1}{(2\pi)^{2n}} \int a(x,\xi') e^{ix\cdot(\xi'-\xi)} e^{ix\cdot\xi} \widehat{\psi}(\xi'-\xi) \widehat{u}(\xi) \, d\xi' \, d\xi.$$

Consequently,

$$b(x,\xi) = \frac{1}{(2\pi)^n} a(x,.) * (e^{-ix\cdot} \widehat{\psi}(-.))(\xi).$$

Let K'' be the symbol of b. Then we have $K''(x,y) = \mathcal{F}^{-1}(b(x,.))(x-y)$ and therefore:

$$K''(x,y) = \mathcal{F}^{-1}(a(x,.))(x-y) \mathcal{F}^{-1}(e^{-ix\cdot} \widehat{\psi}(-.))(x-y)$$

$$= K(x,y) \psi(y).$$

A similar argument (but much simpler) shows that we also have $K'(x, y) = \phi(x)K''(x, y)$, which indeed proves:

$$K'(x, y) = \phi(x)K(x, y)\psi(y).$$

5. According to question (3), K' is the kernel of a pseudo-differential operator of order $-\infty$. According to question (2), it is a C^∞ function. Since K is equal to K' in the vicinity of (x, y), K is of class C^∞ in the vicinity of (x, y).

Solution 20 (Solution to Exercise 10.4)

1. (a) Let $u \in C_0^\infty(\Omega)$. As it is a function of the Schwartz space, $\phi \operatorname{Op}_\Omega(a)u = \operatorname{Op}(\phi a)u$ is a function of the Schwartz space for all $\phi \in C_0^\infty(\Omega)$. Therefore, $\operatorname{Op}_\Omega(a)u$ is of class C^∞ on Ω.
   (b) We have

$$\operatorname{Op}(\phi\tilde{\phi}a)^*v = (M_\phi \operatorname{Op}(\tilde{\phi}a))^*v = (\operatorname{Op}(\tilde{\phi}a))^* M_\phi^* v$$
$$= (\operatorname{Op}(\tilde{\phi}a))^*(\bar{\phi}v) = (\operatorname{Op}(\tilde{\phi}a))^*v.$$

Similarly, $(\operatorname{Op}(\phi\tilde{\phi}a))^*v = (M_{\tilde{\phi}} \operatorname{Op}(\phi a))^*v = (\operatorname{Op}(\phi a))^*v$. This leads to the requested equality.
   (c) Let $u \in S'(\Omega)$. We define $\operatorname{Op}_\Omega(a)u \in D'(\Omega)$ by:

$$\forall v \in C_0(\Omega) \quad \langle \operatorname{Op}_\Omega(a)u, v \rangle = \langle u, (\operatorname{Op}(\phi a))^*v \rangle,$$

where ϕ is a function in $C_0^\infty(\Omega)$ equal to 1 on the support of v. This definition does not depend on the choice of ϕ, according to the previous question. The continuity properties of $(\operatorname{Op}(\phi a))^*$ ensure that this indeed defines a distribution.
   Moreover, we verify that this definition coincides with the previous one for $u \in C_0(\Omega)$.
2. (a) Let $u \in C_0^\infty(\Omega)$.
   Let K be the (finite) set of indices k such that $\operatorname{supp}(u) \cap \operatorname{supp}(\psi_k) \neq \emptyset$. We have $u = \sum_{k \in K} \psi_k u$ and therefore $Au = \sum_{k \in K} AM_{\psi_k}u$.
   Let $x \in \Omega$ be fixed. Let J be the (finite) set of indices j such that $x \in \operatorname{supp}(\psi_j)$. Then:

$$Au(x) = \sum_{j \in J} \psi_j(x)Au(x) = \sum_{j \in J, k \in K} (M_{\psi_j}AM_{\psi_k})u(x) = \sum_{j \in J, k \in K} A_{jk}u(x).$$

If $j \notin J$ or $k \notin K$, $A_{jk}u(x) = 0$, so:

$$Au(x) = \sum_{j, k \in \mathbb{N}} A_{jk}u(x).$$

   (b) By hypothesis, for all j, k, A_{jk} is a pseudo-differential operator of order m, meaning there exists a symbol $a_{jk} \in S^m$ such that $A_{jk} = \operatorname{Op}(a_{jk})$.
   For all $x \notin \operatorname{supp}(\psi_j)$, $a_{jk}(x, .) = 0$ (since $A_{jk}u(x) = 0$ for all $u \in S(\mathbb{R}^n)$).

If we set $a = \sum_{(j,k)\in I} a_{jk}$, the operator a is well defined since at each point, the sum is finite. Moreover, for all $\phi \in C_0^\infty(\Omega)$, $\phi a_{jk} = 0$ for all except finitely many $(j,k) \in I$. Therefore, ϕa is a finite sum of symbols of order m, which implies that ϕa is a symbol of order m. So $a \in S_{\mathrm{loc}}^m(\Omega)$. We now need to verify that with this definition, we indeed have $\sum_{(j,k)\in I} A_{jk} = \mathrm{Op}_\Omega(a)$. For all $u \in \mathcal{E}(\mathbb{R}^n)$ and for all $\phi \in C_0^\infty(\Omega)$:

$$\phi \sum_{(j,k)\in I} A_{jk}u = \sum_{\substack{(j,k)\in I \\ \mathrm{supp}(\psi_j)\cap\mathrm{supp}(\phi)\neq\varnothing}} \phi A_{jk}u = \sum_{\substack{(j,k)\in I \\ \mathrm{supp}(\psi_j)\cap\mathrm{supp}(\phi)\neq\varnothing}} \phi\,\mathrm{Op}(a_{jk})u$$

$$= \mathrm{Op}\left(\phi \sum_{\substack{(j,k)\in I \\ \mathrm{supp}(\psi_j)\cap\mathrm{supp}(\phi)\neq\varnothing}} a_{jk}\right)u$$

$$= \mathrm{Op}(\phi a)u = \phi\,\mathrm{Op}_\Omega(a)u.$$

Since this is true for all functions ϕ, $\sum_{(j,k)\in I} A_{jk}u = \mathrm{Op}_\Omega(a)u$.

(c) Similarly to question (3) of Exercise 10.3, $M_{\psi_j}AM_{\psi_k} \in \mathrm{Op}(S^{-\infty})$ if $\mathrm{supp}(\psi_j)\cap \mathrm{supp}(\psi_k) = \varnothing$. According to question (2) of Exercise 10.3, it is therefore a C^∞ kernel operator.

Let us denote the kernel by K and show that it is supported within $\mathrm{supp}(\psi_j) \times \mathrm{supp}(\psi_k)$. For all $x \notin \mathrm{supp}(\psi_j)$, for all $u \in \mathcal{S}(\mathbb{R}^n)$, $A_{jk}u(x) = 0$ so:

$$\int_{\mathbb{R}^n} K(x,y)u(y)\,\mathrm{d}y = 0.$$

This is equivalent to $K(x,y) = 0$ for all $y \in \mathbb{R}^n$. Therefore, $\mathrm{supp}(K) \subset \mathrm{supp}(\psi_j) \times \mathbb{R}^n$.

Let us now show that $K(x,y) = 0$ for all x, y such that $y \notin \mathrm{supp}(\psi_k)$. Assume, by contradiction, that this is not the case: $K(x,y) \neq 0$ for a certain $(x,y) \in \mathbb{R}^n \times \mathbb{R}^n$ such that $y \notin \mathrm{supp}(\psi_k)$. Then there exists a function u, of class C^∞, supported within an arbitrarily small neighborhood of y, such that:

$$\int_{\mathbb{R}^n} K(x,y)u(y)\,\mathrm{d}y \neq 0.$$

We then have $A_{jk}u(x) \neq 0$. But, if the support of u is small enough, $\mathrm{supp}\,u \cap \mathrm{supp}(\psi_k) = \varnothing$ and $A_{jk}u = M_{\psi_j}A(\psi_k u) = M_{\psi_j}A(0) = 0$. This is a contradiction.

(d) For all $(j,k) \notin I$, let us denote the kernel of A_{jk} by K_{jk}. Since $\mathrm{supp}(K_{jk}) \subset \mathrm{supp}(\psi_j) \times \mathrm{supp}(\psi_k)$, the sum $K = \sum_{(j,k)\notin I} K_{jk}$ is well defined (each point admits a neighborhood on which only a finite number of terms are non-zero); it is of class C^∞.

We also verify that, for any function $u \in C_0^\infty(\Omega)$,

$$\left(\sum_{(j,k)\notin I} A_{jk}\right)u = \int_{\mathbb{R}^n} K(.,y)u(y)\,\mathrm{d}y.$$

This implies that $\sum_{(j,k)\notin I} A_{jk}$ is an operator with a C^∞ kernel.

By setting $R = \sum_{(j,k)\notin I} A_{jk}$ and defining a as in question (b), we have, according to question (a):

$$A = \sum_{(j,k)\in I} A_{jk} + \sum_{(j,k)\notin I} A_{jk} = \mathrm{Op}_\Omega(a) + R.$$

Solution 21 (Solution to Exercise 10.5)

1. We will show that

$$b(x, \xi) = \frac{1 - \chi(\xi)}{P(\xi)} u(x)$$

satisfies the following property: for all multi-indices α and β, there exists a certain constant C' such that:

$$\partial_\xi^\alpha \partial_x^\beta b(x, \xi) \leqslant C'(1 + |x| + |\xi|)^{-m}.$$

Note that

$$\partial_\xi^\alpha \partial_x^\beta b(x, \xi) = \partial_\xi^\alpha \left(\frac{1 - \chi(\xi)}{P(\xi)} \right) \partial_x^\beta u(x).$$

The first term of the product is bounded by $C(1 + |\xi|)^{-m-|\alpha|}$ for a certain constant $C > 0$ (due to the ellipticity condition). The second is bounded by $C_k(1 + |x|)^k$ for all k, since u has compact support. This leads to the result.

2. For all s and for all u with support in U:

$$T(u) = \frac{(-1)^s}{(2\pi)^n} \int (-1)^s (x_1^2 + \cdots + x_n^2)^s e^{ix\cdot\xi} \frac{u(x)}{(x_1^2 + \cdots + x_n^2)^s} \frac{1 - \chi(\xi)}{P(\xi)} \, dx \, d\xi$$

$$= \frac{(-1)^s}{(2\pi)^n} \int (\partial_{\xi_1}^2 + \cdots + \partial_{\xi_n}^2)^s \left[e^{ix\cdot\xi} \frac{u(x)}{(x_1^2 + \cdots + x_n^2)^s} \right] \frac{1 - \chi(\xi)}{P(\xi)} \, dx \, d\xi$$

$$= \frac{(-1)^s}{(2\pi)^n} \int e^{ix\cdot\xi} \frac{u(x)}{(x_1^2 + \cdots + x_n^2)^s} (\partial_{\xi_1}^2 + \cdots + \partial_{\xi_n}^2)^s \left[\frac{1 - \chi(\xi)}{P(\xi)} \right] dx \, d\xi.$$

The function $(1 - \chi)P^{-1}$ belongs to S^{-m} so, for all s large enough, $(\partial_{\xi_1}^2 + \cdots + \partial_{\xi_n}^2)^s \left[\frac{1-\chi(\xi)}{P(\xi)} \right]$ is integrable.

We then set, for all $x \in U$:

$$t(x) = \frac{(-1)^s}{(x_1^2 + \cdots + x_n^2)^s} \frac{1}{(2\pi)^n} \int e^{ix\cdot\xi} (\partial_{\xi_1}^2 + \cdots + \partial_{\xi_n}^2)^s \left[\frac{1 - \chi(\xi)}{P(\xi)} \right] d\xi.$$

The function t is bounded on U; therefore, it belongs to $L^2(U)$. Moreover, for all $u \in C_0^\infty(U)$:

$$T(u) = \int u(x) t(x) \, dx.$$

3. We have, for any function u:

$$\partial^\alpha T(u) < = \frac{(-1)^{|\alpha|}}{(2\pi)^n} \int e^{ix\cdot\xi} \frac{1-\chi(\xi)}{P(\xi)} \partial^\alpha u(x)\, dx\, d\xi$$

$$= \frac{i^{|\alpha|}}{(2\pi)^n} \int e^{ix\cdot\xi} \xi^\alpha \frac{1-\chi(\xi)}{P(\xi)} u(x)\, dx\, d\xi,$$

which identifies as a function of $L^2(U)$ for the same reason as before. On any bounded open set U not containing 0, T identifies as an L^2 function, which admits L^2 derivatives of any order; T therefore identifies as a C^∞ function. On $\mathbb{R}^n \setminus \{0\}$, T therefore identifies as a C^∞ function.

4. We write

$$P(D)Tu = T(P(-D)u)$$

$$= \frac{1}{(2\pi)^n} \int e^{ix\cdot\xi} \frac{1-\chi(\xi)}{P(\xi)} [P(-D)u](x)\, dx\, d\xi$$

$$= \frac{1}{(2\pi)^n} \int P(D_x) \left[e^{ix\cdot\xi} \frac{1-\chi(\xi)}{P(\xi)} \right] u(x)\, dx\, d\xi$$

$$= \frac{1}{(2\pi)^n} \int e^{ix\cdot\xi} P(\xi) \frac{1-\chi(\xi)}{P(\xi)} u(x)\, dx\, d\xi$$

$$= \frac{1}{(2\pi)^n} \int e^{ix\cdot\xi} u(x)\, dx\, d\xi - \frac{1}{(2\pi)^n} \int e^{ix\cdot\xi} \chi(\xi) u(x)\, dx\, d\xi$$

$$= u(0) - \frac{1}{(2\pi)^n} \int e^{ix\cdot\xi} \chi(\xi) u(x)\, dx\, d\xi.$$

Since the function $r: x \mapsto (2\pi)^{-n} \int e^{ix\cdot\xi} \chi(\xi)\, d\xi$ is C^∞ (the Fourier transform of a function with compact support), we deduce that $P(D)T = \delta_0 + r$.

5. Let $\varepsilon > 0$ be arbitrary. Let ϕ be a function with support in $B(0, \varepsilon)$, which is 1 in the neighborhood of 0. Let us define $T_\varepsilon : u \mapsto T(\phi u)$. This is a distribution with support in $B(0, \varepsilon)$.

Moreover, for all j, $\partial_j T_\varepsilon(u) = \partial_j T(\phi u) + T((\partial_j \phi)u)$. The distribution $u \mapsto T((\partial_j \phi)u)$ identifies as a function of $C^\infty(\mathbb{R}^n)$ because $\partial_j \phi$ vanishes on a neighborhood of 0.

We deduce that there exists a function $\tilde{r} \in C^\infty(\mathbb{R}^n)$ such that, for all u,

$$P(D)T_\varepsilon(u) = (P(D)T)(\phi u) + \int \tilde{r} u = u(0) + \int (\tilde{r} + \phi r) u.$$

So $P(D)T_\varepsilon = \delta_0 + (\tilde{r} + \phi r)$.

Solution 22 (Solution to Exercise 11.1)

1. We have already seen that if $a = a(x, \xi)$ and $b = b(\xi)$ (symbol independent of x) then $\mathrm{Op}(a) \circ \mathrm{Op}(b) = \mathrm{Op}(ab)$. We deduce that, for all $\varepsilon > 0$, $\mathrm{Op}(a)J_\varepsilon = \mathrm{Op}(a^\varepsilon)$ where $a_t^\varepsilon(x, \xi) = a_t(x, \xi)\chi(\varepsilon\xi)$.

2. For all t and all $\varepsilon > 0$, the symbol a_t^ε has compact support in ξ and therefore belongs to $S^{-\infty}$ and in particular to $S^0(\mathbb{R}^n)$. The continuity theorem of pseudo-differential operators of order 0 on Sobolev implies that $\mathrm{Op}(a_t^\varepsilon)$ is a continuous operator on $H^s(\mathbb{R}^n)$. Then the equation $\partial_t u + \mathrm{Op}(a^\varepsilon)u = f$ is an ordinary differential equation, for which the Cauchy–Lipschitz theorem applies (we use the version of this result given by Exercise 2.1).

3. Note that the symbol $a_t^\varepsilon(x,\xi) = a_t(x,\xi)\chi(\varepsilon\xi)$ is bounded in $S^1(\mathbb{R}^n)$ uniformly in ε, in the sense that $\{a_t^\varepsilon : \varepsilon \in (0,1], t \in [0,T]\}$ is a bounded subset of $S^1(\mathbb{R}^n)$, which means that,

$$\forall(\alpha,\beta) \in \mathbb{N}^n, \quad \sup_{\varepsilon\in(0,1]}\ \sup_{t\in[0,T]}\ \sup_{x,\xi}\ \langle\xi\rangle^{|\beta|-1}\left|\partial_x^\alpha\partial_\xi^\beta a_t^\varepsilon(x,\xi)\right| < +\infty.$$

Furthermore, $\mathrm{Re}\,a_t^\varepsilon$ is uniformly bounded in $S^0(\mathbb{R}^n)$. The desired inequality is therefore a consequence of (11.2).

By applying the previous inequality to $v = u_\varepsilon$, we obtain that there exists a constant C such that for all $\varepsilon > 0$ and all $t \in [0,T]$,

$$\|u_\varepsilon\|_{C^0([0,T];H^s)} \leqslant C\|u_0\|_{H^s} + C\int_0^T \|f(t)\|_{H^s}\, dt. \tag{19.4}$$

This implies that $(u_\varepsilon)_{\varepsilon\in(0,1]}$ is bounded in $C^0([0,T];H^s(\mathbb{R}^n))$. Using the equation, we verify that $(u_\varepsilon)_{\varepsilon\in(0,1]}$ is a bounded family in $C^1([0,T];H^{s-1}(\mathbb{R}^n))$.

4. Let ε and ε' be in $(0,1]$. Starting from

$$\partial_t u_\varepsilon + \mathrm{Op}(a)J_\varepsilon u_\varepsilon = f,$$
$$\partial_t u_{\varepsilon'} + \mathrm{Op}(a)J_{\varepsilon'} u_{\varepsilon'} = f,$$

we deduce that $v = u_\varepsilon - u_{\varepsilon'}$ satisfies

$$\partial_t v + \mathrm{Op}(a^\varepsilon)v = f_\varepsilon \quad \text{with} \quad f_\varepsilon = \mathrm{Op}(a)(J_{\varepsilon'} - J_\varepsilon)u_{\varepsilon'}.$$

Since u_ε and $u_{\varepsilon'}$ coincide for $t = 0$, we have $v(0) = 0$ and we can then use the inequality of the previous lemma to obtain that

$$\|v\|_{C^0([0,T];H^{s-2})} \leqslant C\int_0^T \|f_\varepsilon(t)\|_{H^{s-2}}\, dt.$$

However

$$\|f^\varepsilon(t)\|_{H^{s-2}} = \left\|\mathrm{Op}(a)(J_{\varepsilon'} - J_\varepsilon)u_{\varepsilon'}(t)\right\|_{H^{s-2}} \leqslant K\left\|(J_{\varepsilon'} - J_\varepsilon)u_{\varepsilon'}(t)\right\|_{H^{s-1}}.$$

By definition,

$$\left\|(J_{\varepsilon'} - J_\varepsilon)u_{\varepsilon'}(t)\right\|_{H^{s-1}}^2 = \frac{1}{(2\pi)^{2n}}\int \langle\xi\rangle^{2(s-2)}\,|\chi(\varepsilon\xi) - \chi(\varepsilon'\xi)|^2\,|\widehat{u}_{\varepsilon'}(t,\xi)|^2\,d\xi.$$

We use the elementary inequality $|\chi(\varepsilon\xi) - \chi(\varepsilon'\xi)| \leqslant K |\varepsilon - \varepsilon'| |\xi|$ to conclude that

$$\int_0^T \|f_\varepsilon(t)\|_{H^{s-2}} \, dt \leqslant K' |\varepsilon - \varepsilon'| \int_0^T \|u_{\varepsilon'}(t)\|_{H^s} \, dt.$$

Since $\|u_{\varepsilon'}\|_{C^0([0,T];H^s)}$ is uniformly bounded according to (19.4), we obtain

$$\|v\|_{C^0([0,T];H^{s-2})} = O(|\varepsilon - \varepsilon'|),$$

which is the desired result.

5. We have

$$\|u\|_{H^s}^2 = (2\pi)^{-n} \int \langle\xi\rangle^{2s} |\widehat{u}(\xi)|^2 \, d\xi$$

$$= (2\pi)^{-n} \int \langle\xi\rangle^{2\alpha s_1} |\widehat{u}(\xi)|^{2\alpha} \langle\xi\rangle^{2(1-\alpha)s_2} |\widehat{u}(\xi)|^{2(1-\alpha)} \, d\xi,$$

so that the desired inequality is a consequence of Hölder's inequality.

We have seen that the family $(u_\varepsilon)_{\varepsilon\in(0,1]}$ is bounded in $C^0([0,T];H^s)$ and that it is moreover Cauchy in $C^0([0,T];H^{s-2})$. Given $\sigma \in [s-2,s)$, the inequality (11.7), applied with $(s_1, s_2, \sigma) = (s-2, s, \sigma)$, implies that $(u_\varepsilon)_{\varepsilon\in(0,1]}$ is Cauchy in $C^0([0,T];H^\sigma)$. Using the equation, we also find that $(\partial_t u_\varepsilon)_{\varepsilon\in(0,1]}$ is Cauchy in $C^1([0,T];H^{\sigma-1})$. Therefore, u_ε converges in $C^0([0,T];H^\sigma) \cap C^1([0,T];H^{\sigma-1})$ to a limit denoted u.

By taking the limit, we find that u is a solution to the Cauchy problem

$$\partial_t u + \mathrm{Op}(a)u = f, \quad u(0) = u_0.$$

6. We have seen at the end of Section 11.2 how to show that u belongs to $C^0([0,T];H^s)$. We then show that $\partial_t u$ belongs to $C^0([0,T];H^{s-1})$ using the equation.

Solution 23 (Solution to Exercise 13.1) Let us set $h = \gamma - 1$ and $w = u - v$. We then have

$$Q_\gamma(f) = \int_\Omega (1+h)|\nabla v + \nabla w|^2 \, dx$$

$$= Q_1(f) + 2 \int_\Omega \nabla v . \nabla w \, dx + \int_\Omega |\nabla w|^2 \, dx$$

$$+ \int_\Omega h|\nabla v|^2 \, dx + \int_\Omega h\nabla w \cdot (2\nabla v + \nabla w) \, dx.$$

The function $w \in H^1$ is a solution of

$$\begin{cases} \mathrm{div}(\gamma\nabla(v+w)) = 0 & \text{in } \Omega, \\ v + w = f & \text{on } \partial\Omega. \end{cases}$$

Given the equation satisfied by v, we observe that w satisfies

$$\begin{cases} \operatorname{div}(\gamma\nabla w) = -\operatorname{div}((1+h)\nabla v) = -\nabla h \cdot \nabla v & \text{in } \Omega, \\ w = 0 & \text{on } \partial\Omega. \end{cases}$$

We will use this equation to bound $\|\nabla w\|_{L^2}$:

$$\int_\Omega \gamma|\nabla w|^2\,dx = -\int_\Omega \operatorname{div}(\gamma\nabla w)w\,dx = \int_\Omega w\nabla h \cdot \nabla v\,dx$$
$$\leqslant \|w\|_{L^2}\|\nabla h\|_{L^\infty}\|\nabla v\|_{L^2}.$$

(We used the boundary condition $w_{|\partial\Omega} = 0$ in the integration by parts.) According to Poincaré's inequality, since $w \in H_0^1$, we have $\|w\|_{L^2} \leqslant C_\Omega\|\nabla w\|_{L^2}$ for a certain constant C_Ω that does not depend on w. As $\|\gamma\|_{L^\infty} \geqslant 1 - \|h\|_{L^\infty}$, we deduce from the previous inequalities that

$$(1 - \|h\|_{L^\infty})\|\nabla w\|_{L^2}^2 \leqslant C_\Omega\|\nabla w\|_{L^2}\|\nabla h\|_{L^\infty}\|\nabla v\|_{L^2}$$

and therefore, as soon as $\|h\|_{L^\infty} < 1$,

$$\|\nabla w\|_{L^2} \leqslant \frac{C_\Omega}{1 - \|h\|_{L^\infty}}\|\nabla v\|_{L^2}\|\nabla h\|_{L^\infty}.$$

Thus,

$$Q_\gamma(f) = Q_1(f) + 2\int_\Omega \nabla v.\nabla w\,dx + \int_\Omega h|\nabla v|^2\,dx + O((\|h\|_{L^2} + \|\nabla h\|_{L^2})^2).$$

However, since $w_{|\partial\Omega} = 0$ and $\Delta v = 0$, we have

$$\int_\Omega \nabla v.\nabla w\,dx = -\int_\Omega w\Delta v\,dx = 0,$$

from which we conclude that

$$Q_\gamma(f) = Q_1(f) + \int_\Omega h|\nabla v|^2\,dx + O((\|h\|_{L^2} + \|\nabla h\|_{L^2})^2),$$

which proves the result.

Solution 24 (Solution to Exercise 15.1)

1. If u is of class C^1, we have, for all $h \in \mathbb{R}^n$ such that $d(\overline{\Omega''}, \partial\Omega) < h$:

$$\forall x \in \Omega'', \quad \tau_h u(x) - u(x) = u(x+h) - u(x) = \int_0^1 h \cdot \nabla u(x+th)\,dt$$
$$\leqslant |h|\int_0^1 |\nabla u(x+th)|\,dt \leqslant |h|\left(\int_0^1 |\nabla u(x+th)|^2\,dt\right)^{1/2}.$$

The last inequality is a consequence of the Cauchy–Schwarz inequality. It follows that

$$\int_{\Omega''} |\tau_h u(x) - u(x)|^2 \, dx \leqslant |h|^2 \int_0^1 \int_{x \in \Omega''} |\nabla u(x + th)|^2 \, dx \, dt$$

$$\leqslant |h|^2 \int_0^1 \int_{x \in \Omega} |\nabla u(x)|^2 \, dx \, dt,$$

and we extend to all $H^1(\Omega)$ by density.

2. We take h such that $\Omega'' + h \subset \Omega$. In this question, we denote the scalar product of two elements of $\mathbb{R}^n$ by $\langle x, y \rangle$. Then, we have

$$\int_\Omega \langle (\tau_h A) \nabla (\Delta_h u), \nabla \phi \rangle$$

$$= \frac{1}{|h|} \int_\Omega \langle (\tau_h A) (\tau_h \nabla u - \nabla u), \nabla \phi \rangle$$

$$= \frac{1}{|h|} \int_\Omega \langle (\tau_h A) \tau_h \nabla u, \nabla \phi \rangle - \frac{1}{|h|} \int_\Omega \langle (\tau_h A) \nabla u, \nabla \phi \rangle$$

$$= \frac{1}{|h|} \int_\Omega \langle A \nabla u, \tau_{-h} \nabla \phi \rangle - \int_\Omega \langle (\Delta_h A) \nabla u, \nabla \phi \rangle - \frac{1}{|h|} \int_\Omega \langle A \nabla u, \nabla \phi \rangle$$

$$= \int_\Omega \langle A \nabla u, \nabla \Delta_{-h} \phi \rangle - \int_\Omega \langle (\Delta_h A) \nabla u, \nabla \phi \rangle$$

$$= \int_\Omega f(\Delta_{-h} \phi) - \int_\Omega \langle (\Delta_h A) \nabla u, \nabla \phi \rangle = \int_\Omega (\Delta_h f) \phi - \int_\Omega \langle (\Delta_h A) \nabla u, \nabla \phi \rangle,$$

which is the desired result.

3. Let γ be a C^∞ function that equals 1 on Ω' and whose support is included in Ω''. With $\phi = \gamma^2 (\Delta_h u)$, we have:

$$\int_\Omega \langle (\tau_h A) \nabla (\Delta_h u), \nabla (\gamma^2 (\Delta_h u)) \rangle$$

$$= \int_\Omega \gamma^2 \langle (\tau_h A) \nabla (\Delta_h u), \nabla (\Delta_h u) \rangle + 2 \int_\Omega \langle (\tau_h A) \nabla (\Delta_h u), (\nabla \gamma) \rangle \gamma \Delta_h u$$

$$\geqslant \lambda \int_\Omega \gamma^2 \|\nabla (\Delta_h u)\|^2 + 2 \int_\Omega \langle (\tau_h A) \nabla (\Delta_h u), (\nabla \gamma) \rangle \gamma \Delta_h u.$$

Using the previous question:

$$\lambda \int_\Omega \gamma^2 |\nabla (\Delta_h u)|^2$$

$$\leqslant -2 \int_\Omega \langle (\tau_h A) \nabla (\Delta_h u), (\nabla \gamma) \rangle \gamma \Delta_h u + \int_\Omega \gamma^2 (\Delta_h f)(\Delta_h u)$$

$$- \int_\Omega \gamma^2 \langle (\Delta_h A) \nabla u, \nabla \Delta_h u \rangle - 2 \int_\Omega \langle (\Delta_h A) \nabla u, \nabla \gamma \rangle \gamma \Delta_h u$$

$$\leqslant 2\|A\|_{L^\infty}\|\gamma\nabla(\Delta_h u)\|_{L^2}\|\,|\nabla\gamma|\Delta_h u\|_{L^2}$$
$$+\|f\|_{L^2}\left(\|\gamma\|_{L^\infty}\|\gamma\nabla(\Delta_h u)\|_{L^2}+2\|\nabla\gamma\|_{L^\infty}\|\gamma\Delta_h u\|_{L^2}\right)$$
$$+\|\gamma\nabla(\Delta_h u)\|_{L^2}\|\Delta_h A\|_{L^\infty}\|\gamma\nabla u\|_{L^2}$$
$$+2\|\Delta_h A\|_{L^\infty}\|\,|\nabla u|.|\nabla\gamma|\,\|_{L^2}\|\gamma\Delta_h u\|_{L^2},$$

where the penultimate inequality comes from question (15.1).
Since $A \in C^{0,1}$, $\|\Delta_h A\|_{L^\infty}$ is bounded by a constant independent of h. Moreover, we have

$$\|\gamma\Delta_h u\|_{L^2} \leqslant \|\gamma\|_{L^\infty}|h|^{-1}\|\tau_h u - u\|_{L^2(\mathrm{supp}\,\gamma)} \leqslant \|\gamma\|_{L^\infty}\|\nabla u\|_{L^2(\Omega'')}.$$

Similarly, $\|\,|\nabla\gamma|\Delta_h u\|_{L^2} \leqslant \|\nabla\gamma\|_{L^\infty}\|\nabla u\|_{L^2(\Omega'')}$. It follows that, for constants C_1, C_2 independent of h:

$$\lambda\|\gamma\nabla(\Delta_h u)\|_{L^2}^2 \leqslant C_1\|\gamma\nabla(\Delta_h u)\|_{L^2}\left(\|\nabla u\|_{L^2(\Omega'')}+\|f\|_{L^2}\right)$$
$$+C_2\|\nabla u\|_{L^2(\Omega'')}\left(\|f\|_{L^2}+\|\nabla u\|_{L^2(\Omega'')}\right).$$

For all $X \in \mathbb{R}$, if $X^2 \leqslant aX + b$ with $a, b \geqslant 0$, we see (by solving the associated polynomial equation) that

$$X \leqslant \sqrt{b} + a.$$

Applying this inequality with:

$$a = C_1\lambda^{-1}\left(\|\nabla u\|_{L^2(\Omega'')}+\|f\|_{L^2}\right),$$
$$b = C_2\lambda^{-1}\|\nabla u\|_{L^2(\Omega'')}\left(\|f\|_{L^2}+\|\nabla u\|_{L^2(\Omega'')}\right),$$

we obtain:

$$\|\gamma\nabla(\Delta_h u)\|_{L^2} \leqslant C_3\left(\|f\|_{L^2}+\|\nabla u\|_{L^2(\Omega'')}\right).$$

By squaring and using the hint, we get:

$$\|\gamma\nabla(\Delta_h u)\|_{L^2}^2 \leqslant 2C_3\left(\|f\|_{L^2}^2+\|\nabla u\|_{L^2(\Omega'')}^2\right)$$
$$\leqslant 2C_3\left(\int_\Omega f^2 + c\int_\Omega(u^2+f^2)\right) \leqslant C\int_\Omega(u^2+f^2).$$

Since γ equals 1 on Ω', we obtain

$$\int_{\Omega'}\|\nabla\Delta_h u\|^2 \leqslant \|\gamma\nabla(\Delta_h u)\|_{L^2}^2,$$

from which the desired result follows.

4. Let us consider two indices $1 \leqslant i, j \leqslant n$ and denote by e_j the j-th vector of the canonical basis. For any function $\phi \in C_0^\infty(\Omega')$, we have

$$\int_{\Omega'} (\partial_i u)(\partial_j \phi) = \lim_{t \to 0} \int_{\Omega'} (\partial_i u)(\Delta_{te_j} \phi) = \lim_{t \to 0} \int_{\Omega'} (\Delta_{-te_j} \partial_i u) \phi$$

$$\leqslant \limsup_{t \to 0} \|\Delta_{-te_j} \partial_i u\|_{L^2(\Omega')} \|\phi\|_{L^2(\Omega')}$$

$$\leqslant \limsup_{t \to 0} \|\nabla \Delta_{-te_j} u\|_{L^2(\Omega')} \|\phi\|_{L^2(\Omega')}$$

$$\leqslant C^{1/2} \left(\int_{\Omega} u^2 + f^2 \right)^{1/2} \|\phi\|_{L^2(\Omega')}.$$

According to Proposition 7.24, this implies that $\partial_i u$ is weakly differentiable with respect to x_j. This proves that u belongs to the space $H^2(\Omega')$ and the desired inequality is a direct consequence of the above.

5. Let $u \in H_0^1(R)$ be a solution of the problem. We extend u over $[-1;0] \times [-1;1]^{n-1}$ by:

$$u(x_1, \ldots, x_n) = -u(-x_1, x_2, \ldots, x_n) \qquad \text{if } x_1 < 0.$$

Let $\overline{R} = R \cup \left([-1;0] \times [-1;1]^{n-1}\right)$. The extended function u belongs to $H_0^1(\overline{R})$ and satisfies:

$$\nabla u(x_1, \ldots, x_n) = \nabla u(-x_1, \ldots, x_n) \qquad \text{if } x_1 < 0.$$

To show this last property, it is enough to observe that it is true on the set of functions of class C^∞ with compact support included in R and then use a continuity argument. We note here the use of the boundary condition: the same extension method, applied to any element of $H^1(R)$, would not give an element of $H^1(\overline{R})$.

We also extend A, by:

$$A(x_1, \ldots, x_n) = A(-x_1, x_2, \ldots, x_n) \qquad \text{if } x_1 < 0.$$

The function, thus extended, remains Lipschitz continuous.

For any function $\phi \in H_0^1(\overline{R})$:

$$\int_{\overline{R}} \langle A \nabla u, \nabla \phi \rangle = \int_R \langle A \nabla u, \nabla \phi \rangle - \int_R \langle A \nabla u, \nabla \widetilde{\phi} \rangle$$

$$= \int_R \langle A \nabla u, \nabla (\phi - \widetilde{\phi}) \rangle,$$

where $\widetilde{\phi}(x_1, \ldots, x_n) = \phi(-x_1, x_2, \ldots, x_n)$ on $\overline{R} - R$. The function $\phi - \widetilde{\phi}$ belongs to $H_0^1(R)$ (because its trace on the boundary of R is zero). Therefore, if we extend f by

$$f(x_1, \ldots, x_n) = -f(-x_1, x_2, \ldots, x_n) \qquad \text{if } x_1 < 0,$$

we conclude that

$$\int_{\overline{R}} \langle A \nabla u, \nabla \phi \rangle = \int_R \langle A \nabla u, \nabla (\phi - \widetilde{\phi}) \rangle = \int_R f(\phi - \widetilde{\phi}) = \int_{\overline{R}} \phi f.$$

Therefore, on $\overline{R}$, $-\operatorname{div}(A\nabla u) = f$. The rectangle R' is included in the interior of $\overline{R}$ so we can apply the previous question, which gives the desired result.

Solution 25 (Solution to Exercise 16.1)

1. Without loss of generality, we can assume that $\lambda = 1$. Using the identity

$$e^{i\phi} = \frac{1}{i\phi'}\frac{d}{dx}\left(e^{i\phi}\right),$$

and integrating by parts, we find that

$$\int_a^b e^{i\phi(x)}\,dx = \frac{e^{i\phi(b)}}{i\phi'(b)} - \frac{e^{i\phi(a)}}{i\phi'(a)} - \frac{1}{i}\int_a^b \frac{d}{dx}\left(\frac{1}{\phi'(x)}\right)e^{i\phi(x)}\,dx,$$

and therefore, as ϕ is real-valued (note that $\operatorname{Im}\phi \geqslant 0$ would suffice),

$$\left|\int_a^b e^{i\phi(x)}\,dx\right| \leqslant \frac{1}{|\phi'(b)|} + \frac{1}{|\phi'(a)|} + \int_a^b \left|\frac{d}{dx}\left(\frac{1}{\phi'(x)}\right)\right|\,dx. \qquad (19.5)$$

The result follows directly from this.

2. We can assume without loss of generality that $\lambda = 1$ and that ϕ' is increasing. The increase of ϕ' implies that

$$\left|\frac{d}{dx}\left(\frac{1}{\phi'(x)}\right)\right| = -\frac{d}{dx}\left(\frac{1}{\phi'(x)}\right).$$

By substituting this identity into (19.5), we deduce the upper bound

$$\left|\int_a^b e^{i\phi(x)}\,dx\right| \leqslant \frac{1}{|\phi'(b)|} + \frac{1}{|\phi'(a)|} + \frac{1}{\phi'(a)} - \frac{1}{\phi'(b)}.$$

From this, we deduce the desired estimate (note that if $\phi'(a) > 0$ then $\phi'(b) > 0$).

3. We will show that, for all $(a, b) \in \mathbb{R}^2$ and for any phase $\phi \in C^2(\mathbb{R})$ such that ϕ'' does not vanish on $[a, b]$, we have

$$\left|\int_a^b e^{i\phi(x)}\,dx\right| \leqslant \frac{8}{\sqrt{\inf_{a\leqslant x\leqslant b}|\phi''(x)|}}.$$

It suffices to treat the case $\phi'' > 0$ on $[a, b]$, the case $\phi'' < 0$ on $[a, b]$ follows directly by writing $\overline{e^{i\phi}} = e^{-i\phi}$.

For $\alpha > 0$, let us introduce the set

$$J_\alpha := \{x \in [a, b] : |\phi'(x)| \leqslant \alpha\}.$$

By hypothesis, ϕ' is strictly increasing and continuous on $[a, b]$, and therefore J_α is a closed interval (possibly empty) which we will denote by $J_\alpha = [a_\alpha, b_\alpha]$. Let us decompose the integral according to

$$\int_a^b e^{i\phi(x)}\,dx = \int_a^{a_\alpha} e^{i\phi(x)}\,dx + \int_{a_\alpha}^{b_\alpha} e^{i\phi(x)}\,dx + \int_{b_\alpha}^b e^{i\phi(x)}\,dx.$$

Since ϕ' is strictly increasing over $[a,b]$, the result of the previous question implies that

$$\left| \int_a^{a_\alpha} e^{i\phi(x)}\,dx \right| \leqslant \frac{3}{\inf_{a \leqslant x \leqslant a_\alpha}|\phi'(x)|} \leqslant \frac{3}{\alpha}.$$

Similarly, the modulus of the last integral is bounded by $3/\alpha$. The integral over $[a_\alpha, b_\alpha]$ is smaller than $b_\alpha - a_\alpha$, which we bound as follows

$$2\alpha \geqslant \phi'(b_\alpha) - \phi'(a_\alpha) = \int_{a_\alpha}^{b_\alpha} \phi''(x)\,dx,$$

which implies

$$b_\alpha - a_\alpha \leqslant \frac{2\alpha}{\inf_{a \leqslant x \leqslant b} \phi''(x)}.$$

We have therefore shown that, for all $\alpha > 0$,

$$\left| \int_a^b e^{i\phi(x)}\,dx \right| \leqslant \frac{6}{\alpha} + \frac{2\alpha}{\inf_{a \leqslant x \leqslant b} \phi''(x)}.$$

We conclude by applying this inequality with $\alpha = \sqrt{\inf_{a \leqslant x \leqslant b} \phi''(x)}$.

Solution 26 (Solution to Exercise 16.2)

1. This is proved directly by integrating by parts.
2. Follows from Exercise 16.1.
3. We will prove a stronger result. Specifically, we will show that there exists a constant C such that, for all $t \in \mathbb{R}$, all $x \in \mathbb{R}^*$ and all $R > 0$,

$$\left| \int_{-R}^R e^{i(x\xi + t\xi^2)}|\xi|^{-1/2}\,d\xi \right| \leqslant C|x|^{-1/2}.$$

Already the integral is well defined because $|\xi|^{-1/2}$ is integrable in the neighborhood of the origin. To estimate it, we make the change of variables $\eta = x\xi$, which gives

$$\left| \int_{-R}^R e^{i(x\xi + t\xi^2)}|\xi|^{-1/2}\,d\xi \right| = \frac{|x|^{1/2}}{x} \int_{-\rho}^\rho e^{i(\eta + \tau\eta^2)}|\eta|^{-1/2}\,d\eta,$$

with $\tau = t/x^2$ and $\rho = xR$. It is therefore sufficient to show that there exists a $C > 0$ such that, for all $\rho \in \mathbb{R}$ and all $\tau \in \mathbb{R}$, we have

$$|I(\rho, \tau)| = \left| \int_{-\rho}^\rho e^{i(\eta + \tau\eta^2)}|\eta|^{-1/2}\,d\eta \right| \leqslant C.$$

Since $I(\rho, \tau) = \overline{I(\rho, -\tau)}$, we note that we can without loss of generality assume that $\tau \geqslant 0$. Also, we can always assume $\rho > 0$.

To deal with the singularity at the origin of the amplitude, we introduce a plateau function $\chi \in C_0^\infty(\mathbb{R})$ such that

$$0 \leqslant \chi \leqslant 1, \quad \chi(\eta) = 1 \text{ if } x \in [-1, 1], \quad \chi(\eta) = 0 \text{ if } \eta \notin [-2, 2].$$

Since $\chi(\eta)|\eta|^{-1/2} \in L^1(\mathbb{R})$ it is enough to conclude to estimate

$$\left| \int_{-\rho}^{\rho} e^{i(\eta + \tau \eta^2)} (1 - \chi(\eta))|\eta|^{-1/2} \, d\eta \right|.$$

Let us set

$$\varphi(\eta) = \eta + \tau \eta^2, \quad \psi(\eta) = (1 - \chi(\eta))|\eta|^{-1/2}.$$

The discussion is then organized according to the cancellation of the derivative from the phase: we decompose the integration interval into three intervals (possibly empty),

$$\begin{aligned}
I_1 &:= \{\eta \in [-\rho, \rho] : 1 + \tau\eta \leqslant -1/2\}, \\
I_2 &:= \{\eta \in [-\rho, \rho] : -1/2 < 1 + \tau\eta < 1/2\}, \\
I_3 &:= \{\eta \in [-\rho, \rho] : 1 + \tau\eta \geqslant 1/2\}.
\end{aligned}$$

The reason for this decomposition is that on I_1 and on I_3 the phase φ is strictly increasing and does not cancel out, so the results of Exercise 16.1 imply that

$$\left| \int_{I_1 \cup I_3} e^{i(\eta + \tau\eta^2)} \psi(\eta) \, d\eta \right| \leqslant 6 \left(|\psi(\sup I_3)| + |\psi(\sup I_1)| + 2 \int_{-\rho}^{\rho} |\psi'(\eta)| \, d\eta \right).$$

Since $0 \leqslant \psi \leqslant 1$ and $\psi' \in L^1(\mathbb{R})$, we verify that the right-hand side is uniformly bounded.

It remains to estimate the integral over I_2. For this, let us rewrite the integral in the form

$$\tau^{1/2} \int_{I_2} e^{i(\eta + \tau\eta^2)} \frac{(1 - \chi(\eta))}{|\tau\eta|^{1/2}} \, d\eta.$$

Since on I_2 we have $|\tau\eta| \geqslant 1/2$, we deduce,

$$\left| \int_{I_2} e^{i(\eta + \tau\eta^2)} \psi(\eta) \, d\eta \right| \leqslant 2\tau^{1/2} \left| \int_{I_2} e^{i(\eta + \tau\eta^2)} (1 - \chi(\eta)) \, d\eta \right|.$$

It follows from Exercise 16.1 that

$$\left| \int_{I_2} e^{i(\eta + \tau\eta^2)} \psi(\eta) \, d\eta \right| \leqslant 20 \left(|1 - \chi(\sup I_2)| + \int_{-\rho}^{\rho} |\chi'(\eta)| \, d\eta \right).$$

This concludes the proof.

Solution 27 (Solution of Exercise 17.3)

1. Using Fubini's theorem, we can write

$$
\int_\Omega \operatorname{div}(f)\, dx = \sum_{i=1}^n \int_\Omega \frac{\partial f_i}{\partial x_i}\, dx
$$

$$
= \sum_{i=1}^n \int_{a_1}^{b_1} \cdots \int_{a_{i-1}}^{b_{i-1}} \int_{a_{i+1}}^{b_{i+1}} \cdots \int_{a_n}^{b_n} \int_{a_i}^{b_i} \frac{\partial f_i}{\partial x_i}\, dx_i\, dx_n \cdots dx_{i+1}\, dx_{i-1} \cdots dx_1.
$$

This leads to $\int_\Omega \operatorname{div}(f)\, dx = 0$ because

$$
\int_{a_i}^{b_i} \frac{\partial f_i}{\partial x_i}\, dx_i = \big[f_i(x_1, \dots, x_{i-1}, ., x_{i+1}, \dots, x_n) \big]_{x_i=a_i}^{x_i=b_i} = 0,
$$

by hypothesis on the support of f.

2. We will reduce the calculation of the two integrals to an integration over $\mathcal{U}$ by setting $g = f \circ \phi$ and $\mathcal{U}^- = \{(x_1, x_2, \dots, x_n) \in \mathcal{U} : x_1 < 0\}$.

$$
\int_\Omega \operatorname{div}(f)\, dx = \int_\Omega \operatorname{div}(g \circ \phi^{-1})\, dx
$$

$$
= \int_\Omega \sum_{i,j} \Big(\frac{\partial g_i}{\partial x_j} \circ \phi^{-1} \Big) \frac{\partial (\phi^{-1})_j}{\partial x_i}\, dx
$$

$$
= \int_{\mathcal{U}^-} \sum_{i,j} \frac{\partial g_i}{\partial x_j} \Big(\frac{\partial (\phi^{-1})_j}{\partial x_i} \circ \phi \Big) |\det d\phi|\, dx.
$$

Using the fact that $\mathcal{U}^- = [a_1; 0) \times [a_2; b_2] \times \cdots \times [a_n; b_n]$ and integrating by parts (separately treating the cases $j = 1$ and $j \ne 1$), we obtain:

$$
\int_\Omega \operatorname{div}(f)\, dx = -\int_{\mathcal{U}^-} \sum_{i,j} g_i \frac{\partial}{\partial x_j} \Big(\Big(\frac{\partial (\phi^{-1})_j}{\partial x_i} \circ \phi \Big) |\det d\phi| \Big)\, dx
$$

$$
+ \int_{\widetilde{\mathcal{U}}} \sum_i g_i(0, \tilde{x}) \Big(\frac{\partial (\phi^{-1})_1}{\partial x_i} \circ \widetilde{\phi}(\tilde{x}) \Big) |\det d\phi(0, \tilde{x})|\, d\tilde{x},
$$

where $\widetilde{\mathcal{U}}$ and $\widetilde{\phi}$ are defined as in the statement.

According to the first case we treated, $\int_\Omega \operatorname{div}(f)\, dx$ only depends on the values of f on a neighborhood of $\partial\Omega$: if two functions f and $\widetilde{f}$ coincide on a neighborhood of $\partial\Omega$, then $f - \widetilde{f}$ is a sum of functions as in the first case and we have $\int_\Omega \operatorname{div}(f - \widetilde{f})\, dx = 0$. This implies that the first term of the sum we just obtained is equal to 0, hence

$$
\int_\Omega \operatorname{div}(f)\, dx = \int_{\widetilde{\mathcal{U}}} \sum_i g_i(0, \tilde{x}) \Big(\frac{\partial (\phi^{-1})_1}{\partial x_i} \circ \widetilde{\phi}(\tilde{x}) \Big) |\det d\phi(0, \tilde{x})|\, d\tilde{x}.
$$

It remains to show that

$$\int_{\widetilde{\mathcal{U}}} \sum_i g_i(0,\widetilde{x})\left(\frac{\partial(\phi^{-1})_1}{\partial x_i} \circ \widetilde{\phi}(\widetilde{x})\right) |\det d\phi(0,\widetilde{x})| \, d\widetilde{x}$$
$$= \int_{\widetilde{\mathcal{U}}} (f \cdot n)(\widetilde{\phi}(\widetilde{x})) |\det d\widetilde{\phi}(\widetilde{x})| \, d\widetilde{x}.$$

Since $f \circ \widetilde{\phi} = g(0,\cdot)$, it suffices to show:

$$(n \circ \widetilde{\phi}(\widetilde{x}))| \det d\widetilde{\phi}(\widetilde{x})| = |\det d\phi(0,\widetilde{x})| \left(\frac{\partial(\phi^{-1})_1}{\partial x_i} \circ \widetilde{\phi}(\widetilde{x})\right)_{i=1,\ldots,n}. \qquad (19.6)$$

The vector $\left(\frac{\partial(\phi^{-1})_1}{\partial x_i} \circ \widetilde{\phi}(\widetilde{x})\right)_{i=1,\ldots,n}$ belongs to $\left(d\phi_{(0,\widetilde{x})}(\{0\} \times \mathbb{R}^{n-1})\right)^{\perp}$. This is a consequence of the equality $(d(\phi^{-1}) \circ \phi) \, d\phi = I_{\mathbb{R}^n}$. It is therefore collinear to $n(\widetilde{\phi}(\widetilde{x}))$. We need to calculate the proportionality coefficient.

Moreover, if $d\phi_{(0,\widetilde{x})}(1,0,\ldots,0) = \alpha(\widetilde{x})n(\widetilde{\phi}(\widetilde{x}))+u(\widetilde{x})$ with $u(\widetilde{x}) \in d\phi_{(0,\widetilde{x})}(\{0\}\times \mathbb{R}^{n-1})$ and $\alpha(\widetilde{x}) \in \mathbb{R}_+^*$, the equality $(d(\phi^{-1}) \circ \phi) \, d\phi = I_{\mathbb{R}^n}$ gives:

$$\left\langle \left(\frac{\partial(\phi^{-1})_1}{\partial x_i} \circ \widetilde{\phi}(\widetilde{x})\right)_{i=1,\ldots,n}, d\phi_{(0,\widetilde{x})}(1,0,\ldots,0) \right\rangle = ((d(\phi^{-1}) \circ \phi) \, d\phi)_{1,1} = 1$$
$$\implies \left(\frac{\partial(\phi^{-1})_1}{\partial x_i} \circ \widetilde{\phi}(\widetilde{x})\right)_{i=1,\ldots,n} = \frac{1}{\alpha(\widetilde{x})} n(\widetilde{\phi}(\widetilde{x})).$$

Furthermore, $|\det d\phi(0,\widetilde{x})| = \alpha(\widetilde{x}) \, | \det d\widetilde{\phi}(\widetilde{x})|$. By combining the last two equations, we obtain (19.6).

3. The general result follows by using a suitable partition of unity (see Proposition 17.35).

4. If we set $f = v\nabla u$, the Green's formula is exactly the equality of question (1), since $\operatorname{div}(f) = \nabla u \cdot \nabla v + v\Delta u$.

Solution 28 (Solution to Exercise 18.1) We use the classical example of the sliding bump, the construction of which we recall here. We proceed in several steps. At the first step, we define f_0 as the indicator function of $[0,1]$. At the second step, we define two functions: f_1 is the indicator function of $[0,1/2]$ and f_2 is the indicator function of $[1/2,1]$. We then define at the nth step 2^n functions by dividing the interval $[0,1]$ into 2^n intervals of size 2^{-n} and by considering the 2^n indicator functions of these intervals. Then there is a way to order these functions so that, on the one hand, the sequence converges to 0 in $L^p(\mathbb{R})$ for all finite p and, on the other hand, it does not converge at any point in $[0,1]$.

Solution 29 (Solution to Exercise 18.2)

1. Let us denote by $X = (x_1,\ldots,x_k)$ the variable of $(\mathbb{R}^n)^k$ where each variable x_j belongs to $\mathbb{R}^n$.

The hypothesis implies

$$\left|\int_{(\mathbb{R}^n)^k} F(X)G(X)\,\mathrm{d}X\right| \leq C \left(\int_{(\mathbb{R}^n)^k} |F(X)|^p\,\mathrm{d}X\right)^{1/p} \left(\int_{(\mathbb{R}^n)^k} |G(X)|^q\,\mathrm{d}X\right)^{1/q}$$

for a certain universal constant C independent of k. It follows that

$$\left(\int_{\mathbb{R}^n} f(x)g(x)\,\mathrm{d}x\right)^k \leq C\left(\int_{\mathbb{R}^n} |f(x)|^p\,\mathrm{d}x\right)^{k/p} \left(\int_{\mathbb{R}^n} |g(x)|^q\,\mathrm{d}x\right)^{k/q},$$

from which we get

$$\left|\int_{\mathbb{R}^n} f(x)g(x)\,\mathrm{d}x\right| \leq C^{1/k}\left(\int_{\mathbb{R}^n} |f(x)|^p\,\mathrm{d}x\right)^{1/p} \left(\int_{\mathbb{R}^n} |g(x)|^q\,\mathrm{d}x\right)^{1/q}.$$

By letting k tend to $+\infty$, we obtain the Hölder inequality with $C = 1$.

2. We can assume without loss of generality that $\|f\|_{L^p} = 1$ and $\|g\|_{L^q} = 1$. Lemma 18.8 implies that

$$\|fg\|_{L^1} = \int_0^\infty |\{x \in \mathbb{R}^n : |f(x)g(x)| > \lambda\}|\,\mathrm{d}\lambda.$$

Recall that $\{|h| > \mu\}$ denotes the set $\{x \in \mathbb{R}^n : |h(x)| > \mu\}$.

Note that if a product ab of two positive real numbers is strictly greater than another product cd of two positive real numbers, then we necessarily have $a > c$ or $b > d$. As

$$\lambda = \lambda^{1/p}\lambda^{1/q},$$

this observation implies that

$$\{|f(x)g(x)| > \lambda\} \subset \left\{|f(x)| > \lambda^{1/p}\right\} \cup \left\{|g(x)| > \lambda^{1/q}\right\}.$$

Then, by measuring these sets,

$$|\{|f(x)g(x)| > \lambda\}| \leq \left|\left\{|f(x)| > \lambda^{1/p}\right\}\right| + \left|\left\{|g(x)| > \lambda^{1/q}\right\}\right|.$$

By integrating this inequality, we find that

$$\|fg\|_{L^1} \leq \int_0^{+\infty} \left|\left\{|f(x)| > \lambda^{1/p}\right\}\right|\,\mathrm{d}\lambda + \int_0^{+\infty} \left|\left\{|g(x)| > \lambda^{1/q}\right\}\right|\,\mathrm{d}\lambda.$$

Then, by changing variables,

$$\|fg\|_{L^1} \leq \int_0^{+\infty} p\mu^{p-1}\,|\{|f(x)| > \mu\}|\,\mathrm{d}\mu + \int_0^{+\infty} q\zeta^{q-1}\,|\{|g(x)| > \zeta\}|\,\mathrm{d}\zeta.$$

This gives
$$\|fg\|_{L^1} \leqslant \|f\|_{L^p} + \|g\|_{L^q} = 2 = 2\|f\|_{L^p}\|g\|_{L^q}.$$

This is Hölder's inequality, but with a bad constant (2 instead of 1). We deduce Hölder's inequality with the correct constant 1 by using the first question.

Solution 30 (Solution of Exercise 18.3)

1. The integral is well defined by continuity and compactness of the support of f.
2. By homogeneity of K, we have
$$T(f)(x) = \int_0^{+\infty} K(x,y)f(y)\,dy = \int_0^{+\infty} K(x, x \cdot y/x)\, f(y)\,dy$$
$$= \int_0^{+\infty} K(1, y/x)\frac{1}{x} f(y)\,dy$$
$$= \int_0^{+\infty} K(1,r)\, f(rx)\,dr.$$

According to Minkowski's inequality, we have
$$\left(\int_0^{+\infty}\left|\int_0^{+\infty} K(1,r)f(rx)\,dr\right|^p dx\right)^{1/p}$$
$$\leqslant \int_0^{+\infty}\left(\int_0^{+\infty} |K(1,r)|^p\,|f(rx)|^p\,dx\right)^{1/p} dr$$
$$\leqslant \int_0^{+\infty} |K(1,r)|\,r^{-1/p}\left(\int_0^{+\infty} |f(z)|^p\,dz\right)^{1/p} dr.$$

This proves that
$$\left(\int_0^{+\infty} |T(f)|^p\,dx\right)^{1/p} \leqslant C\left(\int_0^{+\infty} |f(z)|^p\,dz\right)^{1/p},$$
with
$$C := \int_0^{+\infty} |K(1,r)|\,r^{-1/p}\,dr.$$

This constant is finite by hypothesis on the kernel K.
3. We apply the above with $K(x,y) = \frac{1}{x}\mathbf{1}_{[0,x]}(y)$, where $\mathbf{1}_{[0,x]}$ is the indicator function of $[0,x]$. It is easy to verify that K satisfies the two hypotheses (homogeneity and integrability). In this case, the constant is given by
$$C := \int_0^{+\infty} |K(1,r)|\,r^{-1/p}\,dr = \int_0^1 r^{-1/p}\,dr = \frac{p}{p-1}.$$

This proves the desired inequality if f is a continuous function with compact support. The result follows by the density of $C_0^0((0,+\infty))$ in $L^p(0,+\infty)$.

References

1. Thomas Alazard and Claude Zuily. *Tools and problems in partial differential equations.* Universitext. Springer, Cham, 2020.
2. Serge Alinhac. *Hyperbolic partial differential equations.* Universitext. Springer, Dordrecht, 2009.
3. Serge Alinhac and Patrick Gérard. *Opérateurs pseudo-différentiels et théorème de Nash-Moser.* Savoirs Actuels. InterEditions/EDP Sciences, Paris, 1991.
4. Luigi Ambrosio, Alessandro Carlotto, and Annalisa Massaccesi. *Lectures on elliptic partial differential equations,* volume 18 of *Appunti. Scuola Normale Superiore di Pisa (Nuova Serie).* Edizioni della Normale, Pisa, 2018.
5. Silvia Annaratone. Les premières démonstrations de la formule intégrale de Fourier. *Rev. Histoire Math.,* 3(1):99–136, 1997.
6. Juan Arias de Reyna. *Pointwise convergence of Fourier series,* volume 1785 of *Lecture Notes in Math.* Springer-Verlag, Berlin, 2002.
7. Vladimir I. Arnold. Small denominators. I. Mapping the circle onto itself. *Izv. Akad. Nauk SSSR Ser. Mat.,* 25:21–86, 1961.
8. Hajer Bahouri. The Littlewood-Paley theory: a common thread of many works in nonlinear analysis. *Eur. Math. Soc. Newsl.,* (112):15–23, 2019.
9. Hajer Bahouri, Jean-Yves Chemin, and Raphaël Danchin. *Fourier analysis and nonlinear partial differential equations,* volume 343 of *Grundlehren der Math. Wissenschaften.* Springer, Heidelberg, 2011.
10. Lennart Bondesson, Jan Grandell, and Jaak Peetre. The life and work of Olof Thorin (1912–2004). *Proc. Est. Acad. Sci.,* 57(1):18–25, 2008.
11. Jean-Michel Bony. Calcul symbolique et propagation des singularités pour les équations aux dérivées partielles non linéaires. *Ann. Sci. École Norm. Sup. (4),* 14(2):209–246, 1981.
12. Jean-Michel Bony. *Cours d'analyse: théorie des distributions et analyse de Fourier.* Éditions de l'École Polytechnique, Palaiseau, 2001.
13. Jean Bourgain. Fourier transform restriction phenomena for certain lattice subsets and applications to nonlinear evolution equations. I. Schrödinger equations. *Geom. Funct. Anal.,* 3(2):107–156, 1993.
14. Jean Bourgain. Fourier transform restriction phenomena for certain lattice subsets and applications to nonlinear evolution equations. II. The KdV-equation. *Geom. Funct. Anal.,* 3(3):209–262, 1993.
15. Haim Brezis. *Functional analysis, Sobolev spaces and partial differential equations.* Universitext. Springer, New York, 2011.
16. Haïm Brézis and Thierry Gallouet. Nonlinear Schrödinger evolution equations. *Nonlinear Anal.,* 4(4):677–681, 1980.
17. Luitzen Egbertus Jan Brouwer. Beweis der Invarianz des n-dimensionalen Gebiets. *Math. Ann.,* 71(3):305–313, 1911.

T. Alazard, *Analysis and Partial Differential Equations,* Universitext,
https://doi.org/10.1007/978-3-031-70909-8

18. Elia Brué and Quoc-Hung Nguyen. On the Sobolev space of functions with derivative of logarithmic order. *Adv. Nonlinear Anal.*, 9(1):836–849, 2020.

19. Claudia Bucur and Enrico Valdinoci. *Nonlocal diffusion and applications*, volume 20 of *Lecture Notes of the Unione Matematica Italiana*. Springer, 2016.

20. Almut Burchard. A short course on rearrangement inequalities. *Lecture notes, IMDEA Winter School, Madrid*, 2009.

21. Victor I. Burenkov. *Sobolev spaces on domains*, volume 137 of *Teubner-Texte zur Mathematik [Teubner Texts in Mathematics]*. B. G. Teubner Verlagsgesellschaft mbH, Stuttgart, 1998.

22. Alberto-P. Calderón. Lebesgue spaces of differentiable functions and distributions. In *Proc. Sympos. Pure Math., Vol. IV*, pages 33–49. Amer. Math. Soc., Providence, RI, 1961.

23. Alberto-P. Calderón. Intermediate spaces and interpolation, the complex method. *Studia Math.*, 24:113–190, 1964.

24. Alberto-P. Calderón. On an inverse boundary value problem. In *Seminar on Numerical Analysis and its Applications to Continuum Physics (Rio de Janeiro, 1980)*, pages 65–73. Soc. Brasil. Mat., Rio de Janeiro, 1980.

25. Alberto-P. Calderón and Rémi Vaillancourt. On the boundedness of pseudo-differential operators. *J. Math. Soc. Japan*, 23:374–378, 1971.

26. Alberto-P. Calderón and Rémi Vaillancourt. A class of bounded pseudo-differential operators. *Proc. Nat. Acad. Sci. U.S.A.*, 69:1185–1187, 1972.

27. Sergio Campanato. Proprietà di hölderianità di alcune classi di funzioni. *Ann. Scuola Norm. Sup. Pisa Cl. Sci. (3)*, 17:175–188, 1963.

28. Sergio Campanato. Equazioni ellittiche del II deg ordine espazi $\mathfrak{L}^{(2,\lambda)}$. *Ann. Mat. Pura Appl. (4)*, 69:321–381, 1965.

29. James Caristi. Fixed point theorems for mappings satisfying inwardness conditions. *Trans. Amer. Math. Soc.*, 215:241–251, 1976.

30. Lennart Carleson. On convergence and growth of partial sums of Fourier series. *Acta Math.*, 116:135–157, 1966.

31. Augustin-Louis Cauchy. *Cours d'analyse de l'École Royale Polytechnique*. Cambridge Library Collection. Cambridge University Press, Cambridge, 2009.

32. Thierry Cazenave and Alain Haraux. *An introduction to semilinear evolution equations*, volume 13 of *Oxford Lecture Series in Math. and its Applications*. The Clarendon Press, Oxford University Press, New York, 1998.

33. Jacques Chazarain and Alain Piriou. *Introduction à la théorie des équations aux dérivées partielles linéaires*. Gauthier-Villars, Paris, 1981.

34. Jean-Yves Chemin and Chao-Jian Xu. Inclusions de Sobolev en calcul de Weyl-Hörmander et champs de vecteurs sous-elliptiques. *Ann. Sci. École Norm. Sup. (4)*, 30(6):719–751, 1997.

35. Ole Christensen. *An introduction to frames and Riesz bases*. Applied and Numerical Harmonic Analysis. Birkhäuser/Springer, 2016.

36. Ronald R. Coifman and Yves Meyer. *Au delà des opérateurs pseudo-différentiels*, volume 57 of *Astérisque*. Société Mathématique de France, Paris, 1978.

37. Antonio Córdoba and Charles Fefferman. Wave packets and Fourier integral operators. *Comm. Partial Differential Equations*, 3(11):979–1005, 1978.

38. Richard Courant and David Hilbert. *Methods of mathematical physics. Vol. I*. Interscience Publishers, Inc., New York, N.Y., 1953.

39. Richard Courant and David Hilbert. *Methods of mathematical physics. Vol. II: Partial differential equations*. Interscience Publishers, New York-London, 1962.

40. Walter Craig. *Problèmes de petits diviseurs dans les équations aux dérivées partielles*, volume 9 of *Panoramas et Synthèses*. Société Mathématique de France, Paris, 2000.

41. George Cybenko. Approximation by superpositions of a sigmoidal function. *Math. Control Signals Systems*, 2(4):303–314, 1989.

42. Raphaël Danchin. Cours de topologie et d'analyse fonctionnelle Master première année. 2013.

43. Raphaël Danchin. Fourier analysis methods for the compressible Navier-Stokes equations. In *Handbook of mathematical analysis in mechanics of viscous fluids*, pages 1843–1903. Springer, Cham, 2018.

44. Guy David and Jean-Lin Journé. A boundedness criterion for generalized Calderón-Zygmund operators. *Ann. of Math. (2)*, 120(2):371–397, 1984.
45. Ennio De Giorgi. Sulla differenziabilità e l'analiticità delle estremali degli integrali multipli regolari. *Mem. Accad. Sci. Torino. Cl. Sci. Fis. Mat. Nat. (3)*, 3:25–43, 1957.
46. Jacques Dixmier. *Topologie générale*. Mathématiques. Presses Universitaires de France, Paris, 1981.
47. Asen L. Dontchev and R. Tyrrell Rockafellar. *Implicit functions and solution mappings*. Springer Series in Operations Research and Financial Engineering. Springer, New York, second edition, 2014. A view from variational analysis.
48. Ivar Ekeland. On the variational principle. *J. Math. Anal. Appl.*, 47:324–353, 1974.
49. Lawrence C. Evans. *Partial differential equations*, volume 19 of *Graduate Studies in Math.* American Mathematical Society, Providence, RI, 2010.
50. Lawrence C. Evans and Te Zhang. Weak convergence and averaging for ODE. *Nonlinear Anal.*, 138:83–92, 2016.
51. Charles Fefferman. L^p bounds for pseudo-differential operators. *Israel J. Math.*, 14:413–417, 1973.
52. Jean Baptiste Joseph Fourier. *Théorie analytique de la chaleur*. Cambridge Library Collection. Cambridge University Press, Cambridge, 2009.
53. Kurt Otto Friedrichs. The identity of weak and strong extensions of differential operators. *Trans. Amer. Math. Soc.*, 55:132–151, 1944.
54. Dennis Gabor. Theory of communication. Part 1: The analysis of information. *Journal of the Institution of Electrical Engineers-part III: radio and communication engineering*, 93(26):429–441, 1946.
55. David Gale. The game of Hex and the Brouwer fixed-point theorem. *Amer. Math. Monthly*, 86(10):818–827, 1979.
56. Lars Gårding. *Some points of analysis and their history*, volume 11 of *University Lecture Series*. American Mathematical Society, Providence, RI, 1997.
57. Patrick Gérard. Microlocal defect measures. *Comm. Partial Differential Equations*, 16(11):1761–1794, 1991.
58. Patrick Gérard. Nonlinear Schrödinger equations on compact manifolds. In *European Congress of Mathematics*, pages 121–139. Eur. Math. Soc., Zürich, 2005.
59. Étienne Ghys. Resonances and small divisors. In *Kolmogorov's heritage in mathematics*, pages 187–213. Springer, Berlin, 2007.
60. Jean Ginibre and Giorgio Velo. Generalized Strichartz inequalities for the wave equation. *J. Funct. Anal.*, 133(1):50–68, 1995.
61. François Golse, Yves Laszlo, Frank Pacard, and Claude Viterbo. *Analyse réelle et complexe*. Éditions de l'École Polytechnique, Palaiseau, 2013.
62. Loukas Grafakos. *Classical Fourier analysis*, volume 249 of *Graduate Texts in Mathematics*. Springer, New York, third edition, 2014.
63. Alain Grigis and Johannes Sjöstrand. *Microlocal analysis for differential operators*, volume 196 of *London Mathematical Society Lecture Note Series*. Cambridge University Press, Cambridge, 1994.
64. Alexander Grossmann and Jean Morlet. Decomposition of Hardy functions into square integrable wavelets of constant shape. *SIAM J. Math. Anal.*, 15(4):723–736, 1984.
65. Alfred Haar. Zur Theorie der orthogonalen Funktionensysteme. *Math. Ann.*, 69(3):331–371, 1910.
66. Peter Hähner. A periodic Faddeev-type solution operator. *J. Differential Equations*, 128(1):300–308, 1996.
67. Richard S. Hamilton. The inverse function theorem of Nash and Moser. *Bull. Amer. Math. Soc. (N.S.)*, 7(1):65–222, 1982.
68. Qing Han and Fanghua Lin. *Elliptic partial differential equations*, volume 1 of *Courant Lecture Notes in Math.* Courant Institute of Mathematical Sciences, New York; American Mathematical Society, Providence, RI, 2011.
69. Godfrey H. Hardy. Weierstrass's non-differentiable function. *Trans. Amer. Math. Soc.*, 17(3):301–325, 1916.

70. Godfrey H. Hardy and John E. Littlewood. Some properties of fractional integrals. I. *Math. Z.*, 27(1):565–606, 1928.
71. Godfrey H. Hardy and John E. Littlewood. A maximal theorem with function-theoretic applications. *Acta Math.*, 54(1):81–116, 1930.
72. Bernard Helffer and Francis Nier. *Hypoelliptic estimates and spectral theory for Fokker-Planck operators and Witten Laplacians*, volume 1862 of *Lecture Notes in Math.* Springer-Verlag, Berlin, 2005.
73. Michael-Robert Herman. Sur la conjugaison différentiable des difféomorphismes du cercle à des rotations. *Publ. Math. Inst. Hautes Études Sci.*, (49):5–233, 1979.
74. Lars Hörmander. Hypoelliptic second order differential equations. *Acta Math.*, 119:147–171, 1967.
75. Lars Hörmander. Pseudo-differential operators and hypoelliptic equations. In *Singular integrals (Proc. Sympos. Pure Math., Vol. X, Chicago, Ill., 1966)*, pages 138–183. American Mathematical Society, Providence, RI, 1967.
76. Lars Hörmander. Oscillatory integrals and multipliers on FL^p. *Ark. Mat.*, 11:1–11, 1973.
77. Lars Hörmander. On the Nash-Moser implicit function theorem. *Ann. Acad. Sci. Fenn. Ser. A I Math.*, 10:255–259, 1985.
78. Lars Hörmander. *Lectures on nonlinear hyperbolic differential equations*, volume 26 of *Mathématiques & Applications*. Springer-Verlag, Berlin, 1997.
79. Lars Hörmander. *The analysis of linear partial differential operators. I.* Classics in Math. Springer-Verlag, Berlin, 2003.
80. Lars Hörmander. *The analysis of linear partial differential operators. III.* Classics in Mathematics. Springer, Berlin, 2007.
81. Kurt Hornik. Approximation capabilities of multilayer feedforward networks. *Neural networks*, 4(2):251–257, 1991.
82. Kurt Hornik, Maxwell Stinchcombe, and Halbert White. Multilayer feedforward networks are universal approximators. *Neural networks*, 2(5):359–366, 1989.
83. Ing-Lung Hwang. The L^2-boundedness of pseudodifferential operators. *Trans. Amer. Math. Soc.*, 302(1):55–76, 1987.
84. Stéthane Jaffard. Yves Meyer, prix Abel 2017. *Gaz. Math.*, (153):20–26, 2017.
85. F. John and L. Nirenberg. On functions of bounded mean oscillation. *Comm. Pure Appl. Math.*, 14:415–426, 1961.
86. Fritz John. *Partial differential equations*, volume 1 of *Applied Math. Sciences*. Springer-Verlag, New York, 1982.
87. Peter W. Jones. Quasiconformal mappings and extendability of functions in Sobolev spaces. *Acta Math.*, 147(1-2):71–88, 1981.
88. Jürgen Jost. *Partial differential equations*, volume 214 of *Graduate Texts in Math.* Springer, New York, 2013.
89. Jean-Pierre Kahane. À partir et autour de Wiener. In *L'émergence de l'analyse harmonique abstraite (1930–1950) (Paris, 1991)*, volume 2 of *Cahiers Sém. Hist. Math. Sér. 2*, pages 65–78. Univ. Paris VI, Paris, 1992.
90. Jean-Pierre Kahane. Quelques points d'histoire des séries de Fourier. In *Advances in mathematical sciences: CRM's 25 years (Montreal, PQ, 1994)*, volume 11 of *CRM Proc. Lecture Notes*, pages 201–213. American Mathematical Society, Providence, RI, 1997.
91. Jean-Pierre Kahane. The return of Fourier. *Gac. R. Soc. Mat. Esp.*, 10(3):678–688, 2007.
92. Jean-Pierre Kahane. Analyse et synthèse harmoniques. In *Histoire de mathématiques*, Journées X-UPS, pages 17–53. Éditions de l'École Polytechnique, Palaiseau, 2012.
93. Jean-Pierre Kahane. Qu'est-ce que Fourier peut nous dire aujourd'hui? *Gaz. Math.*, (141):69–75, 2014.
94. Jean-Pierre Kahane and Pierre Gilles Lemarié-Rieusset. *Séries de Fourier et ondelettes*, volume 3 of *Nouvelle Bibliothèque Math.* Cassini, Paris, deuxième edition, 2016.
95. Jean-Michel Kantor. Mathématiques d'Est en Ouest. Théorie et pratique: l'exemple des distributions. *Gaz. Math.*, (100):33–43, 2004.
96. Pytor Kapitza. Pendulum with an oscillating pivot. *Sov. Phys. Uspekhi*, 44:7–20, 1951.

97. Yitzhak Katznelson. *An introduction to harmonic analysis*. Cambridge Math. Library. Cambridge University Press, Cambridge, 2004.

98. Markus Keel and Terence Tao. Endpoint Strichartz estimates. *Amer. J. Math.*, 120(5):955–980, 1998.

99. John L. Kelley. *General topology*, volume 27 of *Graduate Texts in Math.* Springer-Verlag, New York-Berlin, 1975.

100. Sergiu Klainerman. Lecture notes. Introduction to analysis. 2011.

101. Joseph J. Kohn. Lectures on degenerate elliptic problems. In *Pseudodifferential operator with applications (Bressanone, 1977)*, pages 89–151. Liguori, Naples, 1978.

102. Joseph J. Kohn and Louis Nirenberg. An algebra of pseudo-differential operators. *Comm. Pure Appl. Math.*, 18:269–305, 1965.

103. Andreï N. Kolmogorov. On conservation of conditionally periodic motions for a small change in Hamilton's function. *Dokl. Akad. Nauk SSSR (N.S.)*, 98:527–530, 1954.

104. Andreï N. Kolmogorov. On the representation of continuous functions of many variables by superposition of continuous functions of one variable and addition. *Dokl. Akad. Nauk SSSR*, 114:953–956, 1957.

105. Steven G. Krantz and Harold R. Parks. *The implicit function theorem*. Birkhäuser Boston, Inc., Boston, MA, 2002. History, theory, and applications.

106. Nicolai V. Krylov. *Lectures on elliptic and parabolic equations in Hölder spaces*, volume 12 of *Graduate Studies in Mathematics*. American Mathematical Society, Providence, RI, 1996.

107. Sergei B. Kuksin. *Analysis of Hamiltonian PDEs*, volume 19 of *Oxford Lecture Series in Mathematics and its Applications*. Oxford University Press, Oxford, 2000.

108. Wladyslaw Kulpa. Poincaré and domain invariance theorem. *Acta Univ. Carolin. Math. Phys.*, 39(1-2):127–136, 1998.

109. Klaas Landsman. Lecture notes on Hilbert spaces and quantum mechanics. Course notes available at https://www.math.ru.nl/~landsman/HSQM.pdf.

110. Klaas Landsman. *Foundations of quantum theory: from classical concepts to operator algebras*. Springer Nature, 2017.

111. Peter D. Lax. Change of variables in multiple integrals. *Amer. Math. Monthly*, 106(6):497–501, 1999.

112. Peter D. Lax. Change of variables in multiple integrals. II. *Amer. Math. Monthly*, 108(2):115–119, 2001.

113. Peter D. Lax. *Functional analysis*. Pure and Applied Mathematics. Wiley-Interscience [John Wiley & Sons], New York, 2002.

114. Pierre Gilles Lemarié-Rieusset. Ondelettes et fonction de Weierstrass. Personal communication.

115. Jean Leray. Sur le mouvement d'un liquide visqueux emplissant l'espace. *Acta Math.*, 63(1):193–248, 1934.

116. Jean Leray. My friend Julius Schauder. In *Numerical solution of highly nonlinear problems (Sympos. Fixed Point Algorithms and Complementarity Problems, Univ. Southampton, Southampton, 1979)*, pages 427–439. North-Holland, Amsterdam-New York, 1980.

117. Jean Leray and Jules Schauder. Topologie et équations fonctionnelles. *Ann. Sci. École Norm. Sup. (3)*, 51:45–78, 1934.

118. Nicolas Lerner. The Wick calculus of pseudo-differential operators and some of its applications. *Cubo Mat. Educ.*, 5(1):213–236, 2003.

119. Nicolas Lerner. *Carleman inequalities*, volume 353 of *Grundlehren der Math. Wissenschaften*. Springer, Cham, 2019.

120. Moshe Leshno, Vladimir Ya Lin, Allan Pinkus, and Shimon Schocken. Multilayer feedforward networks with a nonpolynomial activation function can approximate any function. *Neural networks*, 6(6):861–867, 1993.

121. Leon Lichtenstein. Eine elementare Bemerkung zur reellen Analysis. *Math. Z.*, 30(1):794–795, 1929.

122. Elliott H. Lieb. Sharp constants in the Hardy-Littlewood-Sobolev and related inequalities. *Ann. of Math. (2)*, 118(2):349–374, 1983.

123. Elliott H. Lieb and Michael Loss. *Analysis*, volume 14 of *Graduate Studies in Math*. American Mathematical Society, Providence, RI, 2001.

124. Gary M. Lieberman. *Second order parabolic differential equations*. World Scientific Publishing Co., Inc., River Edge, NJ, 1996.

125. Felipe Linares and Gustavo Ponce. *Introduction to nonlinear dispersive equations*. Universitext. Springer, New York, second edition, 2015.

126. Elaine Machtyngier and Enrique Zuazua. Stabilization of the Schrödinger equation. *Portugal. Math.*, 51(2):243–256, 1994.

127. Lech Maligranda. On interpolation of nonlinear operators. *Comment. Math. Prace Mat.*, 28(2):253–275, 1989.

128. Stéphane Mallat. *Une exploration des signaux en ondelettes*. Éditions de l'École Polytechnique, Palaiseau, 2000.

129. Józef Marcinkiewicz. Sur l'interpolation d'operations. *CR Acad. Sci. Paris.*, 208:1272–1273, 1939.

130. Jean Mawhin. Le théoreme du point fixe de brouwer: un siècle de métamorphoses. *Sci. Tech. Perspect*, 2(10):1–2, 2007.

131. Jean Mawhin. A tribute to Juliusz Schauder. *Antiq. Math.*, 12:229–257, 2018.

132. Warren S McCulloch and Walter Pitts. A logical calculus of the ideas immanent in nervous activity. *The bulletin of mathematical biophysics*, 5(4):115–133, 1943.

133. Guy Métivier. Intégrales singulières, cours de DEA, Rennes 1981. Course notes available at https://www.math.u-bordeaux.fr/~gmetivie/cours.html.

134. Guy Métivier. *Para-differential calculus and applications to the Cauchy problem for nonlinear systems*, volume 5 of *Centro di Ricerca Matematica Ennio De Giorgi (CRM) Series*. Edizioni della Normale, Pisa, 2008.

135. Yves Meyer. *Ondelettes et opérateurs. I*. Actualités Mathématiques. Hermann, Paris, 1990. Ondelettes.

136. Yves Meyer. *Ondelettes et opérateurs. II*. Actualités Mathématiques. Hermann, Paris, 1990. Opérateurs de Calderón-Zygmund.

137. Yves Meyer. Le traitement du signal et l'analyse mathématique. *Ann. Inst. Fourier (Grenoble)*, 50(2):593–632, 2000.

138. Yves Meyer and Ronald R. Coifman. *Ondelettes et opérateurs. III*. Actualités Mathématiques. Hermann, Paris, 1991. Opérateurs multilinéaires.

139. John Milnor. Analytic proofs of the "hairy ball theorem" and the Brouwer fixed-point theorem. *Amer. Math. Monthly*, 85(7):521–524, 1978.

140. C. B. Morrey, Jr. Second-order elliptic systems of differential equations. In *Contributions to the theory of partial differential equations*, volume no. 33 of *Ann. of Math. Stud.*, pages 101–159. Princeton Univ. Press, Princeton, NJ, 1954.

141. Jürgen Moser. A new proof of De Giorgi's theorem concerning the regularity problem for elliptic differential equations. *Comm. Pure Appl. Math.*, 13:457–468, 1960.

142. Jürgen Moser. A rapidly convergent iteration method and non-linear partial differential equations. I. *Ann. Scuola Norm. Sup. Pisa (3)*, 20:265–315, 1966.

143. Jürgen Moser. A rapidly convergent iteration method and non-linear differential equations. II. *Ann. Scuola Norm. Sup. Pisa (3)*, 20:499–535, 1966.

144. François Murat. Compacité par compensation. *Ann. Scuola Norm. Sup. Pisa Cl. Sci. (4)*, 5(3):489–507, 1978.

145. Camil Muscalu and Wilhelm Schlag. *Classical and multilinear harmonic analysis. Vol. I*, volume 137 of *Cambridge Studies in Advanced Math*. Cambridge University Press, Cambridge, 2013.

146. Camil Muscalu and Wilhelm Schlag. *Classical and multilinear harmonic analysis. Vol. II*, volume 138 of *Cambridge Studies in Advanced Math*. Cambridge University Press, Cambridge, 2013.

147. John Nash. The imbedding problem for Riemannian manifolds. *Ann. of Math. (2)*, 63:20–63, 1956.

148. John Nash. Parabolic equations. *Proc. Nat. Acad. Sci. U.S.A.*, 43:754–758, 1957.

149. Donald J. Newman. A simple proof of Wiener's $1/f$ theorem. *Proc. Amer. Math. Soc.*, 48:264–265, 1975.

150. Louis Nirenberg. *Topics in nonlinear functional analysis*. Courant Institute of Mathematical Sciences New York University, New York, 1974.

151. Sehie Park. Ninety years of the Brouwer fixed point theorem. *Vietnam J. Math.*, 27(3):187–222, 1999.

152. Lawrence E. Payne and Hans F. Weinberger. An optimal Poincaré inequality for convex domains. *Arch. Rational Mech. Anal.*, 5:286–292 (1960), 1960.

153. Gideon Peyser. On the Cauchy-Lipschitz theorem. *Amer. Math. Monthly*, 65:760–762, 1958.

154. Allan Pinkus. Approximation theory of the MLP model in neural networks. In *Acta numerica, 1999*, volume 8 of *Acta Numer.*, pages 143–195. Cambridge Univ. Press, Cambridge, 1999.

155. Augusto C. Ponce. *Elliptic PDEs, measures and capacities*, volume 23 of *EMS Tracts in Math.* European Mathematical Society (EMS), Zürich, 2016.

156. Jürgen Pöschel. A lecture on the classical KAM theorem. In *Smooth ergodic theory and its applications (Seattle, WA, 1999)*, volume 69 of *Proc. Sympos. Pure Math.*, pages 707–732. Amererican Mathematical Society, Providence, RI, 2001.

157. Christophe Prange. Weak and strong convergence methods for partial differential equations. Course notes available at http://prange.perso.math.cnrs.fr.

158. Michael Reed and Barry Simon. *Methods of modern mathematical physics. II. Fourier analysis, self-adjointness.* Academic Press, New York-London, 1975.

159. Michael Reed and Barry Simon. *Methods of modern mathematical physics. I.* Academic Press, Inc., New York, 1980. Functional analysis.

160. Marcel Riesz. Sur les maxima des formes bilinéaires et sur les fonctionnelles linéaires. *Acta Math.*, 49(3-4):465–497, 1927.

161. Didier Robert. Analyse fonctionnelle. Course notes available at https://www.math.sciences.univ-nantes.fr/~robert/.

162. Frank Rosenblatt. *The perceptron, a perceiving and recognizing automaton Project Para.* Cornell Aeronautical Laboratory, 1957.

163. Walter Rudin. *Real and complex analysis*. McGraw-Hill Book Co., New York, 1987.

164. Walter Rudin. *Functional analysis*. International Series in Pure and Applied Math. McGraw-Hill, Inc., New York, 1991.

165. Laure Saint-Raymond. Analyse fonctionnelle. Course notes available at https://www.math.ens.psl.eu/shared-files/10357/analyse-fonctionnelle2013-2.pdf, 2013.

166. Xavier Saint Raymond. *Elementary introduction to the theory of pseudodifferential operators*. Studies in Advanced Math. CRC Press, Boca Raton, FL, 1991.

167. Mikko Salo. Calderón problem, lecture notes. 2018.

168. J. Schauder. Über lineare elliptische Differentialgleichungen zweiter Ordnung. *Math. Z.*, 38(1):257–282, 1934.

169. Julius Schauder. Zur Theorie stetiger Abbildungen in Funktionalräumen. *Math. Z.*, 26(1):47–65, 1927.

170. Laurent Schwartz. *Théorie des distributions*, volume IX-X of *Publications de l'Institut de Mathématique de l'Université de Strasbourg*. Hermann, Paris, 1966.

171. Robert T. Seeley. Extension of C^∞ functions defined in a half space. *Proc. Amer. Math. Soc.*, 15:625–626, 1964.

172. Hourya Sinaceur. Cauchy et Bolzano. *Rev. Histoire Sci. Appl.*, 26(2):97–112, 1973.

173. Didier Smets. Régularité pour les problèmes elliptiques. Course notes available at https://www.ljll.fr/smets/DEA/Cours_DEA.pdf, 2008.

174. Sergej Lvovich Sobolev. On a theorem of functional analysis. *Mat. Sbornik*, 4:471–497, 1938.

175. Alan D. Sokal. A really simple elementary proof of the uniform boundedness theorem. *Amer. Math. Monthly*, 118(5):450–452, 2011.

176. Elias M. Stein. *Singular integrals and differentiability properties of functions*, volume 30 of *Princeton Math. Series*. Princeton University Press, Princeton, NJ, 1970.

177. Elias M. Stein. *Harmonic analysis: real-variable methods, orthogonality, and oscillatory integrals*, volume 43 of *Princeton Math. Series*. Princeton University Press, Princeton, NJ, 1993.

178. Marshall H. Stone. The generalized Weierstrass approximation theorem. *Math. Mag.*, 21:167–184, 237–254, 1948.

179. Robert S. Strichartz. Restrictions of Fourier transforms to quadratic surfaces and decay of solutions of wave equations. *Duke Math. J.*, 44(3):705–714, 1977.

180. John Sylvester and Gunther Uhlmann. A global uniqueness theorem for an inverse boundary value problem. *Ann. of Math. (2)*, 125(1):153–169, 1987.

181. Terence Tao. *Nonlinear dispersive equations*, volume 106 of *CBMS Regional Conference Series in Math.* American Mathematical Society, Providence, RI, 2006. Local and global analysis.

182. Terence Tao. *An introduction to measure theory*, volume 126 of *Graduate Studies in Math.* American Mathematical Society, Providence, RI, 2011.

183. Terence Tao. *Hilbert's fifth problem and related topics*, volume 153 of *Graduate Studies in Math.* American Mathematical Society, Providence, RI, 2014.

184. Luc Tartar. Compensated compactness and applications to partial differential equations. In *Nonlinear analysis and mechanics: Heriot-Watt Symposium, Vol. IV*, volume 39 of *Res. Notes in Math.*, pages 136–212. Pitman, Boston, Mass.-London, 1979.

185. Luc Tartar. *H*-measures, a new approach for studying homogenisation, oscillations and concentration effects in partial differential equations. *Proc. Roy. Soc. Edinburgh Sect. A*, 115(3-4):193–230, 1990.

186. Michael E. Taylor. *Pseudodifferential operators*, volume 34 of *Princeton Math. Series*. Princeton University Press, Princeton, NJ, 1981.

187. Michael E. Taylor. *Pseudodifferential operators and nonlinear PDE*, volume 100 of *Progress in Math.* Birkhäuser Boston, Inc., Boston, MA, 1991.

188. Michael E. Taylor. *Tools for PDE*, volume 81 of *Math. Surveys and Monographs*. American Mathematical Society, Providence, RI, 2000. Pseudodifferential operators, paradifferential operators, and layer potentials.

189. Michael E. Taylor. *Partial differential equations III. Nonlinear equations*, volume 117 of *Applied Math. Sciences*. Springer, New York, 2011.

190. G Olof Thorin. An extension of a convexity theorem due to m. *Riesz, Kungl. Fysiografiska Saellskapet i Lund Forhaendlinger*, 8:14, 1939.

191. François Trèves. *Introduction to pseudodifferential and Fourier integral operators. Vol. 1*. University Series in Math. Plenum Press, New York-London, 1980. Pseudodifferential operators.

192. Gunther Uhlmann. Complex geometrical optics and Calderón's problem. In *Control and inverse problems for partial differential equations*, volume 22 of *Ser. Contemp. Appl. Math. CAM*, pages 107–169. Higher Ed. Press, Beijing, 2019.

193. André Unterberger and Juliane Bokobza. Les opérateurs de Calderon-Zygmund précisés. *C. R. Acad. Sci. Paris*, 260:34–37, 1965.

194. André Unterberger and Juliane Bokobza. Les opérateurs pseudo-différentiels d'ordre variable. *C. R. Acad. Sci. Paris*, 261:2271–2273, 1965.

195. C. Eugene Wayne. An introduction to KAM theory. In *Dynamical systems and probabilistic methods in partial differential equations (Berkeley, CA, 1994)*, volume 31 of *Lectures in Appl. Math.*, pages 3–29. American Mathematical Society, Providence, RI, 1996.

196. Norbert Wiener. Tauberian theorems. *Ann. of Math. (2)*, 33(1):1–100, 1932.

197. Michael M Wolf. Mathematical foundations of supervised learning. 2023.

198. Kenji Yajima. Existence of solutions for Schrödinger evolution equations. *Comm. Math. Phys.*, 110(3):415–426, 1987.

199. Eduard Zehnder. An implicit function theorem for small divisor problems. *Bull. Amer. Math. Soc.*, 80:174–179, 1974.

200. Claude Zuily. *Problèmes de distributions. Avec solutions détaillées*. Méthodes. Hermann, Paris, 1978.

201. Claude Zuily. *Distributions et équations aux dérivées partielles*. Dunod, Paris, 2002.
202. Maciej Zworski. *Semiclassical analysis*, volume 138 of *Graduate Studies in Math.* American Mathematical Society, Providence, RI, 2012.
203. Antoni Zygmund. *Trigonometric series. Vol. I, II.* Cambridge Mathematical Library. Cambridge University Press, Cambridge, third edition, 2002. With a foreword by Robert A. Fefferman.

Notation

Function Spaces

$C_0^\infty(\Omega)$:	C^∞ functions with compact support in Ω, page 371
$C_b^\infty(\Omega)$:	space of bounded functions whose derivatives are all bounded, page 216
$C^{0,\alpha}(\mathbb{R}^n)$:	Hölder space of order $\alpha \in (0,1]$, page 127
$C^{k,\alpha}(\mathbb{R}^n)$:	Hölder space of order $k+\alpha$, page 127
$C_*^r(\mathbb{R}^n)$:	Zygmund space of order $r \in \mathbb{R}$, page 129
$H^1(\Omega) = W^{1,2}(\Omega)$:	Sobolev space, page 168
$H_0^1(\Omega)$:	closure of $C_0^\infty(\Omega)$ in $H^1(\Omega)$, page 171
$H^s(\mathbb{R}^n)$:	Sobolev space of order $s \in \mathbb{R}$, page 181
$L^p(\Omega)$:	Lebesgue space, page 377
$L_{\mathrm{per}}^p(\mathbb{R}^n)$:	periodic Lebesgue space, page 95
$L_w^p(\Omega)$:	weak Lebesgue space, page 382
$S^{-\infty}$:	class of symbols of order $-\infty$, page 216
S^m:	class of symbols of order m, page 216
$S(\mathbb{R}^n)$:	Schwartz space, page 116
$S'(\mathbb{R}^n)$:	space of tempered distributions, page 121
$W^{k,p}(\Omega)$:	Sobolev space, page 169
$W^{1,p}(\Omega)$:	Sobolev space, page 167
$X^{s,b}(\mathbb{T}^n)$:	Bourgain space, page 348

Symbols

$\langle \cdot \rangle$:	Japanese bracket, page 181
supp:	support of a function, page 139
a^*:	adjoint symbol, page 232
$a \# b$:	composed symbol, page 245
$A \lesssim B$:	$A \leq CB$ for a constant C depending on fixed parameters, page 61
Δ:	Laplacian, page 173
$\sum \Delta_p$:	Littlewood–Paley decomposition, page 126
$\widehat{f}$ or $\mathcal{F}(f)$:	Fourier transform of f, page 114

T. Alazard, *Analysis and Partial Differential Equations*, Universitext,
https://doi.org/10.1007/978-3-031-70909-8

$\widehat{f}(k)$ or $\widehat{f_k}$:	Fourier coefficient of f, page 96
$\partial_j f = \partial_{x_j} f = \partial f / \partial x_j$:	derivative of f with respect to x_j, page 167
$f * g$:	convolution product, page 136
I_α:	Riesz potential, page 149
$\mathbb{K}$:	denotes either $\mathbb{R}$ or $\mathbb{C}$, page 4
$m(D_x)$:	Fourier multiplier, page 125
Mf:	maximal function of f, page 143
$\mathrm{Op}(a)$:	pseudo-differential operator of symbol a, page 218
$\mathrm{WF}(f)$:	wave front set of f, page 273

Topology

$f_n \rightharpoonup f$ weak-$*$:	weak-$*$ convergence, page 86
$x_n \rightharpoonup x$:	weak convergence, page 84
$\|\cdot\|$:	norm, page 4
$\overline{A}$:	closure of the set A, page 7
$A \Subset B$:	means $\overline{A} \subset B$, page 268
$B(x, r)$:	open ball centered at x with radius r, page 4
E':	topological dual of E, often denoted by E^*, page 6

Index

accumulation point, 360
adherence, 360
approximation of the identity, 140

balanced, 7
ball
  closed, 4, 361
  open, 4, 361
basis of open neighborhoods, 360
boundary, 360

characteristic variety, 274
closure, 360
continuity
  global, 360
  local, 360
convergence
  strong, 70
  weak, 70, 81
  weak-*, 81
convex, 7
convolution (product of), 98, 135
cover, 366
  countable, 366
  finite, 366

diffeomorphism, 39
differentiable function, 37
differential, 37
Diophantine number, 371
distance, 4, 361
dual
  topological, 6
dyadic decomposition, 125

equicontinuity, 22
exhaustive sequence of compact sets, 13

family
  equicontinuous, 18
  graded of semi-norms, 13
  pointwise bounded, 18
  separating of semi-norms, 13
Fourier
  coefficient of, 96
  series of, 96
  transform of, 114
Fourier multiplier, 125, 230
Fourier transform, 114
function
  contraction, 38
  distribution, 381
  Hardy–Littlewood maximal, 143
  harmonic, 199
  Lipschitzian, 38
  of class C^k, 38
  proper, 366
  Weierstrass, 154

Gaussian, 118
graph, 21

heat equation, 110
Hilbert basis, 71
homeomorphism, 5, 361
hyperbolic equation, 264

inequality
  Bessel, 71
  Caccioppoli, 206
  Cauchy–Schwarz, 66
  Chebyshev, 382
  Gårding, 251
  Gagliardo–Nirenberg, 186
  Hardy, 381